# Universitext

T0202713

Springer
*Berlin*
*Heidelberg*
*New York*
*Hong Kong*
*London*
*Milan*
*Paris*
*Tokyo*

Roger Godement

# Analysis I

## Convergence, Elementary functions

Translated from the French by Philip Spain

 Springer

*Roger Godement*
Université Paris VII
Département de Mathématiques
2, place Jussieu
75251 Paris Cedex 05
France
*e-mail:* rgodement@aol.com

*Translator:*
Philip Spain
Mathematics Department
University of Glasgow
Glasgow G12 8QW
Scotland
*e-mail:* pgs@maths.gla.ac.uk

Cataloging-in-Publication Data applied for

A catalog record for this book is available from the Library of Congress.

Bibliographic information published by Die Deutsche Bibliothek
Die Deutsche Bibliothek lists this publication in the Deutsche Nationalbibliografie;
detailed bibliographic data is available in the Internet at <http://dnb.ddb.de>.

Originally published by Springer-Verlag as *Analyse mathématique I. Convergence, fonctions élémentaires*, 1998: ISBN 3-540-42057-6

ISBN 3-540-05923-7 Springer-Verlag Berlin Heidelberg New York

Mathematics Subject Classification (2000): 26-01, 26A03, 26A06, 26A12, 26A15, 26A24, 26A42, 26B05, 28-XX, 30-XX, 30-01, 31-XX, 41-XX, 42-XX, 42-01, 43-XX, 54-XX

Springer-Verlag is a part of Springer Science+Business Media
springeronline.com

© Springer-Verlag Berlin Heidelberg 2004
Printed in Germany

Cover design: *design & production,* Heidelberg
Typesetting by the Translator using a TₑX macro package
Printed on acid-free paper                    41/3142ck-5 4 3 2 1 0

# Preface

## Analysis and its Adhesions

Between 1946 and 1990 I had thousands of students; in the very economical French system with its auditoria for two hundred people or more, this was not difficult. On several occasions I felt the desire to write a book which, presupposing only a minimal level of knowledge and a taste for mathematics, would lead the reader to a point from which he (or she) could launch himself without difficulty into the more abstract or more complicated theories of the $XX^{th}$ century. After various attempts I began to write it for Springer-Verlag in the Spring of 1996.

A long-established house, with unrivalled experience in scientific publishing in general and mathematics in particular, Springer seemed to be by far the best possible publisher. My dealings with their mathematical department over six years have quite confirmed this. As, furthermore, Catriona Byrne, who has responsibility for author relations in this sector, has been a friend of mine for a long time, I had no misgivings at confiding my francophone production to a foreign publisher who, though not from our parish, knows its profession superlatively.

My text has been prepared in French on computer, in DOS, with the aid of *Nota Bene*, a perfectly organized, simple and rational American word processor; but it is hardly more adapted to mathematics than the traditional typewriters of yesteryear: greek letters, $\in$, $\int$, $\Sigma$ have to be written by hand on the printout, something I had been doing anyway since my first 1946 typewriter. I eventually devised a coding system, for instance [[alpha]] for greek letters, that made it easier to translate the NB files into TEX by using global commands. But apart from simple formulae in the main text, most of the others had to be typeset again for the French version.

The excellent English translation has been much easier to do since Dr Spain, who types in TEX, had the TEX version of the French edition. I have taken this opportunity to make some small changes to the French version.

* * *

This is not a standard textbook geared to those many students who have to learn mathematics for other purposes, although it may help them; it is the reader interested in mathematics for its own sake of whom I have thought while writing. To many of the French students and particularly to many of the brightest, mathematics is merely a lift to the upper strata of society[1]. My goal is not to help bright young people to arrive among the first few in the entry competition for the French *Ecole polytechnique* so as to find themselves thirty years later at the service or at the head of a public or private enterprise producing possibly war planes, missiles, military electronics, or nuclear weapons[2], or who will devise all kinds of financial stunts to make their company grow beyond what they can control, and who, in both cases, will make at least twenty times as much money as the winner of a Fields Medal does.

The sole aim of this book thus is mathematical analysis as it was and as it has become. The fundamental ideas which anyone must know – convergence, continuity, elementary functions, integrals, asymptotics, Fourier series and integrals – are the subject of the first two volumes. Volume II also deals with that part (Weierstrass) of the classical theory of analytic functions which can be explained with the use of Fourier series, while the other part (Cauchy) will be found at the beginning of Volume III. I have not hesitated to introduce, sometimes very early, subjects considered as relatively advanced when they can be explained without technical complications: series indexed by arbitrary countable sets, the definition and elementary properties of Radon measures in $\mathbb{R}$ or $\mathbb{C}$, integrals of semi-continuous functions and even, in an Appendix to Chap. V, a short account of the basic theorems of Lebesgue's theory for those who may care to read it at this early stage, analytic functions, the construction of Weierstrass elliptic functions as a beautiful and useful example of a sophisticated series, etc.

I have tried to give the reader an idea of the axiomatic construction of set theory while hoping that he will take Chap. I for what it is: a contribution to his mathematical culture aiming at showing that the whole of mathematics can, in principle, be built from a small number of axioms and definitions. But a full understanding of this Chapter *is not an obligatory prerequisite to an apprenticeship in analysis*. The only thing the reader will have to retain is the naive version of set theory – standard operations on sets and functions to which, anyway, he will get used by merely reading the next chapters – as well as the fact that, even at the simplest level, mathematics rests upon *proofs* of statements, an old art which, in French high schools and probably elsewhere as well, is in the process of becoming obsolete because, we are told, learning to use formulae is much more useful to most people, or because it is too difficult for the many children of the lower strata of society (I was one in the 1930s) who now flood the high schools ...

---

[1] In XIX[th] century Cambridge, the winners of the Math Tripos would far more often become judges or bishops than scientists.

[2] One of the brightest students I have known in thirty-five years is today the head of a holding company that controls, among other things, a chain of supermarkets. He sells Camembert, shrink-wrapped meat, Tampax, orange juice, noodles, mustard, etc. If you have to choose, this is a more civilised way to squander your grey matter.

The sequel, in Volumes III and IV, explains subjects which require either a much higher level of abstraction (short introductions to differential varieties and Riemann surfaces, general integration, Hilbert spaces, general harmonic analysis), or, in the last Chap. XII, a much higher level in computation techniques: Dirichlet series of number theory, elliptic and modular functions, connection with Lie groups. While the choice of material in Volumes I to IV represents a coherent and nearly selfcontained block of mathematics, it constitutes nothing more than one particular view of analysis. Other authors could have chosen other views and, for instance, tried to lead their readers into the theory of partial differential equations. I have not even treated differential equations in one variable: one can learn all about them in a myriad of books, and the classical results of the theory, direct applications of the general principles of analysis, should pose no serious problem to the student who has assimilated these reasonably well.

In the two first volumes – Volumes III and IV are written in a much more orthodox fashion – I have firmly emphasised, sometimes with the aid of out of fashion excurses in ordinary language, the ideas at the basis of analysis, and, in some cases, their historical evolution. I am not, far from it, an expert in the history of mathematics; some mathematicians, sensing their end coming, devote themselves to it late in life; others, younger, consider the subject sufficiently interesting to devote a substantial part of their activity to it; they perform a most useful task even from the pedagogical point of view[3] since, at twenty, which I once was, one thinks only of forging ahead without looking behind, and almost always without knowing where one is going: where and when will one learn? I have myself preferred for a quarter of a century to take an interest in a kind of history – science, technology, and armaments in the XX[th] century – for which mathematics does not prepare one, though there are some indirect connections. Nevertheless I have made some effort to convey to the reader that the ideas and the techniques have evolved, and that it took between one and two centuries for the intuitions of the Founding Fathers to be transformed into perfectly clear concepts founded on unassailable arguments, awaiting the great generalisations of the XX[th] century.

Adopting this point of view has led me, in these first two volumes, systematically to eschew a perfectly linear exposition, organised like a clockwork and only presenting to the reader the dominant or à la mode point of view, with assorted *Blitzbeweise*, lightning proofs in the sense in which we speak of *Blitzkrieg*[4]: one ratifies the result but does not comprehend the strategy until six months after the battle. At the cost of proving the same classical results several times over I have tried to present several methods of arguing to the

---

[3] E. Hairer and G. Wanner, *Analysis by Its History* (Springer-New York, 1996), is a prime example.

[4] René Etiemble, a great French specialist of comparative literature, once made a study of the styles prevailing in various kinds of activities. He came to the conclusion that mathematical style was the closest there was to the military.

reader, and to make clear the necessity of rigour by evidencing the doubtful arguments, and sometimes false results, due to mathematicians like Newton, the Bernoullis, Euler, Fourier, or Cauchy. Adopting this point of view lengthens the text palpably, but one of the ground principles of N. Bourbaki – no economies of paper – is, I think, mandatory when one addresses students embarking on a subject.

The other principle of this same author – to substitute ideas for computations – appears even more commendable to me whenever it can be applied. All the same, one will, inevitably, find calculations in this book; but I have essentially confined myself to those which, inherited from the great mathematicians of the past, form an integral part of the theory and can be considered as ideas.

Except occasionally, to round off the text, one will find no exercises here. Working at exercises is indispensable when one learns mathematics, and one will find them in profusion in many other books and specialised collections. The majority of French students, obsessed by the string of examinations imposed on them, have a very exaggerated tendency to consider the "lectures" of little use and that only "practical work" and "formulae" count or pay. The result is that the majority of them are able, up to errors in calculation, to integrate a rational function but incapable of answering questions of a general nature, e.g. why is a rational function integrable? To understand a theorem is to be able to reconstruct its proof. To understand a block of mathematics does not reduce to knowing how to apply its results; to understand a theory is to be able to reconstruct its logical structure. Every mathematician knows this.

One does not learn analysis or anything else from one single book; there is neither Bible, nor Gospel nor Koran in Mathematics. The fact that the spirit of my book is radically different from that of Serge Lang, *Undergraduate Analysis* (Springer, 2nd. ed., 1997) for example, should not dissuade from reading it, quite the contrary; even less the books of E. Hairer and G. Wanner, *Analysis by Its History*, Wolfgang Walter, *Analysis I* (Springer, 1992, in German) or Reinhold Remmert, *Theory of Complex Functions* (Springer-New York, 1991, translation of *Funktionentheorie 1*, 4. Auflage, 1995), which I have often used, and cite when I do so. These excellent books present numerous exercises, as does Jean Dieudonné's *Calcul Infinitésimal* (Hermann, 1968) though his style enthuses me less.

I have not acceded to the new fashion which likes to decorate elementary analysis textbooks with numerical calculations to fifteen decimal places under the pretext they will be useful to future computer scientists or applied mathematicians. Everyone knows that the mathematicians of the XVII[th] and XVIII[th] centuries loved numerical computations – done by hand, not by tapping the keys of an electronic gadget – that enabled them to verify their theoretical results or to demonstrate the power of their methods. This childhood sickness of analysis disappeared when in the XIX[th] century one began addressing rigour of proof and generality of formulations, rather than formulae.

This does not mean that numerical calculations have become pointless: thanks to computers, one can do more and more of them, for better or worse, in all scientific and technical areas that, from medical imaging to the perfecting of nuclear weapons[5], use mathematics. One does the same in certain branches of mathematics too; for example, displaying a large number of curves may open the way to a general theorem or to understanding a topological situation, not to speak of the traditional number theory where numerical experiment always was, and still is, used to formulate or verify conjectures.

This only means that the aim of an exposition of the *principles* of analysis is not to teach numerical techniques. Moreover, the partisans of applied mathematics, of numerical analysis and of computer science in all the universities of the world manifest their imperialist tendencies far too clearly for real mathematicians to take on in their stead a task for which they generally lack both taste and competence.

* * *

The innocent reader and many confirmed mathematicians will probably be surprised, possibly even shocked, to find in my book some very heavy allusions to extramathematical subjects and particularly to the relations between science and weaponry. This is neither politically nor scientifically correct: Science is politically neutral[6], even when someone lets it fall inadvertently on Hiroshima while the future winner of a Nobel Prize in physics is recording the results in a B-29 trailing the *Enola Gay*[7]. Nor is it part of the curriculum: a

---

[5] Nuclear powers agreed a dozen years ago to stop testing. The reason why was that testing was made unnecessary by the improvement of numerical analysis. The most immediate consequence of this "progress" is that everything is now done in full secrecy, which was not the case when they had to propel into the stratosphere two million tons of radioactive rock and sand in order to check their "gadgets".

[6] An assertion long since demolished by countless studies, notably American, either of particular facets of scientific activity, or of Science in a general way, e.g. in Bernard Barber, *Science and the Social Order* (Collier Books, 1952) and Jean-Jacques Salomon, *Science et Politique* (Paris, Ed. du Seuil, 1970, reed. Economica). Disclosing the influence of politics, for instance of WW II and the Cold War, upon Science and Technology is not the same as going in politics as so many scientists believe without ever having read any serious historical work. And I do not see why being opposed to the military exploitation of mathematics and science should be considered as a more political stand than, for instance, helping Los Alamos or Arzamas to develop their "weapons of genocide" was.

[7] In *Alvarez: Adventures of a Physicist* (Basic Books, 1987), Luis Alvarez's trip to Hiroshima is the very first thing he relates in his book. He was also one of the main proponents of the H-bomb and, at the end of October, 1949, went to Washington to lobby in favor of it. In 1954, he testified at the Oppenheimer security hearing that Oppenheimer's opposition to the H-bomb was proof of an *exceedingly poor judgment*. Alvarez is one of many similar counterexamples to the "neutrality of Science" theory.

scientist's business is to provide his students or readers, without commentary, the knowledge they will later use, for better or for worse, as suits them. It will be up to them to discover by themselves, possibly years after graduating, that which "has no place" (why, please?) in scientific books or lectures and which was not told them by older scientists well aware of it, or who should have been. Let me give you a few French examples.

As a dozen of people who mostly work in particle physics assured me, you can spend five or eight years learning physics without ever having heard anything about nuclear weapons. I once checked the chemistry library of my Paris university for books by Louis Fieser, a most eminent Harvard chemist who, during WW II, was in charge of improving incendiary weapons and developed napalm; all of his chemistry books are there, but not his account of war work[8].

I found another, particularly caricatural, example in a textbook of physics for high-school finishing students; as required by the French official instructions in 1995, the concluding chapter, on the laser (never mind that an eighteen year old boy or girl can understand practically nothing to it), mentioned a number of civilian applications – ophthalmology, measure of atmospheric pollution, compact discs, energy production by laser-induced thermonuclear fusion[9], etc. – but not a single military use of lasers, a domain in which French industry was always very strong. This is not only dishonest; it is a foolish way to hide the truth since students read newspapers, look at TV, and if they type "laser military history" on www.google.com, they will get about 164,000 documents!

Thirty years ago, in part under the influence of what I had seen on American campuses and read in American newspapers and such reviews as *Science*

---

[8] Louis Fieser, *The Scientific Method. A Personal Account of Unusual Projects in War and in Peace* (Reinhold, 1964). In the *Biographical Memoirs*, v. 65, 1994, of the (American) National Academy of Science, Fieser's biographer has this to say (p. 165) about his work during WW II: "*With the approach of World War II, Fieser was drawn increasingly into war-related projects. A brief excursion into the area of mixed aliphatic-aromatic polynitro compounds for possible use as exotic explosives was followed by studies of alkali salts of long chain fatty acids as incendiaries, but by far the most important of his war-related work was his long and intensive study of the quinone antimalarials*", to which the author devotes one full page. The word "napalm" is nowhere to be found in this fourteen-page biography, a beautiful example of the art of fooling the reader with opaque technical jargon. All the more remarkable since Fieser was strongly criticised during the Vietnam War for his development of napalm. In his long biography of von Neumann in the *Dictionary of Scientific Biography*, J. Dieudonné devotes two lines to what he calls his "government" work without telling us whether it had to do with, say, the H-bomb or cancer research, two strongly supported domains of "government" work.

[9] This is a very long term project, but the French and American military have justified this very expensive enterprise by pointing out that the new knowledge of fusion processes it will provide will be used to improve nuclear weapons, a fact that is of course not mentioned in the textbook.

and the *Bulletin of the Atomic Scientists*, I succeeded, to my great aston-
ishment, to convince the head of my Paris university library to start a new
section that would be devoted to what was then called in America Science
and Society studies. Although it received little money, you can now find there
several thousands of (mostly American) books and the main reviews in the
history of science and technology, including the military side of it, the arms
race, economics of research and development, science policy, etc.: no exclusive. But almost all the readers are people who specialise in that field, while
most of the 5,000 scientists working at the university don't even know the
existence of this library. Since their specialised libraries are practically empty
in this respect, the conclusion is inescapable: their only sources of information
are their generally narrow personal experience[10], perhaps some historical articles written in scientific reviews by scientists who have no idea of historical
writing[11], and cafeteria conversations:

> The humanist who looks at science from the point of view of his own
> endeavours is bound to be impressed, first of all, by its startling lack of
> insight into itself. Scientists seem able to go about their business in a state
> of indifference to, if not ignorance of, anything but the going, currently
> acceptable doctrines of their several disciplines ... The only thing wrong
> with scientists is that they don't understand science. They don't know
> where their institutions come from, what forces shaped and are still shaping
> them, and they are wedded to an antihistorical way of thinking which
> threatens to deter them from ever finding out[12].

It appears more honest to me to violate these miserable and far too comfortable taboos and to put on their guard those innocents who leap into the dark
into careers of which they know nothing. Because of their past and potential

---

[10] It is not always that narrow. As in the USA – the model – there are in
France scientists who have been for a long time in top government committees
or who have cooperated with industry. They obviously know a lot more than the
average researcher, let alone student. But they mostly don't speak, much less
write, particularly when defence activities are involved. This striking difference
between French and American "Statesmen of Science" can perhaps be explained
by the fact that the political spectrum extends much farther to the left in France
than in the USA, so that defence work was, at least during most of the Cold
War, much more controversial here than on the other side of the Atlantic.

[11] One of the books I have recently read is Gregg Herken, *Brotherhood of the Bomb:
The Tangled Lives and Loyalties of Robert Oppenheimer, Ernest Lawrence, and
Edward Teller* (Henry Holt, 2002), a superb though very concentrated book.
The main text, 334 pages, is followed by over 2,000 notes: an average of six
references to sources per page (and a lot more on Internet). No active scientist
could spend ten years reading two hundred books and papers already published,
interviewing at length eighty colleagues, discovering and reading hundreds of
recently declassified government files, and organizing this amount of information
into a coherent book.

[12] Eric Larrabee, *Science and the Common Reader* (Commentary, June 1966). As I
said above, old scientists who have long been top consultants to their government
are not as innocent as Larrabee puts it, but the new generation has not their
experience of science politics.

catastrophic consequences, the connections between science, technology and armaments concern all who go into science or technology or who practice them. They have been governed for half a century by the existence of public organisations and private enterprises whose function is *the systematic transformation of scientific and technological progress into military progress* within the limits, often elastic, of the economic capacities of the various countries which take part in it:

> With the attention which is paid in these days to weapons of war, there is probably no known scientific principle that has not already been carefully scrutinized to see whether it is of any significance for defence[13].

In countries – France is a prime example – where discussions on the relations between Science and Defence have been dominated for decades first by silence, then by a thick consensus[14], and have been totally absent from university teaching[15], the thing to say to young people is that *one of the forms of intellectual liberty is not to let oneself be dominated by the dominant ideas.*

But this requires access to other information sources. It would be impossible to thoroughly discuss this subject and its history within the framework of a mathematical treatise. I nevertheless decided to write a few dozen pages – the Postface to Volume II – in order to give the interested reader an idea of it and, in particular, to show that the question and the subject do exist. I have not balked at citing a good number of important bibliographical references

---

[13] Sir Solly Zuckerman, *Scientists and War* (London, Hamish Hamilton, 1962, p. 80); the author was at the time the British Government chief scientist and had formerly been the head of British military research. There is no reason to believe that Zuckerman's statement is no longer valid, particularly in America.

[14] "Science et Défense" is the title of a French association founded in 1983 by Charles Hernu, then the (socialist) Secretary of Defence and future hero of the Greenpeace affair – the clumsy sinking in Auckland harbour by French agents of a ship that would have interfered with a French nuclear test in the Pacific. Supported by the Armament branch of Defence, the association organises a yearly congress, where, over two days, engineers and scientists lecture on the technical problems of armaments and the closely related sciences. Several hundreds of people attend: military, engineers, industrialists, scientists, and, inevitably, political scientists and metaphysicians of strategy. France is, to my knowledge, the only country where what a number of American historians now call the scientific-military-industrial complex dares to exhibit itself so publicly and without provoking the least reaction. This would not have been possible before the conversion of the Socialist and Communist parties to nuclear weapons when, at the end of the 1970s, they saw a good prospect of winning the 1981 presidential election.

[15] America was, in the 1970s, a notable exception to this general statement: student protests against the Vietnam war and the cooperation of many university departments or laboratories with the DoD led some universities to add to their curriculum lectures on various aspects of "Science and Society" that attracted a sizeable number of science students, while some teachers in the history of science saw their audience suddenly grow. Although the traditional back to normal process did not take very long, many of the present generation of specialists found their calling during this period.

– there are plenty more – which will allow those who so desire to complete, verify, or discuss this text. I do not have the naive hope that a twenty year old student of mathematics will plunge into this ocean of literature; it would hardly even be a very good service to encourage him to do so. But maybe this text will find readers who are not so young and no longer have to submit to examinations or competitions for success. Although the French version of this Postface devoted a good deal of space to the French situation, I thought it better, in the English version, to emphasise the American situation more than I did in French, and this for several good reasons.

From Pearl Harbor to the present day, America has been the world leader in this domain – a leader which, for a dozen years, has no longer had any competitor worth naming and seems to be in a technological arms race against itself[16], as was already the case when it spent $2 000 000 000 (one percent of its 1945 GNP) during WW II in order to get the atomic bomb before the Nazis, who did not believe it could be available in time and devoted very little resources to it. This American polarisation on scientific weapons, more or less faithfully imitated in the Soviet Union, Britain and France, had enormous political consequences; among others, it compelled the much weaker Soviet Union to devote to defence a proportion of its resources which must have strongly contributed to its downfall and to the present American hegemony. On the other hand, the civilian uses of mostly American military innovations in electronics, informatics, aviation, space, telecommunications, nuclear power, etc., had a deep influence on the daily life of people everywhere. Without WW II and the arms race, most of these innovations would have come much later, or never, because the financing of research, development and initial production by defence organisations made it possible for private enterprises to take risks which, otherwise, would have been barred by the return on investment principle that governs civilian innovations. Without World War II, no V-2 missiles and no atomic weapons; without these and the Cold War, no intercontinental ballistic missiles; without ICBMs and the need of the central military authorities for instant worldwide command, control, communication, and intelligence – C$^3$I as they call it – no satellites; and without satellites and many other innovations propelled by the military - computers, integrated circuits, Arpanet, etc. – then no Internet, to mention only this most spectacular spin-off of the arms race. The idea that civilian industry could have, by itself, spent tens or hundreds of billions in order to invent, produce and market such gigantic amounts of hardware and software at a time when nobody but the military had any proven need for it is foolish. Civilian business does not deal in science fiction.

WW II and the arms race also contributed to propelling the funding of scientific research proper to levels which, before 1939, would have seemed

---

[16] It has been recently disclosed that America will develop in the next 10 or 15 years an hypersonic cruise missile that will be able to strike anywhere on the Earth in less than two hours from bases in continental America.

unrealistic in the utmost, a fact of which scientists everywhere were the first beneficiaries although never nearly as much as American ones[17]. This not only made it possible for many more young Americans to choose scientific careers than was the case before WW II, it also attracted to America many scientists (and still more engineers) who had been educated elsewhere, a process that is continuing to this day – the famous *brain drain* that was first noticed in the 1950s, not to mention some Russian immigrants after 1917 and the European Jews in the 1930s.

\* \* \*

The French version of this book included many citations and references in English, particularly in the Postface to Volume II, this in order to encourage the reader to use a language that is absolutely indispensable if one wants to inform oneself on anything at all: for clear demographic reasons France accounts for only a small proportion of the literature, for instance from 3% (technology) to 7% (mathematics) in the sciences on the world scale, and although French authors publish excellent books in many domains, scientific or not, they cannot be leaders everywhere. There is for instance nothing of any value on the history of nuclear weapons, not even of French ones, and none of the best American books have been translated. Almost all I know in the Science and Defence domain has been learned from American authors, although a few French historians of Science and Technology are beginning to deal with it.

There is no need to suggest to readers of the English translation of the present book to learn English. One should however warn the beginner that, even though well over 60% of the mathematical literature is now in English, an ability to read French is, at the research level, still needed. Since 1945, the Fields Medal has been awarded to 44 people worldwide; seven of them were French, and two more, although they are not, did all their previous work in France; the first Abel Prize (a recently created substitute to the nonexistent Nobel Prize for Mathematics) has been awarded in 2003 to Jean-Pierre Serre, who won a Fields Medal in 1954, and others won for instance the Wolf Prize. There are in France many more excellent mathematicians than these stars; although some publish in English, still many write in French. And there are of course German and Russian authors, among others, who still publish in the one language they learned as infants, as anglophone authors always did.

---

[17] In 1965, Isidor Rabi, a winner of the Nobel Prize in physics, pointed out that the budget of the Columbia University physics lab had grown from 15,000 dollars before the war to three millions and attributes this to the war which *"did wonderful things in some respects"*. Hans Bethe, another Nobel Laureate, remembered in 1962 that before WW II he found it difficult to get some $3,000 for a cyclotron at Cornell, but that, *"Today, $3,000 is pin money. We use it in this laboratory in a day"*. To be objective, one should also note the fantastic increase of civilian research funds allocated to Life Sciences, mainly Biology and Medicine; but even in this case, it was WW II, especially the development of penicillin, which at the start demonstrated what could be done in these fields with enough money and a concerted effort.

You fortunately don't need to learn Japanese: Japanese authors do not use it at the international level, a most courteous stand when you think that, for them, learning English is a lot more work than learning French is for the American, or English for the French.

The fact that English has acquired almost the status of an international common language, or lingua franca, has of course its upside, and any other reasonably widespread language, as Latin was three centuries ago, would do. The prevalence of English is often explained by the fact that it is supposedly simpler than, say, German, French, or Russian, and that anyway anglophones now form a large proportion of scientists at the world level. As suggested above, this preponderance of English, which goes far beyond Science, is also, and possibly mainly, a corollary of the enormous resources American government, industry and private foundations have devoted to Science and Technology since the 1940s and more generally of the overwhelming superiority of the American economy[18].

There is in France, and probably elsewhere too, a theory according to which, thanks to the overwhelming power America acquired in 1945 and still more in 1990, the result, or even purpose, of the "invasion of English" is to spread across the whole world the American conceptions of society, politics, economy, technology, mass media, etc. and to help American enterprises to acquire larger and larger parts of foreign markets everywhere, a process that, although or because successful, meets strong opposition in many countries.

Although greatly reinforced by WW II, it started much sooner. The use in America of such typical expressions as "richest in the world", "greatest in the world", "tallest in the world", "fastest in the world", "first in the world", etc. was already widespread in the 1900s and was a plain enough symptom. Standard Oil, General Electric, Ford were models of multinational companies that European enterprises tried (generally without much success at the time, if you except I.G. Farben in the 1920s) to imitate. American sewing machines, typewriters and accounting machines, agricultural machinery, machine tools and, between WW I and WW II, automobiles were invading Europe long before computers did. In the 1920s, jazz had already its fanatics everywhere, Hollywood's movies had already 60-80% of the French market, most of the best movie theaters were in American hands, and the answer to French attempts to impose import quotas was a near total boycott of French movies in America (19 in 1929, against hundreds of American movies in France), a situation which did not improve after WW II. After 1918 it was Wilson, a U.S. President with the mind and eloquence of a Protestant missionary, who launched the Society of Nations, which Congress rejected. The United States' interventionist policy was already quite plain in the Americas, China and Japan long before the end of the XIX$^{th}$ century, and as a recent book[19]

---

[18] In some French companies, meetings of the Board are in English because of the presence of one or two American members. At the present time, about 20% of the total capitalisation of the Paris Stock Exchange is American-owned.

[19] Philippe Roger, *L'ennemi américain. Généalogie de l'antiaméricanisme français* (Paris, Seuil, 2002, 600 pp.) puts everything in historical perspective without

reminded us, French hostility toward America was powerfully increased by the war against Spain in 1898, which was viewed by the French right as a threat to European colonial empires, and by the left as a conclusive proof of the transformation of an already unpalatable American capitalism into outright imperialism or economic colonialism. As to the present American taste for firearms, a unique feature among "civilised" countries as they were called in the 1900s, it was Samuel Colt who, during the war against Mexico, triggered the craze by adopting the American system of manufacture invented in arsenals in order to mass produce his celebrated revolvers. Present inequalities in the distribution of income are not worse than they were at the time John D. Rockefeller was worth over one billion dollars, i.e. about 2% of America's GNP: a proportion which, nowadays, would amount to some 200 billion. And New York bar owners pouring French wine on the street[20] were seen already long before March 2003.

Thus, nothing very new under the sun, except that American international preponderance and unilateralism have now acquired the status of an official doctrine supported by a host of ideologists invoking a fundamentalist Protestant ethic in order to justify interventions which, in the eyes of a vast majority of people everywhere, are nothing but displays of power even when they rid nations of barbaric rulers or religious oppression in the hope of establishing there a (probably very weak) version of Western democracy.

That being said, nobody has to appreciate the barbaric music and violent movies which presently come from America (the American stars of my youth were Charlie Chaplin, Buster Keaton and the Marx Brothers). Americans do not merely dictate the export of these productions through international commercial agreements and by owning big distribution companies; they also sell them by finding indigenous customers (or imitators) who are only too happy to make money by distributing them among a young and most often uncultured public. And how would local television fill its hours of programmes, how could the cinemas function, without the flow of American productions? The work force in France (say) is not large enough to replace American mediocrity with French mediocrity; and no country is capable of producing a new Shakespeare or a new Bartok every day. One therefore broadcasts what is available or imitates American crass "games".

---

himself falling into the trap. It goes without saying that many of the criticisms that some French intellectuals and politicians of the right addressed to America would apply just as well to France. Howard Zinn, *A People's History of the United States, 1492-Present* (HarperCollins, 2003 edition), while or because very one-sided, would be very useful to help understand criticism from the French left, which was never as systematic and well organised as Zinn's, not to mention books by Lewis Mumford, Noam Chomsky, etc.

[20] If a boycott of French wines were to lower prices in France, I, for one, would not shed crocodile tears at the tragic fate of poor American patriots heroically depriving themselves of Chateau Latour at $1,000 a bottle (assuming they don't have a stock of it in their cellar).

Nor is one obliged to approve the Darwinian concepts of economic competition and social relations which, thanks to technologies that have emerged straight from the cold war and arms race, are presently expanding under the name of "globalisation": the extension to the planet of a "liberal", i.e. capitalist, and "modern" economic system founded on the principles isolated by Adam Smith in 1776 and assimilated erroneously by the robber barons who, at the end of the XIX[th] century, erected the great American capitalist enterprises, afterwards revised a little and codified. It is now forbidden to shoot the strikers but not to domesticate the unions; to dismiss thousands of employees to improve the competitiveness of companies and in return to exploit the work force at low pay in developing countries; to push for the dismantlement of European social welfare systems hard won after a century of struggle but now judged too expensive – or smacking of Socialism? – by the alumni of the Harvard Business School and its foreign imitations; to suborn the public markets by handing cheques to political parties as is presently the case in France, Germany, Italy, etc., or, in the Third World, to gangsters in high places in order to inundate the countries they rule or own with killing machines under the pretext of lowering the unit price for the countries that produce them[21], or in order to secure the rights to exploiting their natural resources. It is the reign of money, whose rallying slogan was launched a hundred and fifty years ago by a famous French minister: *Enrichissez-vous*! If you can[22] ...

That said, America possesses, notably in its universities, an intellectual class not to be globally confused with the spokesmen of the Pentagon's warlords or the operators of Wall Street. In particular and as I said above, no one, in France, has revealed the military influence on scientific and technological development since 1940 as a number of American historians, particularly of the younger generation, have done for a quarter of a century with the help of massive documentation; if you are interested in, say, the history of the Cold War, you will find in the American literature all the information, points of view and opinions you want. There is no need either to point out that many American novelists did not wait until 2003 to disseminate unorthodox descriptions of the American society. As to the mathematicians, many of whom have always been very critical of official policy, the years I spent with my family in the 1950s and 1960s at Urbana, Berkeley and Princeton were among the happiest of my life. And when, at the end of October 1961,

---

[21] A few years ago, the sale to Taiwan by the French company Thomson-CSF of very sophisticated frigates generated a $500 million return for (mostly unknown) Taiwanese and French politicians or political parties and go-betweens. An investigation of the case by the French judiciary was stopped in a most elegant way: the Department of Defence classified all Thomson-CSF documents pertaining to it. The company decided a few months later to change its name into Thales, a rather unpalatable reference to Mathematics.

[22] What Guizot said is: Get rich, by your work and savings – a cynical precept at a time when the overwhelming majority of people, after working twelve hours a day six days per week, would die as poor as their parents were.

my Paris flat was destroyed[23] because I had, rather mildly in fact, spoken out during a lecture against the savage repression in Paris of a peaceful Algerian demonstration for independence, I received two days later a telegram from J. Robert Oppenheimer inviting me, on very generous terms, to spend the remainder of the academic year at the Princeton Institute; we went two months later, and there my wife recovered her usual balance.

It goes without saying that the facts and opinions to be found in this book are my own and full responsibility. *They do not commit Springer-Verlag to any degree.* Some will perhaps reproach my publisher for not having censored me. Being ill-placed to do so in their place I prefer, for myself, to thank them warmly for having been allowed the liberty to express myself. This is an attitude which I surely would not have encountered everywhere and which I appreciate for its proper worth.

Paris 2003.
Rgodement@aol.com

---

[23] My wife, who was there, was extremely lucky not to be hurt, though she was badly shaken for months afterwards; my three children came home from school fifteen minutes after the bombing. The very first thing I noticed after climbing a ladder to my place was the police searching my papers (there was nothing to be found). Though they certainly identified the authors of this attempt – probably candidates to the French equivalent of West Point school, according to my own students – I was never told anything and I very much doubt that they ever had to bear any unpleasant consequences.

# Contents

# I – Sets and Functions

## §1. Set Theory – §2. Logicians' Logic

It is generally understood that mathematicians concern themselves with objects or concepts about which they establish theorems by applying logical, preferably irrebuttable, arguments. This last detail, which remained theoretical for a long time, apart from in the theory of numbers and in the elementary geometry inherited from the Greeks, has always scared the majority of people, accustomed as they are to produce and hear daily, even in exalted intellectual activities, tens of arguments each more contestable than the other: life in society would be impossible if everyone had to provide incontrovertible proofs of his assertions and to express himself clearly and without ambiguity[1].

Up to the end of the XVII[th] century the objects of mathematics were numbers, geometrical figures, equations or functions more or less directly arising from everyday life, astronomy or mechanics: the triangles and conic sections of the Greeks, the whole numbers, rationals, or irrationals like $\sqrt{2}$, the simplest algebraic equations which one sometimes tried, like Fermat, to solve in terms of integers, the trigonometric functions which had been extensively developed by the Greek and Arab astronomers before the Westerners, Napier's logarithms, Galileo's parabolas, the velocity of a moving body and the calculus of tangents to a curve, which, in the second half of the XVII[th] century, led to the concept of the derivative, the computation of the area bounded by a curve – the circle or the parabola in the case of Archimedes – which in the same period led to the integral calculus etc. Although still very primitive until about 1600, a short while later mathematics exploded thanks to the creation of the infinitesimal calculus by Fermat, Descartes, Huyghens, Wallis, Cavalieri, and, above all, between about 1665 and 1720, by Newton, Leibniz and the Bernoullis. This opened an epoch, when, by the new methods, they solved an amazing number of problems without worrying too greatly about the validity of their proofs; the then Prince of Mathematicians was Leonard Euler (1707–1783), a man who "calculated as he breathed", and was so inventive that he not only discovered innumerable formulae which are still useful, but – and this is far more difficult – also provided proofs, almost

---

[1] See various comments on this point in Chap. II, n° 7.

always shaky, even though the results were themselves correct. This way of doing mathematics reached its apogee at the beginning of the XIX$^{th}$ century with Joseph Fourier and his trigonometric series, without which a large part of present day mathematics and physics would not have been possible. Fourier's "proofs" are not only invalid; they are also unmeaningful, resting as they do, even if not explicitly, on equations as absurd as

$$1 - 3 + 5 - 7 + \ldots = 0,$$
$$1^3 - 3^3 + 5^3 - 7^3 + \ldots = 0,$$
$$1^5 - 3^5 + 5^5 - 7^5 + \ldots = 0$$

etc. His results, here again, are nevertheless correct; his first formula is the series for the "square wave" as used by all electricians. His general theorems on periodic functions were soon correctly proved by quite other methods by more serious mathematicians, Abel and above all Dirichlet, who were the first, with Cauchy, to introduce rigour and precision into analysis in the 1820s; nevertheless Fourier had seen all the simple and fundamental results and had invented a method which, right up to our days, has been unceasingly exploited in situations far more general and difficult than he could have known.

One now enters progressively into a new age, when, particularly in Germany, while waiting for the French and Italians at the end of the XIX$^{th}$ century, mathematicians systematically put to the test all the concepts – limits, convergence, irrational numbers, continuity, differentiation, integration, etc. – which involve, implicitly and more often explicitly, an *infinite number* of operations, and so were not defined in a perfectly clear and unambiguous way; they challenged all the "geometrically obvious" statements which were incorrectly proved and even sometimes false if taken literally, as was finally understood. Sometimes apparently monstrous creatures arrived on the scene, notably in connection with Fourier series, which continued to have surprises in store, and still, in 1997, pose very difficult problems: discontinuous functions which jump brusquely and at all rational values of the variable, continuous curves not admitting even a single tangent, functions which one did not know how to integrate, not because the "formula" was not known, but because they eluded all known definitions of an integral, continuous trajectories which passed through all the points of a square, etc.

On the other hand, there is a branch of mathematics which has always escaped these crises because the infinite plays no rôle there: arithmetic, or algebra, and particularly the classical theory of numbers and of algebraic equations, particularly the study of "algebraic" numbers, that is, roots of polynomial equations with integer coefficients, to which one attempts to generalise such classical results as decomposition into a product of prime factors. Carl Friedrich Gauss, who did many other things as well as mathematics, considered arithmetic so understood to be the "Queen of the Sciences"; he himself was the King of Arithmetic between 1800 and 1830 ... The study of these numbers presents enormous methodological difficulties – it took decades of efforts by several mathematicians of the first rank before Dedekind discovered after 1870 what has become the guiding thread up to our days, the theory of ideals –, but the results obtained, though still partial and limited in scope, and the methods used, left no doubt: the logic of proof was unassailable. One knew exactly what one was speaking of when proving a theorem, even though, to be sure, not all of the very complex mechanism which governs the properties of algebraic numbers had yet been discovered. The problems of arithmetic call for the detective's art; those posed by the fundamentals of analysis are rather a matter for philosophical reflection.

The main part of these developments in arithmetic was due, here again, to the Germans influenced by Gauss; he himself *de facto* founded a dynasty which ruled the subject for most of a century. It is hardly surprising that, in these circumstances, other Germans, sometimes they themselves (Dedekind), little by little elaborated a program which was named *to arithmetise analysis*, as Felix Klein put it in 1895: in other words, to substitute irrebuttable proofs in place of the hazy or wrong arguments of the XVII[th] and XVIII[th] centuries, to substitute completely clear concepts for the vague intuitions founded on deceptive geometric images – no one will ever draw the graph of a continuous nondifferentiable function –, in sum, as much as possible to "replace computation by concepts" as, much later, Bourbaki would write in the Preface to his *Elements of Mathematics*. As we shall see at the beginning of the next chapter, one does not have to go very far to understand the need for this arithmetisation: what is the number $\pi$ ?

There are two fundamental aspects here. In the first place, to elaborate a systematic technique for manipulating approximations to allow one to give perfectly precise definitions of such concepts as convergence, continuity, differentiability etc. This started, still rather vaguely, with Cauchy in about 1820, and was completed about 1870–1880 with Weierstrass and his usage of $\varepsilon$ and $\delta$, *Epsilontik* as our German colleagues term it; vast, more or less abstract generalisations emerged in the XX[th] century, where the $\varepsilon$ and $\delta$ would be replaced by the concept of "neighbourhood", but the ideas remained essentially his own, and his technique remained necessary and often sufficient in the immense majority of branches of analysis. As I shall use them from start to finish of this book (though my typist's habit is to use $r$ and $r'$ where Weierstrass and all contemporary mathematicians use $\varepsilon$ and $\delta$), it is needless to say more here than, in essence, it consists of demonstrating equalities by replacing them by more and more precise inequalities: one shows that $a = b$ by proving that $|a - b| < 1/10^n$ for every integer $n$. This is the fundamental difference between analysis and arithmetic or algebra.

The other aspect, which spread less easily because it constituted a major upset to the modes of thought of mathematicians, was the invention of *Set Theory* by Georg Cantor (1845–1918) between about 1870 and 1890: to a first approximation this consists of conceptualising that the totality of mathematical objects possessing a given property forms, in itself, a new quite distinct

mathematical object, and of reasoning about such objects; the appearance of
the "actual infinity" in mathematics, a concept which for centuries has set
cohorts of metaphysicians and theologians cogitating, and often deliriously
– Cantor knows and quotes them – and which therefore at the beginning
provoked violent opposition from among the mathematicians[2]. Cantor and
Dedekind, with whom he cooperated, were at first interested in more and
more complicated sets of real numbers – here again it was the theory of
trigonometric series which furnished Cantor with his initial motivation and
examples, though the fields of algebraic numbers and Dedekind's "ideals" are
also sets of numbers, even if much simpler[3], – then to sets which, in the eyes
of many mathematicians of the period, were a matter for metaphysics rather
than normal mathematics, a criticism which, at the beginning, was legitimated by the logical paradoxes which arose on applying modes of reasoning
that rely on an abusive extension of ordinary language to them. One should
note that before attempting the assault on "transfinite numbers" which, a
century later, are rarely used in practice, Cantor introduced extraordinarily useful concepts into the study of sets of numbers or points – open and
closed sets, "accumulation points" etc. – which do not seem to pose logical
problems[4], which many mathematicians then revealed in their most orthodox
research, and which would complete the clarification of analysis undertaken
by Weierstrass, of whom, moreover, Cantor had been a student.

Then a period opens which sees the birth of mathematical logic, of the
theory of "abstract" sets, and first attempts to axiomatise all of mathematics, starting with the theory of integers, i.e. arithmetic. Founded by Gottlob
Frege and pursued by Giuseppe Peano, Bertrand Russell and Alfred North
Whitehead, Ernest Zermelo, David Hilbert, etc., to mention only authors
known or famous before 1914, these new theories were intended to codify
on the one hand the rules of construction of mathematical arguments (elementary logical operations, use of "variables", the formulation and calculus
of "propositions", etc.), and on the other hand the rules of construction of
mathematical objects (numbers, functions, sets, etc.), and finally to isolate

---

[2] The opposition came not so much from the fact that the bizarre sets that Cantor
constructed contained an infinite number of elements: no one objected to considering a line, a plane, a curve, as an infinite collection of points, nor to considering,
for example, the intersection of a plane and a surface. The major objection arose
from the fact that Cantor sometimes constructed sets by an infinite number of
intermediate constructions, with arbitrary choices at each stage, inexplicit, and
even impossible to describe explicitly. Present day mathematicians do this every
day, but it was not so around 1870–1880.

[3] In the ring of rational integers $\mathbb{Z}$ (i.e. of arbitrary sign) an ideal is a set $I$ possessing the property that $ux + vy \in I$ for every $x, y \in I$ and $u, v \in \mathbb{Z}$. Such an
ideal is the set of multiples of some integer.

[4] But fifty years it later would be discovered that one of the conjectures which
Cantor and others endeavoured in vain to prove (the Continuum Hypothesis)
is in fact unprovable and one can, at will, accept or reject it, as with Euclid's
Parallel Postulate.

the axioms starting from which one can erect the whole of mathematics in a perfectly coherent way, without, it is to be hoped, stumbling upon an internal contradiction. The rules imposed, while allowing all the standard mathematical arguments and objects, are strict enough for the early paradoxes to be, one hopes, impossible to formulate without infringements; to wit, safety barriers beyond which one ventures at one's own risk and peril. This domain, which has been the subject of a large number of works for a century, still presents very difficult problems and has sometimes given rise to intense debate; most mathematicians observe this from a distance, content with a "naive", i.e. not strictly formalised version, of Logic and Set Theory.

The principal result of this work is that *every mathematical object can be considered as a set* and indeed that the only logically correct way to *define* a mathematical object is to say that it is composed of one or more sets subject to explicitly stated conditions. This method suffers from the inconvenience of making the seemingly simplest mathematical objects, the real numbers for example, appear as extremely complex elaborations of sets; in the definition of the real numbers as "cuts", due to Dedekind, and simplified by Peano and Russell, which we shall give at the beginning of the next chapter, a real number, for example $\pi$, is *by definition* a set of rational numbers (naively: the set of all rational numbers $< \pi$); a rational number, in its turn, is[5] the set of all pairs of integers $(p, q)$ of arbitrary sign such that $x = p/q, q \neq 0$; an integer $n$ of arbitrary sign is itself the set of all pairs $(p, q)$ of whole numbers such that $p - q = n$; a whole number, at last, is a set, the number 3 for example being a set of three elements chosen once and for all; thus the number $\pi$ becomes a set of sets of sets of sets, all infinite except for the last. One might therefore, starting from the whole numbers, establish a kind of *hierarchy of complexity* in the universe of mathematical objects, as do certain logicians. But of course from the moment these objects and the operations to which they may be subjected have been defined by applying the standard operations of set theory to objects already known, one forgets their explicit definitions, their extreme complexity making manipulating them quite impossible; one confines oneself to arguing as was always done, up to one detail: one knows exactly, or one could know, if one put one's mind to it, what the symbol $\pi$ signifies, as it leaves metaphysics and enters into mathematics[6].

Following this historic evolution one can contemplate Set Theory from two different points of view: on the one hand from the "naive" point of view of

---

[5] See for example §§5 and 28 of my *Cours d'algèbre* (Hermann, 1966) or *Algebra* (Addison-Wesley. 1972)

[6] The physicist Emilio Segre explains in his memoirs that, when he was in high-school, he did not understand why his mathematics teacher stressed the need to define the real numbers by means of Dedekind sections, the concept of number seeming to be *sui generis*. This only proves that a Nobel Laureate in physics may not understand the mathematics he has been using during his whole life, or, if one prefers, not understand the difference between Physics and Mathematics. See the beginning of Chapter II.

mathematicians who use it daily while restricting themselves to the imagining of the simplest geometrical and physical images, on the other hand, from the "formalised" point of view of logicians who, by systematically applying rules of argument, and methods of construction stated once and for all, construct everything from nothing "to allow thought to rise above the void by leaning on the void" as wrote a physicist studying Newton's cosmology[7]. We shall confine ourself to expounding an intermediate point of view very succinctly, without using logicians' language, but nevertheless not letting the reader believe that intuitively "evident" results require no justification. The use of ordinary language can make propositions seem intuitive, though professional logicians will consider them by no means evident, and computers will not understand for the simple reason that they lack intuition[8]: to make oneself understood by a stupid though disciplined machine one must explain oneself in a perfectly correct fashion, even if one only wants to receive information[9] on the *Silicon Valley Paedophiles Brotherhood* through the Internet. We shall give some very summary details on the language of logicians later.

----

[7] Loup Verlet, *La malle de Newton* (Gallimard, 1993), p. 292. Highly recommended reading.

[8] Although the invention of computer science has presented several inconveniences to mankind – ask ordinary Iraqis what they think of the computers on a cruise missile –, it offers a great advantage to mathematical pedagogy: a proof is totally correct if a suitably programmed computer could understand it. This point of view does not eject the rôle of intuition from mathematics; it simply shows that one must not confuse an intuition with an argument.

[9] For computer scientists, "information" is a sequence of the digits 0 and 1. It is interesting to note that before becoming a "cultural rubbish bin", as the Chinese call it, the Net was first conceived as a military system with the purpose of assuring American telecommunications in the event of nuclear war, thanks to the total interconnection of the given bases and centres of calculation or command; it was soon linked to laboratories and university departments highly delighted to profit from it (and to profit the military through their expertise) after which the military provided themselves with a system for their exclusive use while the "civil" system passed into the control of the National Science Foundation, from which it has finally been removed.

There is an enormous American literature on the history of computer science; the following titles, by historians who cite their sources, are particularly recommended: Kenneth Flamm, *Creating the Computer: Government, Industry, and High Technology* (Brookings Institution, 1988), Paul Edwards, *The Closed World. Computers and the Politics of Discourse in Cold War America* (MIT Press, 1996), Arthur L. Norberg and Judy O'Neill, *Transforming Computing Technology. Information Processing for the Pentagon 1962-1986* (The Johns Hopkins UP, 1996), very dry, Thomas P. Hughes, *Rescuing Prometheus* (Pantheon Books, 1998) which also treats other similar subjects (missiles, SAGE, systems analysis, etc.), Janet Abbate, *Inventing the Internet* (MIT Press, 1999), from a Ph.D. thesis, Kent Redmond and Thomas M. Smith, *From Whirlwind to Mitre. The R&D Story of the SAGE Air Defense Computer* (MIT Press, 2000), where one can see to what extent the development of the gigantic SAGE net for the air defence of the American continent influenced the progress of computer science and electronics in the 1950s.

# §1. Set Theory

## 1 – Membership, equality, empty set

The concept of a set[10] is a primitive concept in mathematics; one can no more
provide a definition than Euclid could define mathematically what a point is.
In my youth there were those who said that a set is "a collection of objects of
the same nature"; apart from the vicious circle (what indeed is a "collection"?
a set?), to talk of "nature" is empty and means nothing[11]. Certain denigrators
of the introduction of "modern math" into elementary education have been
scandalised to see that in some textbooks they have had the temerity to form
the union of a set of apples with a set of pears; never mind that a normal
child will tell you that this gives a set of fruits, or even of things, and if asked
to count the number of elements of the union any moderately intelligent child
can explain to you that it does not matter that the first set consists of apples
rather than oranges and the second of pears rather than dessert spoons; the
fact that the Louvre Museum combines disparate collections – of pictures,
sculptures, ceramics, gold work, mummies, etc. – has never troubled anyone.
One calls this: to acquire the sense of abstraction.

The logicians have in any case long since invented a radical method of
eliminating questions concerning the "nature" of mathematical objects or
sets (the two terms are synonymous). One can describe this in a figurative
way by saying that a set is a "primary" box containing "secondary" boxes,
its *elements*, no two of which have identical contents, which in their turn
contain "tertiary" boxes themselves containing ... The Louvre is a collection
of collections (of paintings, sculptures, etc.), the collection of paintings is
itself a collection of paintings stolen by Bonaparte, Monge and Berthollet in
Italy (we unfortunately had to return it in 1815), bequeathed by ... private
collectors, bought at sales, etc.

The whole of set theory rests on two sorts of relations. The *membership
relation* $x \in X$ which is read "*x belongs* to $X$" or "*x* is an *element of $X$*"; this
means that $x$ is one of the secondary boxes contained in the box $X$, while
*secondary* means: not contained in any box other than $X$ itself. The negation
of $x \in X$ is denoted $x \notin X$. To express the fact that an object $x$ is an element
of a set which is itself an element of $X$, one might write $x \in\in X$; at the next
level one might write $x \in\in\in X$. These last two notations have not found
common currency, but I will use them occasionally in this chapter. If one
considers the Louvre as a set whose elements are its collection of paintings,
its collection of sculpture etc., then the Mona Lisa $\in\in$ Louvre.

On the other hand there is the *equality relation* $x = y$, whose intuitive
meaning is that the two sets are identical; its negation is written $x \neq y$. The

---

[10] After several tentatives Cantor chose the word *Menge* (quantity, number,
amount, mass, multitude, crowd).

[11] Cantor defined a set as "every assemblage as a whole (Zusammenfassung zu
einem Ganzen) $M$ of defined and distinct objects $m$ of our intuition or thought".

two relations ∈ and = obey axioms which we shall state gradually as needed; they permit us to construct ever more complex sets and relations.

The first, the *axiom of extension*, says that two sets $A$ and $B$ are equal (i.e., for the mathematician, identical or indistinguishable) if and only if they possess the same elements; in other words, if the relations $x \in A$ and $x \in B$ are logically equivalent. In the interpretation of a set $X$ as a nesting of boxes it is thus pointless to place in the primary box $X$ two secondary boxes $A$ and $B$ which, as nestings of boxes, have exactly the same structure: they are not mathematically distinct objects even if physically they appear to be so. If for example you place three empty boxes in a box $X$ (or, more generally, three copies of the same box), you obtain the same set as if you had placed only one, since the axiom of extension shows that the two sets are mathematically identical.

Given two sets $X$ and $Y$ one says that $X$ is *contained in* $Y$ (or that $Y$ contains $X$, or that $X$ is a *subset* of $Y$) if every element of $X$ is an element of $Y$; the notation is $X \subset Y$ or $Y \supset X$. It is clear that if $X \subset Y$ and $Y \subset Z$, then $X \subset Z$. The relation $X = Y$ means that both $X \subset Y$ and $Y \subset X$.

Some authors write $X \subseteq Y$ to maintain the visual analogy with $x \leq y$, and reserve $X \subset Y$ to mean that $X$ is a *proper* subset of $Y$. This is a totally useless notation.

If one considers, as has just been suggested, that mathematics consists essentially of proving theorems about more or less complex nestings of boxes, the simplest box one can imagine is the empty box. One thus needs a particular mathematical object, denoted $\emptyset$, the *empty set*; its existence is the second of the axioms of set theory[12]: there exists a set $\emptyset$ such that the relation $x \in \emptyset$ is false for every $x$.

One meets this set in everyday life. If, when you are travelling by car in the Far West, the police arrest you because you have shot a red light at the intersection of two flat, deserted, absolutely straight, orthogonal roads you might acknowledge that your infraction constituted a mortal danger to the, empty, set of motorists visible within a range of ten miles. (You would pay the same fine, even so.) On 12 August 1997, the atmospheric pollution in Paris having attained too high a level, the Parisian police, relayed by the media, generously announced that Paris residents parking their cars within their authorised perimeter were graciously absolved, the following day, from paying their daily tribute of 15 F for this right[13]; but all the motorists who

---

[12] Some logicians prefer to postulate the existence of *a* set; that of $\emptyset$ follows immediately from the axiom of separation stated below.

[13] The idea is to encourage Parisian motorists to travel to work by public transport. The fact is that normally using your car instead of parking in your street allows you to save 15 F for parking and two metro tickets, i.e. enough money to buy three litres of petrol. The duty of paying 15 F or a fine of 75 F if you do not use your car to get to work before 9 a.m. thus amounts to subsidising the polluters and penalising those who use public transport. This is confirmed by the fact

reside in Paris and park on the public roads know that the set of days of the month of August on which this tax is obligatory is empty, in contrast to the rest of the year.

The most remarkable property of the empty set is that everything one can say about its elements (though not about itself) is simultaneously true and false, and that, further, no logical catastrophe ensues. When I once informed a friend that every man who has passed the age of five hundred years makes love three times a day she replied "that's false, I'm sure you wouldn't be able to"; to this kind of typically feminine *ad hominem* logic – it is well known that women are incapable of reasoning impersonally, objectively and abstractly – I obviously replied "certainly, with you"; she then exclaimed "false, I'd be dead and I don't like old men" and, to finish, threw a fit of nerves when I replied that that did not contradict the initial proposition one whit. To learn to juggle with the innumerable properties of the empty set is an excellent exercise for developing your powers of reasoning; you could in particular set yourself to detect all statements, including those in the present treatise, which, taken literally, are false because the author has forgotten to posit that a certain set is not empty: "every continuous function on a compact set attains its maximum at a point of this set", "every bounded set has a strict upper bound", etc.; these statements are false if the set under consideration is empty because they affirm the existence of an element (possessing certain properties) of the empty set. The perpetrators of such gross errors will generally reply that they have passed the age of priggishness and trust to the good sense of the reader, who is asked, implicitly, to trust the competence of the author.

The empty set is a *subset* of every other set $X$: since $\emptyset$ contains no elements at all, all its elements are also elements of $X$. It is also true that the empty set may be an *element* of another set $X$ as we shall now see.

It is on the empty set that one leans to "lift oneself up from the void"; figure 1 below represents a primary box $X$ containing three secondary boxes $A$, $B$, $C$, which are the elements of $X$; $A$ is the empty set and contains no elements, $B$ is a box containing an empty box, so has one element, namely the empty box in question, while $C$ is a box with two elements: an empty box and a box containing an empty box.

The representation above might lead the reader to believe that there are four distinct empty sets in the schema for $X$; now there is only one empty set in Nature, but, like the Holy Ghost, it is everywhere simultaneously. One can finesse this niggle by replacing the imagery of boxes by the schema of the relation $x \in y$; on writing $x \to y$ to facilitate the graphic representation, figure 1 would be replaced by the following schema:

---

that parking is free between 7 in the evening and 9 in the morning as well as on Saturday and Sunday, i.e. outside working hours. One should teach the abc of formal logic (a few pages of Plato would suffice) to the bureaucrats who hope to fight pollution by taxing the non-polluters. Let us add that in certain countries the residents buy a permit at the beginning of each year, so freeing them from the daily racket which one suffers in Paris, and for a modest sum.

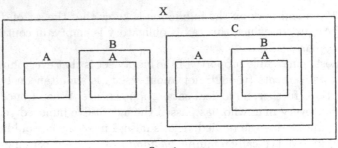

fig. 1.

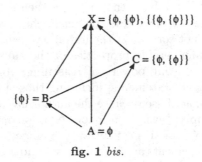

fig. 1 *bis.*

The significance of the signs { and } will be explained in the following n°.

## 2 – The set defined by a relation. Intersections and unions

In practice, a set is most often defined by a characteristic property of its elements: "the set of integers between 2 and 25", "the set of points distant 15 km from a given point O", "the set of rational numbers $< \pi$", "the set of functions $f$ defined on $\mathbb{R}$ with values in $\mathbb{R}$", etc. A seemingly obvious general statement, due to Frege, is that for every proposition or relation $P\{x\}$ in which there appears a variable $x$ symbolising a totally undetermined object one can speak of *the set of those $x$ such that $P\{x\}$ is true*; this set will be unique, by the axiom of extension. If you choose the relation $x = x$ you will thus obtain the set of *all* mathematical objects, a creature which justly evoked great suspicion in Cantor; he spoke of it as a "class" of sets, a concept which the logicians later used and developed.

    In 1903, as Frege was on the point of publishing the second volume of his *Grundgesetze der Arithmetik*, Bertrand Russell[14] demolished Frege's whole

---

[14] On Bertrand Russell, an exceptional personality and formidable source of ideas of all sorts, see Ronald. W. Clark, *The Life of Bertrand Russell* (Penguin Books, 1975). A much more scholarly biography is in course of publication, but the 979 pages of Clark, of which 160 are notes, bibliography and index, already contain much information; sold at £2.95 when it appeared, one cannot ask for much more ... Do not seek mathematical logic there: Clark is a "man of letters", does not

edifice, at least twenty years' work, and a part of his very self, by choosing $x \notin x$ for the relation $P\{x\}$. Suppose that there is indeed a set $A$ such that

(2.1)        for every $x$,   $x \in A$ is equivalent to $x \notin x$.

Since a relation which is true for every $x$ remains so when one substitutes a specific mathematical object for the variable $x$, one sees that the relations $A \in A$ and $A \notin A$ are logically equivalent: contradiction!

The *axiom of separation* (Ernest Zermelo, 1908) obviates "Russell's Paradox": if $P\{x\}$ is a proposition and $X$ is a set one may speak of the set $A$ of those $x$ *which belong to* $X$ and satisfy $P\{x\}$; in logical language[15]:

(2.2)            $(x \in A) \iff (P\{x\} \ \& \ (x \in X))$;

instead of placing oneself in the absurd universe of all possible mathematical objects one places oneself in the specific set $X$: this is one of the guard-rails of the theory[16]. In particular, one may not speak of "the set of all sets", as was done in Cantor's time, for if such a set $X$ existed the relation $x \notin x$ would define, by (2), a set $A \subset X$ satisfying (1), an absurdity. You may certainly think of the "class", "category", "totality" of sets, but this is not a set in the technical sense of the term.

When $X$ and $Y$ are two sets one writes $X - Y$ for the set of elements of $X$ that do not belong to $Y$: the axiom of separation legitimates this definition: $P\{x\}$ here is the relation $x \notin Y$. By far the most frequent case is that when $Y \subset X$; $X - Y$ is then the *complement* of $Y$ in $X$; in this case one obviously has

$$X - (X - Y) = Y.$$

---

understand it, and makes no attempt to appear to understand it. This is a general problem in the history of science: when it is written by scientists who understand the subject the socio-political aspects disappear or reduce to non-documented banalities, and *vice versa*. The exceptions, for instance Loup Verlet's book cited above, are very rare. Dieudonné resolved this dilemma by saying that the socio-political aspects do not explain the scientists' ideas. Apart from the fact that this statement may be false (obvious counterexamples: the logarithms of Napier and Briggs for astronomers and navigators, Lavoisier and the French gunpowder administration, Gauss and geodesy, Liebig and nitrogenous fertilisers, Haber and the direct synthesis of ammonia, von Neumann after 1937, etc.), one is in the right not to be *solely* interested in mathematics or physics in the strict sense.

[15] The sign $\iff$ denotes logical equivalence; the sign $\&$ indicates the conjunction of two statements: the parentheses have the purpose of delimiting relations. See the part of this chapter that treats mathematical logic.

[16] The solution found by Zermelo trivially eliminates Russell's Paradox since then one does not have the right to talk "in the air" of the *set* of $x$ such that $x \notin x$: one must specify at the outset that one places oneself in a given *set*. But the true problem, resolved by Zermelo and his successors, was to show that on adopting their axioms one did not forbid anything commonly done in mathematics. The solution would have been rejected if, for example, it had made it impossible to construct the real numbers.

And $X - X = \emptyset$ for any $X$.

If $X$ and $Y$ are two sets their *intersection* $X \cap Y$ is the set of objects belonging simultaneously to $X$ and to $Y$; this definition is again legitimated by the axiom of separation applied to $X$ and the relation $x \in Y$. One says that $X$ and $Y$ are *disjoint* when $X \cap Y = \emptyset$.

The *union* $A \cup B$ of two sets is, intuitively, the set of objects belonging to $A$ or[17] to $B$. More generally, consider a set $X$ and think of its elements as themselves being sets; the *axiom of union* affirms the existence of a set $Y$ whose elements $y$ are characterised by the following property: there exists an $A \in X$ such that $y \in A$; logicians call this the *union* of $X$, an expression generally eschewed by mathematicians. It is the set of $x$ such that $x \in\in X$. In the imagery of boxes this signifies that one may suppress the various secondary boxes belonging to $X$ and replace them by the tertiary boxes they contain, eliminating double mentions as always.

One ought to transform all this into an exciting game, one might even make a fortune by patenting it. Mr Gates' employees would immediately devise a multicoloured speaking version ("now find the union of the union of the union") for multimedia computers. There would be several degrees of difficulty, characterised by the maximum number of nestings allowed: the BIB (Boxes in Boxes) for babies, the BIBIB (Boxes in Boxes in Boxes) etc., up to BIBIBI ... (Boxes in Boxes in Boxes in ...) at level $\aleph_0$. One could, thanks to the Internet, organise Olympiads on a planetary scale, as in mathematics. The parents of future students of the Polytechnique, of Harvard, or of the Todaï University in Tokyo, could present BIBIBIB to their offspring at the age of six for the gifted, and three for the extra-gifted, the infinite ascensions in the BIBIBI ... being reserved for the Mozarts of Logic.

The axiom of union would allow one to legitimate logically the definition of $A \cup B$ if one knew that there was a set $C$ of which $A$ and $B$ are elements. The existence of $C$ is as "evident" as is, in Euclidean geometry, the existence of a unique straight line joining two given points. As evident naively, as undemonstrable logically. We therefore need another axiom, the *axiom of pairs*: if $A$ and $B$ are two sets there exists a set $C$ of which $A$ and $B$ are the only *elements*. $C$ is unique, by the axiom of extension. One denotes it $\{A, B\}$, or $\{A\}$ if $A = B$; you will have no trouble verifying that $\{A, B\} = \{B, A\}$. Given three sets $A$, $B$, $C$ one puts $\{A, B, C\} = \{A, B\} \cup \{C\}$, etc.

The operations of union and intersection have the following nearly obvious properties:

$$X \cup Y \;=\; Y \cup X,$$

---

[17] In mathematics the conjunction "or" is not disjunctive: if $P$ and $Q$ are statements, "$P$ or $Q$" does not exclude "$P$ and $Q$".

$$X \cup (Y \cup Z) = (X \cup Y) \cup Z,$$
$$X \cap (Y \cup Z) = (X \cap Y) \cup (X \cap Z),$$
$$X - (Y \cup Z) = (X - Y) \cap (X - Z),$$

$$X \cap Y = Y \cap X,$$
$$X \cap (Y \cap Z) = (X \cap Y) \cap Z,$$
$$X \cup (Y \cap Z) = (X \cup Y) \cap (X \cup Z),$$
$$X - (Y \cap Z) = (X - Y) \cup (X - Z),$$

Rather than learn these relations by heart one should be able to reconstruct them on the instant; once the notation has been understood these rules reduce to simple common sense.

The axiom of separation eliminates Russell's Paradox, yet, while the relation $x \notin x$ has a very normal look, the opposite relation, $x \in x$, seems very strange: one has never seen a museum of painting which is an element of its own collection of pictures, and one would be hard put to it to realise $x \in x$ in the Game of Boxes of Boxes; it would be even more strange to consider sets $x$, $y$ and $z$ such that at the same time $x \in y$, $y \in z$ and $z \in x$: as Suppes more or less said "if you do not believe that this is against intuition then try to find an example"; one might add à la Serge Lang: if you succeed you will instantly be *world famous among mathematicians* because you will have demolished a theory painstakingly constructed over a century by excellent or very great mathematicians. The relations which might seem to permit this have been eliminated by means of the *axiom of regularity* or *foundation* (*Fundierung* in German) formulated by von Neumann in 1925 and simplified by Zermelo in 1930: it says that if one considers the elements of a nonempty set $A$ as themselves being sets then there exists a set $X \in A$ such that $X \cap A = \emptyset$; we shall use this in n° 9, but one has no occasion to use it in practical mathematics. To deduce the impossibility of a relation such as $x \in y \in z \in x$ one applies this axiom to the set of three elements $A = \{x, y, z\}$ and deduces a contradiction since then

$$x \in A \cap y, \quad y \in A \cap z, \quad z \in A \cap x,$$

so that the intersections of $A$ with its elements are all nonempty. One can also deduce the impossibility of an unending *descending* chain such that $x_1 \ni x_2 \ni x_3 \ni \ldots$; such a relation contradicts the axiom of regularity for the set $A = \{x_1, x_2, x_3, \ldots\}$, since, for every $p$, one has $x_{p+1} \in x_p \cap A$ and thus $A \cap x \neq \emptyset$ for every $x \in A$. A descending chain of membership relations thus always leads to the empty set if one pursues it far enough.

On the other hand there are unending *ascending* chains, for example

$$0 \in 1 \in 2 \in 3 \ldots$$

as we shall see in the next n°.

# 3 – Whole numbers. Infinite sets

Applied to the innocent empty set the formation of pairs leads to *nonempty* sets, not yet logically guaranteed to exist at this stage of the work: in the first place $\{\emptyset\}$, the set whose only element is the empty set (box containing an empty box), then $\{\{\emptyset\}\}$, the set whose only element is the set whose only element is the empty set (box containing a box containing an empty box) etc. The relation $\emptyset \in \{\{\{\emptyset\}\}\}$ is false; the correct relation is $\emptyset \in\in\in \{\{\{\emptyset\}\}\}$: the empty set belongs to a set which belongs to a set which belongs to $\{\{\{\emptyset\}\}\}$. These sets are pairwise distinct: the relation $\{\{\emptyset\}\} = \{\{\{\{\emptyset\}\}\}\}$ for example would imply $\{\emptyset\} = \{\{\{\emptyset\}\}\}$ by the axiom of extension, then $\emptyset = \{\{\emptyset\}\}$ for the same reason, then $\{\emptyset\} \in \emptyset$, which is false. An empty box contains nothing, not even an empty box.

This type of construction furnishes a possible definition of the primary objects studied in mathematics, namely the *whole numbers* or *natural integers*. According to Zermelo, 1908, and in conformity with the programme of reducing everything to set theory, one defines them by

$$(3.1) \qquad 0 = \emptyset, \qquad 1 = \{\emptyset\}, \qquad 2 = \{\{\emptyset\}\}, \ldots ;$$

simple abbreviations[18] for very particular sets. A computer would understand, but it would take twenty seconds for a machine running at 100 Mhz to write or read the number $10^9$, assuming that one cycle is enough to recognise the signs $\{$ and $\}$.

Another method of defining the integers, equivalent[19] to the preceding, due to von Neumann[20], consists of putting

---

[18] In particular the sign $=$ used in these definitions is not that of set theory; when introducing a definition the logicians (and now some mathematicians) prefer to use the sign $:=$, the sign $:$ warning the reader that one is introducing a definition or new notation, and not a relation to be proved.

[19] The precise mode of definition of the integers (or of any other mathematical object) is of no importance so long as the various possible definitions lead to the same theorems; the "nature" of mathematical objects is irrelevant because one only asks them to be models of real objects. For the same reason the symbol used by computer scientists to designate the number 15 is unimportant on the theoretical plane, so long as the computers are programmed to recognise it.

[20] 1923; he was twenty years old and in the course of spending a few years learning chemistry in Zürich because his father, a banker in Budapest, wanted to direct him to a profession more lucrative than mathematics; he had already read Cantor and Co. at least three years earlier. This definition of the integers, used by N. Bourbaki, made certain French users of mathematics laugh at a certain period. A list, even though incomplete, of von Neumann's activities, from 1937 – classical explosives, operational research and game theory, the A-bomb, the H-bomb, intercontinental missiles, computers – should reassure the philistines of his sense of the concrete. But he was not intellectually cynical. Laurent Schwartz, *Un mathématicien aux prises avec le siècle* (Odile Jacob, 1997), p. 288, translated as *A Mathematician Grappling with his Century* (Birkhäuser, 2001) writes about me that I have "never pardoned von Neumann for having forsaken mathematics to create computer science"; it is true that thanks to a colleague from Princeton we knew vaguely in 1947, at Nancy, that von Neumann "was now doing numerical calculations" though we were ignorant of their purpose; but, at least so far as I am concerned, we know a lot more half a century later ... On von Neumann

(3.2)         $0 = \emptyset$,     $1 = \{\emptyset\} = \{0\}$,     $2 = \{\emptyset, \{\emptyset\}\} = \{0, 1\}$,

$$3 = \{\emptyset, \{\emptyset\}, \{\emptyset, \{\emptyset\}\}\} = \{0, 1, 2\},$$

$$4 = \{\emptyset, \{\emptyset\}, \{\emptyset, \{\emptyset\}\}, \{\emptyset, \{\emptyset\}, \{\emptyset, \{\emptyset\}\}\}\} = \{0, 1, 2, 3\},$$

etc. For example, 3 is the set whose elements are (a) the empty set, (b) the set whose unique element is the empty set, (c) the set whose only elements are the empty set and the set whose unique element is the empty set (fig. 1). Thus $\emptyset \in 4$, $\emptyset \in\in 4$, $\emptyset \in\in\in 4$ and $\emptyset \in\in\in\in 4$. Figure 2 shows the two possible definitions of the whole number 4 in the imagery of boxes.

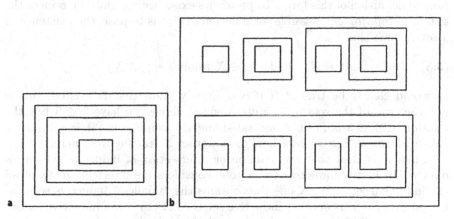

**fig. 2.** a) 4 according to Zermelo; b) 4 according to von Neumann

Given a set $x$ the logicians call the set $s(x) = x \cup \{x\}$ the *successor* of $x$; we have $s(x) \neq x$ since $x \notin x$. One thus obtains the integers by applying this operation repeatedly to the empty set: 0 is the empty set, 1 is the successor of 0, 2 the successor of 1, etc. In other words, if 14 is a primary box containing fourteen secondary boxes, then 15 is the primary box containing identical copies of the fourteen secondary boxes contained in 14 *and* the box 14 itself, which is not identical with any of the fourteen boxes it contains. In what has just been written "fourteen" is what is known to everyone who knows to read, write and count, while 14 is the mathematical or logical number defined by the method of Zermelo or von Neumann; a computer understands 14 but not fourteen; for humans it is usually the opposite.

Von Neumann's definition seems much more complicated than Zermelo's: to write the number $10^9$ explicitly requires $2^{1\,000\,000\,000}$ parentheses, an integer with about three hundred million digits, and $2^{999\,999\,999}$ mentions of $\emptyset$. But in conformity with intuition it defines 14 as a set of fourteen elements. It

and computer science, see William Aspray, *John von Neumann and the Origins of Modern Computing* (MIT Press, 1990).

presents another advantage, because of which it has been adopted almost universally for the construction of the "infinite ordinals" of Cantor, as we shall show in n° 9.

The question of how to know whether the sets used to define the integers are themselves elements of some set, in other words, that of the existence of *the set* $\mathbb{N}$ *of whole numbers*, is not logically evident; it would be, by the axiom of separation, if one already knew of the existence of a set of which at least all the integers are elements, but where to find one? All the sets we have actually constructed up to now, starting from the one set whose existence is guaranteed *a priori*, namely $\emptyset$, have a finite number of elements, while, according to all the evidence, $\mathbb{N}$ must possess infinitely many, whatever the precise definition of this term. To justify its existence one thus introduces the *axiom of infinity*: one possible formulation of this is to posit the existence of a set $X$ such that

$$(3.3) \qquad (\emptyset \in X) \quad \text{and} \quad (x \in X \text{ implies } s(x) \in X),$$

as would clearly be true of $\mathbb{N}$ if one already knew that $\mathbb{N}$ existed; this is not the case of the sets 0, 1, 2 etc. defined above: we have $2 \in 3$ but the relation $s(2) \in 3$ is false. A set satisfying the conditions (3) is sometimes called *inductive*. Let us show how to construct $\mathbb{N}$ starting from here.

First, it is clear that every union or intersection of inductive sets is inductive. If, in any inductive set $X$, one considers the intersection $X_0$ of all the inductive sets $X' \subset X$ one thus obtains the "smallest" inductive set contained in $X$. If $Y$ is another inductive set then $X \cap Y$ is an inductive subset of both $X$ and of $Y$, so contains $X_0$ and $Y_0$; $Y_0$ is thus an inductive subset of $X$, whence $Y_0 \supset X_0$ and *vice versa*; in other words $X_0 = Y_0$. The set $X_0$ defined in each inductive set $X$ is thus the same independently of $X$; it is, by definition, the set $\mathbb{N}$ of whole numbers, and, similarly, the smallest of "all" inductive sets (which are too numerous to be the elements of a set).

Since $\mathbb{N}$ is inductive and contains $\emptyset$ it contains all the whole numbers *à la* von Neumann. This proves that they belong to a common set, and therefore we may speak of the set $E$ of these integers. It is clear that it is inductive and is contained in $\mathbb{N}$. Hence $E = \mathbb{N}$, so the elements of $\mathbb{N}$ are precisely the integers *à la* von Neumann.

From this follows the principle of *proof by induction*: to show that a property $P\{n\}$, in which the letter $n$ symbolises an indeterminate "variable", is true for every $n \in \mathbb{N}$ one shows that

(i)  it is true for $n = \emptyset$, i.e., in ordinary language, for $n = 0$,
(ii) the relation $P\{n\}$ implies $P\{s(n)\}$ i.e., in ordinary language, that $P\{n\}$ implies $P\{n+1\}$.

If this is so, then the set of $n \in \mathbb{N}$ satisfying $P\{n\}$ is inductive and contained in $\mathbb{N}$, so equal to $\mathbb{N}$.

## 4 – Ordered pairs, Cartesian products, sets of subsets

If $a$ and $b$ are mathematical objects one has $\{a, b\} = \{b, a\}$. If, on the other hand, you associate to every point of a plane furnished with coordinate axes its two coordinates $x$ and $y$ and denote the corresponding point by the classical notation $(x, y)$, it is clear that in general $(x, y) \neq (y, x)$. One is therefore led to associate with two objects $x$ and $y$ written in a determinate order a new object $(x, y)$, an *ordered pair*, (or, in French, *couple*) the rule of equality for two ordered pairs being that

(4.1)        $(x, y) = (u, v)$  if and only if $x = u$ and $y = v$.

Sometimes one says that $x$ and $y$ are the *projections* or *coordinates* of the ordered pair $(x, y)$. Similarly one defines *triplets*

$$(x, y, z) = ((x, y), z),$$

quadruplets

$$(x, y, z, t) = ((x, y, z), t),$$

etc. The rules of equality for such objects are obvious.

The axiom of pairs, used in n° 2 to define doubletons (non-ordered pairs) $\{x, y\}$, allows one to introduce ordered pairs without stepping outside the framework of Set Theory, by agreeing, for example, that

(4.2)        $(x, y) = \{\{x\}, \{x, y\}\},$

an astute idea due to the Pole Kasimierz Kuratowski (1921). Suppes tells us it was already to be found in another form in 1914 in the work of the American Norbert Wiener, the "Father of Cybernetics", as journalists knowing nothing else of him have called him since 1950. The relation $(x, y) = (u, v)$, i.e.

(4.3)        $\{\{x\}, \{x, y\}\} = \{\{u\}, \{u, v\}\},$

in fact forces either $\{u\} = \{x\}$, or $\{u\} = \{x, y\}$. In the first case one has $u = x$; in the second, $x = y = u$; so $u = x$ in either case. If $x = y$, one has $\{x, y\} = \{x\}$, hence $\{\{x\}, \{x, y\}\} = \{\{x\}\}$ so that (3) can be written $\{\{x\}\} = \{\{x\}, \{x, v\}\}$, which implies $\{x, v\} = \{x\}$ i.e. $v = x = y$; if $x \neq y$, the second member of (3), which can be written $\{\{x\}, \{x, v\}\}$ since $u = x$, cannot contain the element $\{x, y\} \neq \{x\}$ unless $\{x, v\} = \{x, y\}$, and since $x \neq y$ this forces $v = y$. In conclusion, one sees in every case that the condition (1) is satisfied by the definition (2) of ordered pairs.

*Exercise.* Draw the boxes or graphs that represent the sets $(2, 3)$ and $(3, 2)$.

Given two sets $X$ and $Y$ the set of ordered pairs $(x, y)$ for which $x \in X$ and $y \in Y$ is called the *cartesian product* of $X$ and $Y$, and is denoted $X \times Y$. More generally one defines $X \times Y \times Z = (X \times Y) \times Z$, the set of triplets $(x, y, z)$ with $x \in X$, $y \in Z$, $z \in Z$, etc. If $X$ is a set one defines

$$X^2 = X \times X, \qquad X^3 = X \times X \times X, \quad \text{etc.}$$

The existence of the cartesian product is plain from the naive point of view: for the logicians it demands a proof. Now the elements $z$ of $X \times Y$ are characterised by the following relation $P\{z\}$: there exist an $x \in X$ and a $y \in Y$ such that $z = \{\{x\}, \{x, y\}\}$. The existence of $X \times Y$ can therefore be deduced from the axiom of separation so long as one knows in advance that those $z$ satisfying $P\{z\}$ belong to a common set $Z$; but that precisely is the whole problem.

Instead of just postulating the existence of $X \times Y$ axiomatically the logicians go much further. In the formula $z = \{\{x\}, \{x, y\}\}$, $z$ is a set whose elements $\{x\}$ and $\{x, y\}$ are subsets of the union $X \cup Y$. The axiom which allows one to resolve the problem affirms in a general way that, *for every set $X$, there exists a set $\mathcal{P}(X)$ whose* elements *are the* subsets *of $X$*:

(4.4) $$Y \in \mathcal{P}(X) \iff Y \subset X.$$

If, for example, $X = \{a, b, c\}$ where $a$, $b$, $c$ are pairwise distinct, $\mathcal{P}(X)$ has the elements

$$\emptyset, \quad \{a\}, \quad \{b\}, \quad \{c\}, \quad \{b, c\}, \quad \{a, c\}, \quad \{a, b\}, \quad \{a, b, c\}.$$

Returning to the cartesian product of $X$ and $Y$, its *elements* $z$ are, after definition (2) of Kuratowski, *sets* whose two *elements* are *subsets* of $X \cup Y$, so *elements* of $\mathcal{P}(X \cup Y)$; one thus has $z \subset \mathcal{P}(X \cup Y)$ and so, by definition, $z \in \mathcal{P}(\mathcal{P}(X \cup Y))$. The definition (2) of ordered pairs thus provides (axiom of separation) a *set*

$$X \times Y \subset \mathcal{P}(\mathcal{P}(X \cup Y)).$$

This argument (which you may well forget once you have understood it: the sole thing to retain is the condition for two ordered pairs to be equal) may appear somewhat esoteric and abstract, but it has the merit of showing that the product $X \times Y$ can be constructed by means of the standard operations of Set Theory, and for logicians, further, that it is logically founded, and obviates the risk of internal contradictions of the type of the Russell Paradox. One has to appreciate that Logicians are an even more bizarre lot than Mathematicians: they feel the need to prove everything, including what is plain to see to the "general public". In matters electronical, these are content to press the buttons on the black boxes and check that "it works"; the professionals seek to understand what happens inside them.

The construction of $\mathcal{P}(X)$ for every set $X$ is useful in many other circumstances. The definition of the real numbers proposed by Dedekind amounts to saying, as we shall see at the beginning of Chapter II, that a real number $x$ is a set of rational numbers (intuitively, the set of $\xi \in \mathbb{Q}$ such that $\xi < x$). It is thus an element of $\mathcal{P}(\mathbb{Q})$ so, thanks to the axiom of separation one can speak of the *set* $\mathbb{R} \subset \mathcal{P}(\mathbb{Q})$ of real numbers. *After* having constructed the real numbers as subsets of $\mathcal{P}(\mathbb{Q})$ and proved their fundamental properties one of course forgets their manifestly complicated definition.

One also uses $\mathcal{P}$ in the study of *equivalence relations*. Such is any relation $R$ (denoted for example by $xRy$) between the elements of a set $X$ satisfying the following conditions: (i) $xRy$ and $yRz$ imply $xRz$, (ii) $xRy$ implies $yRx$, (iii) $xRx$ is true for all $x \in X$; consider for instance the relation "$x - y$ is a multiple of 3" between signed integers. If $a \in X$ the axiom of separation allows one to speak of the set $C(a)$ of all $x \in X$ such that $xRa$ is true; this is called the *equivalence class* of $a$. It is immediate that $a \in C(a)$ and that two classes $C(a)$ and $C(b)$ are either identical or disjoint. The axiom of the set of subsets allows us to consider the $C(a)$, $a \in X$, as the elements of a new set, contained in $\mathcal{P}(X)$, denoted $X/R$, and called the *quotient of $X$ by the equivalence relation $R$*. See for example the short §4 of my *Algebra*, where you will find examples and also applications to the construction of the signed integers and of the rational numbers. Observe in passing that, *by explicit construction*, and not just from the general abstract theory, a quotient set is a set of sets.

## 5 – Functions, maps, correspondences

The concept of the cartesian product allows one to introduce the general concept of a *function* or *map*, which is as fundamental as that of a set and which, as we shall see, reduces to it as do all others. In elementary education and in the whole history of mathematics up to the beginning of the XIX[th] century, a function was given by a "formula" such as $f(x) = x^2 - 3$, $f(x) = \sin x$, etc., but starting with Descartes one often also defined a function from a curve whose "equation" one sought. For experimental scientists and engineers a function is very often also given by its *graph*, the geometrical locus of those points $(x, y)$ in the plane such that $y = f(x)$ for a function $f$ which, quite often, one does not really know.

Starting with the XIX[th] century the concept of a function ceased to be associated with a simple or complicated "formula"; the German Dirichlet for example speaks of the function equal to 0 if $x$ is a rational number and to 1 if $x$ is irrational, and one later envisaged much stranger functions, until the general and abstract concept emerged of a *function defined on a set $X$ and having values in a set $Y$*; such a function $f$ associates to every $x \in X$ a well determined $y = f(x) \in Y$ depending on $x$ according to a precise rule. The graph of $f$ is then the set of ordered pairs $(x, y) \in X \times Y$ such that $y = f(x)$ for every $x \in X$. One encounters this in everyday life: if, in a monogamous society, one denotes by $H$ the set of married men and by $F$ the set of women, the relation "$y$ is the wife of $x$" is a function with values in $F$ defined on $H$. Its graph is clearly a set of ... couples.

Conversely, a subset $G$ of $X \times Y$ is the graph of a function $f$ provided that $G$ has the following property: for every $x \in X$ there exists one, and only one, $y \in Y$ such that $(x, y) \in G$; and then one writes $y = f(x)$. This convention allows one to reduce the concept of a function to that of a set: *by definition* a function defined on $X$ with values in $Y$ *is* a subset of $X \times Y$ subject to the preceding condition; no longer is there a "formula".

On suppressing the restriction imposed on $G$ one obtains the concept of a *correspondence* or *relation* between $X$ and $Y$: two elements $x \in X$ and $y \in Y$ correspond under $G$ if $(x, y) \in G$. If one returns to the preceding example and replaces the set $H$ by the set of all men, and does not assume that the society is monogamous, the relation "$y$ is one of the wives of $x$" is a correspondence between $H$ and $F$. One does not insist on the existence for *every* $x \in H$ of a $y \in F$ such that $(x, y) \in G$, nor does one insist that such a $y$ be unique; the $x$ for which there exists such a $y$ constitute the *set of definition* of the correspondence (the owners of a harem); the $y$ such that one has $(x, y) \in G$ for at least one $x$ constitute the *image* or the *set of values* (the women of a harem). If $X = Y = \mathbb{R}$, the relation $x^2 + 3y^2 = 1$, whose graph $G$ in the plane is an ellipse, is a correspondence: its set of definition is the set of $x$ such that $|x| \le 1$, its image the set of $y$ such that $|y| \le 1/\sqrt{3}$; this correspondence is not a function since a real number may have two distinct square roots. The formula $x < y$ is likewise a correspondence (one more often says "relation" in cases of this sort) whose graph the reader will have no trouble finding.

In actual practice one often uses other expressions. Instead of saying

> let $f$ be a function defined on $X$ with values in $Y$,

one often says

> let $f$ be a map of $X$ into $Y$

or

> consider a map $f : X \longrightarrow Y$.

When $f$ is given by a "formula" one also, for example, speaks of

> the map $x \longmapsto x^3$ of $X$ into $Y$,

assuming that this makes sense; do not confuse the signs $\longrightarrow$ and $\longmapsto$; the string $x \longmapsto x^3$ does not denote a map of the set $x$ into the set $x^3$, it denotes the function or map which to each element $x$ of $X$ associates the element $x^3$ of $Y$.

Let us again observe that in mathematics, when speaking of a function or map $f$, one must specify the set $X$ on which $f$ is defined and the set $Y$ in which it takes its values. To speak without further specification of "the function $x^2$" is meaningless[21]. The map $x \longmapsto x^2$ of the interval $0 \le x \le 1$

---

[21] The logicians nevertheless speak of *functional relations* without specifying the sets of departure or arrival: they mean a relation $R\{x, y\}$ between two "variables" $x$ and $y$ such that

$$R\{x, y'\} \quad \& \quad R\{x, y''\} \text{ implies } y' = y''.$$

of $\mathbb{R}$ (the set of real numbers) into the set $\mathbb{R}$ *is not* the same as the map $x \longmapsto x^2$ of $\mathbb{R}$ into $\mathbb{R}$; their graphs are different. Moreover, they do not have the same properties: in the first case the equation $f(x) = b$, for a given $b \in \mathbb{R}$ has one or no solution, while it can have two in the second case. Neglecting these "details" leads to confusion and errors of reasoning.

If $A$ is a subset of a set $X$, the *characteristic function* of $A$ (relative to $X$) is the map $\chi_A : X \longrightarrow \{0,1\}$ given by

$$\chi_A(x) = 1 \text{ if } x \in A, \quad = 0 \text{ if } x \notin A.$$

If $X = \mathbb{R}$, one can sketch its graph easily if $A$ is, for example, the union of a finite number of pairwise disjoint intervals – it consists of horizontal line segments, with "jumps" at the extremities of these intervals – but you will not manage if $A = \mathbb{Q}$, the case Dirichlet had spoken of already in about 1830. The principal interest of these functions is to transform relations between sets into relations between functions, for example:

$$\begin{aligned} \chi_{A \cap B}(x) &= \chi_A(x)\chi_B(x), \\ \chi_{A \cup B}(x) &= \chi_A(x) + \chi_B(x) - \chi_A(x)\chi_B(x), \\ \chi_{X-A}(x) &= 1 - \chi_A(x), \end{aligned}$$

etc.

Instead of speaking of functions one often speaks in mathematics of *families* of numbers, sets, etc. The only difference between these concepts relates to the notation employed: given two sets $I$ and $X$ a family of elements of $X$ *indexed by $I$*, the notation is

$$(x_i)_{i \in I},$$

consists of associating an $x_i \in X$ to each *index $i \in I$*; the preceding notation is thus just another way of speaking of the map $i \longmapsto x_i$ of $I$ into $X$, i.e. of the map $f : I \longrightarrow X$ given by

$$f(i) = x_i \text{ for every } i \in I.$$

One might do entirely without this concept, whose historical origin lies in sequences of real numbers

$$u_1, u_2, \ldots, u_n, \ldots$$

which we will meet from the beginning of the next chapter, for example the sequence

$$1, 1/2, \ldots, 1/n, \ldots;$$

for classical analysts, who concerned themselves only with functions where the variable can take all the real values in an interval, the notation $u_n$ denotes term number $n$ in the sequence; but you can, without the least inconvenience, and sometimes to advantage, as we shall see, write $u(n)$ for what is usually written $u_n$ and declare that a sequence of real numbers is nothing other than a map of the integers $> 0$ into the set of real numbers.

When the terms of a family $(X_i)_{i \in I}$ are considered as sets[22] one can define the union and intersection

$$\bigcup_{i \in I} X_i \ , \qquad \bigcap_{i \in I} X_i$$

of the family: in abbreviated form $\bigcup X_i, \bigcap X_i$; this is the set of $x$ such that $x \in X_i$ for at least one $i \in I$ in the first case, for every $i \in I$ in the second. One recovers the concepts introduced above by choosing $I$ to be a set of two elements. In the general case there are formulae similar to those we have already mentioned in this particular case: if for example $I$ is itself the union of a family $(I_k)_{k \in K}$ of sets, then

$$\bigcup_{i \in I} X_i = \bigcup_{k \in K} \left( \bigcup_{i \in I_k} X_i \right) \qquad \& \qquad \bigcap_{i \in I} X_i = \bigcap_{k \in K} \left( \bigcap_{i \in I_k} X_i \right)$$

(*associativity* of the intersection or of the union): if one wanted to regroup in one hypermuseum all the pictures belonging to all the various European museums one might begin by uniting all the museums of each country, after which one could regroup the national supermuseums so formed; in this example $K$ is the set of European states and, for each $k$, $I_k$ is the set of museums in country $k$. All formulae of this type reduce to common sense despite their abstract and rebarbative appearance.

Given a set $X$, a subset $A$ of $X$, and a family $(E_i)_{i \in I}$ of subsets of $X$, the $E_i$ are said to *cover* $A$, or to be a *covering* of $A$, when $A \subset \bigcup E_i$. For example, the family of intervals $(n - 1, n + 1)$, where $n$ is an integer varying between 0 and $p$, covers the interval $(0, p)$ (in $\mathbb{R}$).

The concept of the union of a family of sets arises notably in the *construction of a function by pasting*. Consider, for example, a function whose graph is piecewise linear on the interval $(0, 1)$; it is not given by a unique "formula" valid everywhere; on certain intervals it might be the function $y = 2x + 5$, on others $y = -x - 1$, etc. More generally, suppose that a set $X$ is the union of a family of sets $(X_i)_{i \in I}$ and that for each $i \in I$ we are given a $f_i : X_i \longrightarrow Y$; does there exist a function $f$ on $X$ such that $f$ coincides with $f_i$ on each $X_i$ ? An obvious necessary condition for the existence of $f$ is that if an $x \in X$ belongs to two sets in the family then the values at $x$ of the two corresponding functions must be equal:

---

[22] This precision may seem superfluous since *all* the objects one studies in mathematics are sets. But, in practice, it can happen that (quite consciously) one forgets this point, and it happens as often that one keeps it present in mind. All depends on context. Simple example: $I$ is the set of points of a circle and $X_i$ is the line (set of points in the plane) tangent to the circle at the point $i \in I$.

$$f_i(x) = f_j(x) \quad \text{for every } x \in X_i \cap X_j.$$

Conversely, if this compatibility condition is satisfied then the function $f$ does exist: to define $f(x)$ at an $x \in X$ one chooses an $i \in I$ such that $x \in X_i$ arbitrarily, and then puts $f(x) = f_i(x)$; one has done the necessary to eliminate any ambiguity in the definition of $f(x)$. Further, the graph $G$ of $f$ is the union of the graphs $G_i \subset X_i \times Y \subset X \times Y$ of the $f_i$.

The situation is particularly simple if one has a *partition* of $X$, i.e. a family of pairwise disjoint sets $(X_i)_{i \in I}$ whose union is $X$. For every $x \in X$, there then exists one and only one $i \in I$ such that $x \in X_i$, so that one may choose the $f_i$ arbitrarily. If, on $\mathbb{R}$, for each integer $n$ of arbitrary sign you have a function $f_n$ defined for $n \leq x < n + 1$ (do not confuse the signs $\leq$ and $<$, see Chap. II, n° 2), then there exists a function $f$ defined on $\mathbb{R}$ which agrees with $f_n$ for $n \leq x < n + 1$. If, on the other hand, the $f_n$ are given for $n \leq x \leq n + 1$, then $f$ exists only if $f_n(n + 1) = f_{n+1}(n + 1)$ for every $n$.

The concept of a family of sets is linked to the *axiom of choice*: given a family of nonempty subsets $(X_i)_{i \in I}$ of a set $X$ there exists a map $f : I \longrightarrow X$ such that $f(i) \in X_i$ for every $i \in I$. Intuitively, one obtains $f$ by "choosing arbitrarily" an element $x_i$ from each $X_i$. Cantor and others used it implicitly until it was identified explicitly (Zermelo, Whitehead and Russell). As we said at the beginning of this chapter, many mathematicians objected to "infinities of random choices" with no precise mathematical sense that could never lead to "explicit" formulae. No matter that it has survived by virtue of its use in all sorts of branches of mathematics, where, most of the time, one uses it without even a mention. Moreover, it was later proved (Paul Cohen, 1963) that the axiom of choice is logically independent of the other axioms of set theory: that if they are themselves not inconsistent, as one hopes (though this has never been proved), then adjoining the axiom of choice will not lead to a contradiction. You can adopt it or reject it. What is more, there are branches of mathematics – arithmetic, for example – which can be constructed without using it.

The axiom of choice comes in when one tries to extend the concept of the cartesian product to an arbitrary family $(X_i)_{i \in I}$ of sets. Their cartesian product, properly, can only be the set of families $(x_i)_{i \in I}$ such that $x_i \in X_i$ for every $i \in I$. The axiom of choice amounts to saying that a cartesian product of nonempty sets is always nonempty.

## 6 – Injections, surjections, bijections

Let us return to maps in general. Given three sets $X$, $Y$, $Z$ and maps $f : X \longrightarrow Y$ and $g : Y \longrightarrow Z$, one can construct the *composed map* $h : X \longrightarrow Z$ by putting

$$h(x) = g[f(x)] \quad \text{for every } x \in X.$$

If $F \subset X \times Y$ and $G \subset Y \times Z$ are the graphs of $f$ and $g$, the graph $H \subset X \times Z$ of $h$ is the set of ordered pairs $(x, z)$ having the following property: there exists a $y \in Y$ such that both $(x, y) \in F$ and $(y, z) \in G$. It is immediate that $H$ is a graph. The composed map $h$ is denoted $g \circ f$: thus

$$(6.1) \qquad (g \circ f)(x) = g[f(x)],$$

but in fact one always writes $g \circ f(x)$ instead of $(g \circ f)(x)$. This concept generalises what one does when speaking of the function $\sin \cos x$ – one ought to write $\sin \circ \cos$ – or, in geometry, when defining the "product" of two homotheties, translations, etc.

Given a map $f : X \longrightarrow Y$, one frequently has to consider those $x \in X$ such that $f(x) = b$ is a given element of $Y$. As regards their existence, all cases are clearly possible. The simplest is that where the equation $f(x) = b$ has at least one solution no matter what $b \in Y$; $f$ is then said to be *surjective* (or is a *surjection*). The map $x \longmapsto x^3$ of $\mathbb{R}$ into $\mathbb{R}$ is surjective, since every real number, whatever its sign, has a cube root. The map $x \longmapsto x^2$ is not, because only positive numbers have square roots.

More generally, if one replaces $X$ by one of its subsets $A$ one is led to introduce the set $B \subset Y$ of $y$ such that the equation $f(x) = b$ has at least one solution in $A$ (and possibly elsewhere[23]). This set, denoted by $f(A)$, is called the *image of $A$ by $f$*, a concept familiar from elementary geometry: the image of a circle by a translation is a circle. Clearly

$$(6.2) \qquad f(A \cup B) = f(A) \cup f(B),$$

but

the relation $f(A \cap B) = f(A) \cap f(B)$ is false

in general, since for $b \in f(A) \cap f(B)$ the equation $f(x) = b$ has at least one solution in $A$ and at least one solution in $B$, but why should they be the same?

The situation is simpler if $f$ is *injective* or is an *injection*, i.e. if the equation $f(x) = b$ has at most one solution for every $b \in Y$. In this case, it is clear, one always has $f(A \cap B) = f(A) \cap f(B)$.

Together with the concept of image – one also says the *direct image*– we have, in the inverse sense, the *inverse image* under $f$ of a subset $B$ of $Y$: this is the set of $x \in X$ such that $f(x) \in B$; the notation is $f^{-1}(B)$. This time one has

$$(6.3) \qquad \begin{aligned} f^{-1}(B' \cup B'') &= f^{-1}(B') \cup f^{-1}(B''), \\ f^{-1}(B' \cap B'') &= f^{-1}(B') \cap f^{-1}(B''); \end{aligned}$$

---

[23] In mathematics one says what one says and does not say what one does not say. Observing this rule in public or private life might eliminate a great number of stupid discussions of the type "You say that the French are racists. Do you believe that the Americans are not?"

in the first case one has to consider the $x \in X$ such that $f(x) \in B' \cup B''$, i.e. such that $f(x) \in B'$ or $f(x) \in B''$; in the second, those $x$ such that $f(x) \in B'$ and $f(x) \in B''$, whence the formulae, which you can extend to the case of unions and intersections of arbitrary families of sets.

It can happen that a map $f : X \longrightarrow Y$ is simultaneously injective and surjective; one then says that $f$ is *bijective* or is a *bijection*; this means that for every $b \in Y$ the equation $f(x) = b$ has one and only one solution $x \in X$. The map $x \longmapsto x^3$ of $\mathbb{R}$ into $\mathbb{R}$ is bijective. The map $x \longmapsto x+1$ of $\mathbb{N}$ into $\mathbb{N}$ is injective but not surjective; it is a bijection of

$$\mathbb{N} = \{0,1,2,3,\ldots\} \quad \text{onto} \quad \{1,2,3,4,\ldots\} = \mathbb{N} - \{0\}.$$

The map $x \longmapsto x^2$ of $\mathbb{R}$ into $\mathbb{R}$ is neither injective nor surjective; it becomes bijective if one replaces $\mathbb{R}$ by $\mathbb{R}_+$, the set of real numbers $\geq 0$.

When a map $f : X \longrightarrow Y$ is bijective one can define the *inverse map* $f^{-1} : Y \longrightarrow X$ as follows: its graph $G \subset Y \times X$ is the set of ordered pairs $(y,x)$ such that $(x,y) \in F$, the graph of $f$. Since one and only one $x \in X$ corresponds to each $y \in Y$, $G$ really is a graph and one sees that

(6.4) $$x = f^{-1}(y) \Longleftrightarrow y = f(x).$$

It comes to the same to say that

$$f^{-1} \circ f(x) = x \text{ for every } x \in X \quad \text{and} \quad f \circ f^{-1}(y) = y \text{ for every } y \in Y.$$

If one writes $id_X$ for the *identity map* $x \longmapsto x$ of $X$ into $X$ then

$$f^{-1} \circ f = id_X, \qquad f \circ f^{-1} = id_Y.$$

For example, the inverse map of $x \longmapsto x^3$ on $\mathbb{R}$ is $x \longmapsto x^{1/3}$, the cube root of $x$.

It is clear that if one composes maps which are all injective, or all surjective, or all bijective, one obtains a map of the same kind: to solve $g(f(x)) = c$, one has to find a $b$ such that $c = g(b)$, then an $x$ such that $b = f(x)$, whence the results.

## 7 – Equipotent sets. Countable sets

It is "obvious" that if $X$ is a finite set – a concept which we have not yet defined strictly – and $f$ is an injective map of $X$ into a set $Y$ then the image $f(X)$ has as many elements as $X$. If, in particular, $Y = X$, then $f$ cannot be injective unless it is surjective too; this is the property which Dedekind used to define a *finite set*; the others being said to be *infinite*. (There are other ways of proceeding, as we shall see.) The example of the map $n \longmapsto n+1$ of $\mathbb{N}$ into $\mathbb{N}$ shows that $\mathbb{N}$ is infinite in Dedekind's sense.

When there is a bijection of a set $X$ onto a set $Y$ one says that $X$ and $Y$ are *equipotent* (or have the *same power*, which assumes that we have defined the difficult concept of the "power" of a set, which generalises that of "number of elements"; see n° 9); since the composition of two bijections is a bijection it is clear that if $X$ is equipotent to $Y$ and $Y$ is equipotent to $Z$ then $X$ is equipotent to $Z$.

The concept of equipotence is familiar as applied to finite sets: it is the foundation of the naive definition of the whole numbers. Its extension to infinite sets was Cantor's first great idea. Seen from 1997 it does not have a very revolutionary air, but when Cantor proved that the set $\mathbb{N}$ of whole numbers is equipotent to the set $\mathbb{Q}$ of rational numbers he created a sensation: there were no more rational numbers $p/q$ than integers, and yet there is an infinity of rationals between 0 and 1, then between 1 and 2, etc.?

Further, Cantor's proof was accessible to anybody. Every rational number can be written in a unique way in the form $p/q$ with $p$ and $q$ having no common divisor (i.e. are relatively prime) and $q > 0$. One can then group the rational numbers according to the value of $|p| + q$; since $q$ is positive there are only finitely many numbers for which $|p| + q$ has a given value $s$. You write on a line of infinite length the numbers for which $s = 0$ (there are none), then those for which $s = 1$, etc.; you obtain

$$0/1, \quad 1/1, \quad -1/1, \quad 1/2, \quad -1/2, \quad 2/1, \quad -2/1, \quad 1/3, \quad -1/3, \quad 3/1, \quad -3/1,$$
$$4/1, \quad -4/1, \quad 3/2, \quad -3/2, \quad 2/3, \quad -2/3, \quad 1/4, \quad -1/4, \quad 5/1, \text{ etc.}$$

In this way one can assign to every irreducible fraction $x = p/q$ an integer $n = f(x)$, its rank in the above order; hence we have a bijection from $\mathbb{Q}$ onto $\mathbb{N}$ after agreeing to assign rank zero to $0/1 = 0$, qed.

A similar argument will show the existence of bijections of $\mathbb{N}$ onto $\mathbb{N} \times \mathbb{N}$, onto $\mathbb{N} \times \mathbb{N} \times \mathbb{N}$ [group the elements $(p, q, r)$ of $\mathbb{N} \times \mathbb{N} \times \mathbb{N}$ according to the value of $p + q + r$], etc., or onto $\mathbb{Q} \times \mathbb{Q}$, $\mathbb{Q} \times \mathbb{Q} \times \mathbb{Q}$, etc.

If there is a bijection of $\mathbb{N}$ onto a set $X$ one says that $X$ is *countable*. Our convention, in contrast to that of some other authors, is that *finite* sets are not reckoned as countable, but in fact we shall often say "countable" when actually meaning "finite or countable".

There are several useful theorems on countable sets; we shall confine ourselves to giving semi-naive proofs of them.

(1) *Every subset $Y$ of a countable set $X$ is finite or countable.* To see this naively one writes the elements of $X$ as a sequence $x_1, x_2, \ldots$, suppresses those $x \notin Y$, and reenumerates the remaining elements, i.e. those of $Y$.

(2) *The image $Y = f(X)$ of a countable set $X$ under a map $f$ is finite or countable.* We enumerate the elements of $X$ as we have just done and put $y'_n = f(x_n)$; one thus obtains all the elements of $Y$, in general several or even infinitely many times. To write the elements of $Y$ once and once only one proceeds as follows: put $y_1 = y'_1$, then let $y_2$ be the first term of the sequence of $y'_n$ which is $\neq y_1$, then $y_3$ the first term of the sequence $y'_n$ different from $y_1$ and $y_2$, and so on indefinitely.

(3) *The cartesian product $X \times Y$ of two countable sets $X$ and $Y$ is countable.* We have essentially proved this in showing above that $\mathbb{Q}$ is countable: one writes the ordered pairs of whole numbers $(p, q)$ "diagonally":

$$(0,0), \quad (0,1), \quad (1,0), \quad (0,2), \quad (1,1), \quad (2,0), \ldots .$$

(4) *The union of a finite or countable family of finite or countable sets is finite or countable.* Indeed, let $(X_i)_{i \in I}$ be such a family, where $I$ is finite or countable like the $X_i$. For each $i$ choose a surjective map $f_i : \mathbb{N} \longrightarrow X_i$ and define a map $f$ of the cartesian product $\mathbb{N} \times I$ onto the union $X$ of the $X_i$ as follows:

$$f((n,i)) = f_i(n)$$

for $n \in \mathbb{N}$ and $i \in I$. For every $x \in X$ there exists an $i \in I$ such that $x \in X_i$, so an $n \in \mathbb{N}$ such that $x = f_i(n)$. The map $f$ is thus surjective, and since the product $\mathbb{N} \times I$ is countable, so is $X$ finite or countable too.

(5) *Every infinite set contains a countable set.* If $X$ is infinite there is a bijection $f$ from $X$ onto a set $Y \subset X$ distinct from $X$. Choose $a \in Y - X$ and put

$$x_0 = a, \quad x_1 = f(x_0), \quad x_2 = f(x_1), \ldots .$$

If one had $x_p = x_q$ for a pair of integers such that $p < q$, one could deduce that $x_{p-1} = x_{q-1}$ since $f$ is injective, whence, continuing this argument, $x_0 = x_{q-p} = f(x_{q-p-1})$, which is impossible since $x_0 \notin Y = f(X)$.

(6) *Let $X$ and $D \subset X$ be two sets; suppose $D$ that is countable and $X - D$ is infinite; then $X$ and $X - D$ are equipotent.* From the preceding result, $X - D$ contains a countable set $D'$ and one has

$$X = Y \cup (D \cup D'), \quad X - D = Y \cup D'$$

where $Y = X - (D \cup D')$ is disjoint from $D$ and $D'$. It is not hard to construct a bijection $g$ of $Y$ onto itself; since $D$ and $D'$ are countable, so is $D \cup D'$; thus there is also a bijection $h$ of $D \cup D'$ onto $D'$. Then one obtains a bijection $f$ of $X$ onto $X - D$ by putting $f(x) = g(x)$ [for example $f(x) = x$] for all $x \in Y$ and $f(x) = h(x)$ for all $x \in D \cup D'$; $f$ is clearly injective and

$$f(X) = f[Y \cup (D \cup D')] = f(Y) \cup f(D \cup D') = Y \cup D' = X - D.$$

(7) *The set of finite subsets of a countable set $X$ is countable.* From point (4) above it is enough to show that, for a given $n$, the $n$-element subsets of a countable set $X$ form a countable subset $P_n$ of $\mathcal{P}(X)$. Now consider the cartesian product $X^n$, the set of systems $(x_1, \ldots, x_n)$ of elements of $X$, and the map $f : X^n \longrightarrow \mathcal{P}(X)$ which transforms $(x_1, \ldots, x_n)$ into the set $\{x_1, \ldots, x_n\} \subset X$. Its image clearly contains $P_n$; and it is countable since $X^n$ is, by (2) and (3); thus $P_n$ is too, since $P_n$ is clearly not finite.

## 8 – The different types of infinity

Since there are "no more" rational numbers than whole numbers one can go further and ask whether there are not also "as many" whole numbers as real numbers (rational or irrational), which would allow one to "enumerate" all the points of a line. The answer is negative (Cantor, 1874).

Let us confine ourselves to considering the set $X = [0,1]$ of numbers $x$ such that $0 \leq x \leq 1$. As we shall see in Chap. II, such a number can be written in decimal notation in the form $x = 0.x_1x_2x_3\ldots$ with "digits" $x_1, x_2, \ldots$ between 0 and 9; this expansion is unique if, for all $n$, one insists that the number $0.x_1 \ldots x_n$ be *strictly* less than $x$, so that, for example, $1/4$ is written as $0.2499999\ldots$ and not $0.2500000\ldots$. That said, consider a map $f$ of $\mathbb{N}$ into $X$; for all $n \in \mathbb{N}$, denote by $a_n$ the $n^{\text{th}}$ digit of $f(n)$ and, for all $n$, let us choose a $b_n \neq a_n$ between 1 and 9. Consider the number $b = 0.b_1b_2\ldots$ whose $n^{\text{th}}$ digit is $b_n$ for every $n$. It belongs to $X$ but not, as we shall see, to the image $f(\mathbb{N})$ of $\mathbb{N}$ of $f$, whence the result: a map $f$ of $\mathbb{N}$ into $X$ is never surjective, let alone bijective.

Indeed, suppose that $b = f(n)$ for an $n \in \mathbb{N}$. Since the digits of $b$ are all $\neq 0$, the decimal expansion $b = 0.b_1b_2\ldots$ cannot terminate in an unending sequence of zeros; thus $0.b_1 \ldots b_p < b$ for all $p$, strict inequality, from which we see that this decimal expansion definitely satisfies the condition imposed above. If one had $b = f(n)$ for a particular $n$, the $n^{\text{th}}$ digit $b_n$ of $b$ would be, from the construction of $b$, different from the $n^{\text{th}}$ digit of $f(n)$, i.e. of $b_n$; which is absurd[24].

Since the set $\mathbb{Q}$ of rational numbers is countable, one sees that $\mathbb{R} - \mathbb{Q}$, the set of irrational numbers, is equipotent to $\mathbb{R}$ (n° 7, point 6).

When there exists a bijection of $\mathbb{R}$, the set of real numbers (geometrically: the set of points of a line), onto a set $X$, one says that $X$ has *the power of the continuum*. One of the most paradoxical of Cantor's results is that $\mathbb{R} \times \mathbb{R}$ has the power of the continuum; in other words, there exist *bijective* maps of the set of points of a line onto the set of points of a plane: there are "no more" points in a plane than on a line. Here again there is a very simple proof[25], for example by using the binary counting of the computer scientists; others would say "of Leibniz", but he had invented a calculating machine and also

---

[24] Cantor's method for for showing that $\mathbb{R}$ is not equipotent to $\mathbb{N}$ is very different but presupposes some knowledge. Suppose that we could write all the real numbers between 0 and 1 as a sequence $u_1, u_2, \ldots$, and let us construct a sequence of compact intervals $I_1 \supset I_2 \supset I_3 \supset \ldots$ of lengths $> 0$ in $[0,1]$ and such that $u_n \notin I_n$ for all $n$ (if $u_n \notin I_{n-1}$, choose $I_n = I_{n-1}$; if $u_n \in I_{n-1}$, choose for $I_n$ an interval contained in $I_{n-1}$ and not containing $u_n$: this is possible since $I_{n-1}$ does not reduce to a single point). The results of Chap. III, n° 9 show that the $I_n$ have a point $x$ in common; if one had $x = u_n$ for some $n$, then one would have $x \notin I_n$, which is absurd.

[25] That of Cantor, much more scholarly, exploits the classical theory of continued fractions.

was one of the precursors of formal logic, which, among other titles, makes him an honorary computer scientist.

In binary counting, every real number between 0 and 1 can be written with the aid of a sequence of digits 0 and 1, and in a unique way, if one insists that the sequence does not consist only of 0 from a certain point on. If one considers only those digits equal to 1, this amounts to writing $x$ in the form

$$x = \left(\frac{1}{2}\right)^{p_1} + \left(\frac{1}{2}\right)^{p_1+p_2} + \left(\frac{1}{2}\right)^{p_1+p_2+p_3} + \ldots$$

with well-determined integers $p_1, p_2, p_3, \ldots > 0$: the digits 1 in the "default" binary description of $x$ are those of rank $p_1, p_1 + p_2$, etc., the others being zeros; if for example one writes $1/2$ in the form $0.01111\ldots$ (and not $0.1000\ldots$), one has $p_1 = 2$, $p_n = 1$ for all $n > 1$. Conversely, such a sequence of integers defines a number between 0 and 1. In other words, there exists a bijection between the interval $I = [0, 1]$ and the set $S$ of sequences[26] $(p_1, p_2, \ldots)$ of integers $> 0$. That said, let $(x, y)$ be a pair of elements of $I$ and let $(p_1, p_2, \ldots)$, $(q_1, q_2, \ldots)$ be the elements of $S$ corresponding to $x$ and $y$; we associate with the pair $(x, y)$ the number $z \in I$ that corresponds to the sequence $(p_1, q_1, p_2, q_2, \ldots)$ obtained by interlacing the sequences corresponding to $x$ and $y$; thus one constructs, as one sees immediately, a bijective map of $I \times I$ into $I$ (or, from the other point of view, of $S \times S$ into $S$), whence Cantor's result.

One would be wrong to believe that he saw all this immediately; to start with, it took him years to surmount the psychological obstacle which the implausibility of the results[27] and the predictable reactions of the majority of his contemporaries presented. The very simple proofs we have presented came later.

Cantor, and others after him, believed for a long time that every infinite subset of $\mathbb{R}$ falls into one of the three categories which we have just defined, the finite, the countable and the power of the continuum; we know now that this "continuum hypothesis" is neither true nor false: one can neither deduce

---

[26] If one denotes the set of integers $> 0$ by $X$ then $S$ is precisely the set of maps from $X$ into $Y = \mathbb{N} - \{0\}$. A map of $X$ into $Y$ is a subset of $X \times Y$, i.e. an element of $\mathcal{P}(X \times Y)$; the set $S$ of these maps thus satisfies

$$S \subset \mathcal{P}(\mathcal{P}(X \times Y)) \subset \mathcal{P}(\mathcal{P}(\mathcal{P}(\mathcal{P}(X \cup Y)))).$$

[27] Peano did much better than Cantor a little later: if one represents $I$ as an interval of a line and $I \times I$ as a square in the plane, Peano constructed a map of $I$ into $I \times I$ which is surjective and continuous. This amounts to the fact that a point moving in the plane in a *continuous* manner can, in a finite time, pass through ALL the points of a square. The Dutchman J. L. E. Brouwer later completed the statement: a map $f$ of $I$ into $I \times I$ can be continuous and surjective, but not continuous and bijective; the point has to pass through all the points of the square an infinite number of times.

it from the axioms of set theory invented by the logicians nor, if one adopts it, deduce a contradiction. You are very unlikely to need this very difficult result; the overwhelming majority of mathematicians die before using it.

In proving these results by methods which, elementary as they are, were totally unknown before him, Cantor showed that *there are different sorts of infinity*, which twenty-five centuries of philosophers and theologians had apparently never discovered. One can even always compare them. A famous theorem (Schröder, 1896 and Bernstein, 1898 – it already appears in Cantor, but his proof leaves much to be desired) says that if $X$ and $Y$ are two sets, then there exists an injection of $X$ into $Y$ (i.e. $X$ is equipotent to a subset of $Y$) or an injection of $Y$ into $X$ and that, if both these cases happen, then $X$ and $Y$ are equipotent. A convenient way of expressing this result is to attach to each set $X$ a symbol $\text{Card}(X)$, the *cardinal* of $X$, agreeing that the relation $\text{Card}(X) = \text{Card}(Y)$ means that $X$ and $Y$ are equipotent and that the relation $\text{Card}(X) < \text{Card}(Y)$ means that $X$ is equipotent to a subset of $Y$, but not to $Y$ itself; the symbol $\text{Card}(X)$ thus plays the rôle of the "number of elements" of $X$. The theorem of Schröder-Bernstein can then be expressed as saying that if[28] $\aleph_1$ and $\aleph_2$ are two cardinals, then one and only one of the three following cases occurs:

$$\aleph_1 < \aleph_2, \quad \aleph_1 = \aleph_2, \quad \aleph_2 < \aleph_1.$$

It is easy to construct infinite sets whose cardinals are increasingly large. If $X$ is a set, the subsets of $X$ are, as we saw above, the elements of a new set $\mathcal{P}(X)$. One can always construct an injection $X \longrightarrow \mathcal{P}(X)$, for example $x \longmapsto \{x\}$, but $X$ and $\mathcal{P}(X)$ are never equipotent, an "obvious" result if $X$ is finite[29]. To see this, consider a map $X \longrightarrow \mathcal{P}(X)$; this associates to each $x \in X$ a set $M(x) \subset X$. Let $A \subset X$ be the set of $x \in X$ such that $x \notin M(x)$

---

[28] The use of Hebrew letters to baptise the cardinals goes back to Cantor and has made many believe that he was a Jew; and with such a name ... His father, a prosperous merchant and cosmopolitan, was protestant and his mother, *née* Marie Böhm, catholic and of a family of musicians. Their son was protestant, but his family link to catholicism "may have made it easier for him to seek, later on, support for his philosophical ideas among Catholic thinkers", his entry in the DSB tells us. In fact, Cantor's father was indeed born to Jewish parents but converted before the birth (in Saint Petersburg) of the mathematician. The "Catholic thinkers" of the DSB were principally Jesuits, Fraenkel tells us in his biography of Cantor. Their interest in his ideas was not the greatest service they could render him ... The *Dictionary of Scientific Biography (DSB)*, Princeton UP, whose publication was directed by Charles Coulton Gillispie, an eminent specialist in the history of science in France of the XVIII[th] century and of the Revolution, comprises about twenty large format volumes in which one can find the essentials, even if the quality and the importance of articles, written by very many authors, varies greatly.

[29] If $X$ has $n$ elements, $\mathcal{P}(X)$ has $2^n$. A subset $Y$ of $X$ is essentially the same as associating to each $x \in X$ the number 1 if $x \in Y$, the number 0 if $x \notin Y$. There are therefore as many subsets of $X$ as ways of constructing a sequence of

and suppose that there exists an $a \in X$ such that $A = M(a)$ [as would be the case if the map $x \longmapsto M(x)$ of $X$ into $\mathcal{P}(X)$ were surjective]. If $a \in A$, one has $a \notin M(a) = A$ by the definition of $A$, absurd; if $a \notin A$, the relation $a \notin M(a)$ is false by the definition of $A$, whence $a \in M(a) = A$, a new contradiction. Thus there cannot be a map of $X$ into $\mathcal{P}(X)$ which is surjective, even less bijective, qed. This proof is due to Cantor, even as formulated here.

The simplest illustration of the preceding result, though not very useful to us, is when $X = \mathbb{N}$: *the set* $\mathcal{P}(\mathbb{N})$ *is equipotent to* $\mathbb{R}$ or, what comes to the same[30], to the interval $X : 0 \leq x \leq 1$ in $\mathbb{R}$. One proves this by associating to each subset $A$ of $\mathbb{N}$ the number $x \in X$ whose $n^{\text{th}}$ binary digit is equal to 1 if $n \in A$ and to 0 if not; one has to take note of the existence of numbers $x$ which have two different binary expansions ($0.1000\ldots = 0.0111\ldots$), but they form a *countable* set since they are of the form $p/2^n$ with $p$ and $n$ integers. The reader may provide the details as an exercise.

However it may be,

$$\mathcal{P}(\mathbb{N}), \quad \mathcal{P}(\mathcal{P}(\mathbb{N})), \quad \mathcal{P}(\mathcal{P}(\mathcal{P}(\mathbb{N}))),$$

etc., are sets whose "powers" become larger and larger, even though negligible in comparison with the analogous sets constructed starting from $\mathbb{R}$. It is not advisable to plunge into contemplation of these vertiginous elaborations. One may also absolve oneself from reading the following n° which, for us, is only a gymnastic exercise in manipulating the symbols $\in$ and $\subset$; but if you read and understand all the proofs, you will be liberated for the rest of your days from any inferiority complex as regards the mysteries of the "transfinite" ...

However one should not believe that these weird creatures have no practical use in mathematics; on the contrary, they are used to prove indispensable theorems in functional analysis (the Hahn-Banach Theorem to mention only one), in general topology (every cartesian product of compact spaces is compact), in algebra (existence of bases in any vector space), etc.

## 9 – Ordinals and cardinals

The von Neumann sets used above in defining the whole numbers have very curious properties. For a start, if $X$ is such a set, then every *element* of $X$ is also a *subset* of $X$; for example, the element $\{\emptyset, \{\emptyset\}, \{\emptyset, \{\emptyset\}\}\}$ of the set 4 has as its elements $\emptyset$, $\{\emptyset\}$ and $\{\emptyset, \{\emptyset\}\}$, which themselves belong to 4. One notes also that if $a$ and $b$ are two elements of $X$, then either $a \in b$, or $a = b$,

---

$n$ numbers equal to 0 or 1, and since there are two possible choices for each of $n$ terms of such a sequence, one obtains $2 \times 2 \times \ldots \times 2$ possibilities (application: coin tossing). More generally, if $X$ has $n$ elements and if $Y$ has $p$, then the set of maps from $X$ into $Y$ has $n^p$ elements (same argument).

[30] Every school-leaver can tell you that $x \longmapsto x/(1 + |x|)$ is a bijection of $\mathbb{R}$ onto the interval $I : -1 < x < 1$. Since the interval $J : -1 \leq x \leq 1$ differs from $I$ by only a countable (even finite) set, it is equipotent to $I$, thus to $\mathbb{R}$. It remains to find a bijection of $J$ onto the interval $0 \leq x \leq 1$.

or $b \in a$ and that the relation $a \in b$ is equivalent to $\{a \subset b$ and $a \neq b\}$. If $a$, $b$, $c$ are three elements of $X$ and both $a \in b$ and $b \in c$, then $a \in c$. Finally one remarks that if $X$ and $Y$ are two integers defined à la von Neumann, the relation $X \leq Y$, which everyone knows, becomes $X \subset Y$, showing that inequalities between integers are reducible to membership relations or set inclusions. One also sees that if $X$ and $Y$ are two von Neumann integers, one always has $X \subset Y$ or $Y \subset X$.

Starting from these properties of von Neumann sets (or whole numbers), one can generalise by calling an *ordinal* (number or set) any set $X$ possessing the following properties :

(O 1)  the relation $x \in X$ implies $x \subset X$;
(O 2)  for any $a, b \in X$, one has either $a \in b$, or $a = b$, or $b \in a$.

The simplest ordinal is naturally $\emptyset$, but there are others, to start with the sets of von Neumann, who gave a slightly different definition of the ordinals, though it is equivalent to the one above.

The set $\mathbb{N}$ is also an ordinal. To check (O 1), one remarks that every $x \in \mathbb{N}$ is a von Neumann integer, from which all its elements are too, so belong to $\mathbb{N}$, whence $x \subset \mathbb{N}$. To check (O 2), one notes that any von Neumann integer is an element of all its successors and that if $b \neq a$ is not one of the successors of $a$, then $a$ is one of the successors of $b$ (one need only be able to read and count ...).

These sets possess rather amusing properties; as in Euclidean geometry and even for the same reason – one constructs an entirely autonomous theory knowing hardly anything –, they are proved in a technically elementary way by applying the definitions.

First note that (O 1) can be written in the form

$$X \subset \mathcal{P}(X)$$

or in the form

$$x \in\in X \Longrightarrow x \in X$$

since an element of a element of $X$ is also an element of a subset of $X$, and so of $X$.

Furthermore, the three cases in (O 2) are pairwise exclusive; for $a \in b$ and $a = b$ would imply $a \in a$, while $a \in b$ and $b \in a$ is, like $a \in a$, forbidden by the providential axiom of regularity at the end of n° 2.

(1) *Every intersection of ordinals is an ordinal.* Clear from the definition.

(2) *If $X$ is an ordinal, then $s(X) = X \cup \{X\}$ is an ordinal.* Verification of (O 1): $x \in s(X)$ implies either $x \in X$, whence $x \subset X \subset s(X)$, or $x = X$ and again $x \subset s(X)$. Verification of (O 2): if $a, b \in s(X)$, one has either $a \in X$ and $b \in X$, whence $a \in b$ or $b = a$ or $b \in a$ since $X$ is an ordinal, or $a \in X$ and $b \in \{X\}$, whence $b = X$ and thus $a \in b$, or $a, b \in \{X\}$, whence $a = b$, qed.

(3) $\emptyset \in X$ *for every nonempty ordinal* $X$. By the axiom of regularity, there exists an $a \in X$ such that $a \cap X = \emptyset$. But $a \subset X$ by (O 1). Thus $\emptyset = a \cap X = a$, qed.

(4) *For* $a, b \in X$, *the relation* $a \in b$ *implies* $a \subset b$. If indeed $x \in a$, one has either $b \in x$, or $b = x$, or $x \in b$ by (O 2). In the first case, one would have $x \in a \in b \in x$, impossible (axiom of regularity). In the second case, one would have $a \in b$ and $b \in a$, impossible. In consequence, $x \in a$ implies $x \in b$, whence $a \subset b$, qed.

(5) Let $A$ be a nonempty subset of $X$; let us show that there exists an $a \in A$ such that $a \subset x$ for all $x \in A$. In other words: as a set of subsets of $X$, *every nonempty subset* $A$ *of* $X$ *possesses a least element*. By the axiom of regularity, there exists an $a \in A$ such that $a \cap A = \emptyset$. For $x \in A$, the relation $x \in a$ would imply $x \in A \cap a = \emptyset$, absurd; since $x \in a$ is impossible, one thus has either $x = a$ or $a \in x$, whence $a \subset x$ in both cases, by (4), qed.

(6) *Every element* $Y$ *of an ordinal* $X$ *is itself an ordinal*. By (4), $x \in Y$ implies $x \subset Y$, whence (O 1). If on the other hand $x, y \in Y$, one has also $x, y \in X$ since $Y \subset X$ by (O 1); since $X$ satisfies (O 2), a *fortiori* so does $Y$, qed.

(7) *Let* $X$ *and* $Y$ *be two ordinals such that* $Y \subset X$ *and* $X \neq Y$; *then* $Y \in X$ *and conversely*. The converse is clear since it implies $Y \subset X$ by (4) and $Y \neq X$ since otherwise one would have $X \in X$.

So suppose that $Y \subset X$ and $X - Y$ is nonempty; $X - Y$ possesses a least element $b$ by (5); we shall see that $b = Y$, which will prove that $Y \in X$ (and even that $Y \in X - Y$, in accordance with the axiom of regularity).

First we shall show that $b \subset Y$. For all $x \in b$, one has either $x \in X - Y$ or $x \in Y$. Since $b$ is the smallest element of $X - Y$, the first eventuality would imply $b \subset x$; but since $b$ is an ordinal, by (6), and since $x \in b$ by hypothesis, one also has $x \subset b$; the relation $x \in X - Y$ would thus imply $x = b$, impossible since $x \in b$. So we see that $x \in b$ implies $x \in Y$, whence $b \subset Y$.

Conversely, for all $y \in Y$, one has either $b \in y$, or $b = y$, or $y \in b$. $Y$ being an ordinal by (6), one has $y \subset Y$. If $b \in y$, one has $b \in Y$, absurd since $b \in X - Y$. If $b = y$, one again has $b \in Y$ since $y \in Y$ implies $y \subset Y$. The only possible case is thus the third, which shows that $Y \subset b$, qed.

(8) *Let* $X$ *and* $Y$ *two ordinals; then either* $X \subset Y$ *or* $Y \subset X$. Let $Z = X \cap Y$, which is an ordinal by (1). If the theorem were false, one would have $Z \subset X$ and $Z \neq X$, so $Z \in X$ by (7), and similarly $Z \in Y$, whence $Z \in X \cap Y$, i.e. $Z \in Z$, qed.

(9) *Let* $X$ *and* $Y$ *two ordinals; one has either* $Y \in X$, *or* $Y = X$, *or* $X \in Y$. If in fact $Y \subset X$ one has either $Y = X$, or $Y \neq X$ and so $Y \in X$ by (7); one finishes the proof with the help of (8).

(10) *Let* $a$ *be an element of an ordinal* $X$; *then either* $s(a) \in X$ *or* $s(a) = X$. Since $s(a) = Y$ is an ordinal by (2), it is enough, by (9), to exclude the possibility that $X \in s(a) = a \cup \{a\}$. But if such were the case, one would

have either $X \in a$ and thus $a \in X \in a$, impossible, or $X = a$ and then $a \in a$, impossible, qed.

The two possibilities $s(a) \in X$ and $s(a) = X$ can very well arise. The second is trivially satisfied if one puts $X = s(a)$ for an ordinal $a$, for example a von Neumann integer, or for $a = \mathbb{N}$. For $X = \mathbb{N}$, the first happens for any $a \in X$.

One can finally prove, but it is much more difficult, that

(11) *Every set is equipotent to an ordinal.* It can be shown that this statement is equivalent to the axiom of choice.

These properties explain the word "ordinal". For consider an ordinal $X$ and, for $x, y \in X$, let us write $x < y$ if $x \in y$. Then we have the following statements, which everyone knows in the case of $\mathbb{N}$:

(i)   $x < y$ and $y < z$ imply $x < z$;
(ii)  for any $x$, $y$, one and only one of the following relations is true:

$$x < y, \quad x = y, \quad y < x;$$

(iii) in every nonempty subset $A$ of $X$, there exists an $a$ such that $a < x$ for all $x \in A$ other than $a$.

(i) follows from (7) since $x < y$, i.e. $x \in y$, is equivalent to $x \subset y$ and $x \neq y$; (ii) and (iii) are the properties (9) and (4).

That said, let us consider an arbitrary set $E$ and, thanks to (11), let us choose an ordinal $X$ and a bijection $f$ of $E$ onto $X$. Given two elements $x$ and $y$ of $E$, let us agree to write that

$$x < y \iff f(x) < f(y).$$

One thus obtains an order relation[31] on $E$, possessing the properties (i), (ii) and (iii): we describe this by saying that $E$ is a *well ordered* set. There does indeed exist such an order relation on the set $\mathbb{R}$ of real numbers; but it clearly cannot be the one that everyone knows – this does not satisfy (iii): the set of numbers $> 0$ does not possess a least element, this is the

---

[31] On a set $E$, an *order relation* is a relation $xRy$ between the elements of $E$ such that (a) one has $xRx$ for all $x \in E$, (b) $xRy$ and $yRz$ imply $xRz$, (c) $xRy$ and $yRx$ imply $x = y$; it is convenient to write $x \leq y$, and to write $x < y$ when also $x \neq y$. Examples: the (nonstrict) inequalities between whole numbers, or between real numbers, the inclusion relation between subsets of a set. When the conditions (i) and (ii) above are satisfied, one speaks of a *total order* – clearly not so in the case of inclusion – and of a *well ordering* when (iii) is also satisfied. This last concept was invented by Cantor whose two fundamental articles are available, with a long and interesting historical introduction in somewhat outdated language by its British translator: Georg Cantor, *Contributions to the Founding of the Theory of Transfinite Numbers* (Open Court Publishing Cy, 1915, reprinted Dover, 1955)

axiom of Archimedes of Chap. II. To "construct" such a well ordering one would have to establish a bijection $f$ between $\mathbb{R}$ and an ordinal $X$ having the power of the continuum. This amounts to writing the elements of $\mathbb{R}$ "one after the other" as one does traditionally for the whole numbers; one might first choose a sequence of elements of $\mathbb{R}$, then an other sequence following the first, then a third following the second, etc., and since the union of a countable infinity of countable sets is again countable, one would not have exhausted $\mathbb{R}$ after having put these sequences "end to end"; one would have to continue *transfinitely*, as Cantor put it. The general theory shows that such an ordinal $X$ and such a bijection exist in the sense of the logicians, but since this result depends on the axiom of choice (and is equivalent to it), it is out of the question to exhibit either $X$ or $f$ by means of a reasonably explicit procedure. One might well think that if an order relation possessing the miraculous property (iii) were easy to see on $\mathbb{R}$, mathematicians would not have waited until the XX[th] century to discover it. No one will ever *see* it.

According to von Neumann, 1923, an ordinal is, by definition, a *well ordered* set subject to a supplementary condition which, apparently, no one had thought of before him:

(iv) every $x \in X$ is the set of $y \in X$ such that $y < x$.

For an ordinal defined by (O 1) and (O 2), the property (iv) is trivial: since $y < x$ is equivalent to $y \in x$ this means that $x$ is the set of $y \in X$ such that $y \in x$. In von Neumann's definition, which employs neither (O 1) nor (O 2), (iv) shows that every *element* of $X$ is, as a set[32], a *subset* of $X$ and thus that $y \in x$ is equivalent to $y < x$. Condition (O 1) follows from this, and (O 2) is none other than the condition (ii) above since $y \in x$ is equivalent to $y < x$. Von Neumann's definition is therefore equivalent to the one we have given.

The fact that the successor of an ordinal is again an ordinal shows that the sets
$$\mathbb{N}, \quad \mathbb{N} \cup \{\mathbb{N}\}, \quad \mathbb{N} \cup \{\mathbb{N}\} \cup \{\mathbb{N} \cup \{\mathbb{N}\}\}, \; etc.$$
are ordinals. On forming the union of the unending sequence formed by $\mathbb{N}$ and its successive successors, if one dares to put it so, one obtains a new ordinal to which one can again apply this process etc. Note in passing that all these sets are countable, so very modest, since there exist ordinals of all possible powers if one believes assertion (11). What distinguishes one from the other is not their cardinal, it is the *order* in which their elements are, at least virtually, written. To be precise, consider two equipotent ordinals $X$ and $Y$ and suppose that there exists a bijection $f$ from $X$ onto $Y$ such that the relation $x' < x''$ implies $f(x') < f(x'')$. Then $X = Y$ (not an obvious result).

---

[32] We again recall that, even if it is not obvious, every mathematical object is a set.

Cantor confined himself to considering well ordered sets, i.e. sets endowed with an order relation satisfying (i), (ii) and (iii), and considered two such sets as equivalent when there exists a bijection of the first onto the second which preserves the order of the elements, which is much more strict than the relation of equipotence; he did not think of imposing the condition (iv), but systematically associated to every element $x$ of a well ordered set $X$ the set of $y \in X$ such that $y < x$; this is not very different, since one can show that, for any well ordered set $E$, there exist an ordinal $X$ and a bijection $f$ of $X$ onto $E$ which transforms the inequalities in $X$ into the inequalities in $E$; further, $X$ and $f$ are determined uniquely by the order relation on $E$.

The condition (iv) of von Neumann thus provides a perfectly determined "standard" in each equivalence class of well ordered sets; his definition of the whole numbers similarly provides, in the "class" of sets of fourteen elements, a Standard-Set of Fourteen Elements, namely the number 14. The physicists do this every day when they compare their folding metres to the iridioplatinum Standard-Metre lodged at the National Bureau of Standards at Washington, D.C.

One should not be surprised that these ideas, introduced by Cantor about 1895 in a quasi-philosophical and very obscure style, did not, at that time, arouse unanimous enthusiasm among his colleagues; but some adopted them immediately and tried to clarify them and to put them on a solid basis; in 1900, at the International Congress of Mathematicians, David Hilbert, who was, with Henri Poincaré, one of the two greatest mathematicians of the time, proposed to his colleagues his famous list of the most important and difficult problems of the age; "to prove the continuum hypothesis" was one; and one knows his opinion on set theory, the Paradise into which Cantor has enabled us to enter and which we shall never leave (I quote from memory). These eulogies did not prevent the unhappy Cantor from spending a large part of his last twenty years in psychiatric establishments. All his *formalisable* ideas have been adopted, but not much of the detail of his definitions and proofs has been retained, conceived as they were in an age when one still lacked a precise language, a convenient notation, and the strict logical discipline introduced by his ... successive successors.

The ordinals may be used to define the cardinals of n° 8 really as *sets* and not just as simple symbols. One cannot use the ordinals themselves since two different ordinals can well be equipotent, for example $\mathbb{N}$ and its successor. But in a given ordinal $X$ the set of ordinals $X' \subset X$ equipotent to $X$ has, by (5), a least element $X_0$ (a nonstandard notation; we remark in passing that the logicians use Greek letters $\alpha$, $\beta$, etc. to denote ordinals, to distinguish them from the Hebrew cardinals); $X_0$ may be $X$ itself, for example if $X = \mathbb{N}$. If $Y \neq X$ is an ordinal equipotent to $X$ and if one supposes for example that $Y \subset X$ in accordance with (8), one has $Y_0 \subset Y \subset X$ and thus $X_0 \subset Y_0$ by the definition of $X_0$. But then one has $X_0 \subset Y$ and since $X_0$ is equipotent to $X$, and so to $Y$, one has $Y_0 \subset X_0$ by the definition of $Y_0$. One finds finally that $X_0 = Y_0$ whenever $X$ and $Y$ are equipotent ordinals. If $E$ is any set whatever, and if one chooses an ordinal $X$ equipotent to $E$, the corresponding ordinal $X_0$ does not depend on the choice of $X$ and is equipotent to $E$. We can thus agree to define $\mathrm{Card}(E) = X_0$; this is the standard-set in the class of sets equipotent to $E$. *Exercise.* An ordinal $X$ is a cardinal if and only if $X \subset Y$ for all ordinals $Y$ equipotent to $X$.

These results lie on the edge of the abyss: one step more and you fall into metaphysics, into mysticism or into contradictions, for example if you speak of "the set of ordinals". Indeed, suppose that such a set existed, and give it the inevitable name: $\Omega$. Let us show that $\Omega$ is again an ordinal. Every $x \in \Omega$ and every $y \in x$ being an ordinal by (6) and so an element of $\Omega$, one sees that $x \in \Omega$ implies $x \subset \Omega$, whence (O 1). If now $a$ and $b$ are two distinct elements of $\Omega$, one has either $a \in b$, or $b \in a$ by (9), whence (O 2). To conclude, it remains to deduce that $\Omega \in \Omega$. *Exercise.* The union of a *set* of ordinals is an ordinal.

We gave above the definition of finite sets according to Dedekind: the set $X$ is finite if every injection $X \longrightarrow X$ is bijective. Many others were found after him, but it is not always easy to establish the equivalence of all these definitions, so we shall not attempt to do so. The Pole Alfred Tarski for example characterised the finite sets in 1924 as follows: every family $(X_i)$ of subsets of $X$ should possess a minimal element, i.e. one not containing any other $X_i$ ("proof": choose an $X_i$ whose number of elements is minimal). $\mathbb{N}$ is not finite, for if one denotes by $X_n$ the set of integers $p \geq n$, one has $X_0 = \mathbb{N} \supset X_1 \supset X_2 \supset \ldots$ with strict inclusions; if $\mathbb{N}$ were finite, one would have $X_n = X_{n+1} = \ldots$ from a certain integer $n$ on.

By reason of (11), this would also suffice to characterise the finite ordinals; for example: an ordinal $X$ is finite if, for any nonempty ordinal $Y \subset X$, including $Y = X$, there exists an ordinal $Z$ such that $Y = Z \cup \{Z\}$. The set $X = \mathbb{N}$ is not finite in this sense: the condition is satisfied for every $Y$ strictly contained in $X$ (since $Y$ is a von Neumann set), but not for $\mathbb{N}$ itself: if one had $\mathbb{N} = Z \cup \{Z\}$, one would have $Z \subset \mathbb{N}$ and $Z \neq \mathbb{N}$, so $Z$ would be an element of $\mathbb{N}$ by property (6), i.e. a von Neumann set, so $s(Z) = \mathbb{N}$ also, and one would be led to the relation $\mathbb{N} \in \mathbb{N}$.

Finally we remark that Cantor and his successors had a lot of fun defining more or less algebraic operations on the cardinals, analogous to those known for the whole numbers; one obtains them starting from the operations of set theory. For example, one defines $\mathrm{Card}(X) + \mathrm{Card}(Y) = \mathrm{Card}(X \cup Y)$ taking the precaution of assuming $X$ and $Y$ disjoint, $\mathrm{Card}(X)\mathrm{Card}(Y) = \mathrm{Card}(X \times Y)$, $\mathrm{Card}(X)^{\mathrm{Card}(Y)} = \mathrm{Card}(F)$ where $F$ is the set of maps from $X$ into $Y$, etc. The formula $\mathrm{Card}(X) + 1 = \mathrm{Card}(X)$ characterises the infinite sets and entrances the mystics, though not the financiers, they who will never

believe that if one possesses an infinite fortune it serves no purpose to continue
to augment it.

# §2. The logic of logicians[33]

In the version of logic more or less universally adopted the basic material comprises the following elements:

(i) expressions which the Greek philosophers and geometers already used in plain language and which one represents by the sign $\Longrightarrow$ ("implies", "thus", "it follows that", "it results that", "if ... then", etc.), the sign $\bigvee$ which resembles the sign $\bigcup$ and which one writes "or" (the nonexclusive logical disjunction of two assertions), and finally the sign $\neg$, "not", the negation of an assertion ("it is false that ..."); to these three fundamental signs one adds two other signs which are also useful but are only convenient abbreviations:

$$P \bigwedge Q \qquad \text{signifies : not[(not } P) \text{ or (not } Q)],$$

$$P \Longleftrightarrow Q \qquad \text{signifies : } (P \Longrightarrow Q) \bigwedge (Q \Longrightarrow P);$$

we always write $P$ or $Q$ instead of $P \bigvee Q$, $P \& Q$ instead of $P \bigwedge Q$ and "not" instead of the sign $\neg$; there are already enough cabbalistic signs in Mathematics not to want to add to them;

(ii) the *quantifiers* $\forall$ ("for all" or "whatever") and $\exists$ ("there exists", "there is at least one"), such as appear in the following statement: for any positive number $x$, there exists an number $y$ such that $x = y^2$ (generally there exist even two ...), which one writes

$$(\forall x[(x \in \mathbb{R}) \ \& \ (x > 0)] \Longrightarrow \{\exists y[(y \in \mathbb{R}) \ \& \ (y^2 = x)]\});$$

by convention,

---

[33] My feeble attempts to find an exposition of logic, either in French or in English, at once accessible and readable, have not been fruitful, with one sole exception: Rene Cori and Daniel Lascar, *Logique mathématique*, transl. as *Mathematical Logic: A Course with Exercises* (OUP, 2001), which also presents the axiomatic theory of sets. Patrick Suppes, *Axiomatic Set Theory* (Van Nostrand, 1960, reprint Dover, 1972), without comparing it to Cori and Lascar, has been of great use, but already assumes the reader to be familiar with the basic principles of logic and, indeed, with the naive theory and its usages. Paul R. Halmos, *Naive Set Theory* (Van Nostrand, 1960 or Springer, 1974) is very readable but ignores Logic completely, which is not necessarily an inconvenience to mathematicians. There are also the volumes on the Theory of Sets by N. Bourbaki; Cori and Lascar write that logic "is not the forte" of Bourbaki, which can be explained by the fact that the treatise was written by mathematicians for mathematicians. Since, before the war, only two people were interested in logic in France – Jacques Herbrand and the philosopher Jean Cavaillès – who both departed prematurely (the first in a mountaineering accident, resistance to German occupation took the second), one should at least ascribe the credit to Bourbaki for having spread the subject widely in France, even if the French Cardinals of logic have a much more elaborate conception of it, as is normal. As far as I am concerned, reading Bourbaki's *Fascicule de résultats* on Set Theory as soon as it was published at the beginning of the 1940s allowed me to learn in a few weeks *everything* I ever needed in this respect.

$$(\exists x P) \qquad \text{signifies} \qquad \text{not}[\forall x(\text{not } P)];$$

(iii) *variables* $x, y, \ldots, X, Y, \ldots, a, b, \ldots$ etc., in unlimited quantity, symbolising totally indeterminate objects;

(iv) logical punctuation signs such as (, ), [, ], etc., whose purpose is to group one or other of the sub-propositions of which a given proposition is composed. The professional logicians use these much less than we have done[34], but we are not addressing them.

In particular, the parentheses allow us to indicate the domains of application of quantifiers clearly. This last point leads to the fundamental distinction between *free variables* and *bound variables*: a variable $x$ is said to be bound if it appears in the domain of application of a quantifier $\forall x$ or $\exists x$, whose domain of application is, in principle, delimited by the parentheses ( and ); a variable is said to be free if it is not bound. Although the professional logicians do not hesitate to write statements in which the same letter $x$ is bound in certain parts of a proposition and free in others, for example

$$(\forall x(x^2 > 0)) \quad \& \quad ((x > 1) \Longrightarrow (x^2 > x)),$$

one can only discourage the reader unfamiliar with logic from this usage; it is much more prudent to write the preceding proposition in the form

$$[\forall x(x^2 > 0)] \quad \& \quad [(y > 1) \Longrightarrow (y^2 > y)]$$

to make clear the fact that it is composed of two unrelated assertions.

The use of these symbols[35] allows one to construct "propositions" or assertions; these are finite sequences of letters and of signs taken from the list above and *a priori* chosen arbitrarily, for example

$$(\forall z((za)\text{or}((\exists y)b) \Longrightarrow \text{not}(x(\exists y)x)$$

It goes without saying that to obtain meaningful sequences (not the case in the example above), one must observe a certain number of generally obvious rules of syntax. To be precise, the only syntactically correct propositions[36] are those one can obtain by repeated application of the following rules or propositions, in which $P$ and $Q$ stand for propositions or assemblages of signs which are already known to be syntactically correct:

---

[34] The parentheses ( and ) are in fact enough, as anyone will know who, for example, has consulted a computerised catalogue in the library using key words.

[35] To respect strict logical formalism is not in my programme – anyway I would be incapable of it – apart from implication arrows and the signs "&" and "or" which will serve *sometimes* to eliminate every ambiguity in a statement. Nor is it to encourage the reader to avail himself of simple stenographic signs which permit him, as one has seen so often, to write gobbledegook instead of expressing himself plainly.

[36] This does not mean "true". The relation $1 = 2$ is syntactically correct. Likewise, the syllogism "every man is immortal, Socrates is a man, therefore Socrates is immortal" is perfectly correct.

(a) not $P$,
(b) $P$ or $Q$,
(c) $P \Longrightarrow Q$,
(d) $(\forall x(P))$ where $x$ appears in $P$ as a free variable.


Some remarks need to be made on (d), the rules of formation (a), (b), (c) presenting no problem other than that of determining the "primitive" propositions starting from which all the others are to be formed with the aid of the four preceding schemata. If $P\{x\}$ is a proposition involving a variable $x$ representing an *a priori* indeterminate object, and maybe other variables, the expressions

$$(\forall x P\{x\}) \ , \qquad (\exists x P\{x\})$$

are to be read, the first "for all $x$, $P\{x\}$" or "one has $P\{x\}$ for all $x$", the second "there exists $x$ such that $P\{x\}$" or "one has $P\{x\}$ for some $x$", "an $x$" signifying "at least one $x$". If $P$ involves other variables $y, z, \ldots$ than $x$, the assertion $(\forall x P\{x, y, z\})$ obtained by applying the quantifier $\forall x$ to $P$ is again an assertion involving the variables $y, z, \ldots$ but in which $x$ has become a bound variable.

Since the free variables represent totally arbitrary objects, you can, in a formula featuring the free variables $y, z, \ldots$, replace them by other distinct variables $u, v, \ldots$, free or bound, apart from the $y, z, \ldots$ which appear in the proposition: the assertions $[(x > y) \Longrightarrow (x+1 > y)]$ and $[(x > z) \Longrightarrow (x+1 > z)]$ are logically equivalent; on the contrary, the propositions $[(x > y) \Longrightarrow (x + 1 > y)]$ and $[(y > y) \Longrightarrow (y + 1 > y)]$ clearly are not; it is similarly obvious that $(\exists x(x \in y))$ is not equivalent to $(\exists x(x \in x))$.

We shall often say that a bound variable $x$ is a *phantom* variable since one can replace the letter $x$ everywhere by a sign with no logical or mathematical meaning, for example the signs \$, %, $\square$, etc., with which at a stroke one can replace $x$ in the quantifier $\forall x$. For example, it is clear that the proposition

$$[(x \in \mathbb{R}) \ \& \ (x^2 = 4) \ \& \ (y + 1 > y)]$$

really contains "arbitrary variables" $x$ and $y$, but that in the assertion

$$(\exists x[(x \in \mathbb{R}) \ \& \ (x^2 = 4) \ \& \ (y + 1 > y)]),$$

the variable $x$ does not play the same rôle as $y$; one could equally well write

$$(\exists \overbrace{\square[(\square \in \mathbb{R}} ) \ \& \ (\square^2 = 4 \ \& \ (y + 1 > y)]$$

as Bourbaki does. In the notation

$$X = \bigcup_{i \in I} X_i$$

used to represent a union of sets, the letter $i$ is, despite the absence of a visible quantifier, a bound variable; the preceding definition is in fact an abbreviated way of writing

$$(\forall x \{(x \in X) \iff (\exists i)[(i \in I) \ \& \ (x \in X_i)]\}).$$

You may therefore replace the letter $i$ with the sign $\square$.

For the rest, if $x$ were again a free variable in the relation $(\forall x P\{x\})$, one could again apply a quantifier to it, which would lead one to idiocies such as

$$(\forall x (\exists x [(x \in \mathbb{R}) \ \& \ (x^2 = 4)]))$$

or, in plain language, "for all $x$, there exists a real number $x$ such that $x^2 = 4$"; this is as if you were to say "for every man, there exists a man called Socrates"; phrases like this have no more meaning in logic than in ordinary language. It is to avoid these mistakes that, in the rule (d), one insists that $x$ be a free variable in $P$: *one has no right to quantify the same variable twice in succession* for the excellent reason that this "right" is not inscribed in the constitution of the Empire.

The "true" propositions are thus those one obtains by repeated application of the rules (a), ... , (d) starting from a small number of explicitly formulated schemata of propositions considered true *a priori*. To obtain the classical syllogisms, it suffices to assume *a priori* that the four following types of relations are true:

$$(P \text{ or } P) \implies P, \quad P \implies (P \text{ or } Q), \quad (P \text{ or } Q) \implies (Q \text{ or } P),$$

$$(P \implies Q) \implies [(P \text{ or } R) \implies (Q \text{ or } R)],$$

where $P$, $Q$, $R$ are any propositions. We can easily deduce other types of valid relations from these, for example that if $P$, $Q$ and $R$ are propositions, the proposition

$$[(P \implies Q) \ \& \ (Q \implies R)] \implies (P \implies R)$$

is true (even if the assertions $P \implies Q$, $Q \implies R$ are not), similarly for the proposition

$$\text{not}(\text{not } P) \iff P$$

for any $P$.

As far as the quantifiers are concerned, the essential point is that if the relation $(\forall x) P\{x, y, z\}$ is true and if $A$ is a mathematical object, then the relation $P\{A, y, z\}$, obtained by substituting the definition of $A$ for the letter $x$ everywhere in $P$, is again true. Correlatively, if $P\{A, y, z\}$ is true for a certain object $A$, then the relation $(\exists x) P\{x, y, z\}$ is true.

To construct the objects – sets – and relations which are the study of mathematics, and not only logic, one needs, as we have seen, the three extra symbols = [which the logicians tend to annex, subjecting it to the axiom $(\forall x)(x = x))$], $\in$ and $\emptyset$, and a certain number of axioms, those set out in the first part of this chapter; some permit the construction of new sets – of unions, of pairs, of sets of sets, etc. – starting from given sets and relations; others permit us to prove relations – equality, membership, inclusion, etc. – between sets. As the use of logical and mathematical signs leads as well to propositions as to sets, one needs to have a *definition of sets*, say, for example:

$$X \text{ is a set } \iff \{(X = \emptyset) \text{ or } [\exists x (x \in X)]\}.$$

To expound all this in strictly formalised language as the logicians do would be pointless and unusable. Indeed, one learns the usage of set theory from practice, and this requires only a little acquaintance with the subject.

It is now time to embark on *true mathematics*, where one has the agreeable illusion of manipulating other things than boxes filled with emptiness, or boxes filled with boxes filled with emptiness, or ... – of mathematics which would never have interested anybody if not for this illusion and its surprising appropriateness to "reality".

# II – Convergence: Discrete variables

*§1. Convergent sequences and series – §2. Absolutely convergent series – §3. First concepts of analytic functions*

## §1. Convergent sequences and series

### 0 – Introduction: what is a real number?

Throughout this book, we shall write $\mathbb{N}$ for the set of "natural" integers, i.e. those $\geq 0$; $\mathbb{Z}$ for that of the "rational" integers, i.e. of any sign; $\mathbb{Q}$ for the set of rational numbers (quotients of two integers); $\mathbb{R}$ for the set of real numbers; and $\mathbb{C}$ for the set of complex numbers $x + iy$ with $x, y \in \mathbb{R}$ and $i^2 = -1$: of course $\mathbb{N} \subset \mathbb{Z} \subset \mathbb{Q} \subset \mathbb{R} \subset \mathbb{C}$.

Since we hope that the reader is relatively familiar with $\mathbb{N}$, $\mathbb{Z}$ and $\mathbb{Q}$, we shall emphasise the construction of real numbers, though without providing all the details, then briefly recall that of $\mathbb{C}$, which is much simpler.

The so-called "real" numbers – it is too late to change the terminology – are truly not to be met in physical reality; they were born in the brains of mathematicians. The event which precipitated this process was the discovery by the Pythagoreans, in the $V^{th}$ century before our era, of the fact that the ratio $\frac{1}{2}(1 + \sqrt{5})$ between the diagonal and the side of a regular pentagon – their emblem – and, later, the numbers $\sqrt{2}$, $\sqrt{3}$, etc. are not rational; if for example one had $\sqrt{2} = p/q$ where $p$ and $q$ are two integers and not both even (if not, simplify!), then the relation $p^2 = 2q^2$ shows that $p$ is even, and so 4 divides the right hand side, whence $q$ is even: contradiction, absolute horror! For mathematicians.

The Greeks of this period, like their Babylonian, Indian or Egyptian predecessors, only knew of fractions: the successors of the Pythagoreans, until Euclid, were forced to develop a very abstruse theory of the (positive) real numbers, grounded in the "measure of magnitudes": to say that the number[1]

---

[1] The notation $\pi$ was introduced unsuccessfully in 1706 by an Englishman, and independently by Euler in 1739; since everyone read him, the usage spread. For

$\pi$ is the ratio between the length of a circumference and that of the diameter, presumes that we have a mathematically exact definition of "lengths", and not just a cadastral or physical one; which is far from obvious. Since, what is more, the modern algebraic notation had not then been developed, everything had to be explained geometrically. The mathematicians of the Renaissance and of the XVII$^{th}$ century inherited this point of view, too rigorous for the age; the invention of the differential and integral calculus about 1665–1700 made it disappear, banished it to a lower level, to the benefit of the prodigiously effective quasimechanical methods of Calculus, whose lack of rigour would certainly have scandalised the Greeks. One had to wait until the second half of the XIX$^{th}$ century to clarify and simplify, by reversing the procedure: the real numbers were then defined by procedures as rigorous as those of arithmetic, and one was then able to define precisely the length of a curve, the area of a surface, etc. In other words, geometry was no longer the foundation of analysis, but the other way round: even though, of course, the first still continues to provide indispensable intuitions.

We spoke above of "so-called" real numbers; why so-called? One might insist that a magnitude to be measured, say the diagonal of a square, or a circumference, is an eminently real object. But it all arises, in the final analysis, from the desire to arrive at an *absolute exactitude* which certainly exists in the minds of mathematicians, if not in physical reality, where, even leaving aside the non-euclidean character of the universe, one never meets "points", "lines", "squares" or "circumferences" in the mathematical sense of the term. Apart perhaps from the whole numbers, mathematical objects, starting from the numbers we call real, are only, at best, idealised models of real objects. Add to this the essential fact that it is impossible to define an irrational number[2] without the intervention of *infinitely* many elementary arithmetical operations or of rational numbers, a situation which does not arise in "reality" or "Nature", and even less, if this were possible, in the experimental sciences.

True, physicists, engineers, etc. constantly use the number $\pi$, with only a bare mention, not bothering to reflect on its exact mathematical meaning.

---

the history and the construction of the rational numbers, real or complex, of $\pi$ and other more advanced subjects, see H. D. Ebbinghaus et al., *Numbers* (Springer, 1991), a book to be recommended from all points of view.

[2] The real and complex numbers divide into two categories. First there are the *algebraic numbers* which, by definition, satisfy algebraic equations with rational coefficients: the rational numbers, and those obtained by extracting roots (for example $i$, $\sqrt[3]{2}$, $\sqrt[7]{5}$ ), the roots of the equation $x^{1848} - 3.14159x^{1789} + 2.718 = 0$, etc. This set is a *field* included between $\mathbb{Q}$ and $\mathbb{C}$. The other numbers are called *transcendental*; $\pi$ is one such. The set of algebraic numbers is countable, but not the set of transcendental numbers, so not $\mathbb{R}$. A simple procedure for constructing transcendental numbers was discovered by Joseph Liouville in 1844: suppose that, for $n$ large, the $n^{th}$ decimal digit is $\neq 0$ if and only there exists a $p \in \mathbb{N}$ such that $n = 1.2.3\ldots p = p!$. See for example Christian Houzel, *Analyse mathématique* (Belin, 1996), p. 64.

But for them, in reality only the first four, ten or twenty-five decimals of $\pi$ matter, since their calculations are intended only to lead to experimentally verifiable formulae. The value of mathematics to its users is to provide systematic procedures for calculating the numbers they need to an arbitrarily high accuracy. At the end of the XVII[th] century, an English astronomer, John Machin, used the formula

$$\pi/4 \;=\; 4.\arctan(1/5) - \arctan(1/239) =$$
$$=\; 4(1/5 - 1/3.5^3 + 1/5.5^5 - 1/7.5^7 + \ldots)$$
$$-\,(1/239 - 1/3.239^3 + 1/5.239^5 - \ldots)$$

to calculate $\pi$ to 100 decimal places: here there are two sums with infinitely many terms, i.e. two series; clearly one cannot calculate their sums exactly: one takes only a finite, though sufficiently large, number of terms in these sums. This formula, mathematically exact if one knows how to give a precise meaning to an infinite sum, provides a systematic procedure for calculating as many decimal places as one wishes of the number $\pi$. In other words, it is *the scheme of a numerical calculation pushed to infinity*, a scheme which, again, exists only in the minds of mathematicians and which the "users" can always interrupt when it has provided the needed places of the complete result. Those who complain of the "pedantry" of mathematicians have simply not understood the problem. Without us, they would drag the letter $\pi$ along in their formulae (or, better still, its 25 first decimals until they need the 150 following – after all, there exist tables of log to 100 decimals, probably not calculated simply for pleasure) without knowing what it represents.

At the present time the simplest way to define the real numbers might be to say that they are "nonterminating decimal expansions", like the number $\pi = 3.14159\ldots$ This supposes that we know all the decimals of the number $\pi$; a vast programme which some pursue, not in the hope of arriving at the end, to be sure, but in the hope, probably also illusory, of showing that the statistical distribution of the decimals of $\pi$ is not the result of a process analogous to a random draw: if such were the case, one could deduce mathematically demonstrable conjectures.

Apart from this particular case, which, after all, has never brought anybody to a halt, what sort of mathematical object does a nonterminating decimal expansion

$$x = x_0.x_1x_2\ldots$$

represent? So long as one has not *defined* the real numbers clearly and precisely and *proved* their *existence* as mathematical objects, such an expansion is only a sequence of numbers $x_0, x_1, \ldots$, where $x_0$ is any rational integer, the "integer part " of the pseudo-number $x$, and the rest are integers between 0 and 9, the "decimals" of $x$; as much to say a pattern on a strip of paper of infinite length. One might also consider the preceding formula as a condensed way of representing an increasing sequence of decimal numbers[3], namely $x_0$; $x_0.x_1$; $x_0.x_1x_2$; ... and agree that, by definition, a real number is such an expansion or such an increasing sequence. To say that it is the *limit* of such an increasing sequence would be a perfect vicious circle: one would first have to define what a limit is, which is precisely the first aim of analysis (and of this chapter), and further to prove that it exists, which assumes that all the problems have been solved. Some mediaeval theologians "proved" the existence of God by observing that existence is one of the divine qualities that figure in His definition. One would not get very far in mathematics by such methods.

This definition of the real numbers brings up other tiresome problems. The first is that one has to explain why, for example, the expansions 1.0000... and 0.9999... define the same number 1. The second, much more serious, is that one has to define the sum and the product of two real numbers.

Everyone knows how to add or multiply decimal numbers that have a finite number of nonzero digits. For the sum, for example, one adds the decimals of the same rank, starting with the last digits to the right, and carrying to the next place when the sum exceeds 9. If one tries to apply this rule of commercial arithmetic to nonterminating expansions, one comes against a little obstacle: *there is no last digit to the right.* One might try to start from the left, but each new addition may have repercussions on all the preceding. In short, it is a practical impossibility to define addition in this way, and even less multiplication; nor to proving the rules of calculation which everyone expects – associativity, distributivity, etc. –, without relying on an "it is obvious that ..." or an "everyone knows that ..."; but mathematics is not based on rumours, no matter how ancient. This will not prevent us, in the sequel, from sometimes using decimal expansions to explain – explain, and not to prove – some theorem or proof, but one cannot deduce anything more without ridiculous contortions – or to resorting to swindles as I myself very deliberately did, to avoid complications, in my course in Algiers in 1964[4].

One mathematically correct method – there are others – for defining the positive real numbers (the negative numbers then come from the usual algebraic procedures) consists of introducing particular sets in the set $\mathbb{Q}_+$ of rational numbers $\geq 0$, the *cuts*: a nonempty subset $X$ of $\mathbb{Q}_+$ is a cut if it satisfies the following conditions[5]:

---

[3] A decimal number is the quotient of an integer by a power of 10, so its decimal expression has only a finite number of nonzero digits. 2/10 is a decimal number, but $2/3 = 0.6666...$ not, even though it is rational.

[4] *Introduction à l'analyse mathématique* (Union nationale des étudiants algériens, 1964). See also *Calcul infinitésimal* in the *Encyclopaedia Universalis*.

[5] The condition (c) is automatically satisfied for every cut that defines an irrational number, but if one omits it one finds that the number 1, for example, corresponds to two different cuts: the set of $x \in \mathbb{Q}_+$ such that $x < 1$, and the set of $x \in \mathbb{Q}_+$ such that $x \leq 1$. The definition adopted here finesses the need to distinguish the

(a) it is bounded above, i.e. there exist numbers in $\mathbb{Q}$ greater than all the $x \in X$,

(b) the relations $x \in X$ and $0 \leq y < x$ imply $y \in X$,

(c) for every nonzero element $x$ of $X$ there exists a $y \in X$ such that $x < y$.

Intuitively, and leaving aside the set $X = \{0\}$ – which satisfies (a), (b) and (c) –, a cut is the set of all the positive rational numbers which are *strictly* less than a given (possibly rational) positive real number, in other words, the set of all its rational positive approximations from below, to an unspecified precision: the number $10^{-123}$ is a rational approximation from below to the number $\pi$ as much as 3.14159 is.

The correct definition of the real numbers to which we have alluded then consists of saying that a positive real number *is* purely and simply a cut; in other words, that there is no difference in nature between a real number and *the set* of all the rational numbers which are strictly less than it. This definition suggests no mystical, metaphysical or physical interpretation of the number $\pi$ for example, but it allows one to argue rationally about all the real numbers, even the irrational ones. For a start, one can very simply define the two fundamental algebraic operations and the inequality relation from this:

(1) the sum $X + Y$ of two cuts is the set of $z = x + y$ with $x \in X$ and $y \in Y$;

(2) the product $XY$ of two cuts is the set of $z = xy$ with $x \in X$, $y \in Y$;

(3) the inequality $X \leq Y$ means simply that $X \subset Y$.

*A posteriori* explanations: (1) and (2) mean that if $a$ and $b$ are two real positive numbers, then every rational positive number $< a + b$ (resp. $ab$) is the sum (resp. the product) of two rational positive numbers $< a$ and $< b$ respectively. (3) means that if $a$ and $b$ are two real numbers, the weak inequality $a \leq b$ is equivalent to the fact that every rational $x < a$ is also $< b$ (strict inequalities).

Naturally one has to verify that these definitions lead to sets satisfying (a), (b) and (c), then that the "obvious" expected properties of addition, of multiplication and of inequalities (see the following n°) are satisfied. One must also show that every *rational* positive number $x$ can be considered as a *real* number: to do this one associates to it the set $C(x)$ of $y \in \mathbb{Q}_+$ such that $y < x$ if $x > 0$, or else the cut $\{0\}$ if $x = 0$. All this requires only most elementary reasoning about the rational numbers, patience, and sometimes a little ingenuity. This done, one can forget this quite abstract construction for defining the real numbers and, as everyone has always done, restrict oneself to reasoning from the fundamental properties that we will state in n° 1. The only interest in this construction is to prove the *existence* of a mathematical object

---

rational numbers from the irrational in the construction of $\mathbb{R}$. The concept of a cut, in Dedekind, has a slightly different meaning: for him, it is a partition of $\mathbb{Q}$ into two nonempty sets $X$ and $Y$ such that $x < y$ for all $x \in X$ and all $y \in Y$; there is then a unique real number "between" $X$ and $Y$. *Exercise*: deduce from (b) that $x < y$ if $x \in X$ and $y \notin X$, $y > 0$.

– the set of real numbers, contained in $\mathcal{P}(\mathbb{Q})$, – possessing these properties (Chap. I, n° 4).

This is a construction of a "modern mathematical" type, very much less intuitive than writing a nonterminating sequence of digits onto a strip of paper of infinite length and keeping on moving always further right along the line in the eternally frustrated hope of arriving one day at the aim of your quest, the Grail: the last digit to the right. It at least has the advantage of being mathematically correct, which could be the reason why Richard Dedekind (1831–1916), a through and through algebraist (theory of fields of algebraic numbers) who would not confuse mathematics and physics, invented it in 1858 and published it in 1872; one might consider this the birth date of modernity in mathematics, which consists of constructing *all* mathematical objects with the aid of logic and set theory and establishing their properties starting from there.

This construction strips the real numbers, for example $\pi$, of all their mystery: reasoning about $\pi$ becomes, in this perspective, reasoning about all the rational numbers $< \pi$. All the calculators, and in particular the physicists and engineers, have always done this, since they replace the nonterminating decimal expansions of $\pi$ by, for example, its 25 first digits. But the genius of Dedekind's idea was to see that instead of privileging one or other more or less arbitrary or artificial procedure of approximating a real number by rational numbers, it was simpler to identify it with the *totality* of its rational approximations from below, i.e. to a subset of the set $\mathbb{Q}$ of rational numbers, and to use this to define the algebraic operations and inequalities in $\mathbb{R}$.

For the reader's convenience we recall briefly the definition of *complex numbers* which we shall use constantly, a much easier enterprise than defining the real numbers.

It is often believed that they were invented to provide roots for the quadratic equations $ax^2 + bx + c = 0$ when $b^2 - 4ac < 0$. This is not so: the Italians of the XVI[th] century invented them because having found miraculous formulae for solving third degree equations, they discovered that these formulae, although sometimes featuring square roots of negative numbers — thus apparently "impossible" – nevertheless provided a real root[6] when one substituted the formula in the equation calculating *à la* Bertrand Russell, i.e. not knowing of what one speaks. They were thus led to introduce new "numbers" of the form $a + b\sqrt{-1}$, where $a$ and $b$ are usual numbers, and to calculate mechanically with them, bearing in mind the "fact" that the square of $\sqrt{-1}$ is equal to $-1$. Later, Euler introduced the convention of denoting this strange number by the letter $i$; it took a long time before everyone, and particularly the "users", adopted it.

---

[6] An equation of odd degree with real coefficients always possesses at least one real root.

Of course one could, as some authors do for $\mathbb{R}$, state axiomatically the existence of a set $\mathbb{C}$ in which are given a particular element denoted by $i$ and two operations, addition and multiplication, satisfying certain conditions:

(C 1): $\mathbb{C}$ is a field, in other words the rules (I.1) to (I.9) of algebraic calculation of the following n° are valid;

(C 2): $\mathbb{R}$ is a subset of $\mathbb{C}$, and the algebraic operations defined in $\mathbb{C}$ coincide, in $\mathbb{R}$, with those already known (in other words, $\mathbb{R}$ is a "sub-field" of $\mathbb{C}$);

(C 3): $i^2 = -1$;

(C 4): every $z \in \mathbb{C}$ can be written as $z = x + iy$ with $x, y \in \mathbb{R}$.

The rule (C 4) is reasonable since the others show that the sum, the product and the quotient of two complex numbers of the form $x + iy$ are again of the same type. Note also that the expression $z = x + iy$ is necessarily unique, since, if this were not the case, one could find, by subtraction, a relation of the form $a + ib = 0$ with $a$ and $b$ real and not both nonzero, whence it would follow, if $a \neq 0$, that $i = 0$, absurd, or else that $i = ab^{-1}$ is a *real* number with square equal to $-1$, also absurd.

All this has for a long time satisfied the mathematicians and *a fortiori* the users, but does not explain whence – Heavens? – this mysterious "number" $i$ comes. Young students have been told hundreds of times during years that no such number exists, but since the teacher and the textbook say it does, why should they try to understand?

Instead of defining complex numbers as if mathematics had made no progress whatsoever during the last 450 years, it is much better, and very easy, to demystify the situation. Now, it has been the usage since the beginning of the XIX[th] century and Gauss to represent a complex number $x + iy$ geometrically by the point of the plane with rectangular coordinates $x$ and $y$; the point which one always identifies with the ordered pair $(x, y)$ of real numbers. Even if, for supposedly pedagogical reasons, you define complex numbers as expressions of the form $x + iy$, you always end up representing them by ordered pairs of real numbers. Why not, then, so define them to begin with? Complex numbers then become mathematical objects obtained from real numbers by a perfectly standard set-theoretic operation.

This method, invented[7] in 1835 by William Rowan Hamilton, though not widely appreciated for a century and still rarely used in elementary textbooks, therefore is to declare that a complex number *is*, by definition, an ordered pair $(a, b)$ of ordinary real numbers, to *define* equality and the two fundamental operations on such pairs by the formulae

(0.1) $$(a, b) \;=\; (c, d) \Longleftrightarrow a = c \;\&\; b = d,$$

(0.2) $$(a, b) + (c, d) \;=\; (a + c, b + d)$$

---

[7] For the history of complex numbers, see the chapter by R. Remmert in H. D. Ebbinghaus et al., *Numbers* (Springer, 1991). Hamilton's method is the simplest case of vastly more general constructions.

(0.3)                $(a, b).(c, d)  =  (ac - bd, ad + bc),$

and to *prove* that conditions (C 1) to (C 4) above are satisfied[8]. This requires only the simplest algebraic calculations and, for beginners, may even be a good exercise in elementary algebra.

Suppose for instance we want to check the associativity of multiplication (I.5), which in $\mathbb{C}$ is the identity

$$[(a, b)(a', b')](a'', b'') = (a, b)[(a', b')(a'', b'')].$$

Applying (0.3) blindly, this first reduces to

$$\cdot \ (aa' - bb', ab' + ba')(a'', b'') = (a, b)(a'a'' - b'b'', a'b'' + b'a''),$$

then, again by (0.3), to

$$((aa' - bb')a'' - (ab' + ba')b'', (aa' - bb')b'' + (ab' + ba')a'') =$$
$$= (a(a'a'' - b'b'') - b(a'b'' + b'a''), a(a'b'' + b'a'') + b(a'a'' - b'b'')).$$

Applying rule (I.9) of n° 1 for real numbers, we are reduced to proving that

$$(aa')a'' - (bb')a'' - (ab')b'' - (ba')b'', (aa')b'' - (bb')b'' + (ab')a'' + (ba')a'') =$$
$$= (a(a'a'') - a(b'b'') - b(a'b'') - b(b'a''), a(a'b'') + a(b'a'') + b(a'a'') - b(b'b'')),$$

and the result then follows from the associativity rules (I.5) for real numbers. Rules (I.1), (I.2), (I.6) and (I.9) for complex numbers are proved in similar ways.

The existence of complex numbers "zero" and "one" is clear: they are the pairs $(0, 0)$ and $(1, 0)$. The "opposite" of $(a, b)$ is obviously $(-a, -b)$. To prove (I.8), i.e. that we can solve $(a, b)(x, y) = (1, 0)$ for any pair $(a, b) \neq (0, 0)$, we write this as

(0.4)                $ax - by = 1, ay + bx = 0;$

if $b = 0$, in which case $a \neq 0$, the pair $(1/a, 0)$ is a solution by (I.8) for $\mathbb{R}$. If $b \neq 0$, the second relation is equivalent to $x = -ay/b$, hence the first to

$$(a^2 + b^2)y = -b :$$

and in $\mathbb{R}$ this can be solved in one and only one way provided that $a^2 + b^2 \neq 0$. But since $b \neq 0$, we may write $a = bc$ for some $c \in \mathbb{R}$, and the relation $a^2 + b^2 = 0$ then is equivalent to

---

[8] These are easy to understand when one sees that the pair $(a, b)$ must after all represent the sought-for symbol $a + ib$:

$$(a + ib) + (c + id)  =  a + b + i(c + d),$$
$$(a + ib)(c + id)  =  ac + ibc + aid + i^2 bd  =  ac - bd + i(ad + bc).$$

$$c^2 + 1 = 0;$$

*it is thus because this equation has no solution in $\mathbb{R}$ that every non-zero complex number has an inverse in $\mathbb{C}$.*

To conclude the construction, we observe that the mapping $x \mapsto (x, 0)$ of $\mathbb{R}$ into $\mathbb{C}$ is injective and transforms addition and multiplication in $\mathbb{R}$ into the corresponding operations in $\mathbb{C}$. We may therefore *agree* to identify each real number $x$ with the couple $(x, 0)$. Since we have

$$(0, 1)(y, 0) = (0, y)$$

by rule (0.3), rule (0.2) then proves that

$$(x, y) = (x, 0) + (0, 1)(y, 0) = x + iy$$

where we *define*

$$i = (0, 1).$$

Again by rule (0.3), we then have

$$(0, 1)(0, 1) = (-1, 0),$$

i.e. $i^2 = -1$. No mysteries anymore.

Having done these constructions and verifications, you may forget them, and start computing mechanically as Euler was doing in 1750.

The geometric representation of complex numbers $x + iy$ by points $(x, y)$ (or by vectors of origin $O$) of a plane allows one to introduce the *modulus* or the *absolute value*

$$|x + iy| = \sqrt{x^2 + y^2}$$

of a complex number. On introducing the *conjugate* $\bar{z} = x - iy$ of $z = x + iy$, one finds immediately that

$$z\bar{z} = |z|^2,$$

whence $|z'z''| = |z'||z''|$ and $1/z = \bar{z}/|z|^2$. These summary indications – plus, of course, the habit of computing complex numbers, which is acquired by practice – will suffice for nearly all our needs.

To conclude, let us state the wonderful property of complex numbers that explains their importance everywhere in mathematics: *any algebraic equation, of any degree, with complex coefficients, has complex roots.* This cannot be proved by purely algebraic methods; analysis is needed, as will be seen in Chap. VII, n° 18.

## 1 – Algebraic operations and the order relation: axioms of $\mathbb{R}$

For those who have faith, one can consider the more fundamental properties of the real numbers as *axioms* to be accepted without worrying about the "nature", concrete or metaphysical, of the objects about which one is reasoning; an eminently "modern" approach: you have probably met it already in Euclid's geometry.

We can class these axioms in four groups.

First of all there is a group of purely algebraic formulae concerning the two fundamental operations; they apply to the rational numbers, to the real numbers and to the complex numbers and state that, endowed with these two operations, $\mathbb{R}$ or $\mathbb{C}$ is, like $\mathbb{Q}$, a commutative *field* (as one says in algebra):

(I.1)  $x + (y + z) = (x + y) + z$ for all $x$, $y$, $z$;
(I.2)  $x + y = y + x$ for all $x$, $y$;
(I.3)  there exists an element $0$ such that $0 + x = x$ for all $x$;
(I.4)  given $x$ there exists a $y$ such that $x + y = 0$;
(I.5)  $x(yz) = (xy)z$ for all $x$, $y$, $z$;
(I.6)  $xy = yx$ for all $x$, $y$;
(I.7)  there exists an element $1 \neq 0$ such that $1.x = x$ for all $x$;
(I.8)  for each $x \neq 0$ there exists a $y$ such that $xy = 1$;
(I.9)  $x(y + z) = xy + xz$ for all $x$, $y$, $z$.

It is not the rôle of an exposition of analysis to develop the consequences of these axioms, but among the "remarkable identities" of algebra there is one which we shall use often, namely the *binomial formula*, which generalises the relation $(x + y)^2 = x^2 + 2xy + y^2$:

$$(1.1) \qquad (x + y)^n \;=\; x^n + nx^{n-1}y/1! + n(n-1)x^{n-2}y^2/2! + \\ + n(n-1)(n-2)x^{n-3}y^3/3! + \ldots + y^n$$

i.e.

$$(1.2) \qquad (x + y)^n = \sum_{p=0}^{n} \binom{n}{p} x^{n-p}y^p$$

where we have put $0! = 1$, $p! = 1.2.\ldots.p$ and

$$\binom{s}{p} = s(s-1)\ldots(s-p+1)/p! \qquad \text{for } s \in \mathbb{C}, \ p \in \mathbb{N},$$

whence

$$\binom{s}{p} = \begin{cases} s!/(s-p)!p! & \text{if } 0 \le p \le s \\ \\ 0 & \text{if } p > s \end{cases} \qquad \text{for } s \in \mathbb{N}.$$

The very simple proof comes from multiplying the right hand side of the formula corresponding to the exponent $n$ by $x + y$ and checking that the result is the formula corresponding to the exponent $n + 1$ (proof by induction).

The binomial formula can be written in the form

$$(1.4) \qquad (x+y)^n/n! = \sum_{p+q=n} x^p y^q /p!q!$$

where the $\sum$ is taken over all the ordered pairs of integers $p, q \in \mathbb{N}$ such that $p + q = n$. If one writes generally

$$(1.5) \qquad x^{[n]} = x^n /n!$$

(*divided powers*), one then has

$$(1.6) \qquad (x+y)^{[n]} = \sum x^{[p]} y^{[q]}.$$

In this form, the relation extends to a sum of any number of terms; for example,

$$(1.7) \qquad (x+y+z+u)^{[n]} = \sum_{p+q+r+s=n} x^{[p]} y^{[q]} z^{[r]} u^{[s]}$$

where, here again, the $\sum$ means that one must give the letters $p$, $q$, $r$, $s$ all positive integer values such that $p + q + r + s = n$ and then calculate the sum of all the corresponding expressions $x^{[p]} y^{[q]} z^{[r]} u^{[s]}$. In a relation such as (6), the letter $n$ denotes a fully determined integer, while the letters $p, \ldots, s$ denote *bound variables*, or phantoms, whose only function is to serve as a logical link between the symbol $\sum$ and the monomial $x^{[p]} y^{[q]} z^{[r]} u^{[s]}$. One can of course denote them by letters other than $p, \ldots, s$ so long as one does not use the letter $n$, which has a totally different sense; an expression such as

$$\sum_{n=1}^{n} x^n$$

is meaningless.

A second group of formulae involves the *order relation* $x \leq y$ in $\mathbb{R}$ or $\mathbb{Q}$:

(II.1) the relations $x \leq y$ and $y \leq z$ imply $x \leq z$;
(II.2) the relation $\{(x \leq y) \ \& \ (y \leq x)\}$ is equivalent to $x = y$;
(II.3) for all $x$ and $y$, one has $x \leq y$ or $y \leq x$;
(II.4) the relation $x \leq y$ implies $x + z \leq y + z$ for all $z$;
(II.5) the relations $0 \leq x$ and $0 \leq y$ imply $0 \leq xy$.

Next comes *Archimedes' axiom*:

(III) given $x, y > 0$ there exists an $n \in \mathbb{N}$ such that $y < nx$.

As we said above, one could, in the preceding, replace $\mathbb{R}$ by the set $\mathbb{Q}$ of rational numbers. The fourth axiom, not true for $\mathbb{Q}$, characterises the real numbers; in this form or in equivalent forms, it is indispensable to proving that there is something nontrivial to analysis. One can state it in several equivalent ways, for example:

(IV) *Let $E$ be a nonempty set of real numbers. Suppose that there exist numbers $M \in \mathbb{R}$ such that $x \leq M$ for all $x \in E$. Then the set of these numbers possesses a least element.*

This least of all the *majorants* or *upper bounds* $M$ of $E$ is called the *least upper bound* of $E$.

If for example $E$ is the set of truncated decimal expansions for $\pi$, this number will be $\pi$ itself. The necessity of axiom (IV) will become clear in n° 9 if it is not yet so at this stage.

If one defines the real numbers by means of cuts as explained at the end of n° 0, then axiom (IV) becomes a theorem, like the others, even almost trivial. For let $E$ be a set of real numbers, positive (for simplicity), and bounded above. By definition, every $x \in E$ is a subset of $\mathbb{Q}_+$ satisfying the conditions (a), (b) and (c) of n° 0, the relation $x \leq y$ being, by definition, equivalent to the inclusion relation $x \subset y$. The least upper bound of $E$ is then the union $u$ in $\mathbb{Q}_+$ of all the sets $x \in E$. It is indeed *obvious* that the set $u$ satisfies the conditions (a), (b), (c) stated at the end of n° 0, that $u \geq x$ (i.e. $u \supset x$) for all $x \in E$, and that $y \geq u$ (i.e. $y \supset u$) for all majorants $y$ of $E$, i.e. for every cut such that $y \supset x$ for all $x \in E$; this is even almost the definition of a union of sets: the smallest set which contains them all.

## 2 – Inequalities and intervals

The handling of inequalities is absolutely fundamental in analysis, since they govern the approximation calculations in constant use. We shall not prove them in detail: everyone knows them, and sceptics, if any there are, may find their proofs in the first volume of Dieudonné's *Treatise on Analysis*, (Academic Press, 1960) for example.

To avoid confusions with the anglo-saxon textbooks, as one calls them in the Quai d'Orsay, let us be very careful to make clear that for us the relation $x \geq 0$ means that $x$ is *positive* (in the wide sense), while the relation $x > 0$ means that $x$ is *strictly positive*. The anglophones say *non negative* and *positive*, the Germans say *positiv* for $x > 0$ and *negativ* for $x < 0$, the number 0 being, for them, neither the one nor the other. Despite the hymns to "the French exception", the French will probably end up imitating the Americans, since roughly 60% of global mathematical production is in English, and America provides about 40% of the total. But as a contemporary sociologist has remarked, one cannot change society by decree.

The first essential point is the *triangle inequality*

$$(2.1) \qquad |x + y| \leq |x| + |y|,$$

valid for all *complex* $x$ and $y$, and obvious geometrically (one can also prove it ...). One can generalise it to

$$(2.2) \qquad |x_1 + \ldots + x_n| \leq |x_1| + \ldots + |x_n|$$

and deduce that

(2.3)   $|(x_1 + \ldots + x_n) - (y_1 + \ldots + y_n)| \leq |x_1 - y_1| + \ldots + |x_n - y_n|$

for all complex $x$ and $y$: if you want to calculate a sum of 100 numbers to within 0.1, it is prudent to calculate each term to within 0.001.

If one defines the *distance* between two complex (or real!) numbers $x$ and $y$ by the formula

$$d(x, y) = |x - y|$$

whose geometric origin is quite clear, the triangle inequality amounts to saying that

(2.1')                    $d(x, y) \leq d(x, z) + d(z, y)$

for all $x, y, z \in \mathbb{C}$. We shall often use the notation $d(a, b)$ to prepare the reader for the extensions of analysis to functions of several variables, i.e. defined on a subset of a real vector space of finite dimension such as $\mathbb{R}^p$, or to much more general spaces (Appendix to Chap. III).

Archimedes' axiom might appear obvious: one can, like John D. Rockefeller, amass a billion 1910 dollars, at five grams of gold to the dollar, by saving one dollar a day for a sufficiently long time. But this does not follow from (I) and (II): the professional mathematicians, in general more competent than the amateurs as in all the sports, have long since invented strange "totally ordered nonarchimedean fields" which satisfy (I) and (II) but not (III). One does not meet them in our usual mathematics.

One of the consequences of (III) is that if a real number $x$ satisfies $x \leq y$ for all *strictly* positive $y$, then $x \leq 0$, for if not there would exist an integer $n$ such that $nx > 1$, whence $x > y$ with $y = 1/n > 0$. Other formulations:

(III')  *for all $a, b \in \mathbb{R}$ with $a < b$, there exists a $u \in \mathbb{Q}$ such that $a < u < b$.*
(III")  *for all $a \in \mathbb{R}$ and $r > 0$, there exists an $x \in \mathbb{Q}$ such that $d(a, x) < r$.*

Let us first prove (III"), which is obvious if one accepts the decimal expression for real numbers. If not, one proceeds as follows. The late lamented Archimedes provides us an integer $p > 0$ such that $1/p < r$; one can thus restrict to the case where $r = 1/p$. The inequality to resolve can be written $|pa - px| < 1$, i.e.

$$b - 1 < y < b + 1,$$

where $b = pa$ and where $y = px$ is neither more nor less rational than $x$. We shall even show that, in this case, one can choose $y$ in $\mathbb{Z}$. If $b > 0$, there are integers $n$ such that $b < n$; the least of these then satisfies $n - 1 < b < n$, whence $b - 1 < n < b + 1$ as desired. If $b = 0$, one takes $y = 0$. If $b < 0$, one reduces to resolving $b' - 1 < y' < b' + 1$ on putting $b' = -b > 0$ and $y' = -y$. Whence (III").

To establish (III'), one puts $c = (a + b)/2$, $r = (b - a)/2$, and any $u \in \mathbb{Q}$ satisfying $|c - u| < r$ will do.

It goes without saying that there always exist not just one, but infinitely many, rational numbers between $a$ and $b$: choose an $x \in \mathbb{Q}$ between $a$ and $b$, then an $x' \in \mathbb{Q}$ between $a$ and $x$, then an $x'' \in \mathbb{Q}$ between $a$ and $x'$, etc.

In the sequel we shall constantly need to speak of *intervals* in the set $\mathbb{R}$ of real numbers. Given two numbers $a$ and $b$, one denotes by

$$[a, b] \quad \text{the interval} \quad a \leq x \leq b,$$
$$[a, b[ \quad \text{the interval} \quad a \leq x < b,$$
$$]a, b] \quad \text{the interval} \quad a < x \leq b,$$
$$]a, b[ \quad \text{the interval} \quad a < x < b.$$

These intervals (which may be empty if $a \geq b$) differ one from the other only in so far as they do, or do not, contain their endpoints. We shall also sometimes employ a notation such as $]a, b)$ to denote an interval "open on the left" but possibly open or closed on the right, optionally. For every real number $a$, one also denotes by

$$[a, +\infty[ \quad \text{the interval} \quad a \leq x,$$
$$]a, +\infty[ \quad \text{the interval} \quad a < x,$$
$$]-\infty, a] \quad \text{the interval} \quad x \leq a,$$
$$]-\infty, a[ \quad \text{the interval} \quad x < a.$$

Finally, one sometimes denotes $\mathbb{R}$ by the analogous notation $]-\infty, +\infty[$.

The intervals of the form $[a, b]$, $[a, +\infty[$, $]-\infty, a]$ are called *closed*; those of the form $]a, b[$, with $a$ and $b$ possibly infinite, are called *open*, the whole interval $\mathbb{R} = ]-\infty, +\infty[$ being simultaneously open and closed. Finally, the intervals of the form $[a, b]$ with $a$ and $b$ finite are called *compact*. Later we shall define much more general open, closed and compact sets.

As to the exact meaning of the symbols $+\infty$ and $-\infty$, used above or elsewhere, one must be clear[9] that (my italics)

(1) $\infty$ *by itself* means nothing, although *phrases containing it* sometimes mean something,
(2) that in every case in which a phrase containing the symbol $\infty$ means something it will do so simply because we have previously attached a meaning to it by means of a *special definition*.

---

[9] Here I quote G. H. Hardy, *Pure Mathematics* (Cambridge University Press, 1908, Tenth ed., 1963, p. 117); it would be difficult to put it better.

## 3 – Local or asymptotic properties

As the reader will observe, one constantly has to examine functions of a single variable – it may vary on an interval of $\mathbb{R}$, or only on $\mathbb{N}$, or on some subset of $\mathbb{R}$ or of $\mathbb{C}$, just to restrict ourselves to functions of a single real or complex variable – when the variable is either "very close" to a fixed value $a$ where the function is or is not defined, or is "very large" i.e. "very close to infinity". One might like to know for example that $x^2$ is equal to 1 to within 0.00001 provided that $x$ is "sufficiently close" to 1, that $(2x+1)/(x-1)$ is equal to 2 to within 0.001 provided that $x$ is "sufficiently large" or that $1/n^3$ is less than $10^{-100}$ when the positive integer $n$ is "sufficiently large"; one also has to know that, for $x$ close to 0, $x^3$ is "negligible" in comparison to $x$, that for $n$ very large $10^{100}n^2 + 10^{100\,000}n$ is "of the same order of magnitude" as $n^2$, etc. It is easy to give a perfectly clear meaning to these *a priori* rather vague expressions.

It is best to consider generally an assertion $P(x)$ in which there appears a letter $x$ (or $y$, or $n$, or $p$, or some other) supposed to represent a number varying in a given set $E$ of real or complex numbers; for example, the relations

$$|x^2 - 1| < 0.00001 \quad \text{where} \ \ x \in E = \mathbb{R},$$

$$|(2x+1)/(x-1) - 2| \le 0.001 \quad \text{where} \ \ x \in E = \mathbb{R} - \{1\},$$

$$1/n^3 < 1/10^{100} \quad \text{where} \ \ n \in E = \mathbb{N} - \{0\},$$

$$n^2 < 10^{100}n^2 + 10^{100\,000}n \le \left(10^{100} + 1\right) n^2 \quad \text{where} \ \ n \in E = \mathbb{Z}.$$

In the case of $\mathbb{R}$, we shall say that $P(x)$ is true *for all sufficiently large positive* $x \in E$ (or, for short, true for $x$ large) if there exists a number $M$ such that, for $x \in E$, the relation $x > M$ implies $P(x)$. There is an analogous definition for $x$ sufficiently large and negative. If one does not specify "positive" or "negative", then we mean that $P(x)$ is true for $|x| > M$. In the case of $\mathbb{C}$, where these inequalities have no meaning, one says that $P(x)$ is true for large $x$ if there exists a number $M > 0$ such that $|x| > M$ implies $P(x)$: in other words if $P(x)$ is true outside a sufficiently large disc.

Given a number $a \in \mathbb{R}$ or $\mathbb{C}$, we say similarly that $P(x)$ *is true for all $x \in E$ sufficiently close* to $a$ or, more briefly, that $P(x)$ is *true on a neighbourhood of $a$*, if there exists a number $r > 0$ such that

(3.1)                       $\{[d(a,x) < r] \ \& \ (x \in E)\} \Longrightarrow P(x);$

in plain language: $P(x)$ is true for all $x \in E$ such that $|x - a| < r$. Another formulation: we call[10] an *open ball* with centre $a$ any set $B(a,r)$ defined by

---

[10] The use of the word "ball" rather than the word "interval" (in the case of $\mathbb{R}$) or of the word "disc" or "circle" (in the case of $\mathbb{C}$) is justified by the case of functions of several variables, i.e. defined on a subset of a Cartesian space $\mathbb{R}^p$.

an inequality $d(a,x) < r$ with a *strictly* positive $r$, in other words, in the case
of $\mathbb{R}$, any interval $]a-r, a+r[$ with $r > 0$ and, in the case of $\mathbb{C}$, any *disc* of
centre $a$, the circumference excluded. Then (3.1) means that there exists an
open ball $B(a,r) = B$ with centre $a$ such that

$$x \in B \cap E \Longrightarrow P(x).$$

The fact that we have chosen "open" balls here, i.e. defined by a strict inequality $d(a,x) < r$, rather than *closed balls* defined by a weak inequality
$d(a,x) \leq r$, is of no importance: an open ball with centre $a$ and radius $r$ contains every closed ball with the same centre and radius $r' < r$, and vice-versa;
a relation valid in an open ball will also be valid in a smaller closed ball and
vice-versa.

Most authors, following a tradition that goes back, at least, to the famous
courses in analysis delivered by Karl Weierstrass in Berlin around 1870, use $\varepsilon$
or $\delta$ for what we call $r$, the psychology of this notation being that these letters
are reserved for "very small" numbers; it is obvious that if one can verify that
the assertion $P(x)$ is true once $d(a,x) < 10^{-4}$ it is unnecessary to examine
what happens in the ball $d(a,x) < 1000$. This usage, however, probably stems
from the fact that $\delta$ is the initial letter of the word "difference" and that $\varepsilon$
immediately follows $\delta$ in the Greek alphabet; we realise this when we define
the continuity of a function at $a$:

for any $\varepsilon > 0$ there exists a $\delta > 0$ such that

$$|x - a| < \delta \Longrightarrow |f(x) - f(a)| < \varepsilon,$$

or again: if the difference between $x$ and $a$ is sufficiently small, i.e. smaller
than a suitably chosen number $\delta > 0$, then the difference between $f(x)$ and
$f(a)$ is also as small as one wishes, i.e. smaller than any number $\varepsilon > 0$
given in advance. The use of the letter $\delta$ (apparently begun by Cauchy) is
thus relatively rational, and that of the letter $\varepsilon$ (introduced by Weierstrass)
probably followed for a reason quite unconnected with mathematics.

In any case, $r$ (or $\varepsilon$, or $\delta$) may also be "very large" since one asks no more
than they exist; further, the concept of a "very small" or "very large" *fixed*
number has no objective meaning. There could be no objection if the reader
followed a preference for the letter $R$, or $\rho$, or whatever he wanted, a $\square$ or $\$$
sign for example: in a statement such as (1), as in the notation $\sum$ in n° 1,
the letter $r$ is a phantom, a "bound variable", whose only rôle is to serve as
the *logical link* between the assertions "there exists $r > 0$" and "$|x - a| < r$
implies $P(x)$". We prefer the letter $r$ because it suggests the radius of a ball;
moreover, it is directly available on all typewriter and computer keyboards.

One might also restrict oneself to powers of 10 and say:

there exists an $n \in \mathbb{Z}$ such that $P(x)$ for all $x \in E$ satisfying
$|x - a| < 10^{-n}$.

In the language of decimal approximations: there exists an $n$ such that the property $P(x)$ is true provided that the first $n$ decimal places of $x$ agree with those of $a$. The equivalence with (1) can be seen from the observation that, first, the powers of 10 are among the numbers $r > 0$, secondly, that if (1) is satisfied for an $r$ which is not a power of 10, it will be satisfied a fortiori if one replaces $r$ by a number $10^{-n}$ with $n$ sufficiently large that $10^{-n} < r$. From time to time we will give translations into this language.

The fundamental point to remember in these modes of expression is that, if one has a *finite number* of assertions $P_1(x), \ldots, P_n(x)$, and if each of these assertions, taken separately, is true on a neighbourhood of a given point $a$, or for $x$ sufficiently large, then so is the logical conjunction of the given relations; in other words, they are *simultaneously* true on a neighbourhood of $a$, or for $x$ sufficiently large. Indeed, in the second case there are numbers $A_1, \ldots, A_n$ such that $P_i(x)$ is true for all $x > A_i$, so that the $n$ assertions considered will simultaneously be true when $x$ exceeds the largest of the numbers $A_i$. In the first case, there are open balls $B(a, r_1), \ldots, B(a, r_n)$ with centre $a$ in which the corresponding assertions are valid; they will thus be simultaneously valid in the intersection of these balls, which is the open ball $B(a, r)$ whose radius is the least of the radii $r_i$ of the balls considered.

Contrariwise, the intersection of infinitely many open balls with centre $a$ (resp. of infinitely many intervals of the form $]A, +\infty[$) can very well reduce to a point $a$ (resp. be empty): this is the case of the balls $B(0, 1/n)$, $n \in \mathbb{N}$, by Archimedes' axiom. We shall prove later that, in $\mathbb{R}$, every intersection of intervals is again an interval, possibly empty, but the intersection of an infinite number of *open* intervals need not again be an open interval; the intervals $] - 1/n, 1/n[$ provide a counterexample.

These concepts are particularly useful when one wants to compare the "orders of magnitude" – a naïve expression having no mathematically precise meaning[11] – of two scalar functions $f(x)$ and $g(x)$ when the variable increases

---

[11] This is not the same as in physics, where an "order of magnitude" means a factor 10. Example: the power of an atomic bomb is "three orders of magnitude" (i.e. $10 \times 10 \times 10 = 10^3$) greater than that of a classical explosive. One of the experts in the subject even claims that the prestige attached to the megatonne, or to a million dollars, is linked to the fact that men have ten fingers; Herbert York, *Race to Oblivion* (Simon & Schuster, 1970), pp. 89–90: "We picked a one-megaton yield for the Atlas warhead for the same reason that everyone speaks of rich men as being millionaires and never as being tenmillionaires or one-hundred-thousandaires. It really was that mystical, and I was one of the mystics. Thus, the actual physical size of the first Atlas warhead and the number of people it would kill were determined by the fact that human beings have two hands with five fingers each and therefore count by tens". The committee entrusted with deciding the characteristics of the Atlas in 1953–1955 was chaired by J. von Neumann. York's book is a sparkling exposition of the American contribution to the arms race before 1970.

indefinitely or approaches indefinitely close to a limit value $a$; what then matters is the ratio $|f(x)/g(x)|$, which may, when $x$ tends to infinity or to $a$, either take values as large as one wants, or remain less than a fixed number, or remain confined between two fixed strictly positive numbers, or approach 1 more and more closely, or approach 0 more and more closely, not to speak of the cases where nothing of this kind happens.

These comparisons can be expressed with the aid of notation which we shall use frequently later and that it is important not to confuse. There are four cases to consider.

(i) The relation

$$f(x) = O(g(x)) \text{ when } x \to +\infty \text{ (or when } x \to a),$$

with an *upper case* $O$, means that there exists a number $M > 0$ independent of $x$ such that one has $|f(x)| \leq M|g(x)|$ for $x$ large (or for $x$ close to $a$) in the sense we have given above to these expressions. The notation $O(g(x))$ is used to denote not only a particular function $f$, but also *an arbitrary function* $O(g(x))$. Experience shows that the ambiguities thus introduced have no unfortunate consequences if one keeps this convention in mind. For example, the relations $f_1 = O(g)$ and $f_2 = O(g)$ do not imply $f_1 = f_2$ despite what one might believe at first glance. Similarly, the obvious relation $O(g(x)) + O(g(x)) = O(g(x))$ – obvious since if two functions are, for $x$ large, majorised by $5|g(x)|$ and $12|g(x)|$ respectively, then their sum is majorised by $17|g(x)|$ – does not imply that $O(g(x)) = 0$. We shall return to these points in detail in Chap. VI.

For example

$$10^{100} n^2 + 10^{100\,000} n = O(n^2) \text{ when } n \to +\infty$$

since for $n \geq 1$ (whence $n \leq n^2$), the left hand side is less than $M n^2$ with

$$M = 10^{100} + 10^{100\,000};$$

this number may appear "very large" to the puny members of the human race, but it is independent of $n$ and one does not demand more.

(ii) The relation

$$f(x) \asymp g(x) \text{ when } x \to +\infty \text{ (or when } x \to a)$$

means that there exist numbers $m > 0$ and $M > 0$ such that

$$m|g(x)| \leq |f(x)| \leq M|g(x)|$$

for $x$ large, or for $x$ close to $a$. One says $f$ and $g$ are *comparable* or have *the same order of magnitude* in these circumstances. This is equivalent to requiring that $f = O(g)$ and $g = O(f)$ simultaneously.

For example,

$$x + \sin x \asymp x \text{ when } x \to +\infty$$

since the left hand side lies between $x - 1$ and $x + 1$, so between $x/2$ and $2x$ for $x \geq 2$.

(iii) The relation

$$f(x) \sim g(x) \text{ when } x \to +\infty \text{ (or when } x \to a)$$

will be defined in the next n°, as will be

(iv) The relation

$$f(x) = o(g(x))$$

with a *lower case o*. These two relations presuppose the concept of limit as already known.

## 4 – The concept of limit. Continuity and differentiability

At our present level the concept of limit applies to complex-valued functions defined on a subset $X$ of $\mathbb{C}$ (and in particular of $\mathbb{R}$) when the variable $x \in X$ increases indefinitely or else approaches a value $a \in \mathbb{C}$ indefinitely closely. If, for example $X \subset \mathbb{R}$, the relation

$$\lim_{x \to +\infty} f(x) = u$$

means that, for all $r > 0$, one has $|f(x) - u| < r$ for $x$ large, in other words that, for every $r > 0$, there exists a number $N$ (depending on $r$) such that

(4.1) $$x > N \Longrightarrow d[f(x), u] < r.$$

The limit when $x$ tends to $-\infty$ is defined analogously: we replace the condition $x > N$ by $x < N$. (One makes no assumption as to the sign of $N$). In the complex case where these inequalities mean nothing one clearly needs to write that

$$|x| > N \Longrightarrow |f(x) - u| < r.$$

Similarly, the relation

$$\lim_{x \to a} f(x) = u$$

means that, for all $r > 0$, one has

(4.2) $$d[f(x), u] < r \text{ for all } x \in X \text{ sufficiently close to } a$$

i.e. that there exists a number $r' > 0$ (depending on $r$) such that, for $x \in X$, the relation

(4.3) $$|x - a| < r' \text{ implies } |f(x) - u| < r.$$

Note in passing that we do not assume that $a \in X$: one can have $X = ]0,1[$ and $a = 0$. For example, in $X = \mathbb{C}$, $1/z$ tends to 0 when $z \to \infty$ since the inequality $|1/z| < r$ is satisfied as soon as $|z| > 1/r$. Similarly, $x^2$ tends to 4 when $x$ tends to 2, since on the one hand

$$|x^2 - 4| = |x - 2|.|x + 2| < 5|x - 2| \text{ for } |x - 2| \leq 1,$$

while on the other hand, $5|x - 2| < r$ for $|x - 2| < r/5$; so $|x^2 - 4| < r$ once $|x - 2| < r' = \min(1, r/5)$.

We shall hardly ever, in this chapter, use the concept of limit except for the case of sequences, i.e. of functions where the independent variable takes only integer values. The general case which we have just mentioned will be covered in detail in Chap. III. Since we shall nevertheless need to speak occasionally of continuity and of differentiability in this chapter, let us now give the definitions of these two fundamental properties.

A scalar (i.e. complex-valued) function $f$ defined on a set $X \subset \mathbb{C}$ is said to be *continuous at a point a of $X$* if

$$\lim f(x) = f(a) \text{ when } x \to a.$$

This means that for all $r > 0$ there exists an $r' > 0$ such that

(4.4)        $\{(x \in X) \ \& \ (|x - a| < r')\} \Longrightarrow |f(x) - f(a)| < r$

or again that, for all $r > 0$, $f(x)$ is *constant to within $r$ on a neighbourhood of $a$* in $X$ or again, in decimal language, that if one wants to calculate $f(a)$ to within $10^{-n}$ it suffices to calculate $f(x)$ for any $x \in X$ having sufficiently many decimal places in common with $a$, for example the number obtained by replacing all the digits of $a$ of sufficiently high rank by 0; this is what all the practitioners of numerical analysis have always done, and this is what all computers do. The calculation above, showing that $x^2$ tends to 4 when $x$ tends to 2, expresses the continuity of the function $x \longmapsto x^2$ at $x = 2$.

More generally, let us show, as a useful exercise, that the functions $x^n$, or, as clearly amounts to the same, $f(x) = x^{[n]} = x^n/n!$, are continuous on $\mathbb{C}$. It suffices to show – the other formulation of continuity – that, for $x$ given, the difference $|f(x + h) - f(x)|$ is $< r$ for $|h|$ sufficiently small. Now the binomial formula (1.6) shows that

(4.5)
$$|f(x + h) - f(x)| =$$
$$= |x^{[n-1]}h + x^{[n-2]}h^2/2! + \ldots + h^n/n!|$$
$$\leq |h|.\{|x|^{[n-1]} + |x|^{[n-2]}|h|/2! + \ldots + |h|^{n-1}/n!\}.$$

The expression between the braces { } differs from the expansion of $(|x| + |h|)^{[n-1]}$ only in the presence, in the term in $|h|^p$, of a denominator $(p + 1)!$ instead of $p!$; since $(p + 1)! > p!$, one concludes that

(4.6)        $$|(x + h)^{[n]} - x^{[n]}| \leq |h|(|x| + |h|)^{[n-1]};$$

since $|h|$ is a factor of the right hand side, continuity is proved, since, for $|h| < 1$ for example, the right hand side is bounded by $M|h|$ where

$M = (|x| + 1)^{[n-1]}$ does not depend on $h$. The inequality (6) will serve us on other occasions in proving the continuity and the differentiability of much more general functions.

The *derivative* of a function $f$ at a point $a$ is defined similarly, as the limit of the ratio

$$\frac{f(x) - f(a)}{x - a}$$

as $x$ tends to $a$ remaining $\neq a$, i.e. tends to $a$ in the set $X' = X - \{a\}$ obtained by omitting the point $a$ from the set of definition $X$ of $f$; or again, by the traditional formula

$$(4.7) \qquad f'(a) = \lim_{h \to 0} \frac{f(a + h) - f(a)}{h}$$

where one clearly has to impose on $h$ the conditions which make the quotient meaningful: $h \neq 0$ and $a + h \in X$. In practice, $X$ is either an interval of $\mathbb{R}$ containing $a$, or, in the case of a function of a complex variable, a subset of $\mathbb{C}$ containing an open ball with centre $a$. This last case, appreciably more subtle than the first, will arise in n° 19 *à propos* so-called analytic functions, but, here, it is no more difficult to understand than that of functions of a real variable.

For example let us calculate the derivative of the function $f(x) = x^{[n]}$ for $n \in \mathbb{N}$. By the binomial formula, one has

$$f(x + h) = x^{[n]} + x^{[n-1]}h + x^{[n-2]}h^{[2]} + \ldots,$$

whence

$$[f(x + h) - f(x)]/h = x^{[n-1]} + \ldots$$

where the terms omitted represent, for $x$ given, a polynomial in $h$ of degree $n - 1$ whose term independent of $h$ is zero. It is almost obvious – and the simpler rules of calculating limits will confirm this if the reader is not yet fully convinced … – that this polynomial tends to 0 with $h$, so that in the limit one obtains the formula

$$(4.8) \qquad \left(x^{[n]}\right)' = x^{[n-1]}$$

or, in more traditional notation,

$$(4.8 \text{ bis}) \qquad (x^n)' = nx^{n-1}.$$

One should note that this calculation is as valid in $\mathbb{C}$ as in $\mathbb{R}$.

Here again, it is helpful to refine the previous calculation a little. If one imitates (5), one obtains

$$|f(x + h) - f(x) - x^{[n-1]}h| = |x^{[n-2]}h^2/2! + \ldots + h^n/n!|$$
$$\leq |h|^{[2]} \left\{ |x|^{[n-2]} + |x|^{[n-3]}|h|/1! + \ldots + |h|^{n-2}/(n-2)! \right\}$$

since $p! \geq 2!(p-2)!$, whence the inequality

(4.9) $$\left|(x+h)^{[n]} - x^{[n]} - hx^{[n-1]}\right| \leq |h|^{[2]} \left(|x| + |h|\right)^{[n-2]}$$

similarly to (6), or

$$\left|\frac{(x+h)^{[n]} - x^{[n]}}{h} - x^{[n-1]}\right| \leq \frac{|h|}{2}(|x| + |h|)^{[n-1]} \leq M|h|$$

for, say, $|h| \leq 1$. We then get (8) by letting $h$ tend to 0. More generally,

(4.10) $$\left|(x+h)^{[n]} - x^{[n]} - hx^{[n-1]} - \ldots - h^{[p]}x^{[n-p]}\right|$$
$$\leq |h|^{[p+1]}(|x| + |h|)^{[n-p-1]} .$$

The concept of a limit now allows us to define the comparison relations

$$f(x) \sim g(x), \qquad f(x) = o(g(x))$$

which we abandoned to their fate at the end of the preceding n°. The first – one says that the functions $f(x)$ and $g(x)$ are *equivalent* at infinity or on a neighbourhood of the point $a$ – means that the ratio $f(x)/g(x)$ tends to 1 when $x$ tends to the limit value considered. The second means that this same ratio tends to 0; one says that $f(x)$ is *negligible with respect to* $g(x)$ in these circumstances.

The second relation means that for all $r > 0$

$$|f(x)| < r|g(x)|$$

for $x$ large (or for $x$ close to $a$), i.e. that there exists an $r' > 0$ such that

(4.11) $$|x - a| < r' \implies |f(x)| < r|g(x)|.$$

For example,

$$x^2 = o(x) \text{ when } x \to 0$$

since $|x| < r$ implies $|x^2| < r|x|$; so for example $|x^2| < 10^{-1000}|x|$ once $|x| < 10^{-1000}$. In $\mathbb{C}$, one has similarly

$$x = o(x^2) \text{ when } x \to \infty$$

since the relation $|x| < r|x^2|$ is satisfied once $|x| > 1/r$.

As to the relation $f \sim g$, it reduces to the one we have just described. One can express this as: for all $r > 0$, one has

$$|f(x)/g(x) - 1| < r$$

for $x$ large (or close to $a$). But this can be written

$$|f(x) - g(x)| < r|g(x)|,$$

in other words $f(x) - g(x) = o(g(x))$, or again

(4.12) $$f(x) = g(x) + o(g(x)) :$$

$f(x)$ is sum of $g(x)$ and of a function negligible with respect to $g$ when $x$ tends to the value considered. Here, as in the case of the notation $O(g(x))$, the notation $o(g(x))$ is used to denote any function negligible with respect to $g(x)$. Some authors introduce indices to avoid confusing functions that are in fact distinct $o_1(g)$, $o_2(g)$, etc.

For example

$$x^2 + x \sim x^2 \text{ when } |x| \to \infty$$

since we have seen above that $x = o(x^2)$.

Another example: to say that a function $f$ defined on a neighbourhood of a point $a$ possesses a derivative at $a$ means that there exists a constant $c$ such that

(4.13) $$f(a + h) = f(a) + ch + o(h) \text{ when } h \to 0;$$

for this relation means that, for all $r > 0$,

$$|f(a + h) - f(a) - ch| < r|h|,$$

i.e.

$$\left| \frac{f(a + h) - f(a)}{h} - c \right| < r,$$

for $|h|$ sufficiently small, in other words that $f$ possesses a derivative $f'(a) = c$ at $a$. For example

$$\sin x = x + o(x) \sim x \text{ when } x \to 0$$

since the ratio $\sin x/x$ tends to the derivative of the function sine for $x = 0$, i.e. to $\cos 0 = 1$.

## 5 – Convergent sequences: definition and examples

Let us return to the principal topic of this chapter: convergent sequences. As we said while explaining Set Theory, a *sequence* of elements of a set $E$ is a function defined for all integers $n \geq 1$ (or, more generally, for all sufficiently large $n \in \mathbb{Z}$) with values in $E$; the value of this function for $x = n$ might, for example, be written $f(n)$, but the tradition has it that it is better to write $u_n$ and to employ the notation $(u_n)$ to denote the succession of values $u_1, u_2, \ldots$ of the terms of the sequence. This of course does not prevent us from using functional notation $u(n)$ when that appears more convenient, particular for typists, a category to which mathematicians have belonged for a long time[12]. Having recalled this, one says that a sequence $(u_n)$ of complex numbers *converges* or is *convergent* if there exists a number $u$, the *limit* of the sequence, such that, for all $r > 0$,

(5.1)          $d(u, u_n) < r$ for all sufficiently large $n$,

in other words if, for all $r > 0$, there exists an integer $N$ (generally depending on $r$, and unless $u_n = u$ for all sufficiently large $n$) such that

(5.1')          $|u - u_n| < r$ for all $n > N$.

This definition is only a particular case of the general concept defined at the beginning of n° 4: $X$ is here the set of $n \in \mathbb{Z}$ for which $u_n$ is meaningful.

It would clearly be enough to verify (1) for numbers $r$ of the form $1/p$, or $1/10^p$ or, if one is a fan of the binary system, $1/2^p$, since one can always choose $p$ so that, for example, $1/10^p < r$. If the sequence has real terms, this would indicate that, for all sufficiently large $n$, the decimal expansion of rank $p$ of $u_n$ is identical to that of $u$. Even though this idea is right, this formulation is not entirely correct because of the eccentricities of representation: the sequence whose terms successive are

$$0.9 \quad 1.1 \quad 0.99 \quad 1.01 \quad 0.999 \quad 1.001 \quad \text{etc.}$$

obviously converges to 1, but contradicts the hypothesis of "asymptotic stability" of the decimal expansions of given order for terms of a sequence.

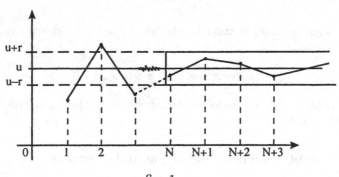

fig. 1.

In the case where all the $u_n$ are real, one can represent the sequence $(u_n)$ by a piecewise linear graph (figure 1) joining the different points $(n, u_n)$ in

---

[12] Indeed, since one can get the majority of scientists to do almost anything by challenging their abilities, the use of very sophisticated mathematical word processors has transformed many mathematicians into voluntary quasi-professional typographers (we repeat: quasi) – to the greater benefit of the true professionals thus displaced ...

the plane. Convergence of $u_n$ to $u$ then means that this graph is "asymptotic" to the horizontal line $y = u$ in the plane (not by the definition of a limit – one has no need to appeal to geometry for this – but, quite the contrary, by the definition of an asymptote). A similar remark applies to the case of a function $f(x)$ defined for sufficiently large $x \in \mathbb{R}$ and which tends to a limit when $x$ increases indefinitely: the fact that the function $1/x$ tends to 0 at infinity shows that its graph is asymptotic to the $x$ axis.

When a sequence $(u_n)$ tends to a limit $u$, one may write

$$\lim_{n \to \infty} u_n = u \quad \text{or} \quad \lim_{n \to +\infty} u_n = u$$

or even simply $\lim u_n = u$. Hardy's admonitions on the significance of the symbol $\infty$ apply here in full. Moreover, like the letter $r$ in the last n°, the letter $n$ is only a phantom and one can replace it by any other sign, on condition that one does so everywhere; one may very well write

$$\lim_{\$ \to \infty} u(\$) = u,$$

the mathematics will not change; however, it is forbidden to keep the first sign $\$$ and to replace the second by the sign $\pounds$ since different letters a priori represent variables independent one of the other: one would then have $\lim u(\pounds) = u(\pounds)$, unless one specifies this by a relation such as $\pounds = f(\$)$.

It is clear from the definition that *for a sequence $(u_n)$ to tend to a limit $u$ it is necessary and sufficient that the sequence with general term $u - u_n$ tends to 0*, which one can write in the form

$$u_n = u + o(1) \text{ when } n \to +\infty$$

since the symbol $o(1)$ represents any sequence or function negligible with respect to the constant function 1, i.e. tending to 0.

Moreover, if $u_n = v_n + iw_n$ is a sequence of complex terms, the inequalities

$$|v_n - v|, \; |w_n - w| \leq |u_n - (v + iw)| \leq |v_n - v| + |w_n - w|$$

show that

$$\lim(v_n + iw_n) = v + iw \iff (\lim v_n = v) \; \& \; (\lim w_n = w).$$

The inequality $\||u_n| - |u|\| \leq |u_n - u|$ shows on the other hand that, for real or complex sequences,

$$\lim u_n = u \implies \lim |u_n| = |u|.$$

The converse is clearly false.

The most obvious example of a convergent sequence is obtained – and it is a long way from being by chance – by starting with a real number $u$ and denoting by $u_p$ its *truncated* or *default decimal expansion of rank $p$*; this is a number of the form $k/10^p$, where $k$ is an integer such that

$$u_p < u \leq u_p + 10^{-p};$$

with this definition, the default decimal expansion of the number 1 is 0.9999&c. Then

$$d(u, u_n) \leq 10^{-p} \text{ for all } n \geq p,$$

from which $u$ is the limit of the $u_n$. This example makes clear the fact that we cannot define the real numbers without using the concept of limit in one way or another. It also shows that if one tries to limit the domain of the analysis to $\mathbb{Q}$ there will be multitudes of sequences which will not converge because of the fact that their limit is irrational.

We mentioned above that when a sequence $(u_n)$ of real numbers converges to a limit $u$, the decimal expansion of the number $u_n$ has a strong tendency to stabilise as $n$ increases indefinitely. One might deduce that a convenient experimental method to ascertain convergence, or to calculate the limit, of a sequence is to examine numerically a sufficiently large number of terms. This can spring several surprises on modern programmed calculators.

If one considers for example the sequence with general term

$$u_n = (1 + 1/n)^n,$$

which, we shall see later, converges to the number

$$e = 2.71828\ 18284\ 590\ldots,$$

the base of the natural logarithms, one finds that

$$u_4 = 2.44141\ldots, \quad u_{64} = 2.69734\ldots, \quad u_{1024} = 2.71696\ldots,$$

which indicates that $u_n$ approaches its limit value only very slowly, and that one would have to choose enormous values of $n$ to obtain even ten or so decimal places of the number $e$; very luckily, the sequence with general term

$$1 + 1/2! + 1/3! + \ldots + 1/n!$$

also converges to $e$, but with prodigious rapidity since the term $1/(n+1)!$ which one adds to $u_n$ to obtain $u_{n+1}$ becomes microscopic very quickly.

Another example of a slowly convergent sequence:

$$\begin{aligned} u_n &= 1/1.2 + 1/3.4 + 1/5.6 + \ldots + 1/(2n-1)2n \\ &= 1 - 1/2 + 1/3 - 1/4 + \ldots + 1/(2n-1) - 1/2n. \end{aligned}$$

It was already known by the end of the XVII$^{\text{th}}$ century that to calculate the limit, namely log 2, exactly to 9 decimal places, quite a modest precision, one must choose $n > 10^8$; see n° 13 for alternating series.

The sequence
$$u_n = 1 + 1/2 + 1/3 + \ldots + 1/n,$$
is divergent or, what comes to the same for an increasing sequence, its terms increase above any bound as we shall see in n° 7. Even on choosing an index $n$ as hyperastronomic as $10^{100}$, one finds a value only of the order of 230, a result which, numerically, is perfectly compatible with the false hypothesis that the sequence converges to 231.

Mathematicians such as Newton, Stirling or Euler who worked to obtain these numerical estimates had methods more ... intellectual than that of churning for an undetermined time (the experts will estimate it for us) the "acres of computers" of the American *National Security Agency*, at the risk of compromising the said security for the idle amusement of mathematicians. They already have trouble in finding two prime numbers $p$ and $q$ when the product $pq$, all one tells them, has a hundred or so digits.

To conclude these generalities, let us observe that the definition of a limit supposes that the limit to be obtained is known. It is nevertheless possible to decide on the convergence of a sequence without knowing the limit in advance. In this direction one notes that if one has $d(u, u_n) < r/2$ for all $n > N$, one will also have, by the triangle inequality,

$$d(u_p, u_q) < r \text{ once } p > N \text{ and } q > N.$$

We shall show in the following chapter that this necessary condition for convergence is also sufficient; this is *Cauchy's general criterion of convergence*, known before him to Bolzano, and which neither really proved. One can make the result seem very plausible by choosing numbers $r$ of the form $10^{-n}$; if decimal numeration did not exhibit the bizarre behaviour to which we have already alluded, the preceding inequality would show that starting from a certain rank, the first $n$ digits of the terms of the sequence would no longer change, and this would demonstrate convergence.

Let us now give some examples of convergent sequences; the first are almost trivial, but those following will be very useful in the sequel.

*Example 1.* The "constant" sequence $u, u, u, \ldots$ converges to $u$.

*Example 2.* One has

(5.2)                              $$\lim 1/n = 0,$$

since the relation $|1/n| < r$ can be written as $nr > 1$ so is satisfied for large $n$ by Archimedes' axiom.

*Example 3.* One has
$$\lim \frac{n}{n+1} = 1$$
since $|1 - n/(n+1)| = 1/(n+1)$ tends to 0 by the preceding example.

*Example 4.* The sequence $u_n = (-1)^n + 1/n$ does not converge; its terms of even order tend to 1, its terms of odd order to $-1$.

We note in this connection that if a sequence $(u_n)$ converges to a limit $u$, then every *subsequence* that one can extract from it will converge, also to $u$. Such a subsequence is obtained by choosing an increasing sequence of integers $p_1$, $p_2$, etc. and omitting the terms of the initial sequence except for those corresponding to this choice. Convergence then follows from the obvious fact that $p_n \geq n$ for all $n$.

For example, the sequence with general term $1/n^3$ tends to 0, since it is a subsequence of the sequence of example 2.

*Example 5.* If $q$ is a complex number, then

$$(5.3) \qquad \lim q^n = 0 \text{ if } |q| < 1.$$

We need to show that, for any $r > 0$, one has $|q^n| < r$ for $n$ large. By replacing $q$ by $|q|$, one reduces to the case where $q \geq 0$, and even to $q > 0$, since the case where $q = 0$ is trivial. As we are now assuming that $q < 1$, we have $1 = q + t$ with $t > 0$. All the terms in the binomial formula

$$1 = (q + t)^{n+1} = q^{n+1} + (n+1)q^n t + \ldots$$

are positive, whence $(n+1)q^n t < 1$, or

$$0 < q^n < 1/t(n+1).$$

The right hand side clearly tends to 0, so $q^n$ does too.

For $q = 1$, the limit is clearly 1. For all other possible values of $q$ the sequence is divergent. For suppose that $\lim q^n = u$ exists for some number $q \in \mathbb{C}$. The sequence with general term $q^{n+1}$ also converges to $u$, since it is a subsequence. But $q^{n+1} = q.q^n$, and it follows from the definition that, for every convergent sequence,

$$\lim u_n = u \text{ implies } \lim q.u_n = q.u$$

for any $q \in \mathbb{C}$, since if $q \neq 0$, the only nontrivial case,

$$|qu_n - qu| < r \Longleftrightarrow |u_n - u| < r/|q|,$$

a relation which is satisfied for large $n$ since $r/|q| > 0$. Returning to the sequence considered, we must have $qu = u$, i.e. either $q = 1$, the trivial case, or $u = 0$, which cannot be the case for $|q| \geq 1$ since then $|q^n| \geq 1$ for all $n$. We therefore have divergence except for the cases $|q| < 1$ and $q = 1$.

*Example 6.* Let us show that

$$(5.4) \qquad \lim z^n/n! = \lim z^{[n]} = 0$$

for any $z \in \mathbb{C}$. By taking absolute values we can restrict to the case where $z > 0$. Writing $u_n$ for the general term, we first remark that, for $n > p$, we have

$$u_n = \frac{z^p z^{n-p}}{p!(p+1)\ldots n} = u_p \cdot \frac{z}{p+1} \cdots \frac{z}{n}.$$

Let us choose for $p$ the integer part of $10z$, so that $p \leq 10z < p+1$, and keep $p$ fixed. We have $z/q < 1/10$ for $q \geq p+1$, and the relation above shows that

$$u_n \leq u_p/10^{n-p} = 10^p u_p/10^n \text{ for all } n > p,$$

whence $u_n < r$ as soon as $n$ is so large that $10^n > 10^p u_p/r$, qed.

*Example 7.* On writing $x^{1/n} = \sqrt[n]{x}$, we have

(5.5)                $\lim x^{1/n} = 1$ for all $x > 0$.

Suppose first that $x > 1$, whence $x^{1/n} = 1 + x_n$ with $x_n > 0$. All the terms in the binomial formula

$$x = (1 + x_n)^n = 1 + n.x_n + \cdots,$$

are $> 0$, whence $0 < x_n < (x-1)/n$ and so $\lim x_n = 0$, which proves (5). The case where $x = 1$ is trivial. If $0 < x < 1$, one puts $x = 1/y$ with $y > 1$, whence $x^{1/n} = 1/y^{1/n}$. It remains to show that, generally,

$$\lim u_n = u \neq 0 \text{ implies } \lim 1/u_n = 1/u,$$

which we shall do in a little while.

This example arose in the construction of the first tables of logarithms by Napier and then Briggs, Kepler, etc. The fundamental relation $\log(xy) = \log x + \log y$ shows that $\log(x^p) = p.\log x$, whence

$$\log(x^{1/n}) = \log(x)/n.$$

Suppose that $x > 1$ and, as above, put

(5.6)          $x^{1/n} = 1 + x_n$, whence $0 < x_n < (x-1)/n < x/n$.

Then $\log x = n \log(1 + x_n)$, whence

$$\log(x)/nx_n = \log(1 + x_n)/x_n.$$

When $n$ increases indefinitely, the right hand side is of the form $\log(1+h)/h$ where $h$ tends to 0. If one assumes that the function log, which clearly satisfies $\log 1 = 0$, is very "regular", it is simplest, so as not to make the calculations too intricate, to assume that

$$\log(1 + h) \sim h \text{ as } h \to 0$$

(or $\sim Mh$ where $M$ is a constant if one works, like Briggs, with logarithms to base 10), in other words, that the function log has a derivative[13] equal to 1 at $x = 1$.

In these circumstances, $\log(1 + x_n)/x_n$ tends to 1 by definition of the derivative, thus $\log(x)/nx_n$ also, and, since $\log x$ does not depend on $n$, one sees finally that $\log x = \lim nx_n$. Taking account of the definition (6) of $x_n$, one thus obtains the fundamental formula

$$(5.7) \qquad\qquad \log x = \lim n(x^{1/n} - 1).$$

This seems to be due to the astronomer Halley (1695), although the essential ideas are already in Napier and Briggs.

*Exercise.* Show that (7) is equivalent to

$$x^{1/n} = 1 + \frac{\log x}{n} + o\left(\frac{1}{n}\right) \quad \text{as } n \to +\infty.$$

To avoid all confusion, we stress the fact that, at this stage of the exposition, the formula (7) is not, in itself, either a definition or a construction of the function log. We have only shown that, *if there exists* a function log satisfying $\log(xy) = \log x + \log y$ and such that $\log(1 + x) \sim x$ when $x$ tends to 0, then it is given by the relation (7). But we have as yet proved neither the existence of the function log, nor the convergence of the sequence (7). This will be the object of Theorem 3 of n° 10.

In fact, as we shall see later, $\log(1 + u)$ lies, for $0 < u < 1$, between $u - u^2/2$ and $u$. Since

$$\log x = n.\log(1 + x_n),$$

we have

$$(5.8) \qquad\qquad 0 < nx_n - \log x < nx_n^2/2 < x^2/2n,$$

the last inequality following from (6). In other words, the error committed in replacing $\log x$ by $nx_n$ is $< x^2/2n$ provided that $x_n < 1$. It only remains, a modest enterprise, to perform the numerical calculations.

---

[13] This is the crucial point in obtaining formula (7). Napier, who worked at his tables from about 1590 to his death in 1617, and Briggs, who transformed them into logarithms to base 10 from about 1615 on, did not argue in terms of "derivatives" for the excellent reason that these would not appear, and then in a rather hazy way, until about twenty years later, with Fermat and Descartes, *à propos* the calculus of tangents to a curve. But Napier imagined a point moving along a segment of line with a speed inversely proportional to its distance $x$ from the origin of the segment, and the concept of "instantaneous speed" was just that of derivative with respect to time. The reader should not believe that the concepts and ideas which, nowadays, appear simple enough to be taught every year to hundreds of thousands of young people of the Earth, were born fully armed from certain brains of genius, like Athena from that of Jupiter ...

Since there is no practical method of extracting $n^{\text{th}}$ roots numerically (by hand ...) apart from the case where $n$ is a power of 2 – then it suffices to extract successive square roots –, one restricts to integers $n$ of the form $2^p$. In this case, the error given by (8) is majorised by $x^2/2^{p+1}$.

Napier, Briggs after his death, and then Kepler, set themselves, among other things, to calculate the logarithms of the first thousand integers exact to 7 or 14 decimal places; starting from this, one could find those of a large number of fractional values of $x$, since $\log(p/q) = \log p - \log q$. (Above all, they needed the logs of the trigonometric functions, but this is another problem again). Of course, they did not have to calculate all these logs directly; it is enough, for a start, to calculate those of the *prime* numbers 2, 3, 5, 7, 11, 13, etc. and even though in this case there are "tricks" to reduce the labour, it remains considerable, to express it mildly.

The first candidate is $x = 2$. One has to extract $p$ successive square roots, choosing $p$ and performing the calculations to sufficiently many places to have a hope of obtaining the required precision. In this particular case the error is less than $2^2/2^{p+1} = 2^{-p+1}$ in addition to those committed in extracting the $p$ successive square roots of 2. To obtain the result to 14 places, it is thus prudent to choose $p$ so that $2^{-p+1} < 10^{-15}$, i.e. $2^p > 2.10^{15}$. Now $2^9 = 512 < 10^3 < 2^{10} = 1024$, whence $2^{50} > 10^{15} > 2^{45}$, which indicates the need to choose for $p$ a number between 45 and 50, in other words to extract at least 45 successive square roots of 2 to ensure having 15 places exact at the end. One can reduce the work with the aid of the following remark.

We know, and they already knew then, that

$$1 + u/2 - u^2/8 < (1+u)^{1/2} < 1 + u/2 \text{ for } 0 < u < 1$$

(square it all), so that, for $u$ small, the error committed in replacing the square root of $1 + u$ by $1 + u/2$ is less than $u^2/8 = u^2/2^3$; the error is thus $< 2^{-50}$ if $u < 2^{-24}$. Now, in calculating the $x_n$, one has

$$1 + x_{n+1} = (1 + x_n)^{1/2} \quad \text{and} \quad x_n < 2^{-n-1} < 2^{-24}$$

once $n > 25$. One thus sees that after having extracted the 24 or 25 first successive square roots of 2 to 15 exact places, one may assume that $(1 + u)^{1/2} = 1 + u/2$ for $p > 25$. In other words, one has to calculate only the first 25 square roots to 15 places, which is still not within reach of everyone.

In the case of logs to base 10, Briggs started by extracting 54 successive square roots of 10, which gave him the number[14]

$$1.00000\ 00000\ 00000\ 12781\ 91493\ 20032\ 35 = 1 + h$$

and allowed him to calculate the number $M$ such that $\log_{10}(1 + h) \sim Mh$ since

---

[14] See E. Hairer and G. Wanner, *Analysis by Its History* (Springer-New York, 1996), p. 30, which also reproduces in facsimile the page where Briggs tabulates the 54 successive square roots of 10 and their logarithms.

$$1 = \log_{10}(10) = 2^{54}\log_{10}(1+h) \sim 2^{54}Mh.$$

To calculate $\log_{10} 2$ for example, he again extracted 54 square roots of 2, and found

$$1.00000\ 00000\ 00000\ 03847\ 73979\ 65583\ 10 = 1 + h'$$

and since

$$\log_{10} 2 = 2^{54}\log_{10}(1+h') \sim 2^{54}Mh',$$

finally he found

$$\log_{10} 2 = h'/h = 0.30102\ 99956\ 63881\ 2.$$

This done, you start again with 3, 5, 7, etc. One may well suspect that, as we shall see, more economical procedures were invented later.

We shall return to all this *à propos* the logarithmic functions.

*Example 8.* We have
(5.9)                                              $$\lim n^{1/n} = 1.$$

Let us put $n^{1/n} = 1 + x_n$ where clearly $x_n > 0$. The binomial formula

$$n = (1 + x_n)^n = 1 + n.x_n + \frac{1}{2}n(n-1)x_n^2 + \dots,$$

in which all the terms are $> 0$, shows that $x_n^2 < 2/(n-1)$, an expression which tends to 0. For all $r > 0$, one thus has $x_n^2 < r^2$ for $n$ large, so $0 < x_n < r$, and thus $x_n$ tends to 0, qed. (A sequence which tends to 0 is $< r^2$ or $r^{624}$ for $n$ large since $r^2 > 0$).

## 6 – The language of series

Though not basically different from that of sequences, the language of *series* is frequently convenient, principally because many individual functions can be represented more naturally by series than by sequences; their theory scarcely appeared before Cauchy's time, the 1820s, while they are ubiquitous after 1650.

The fundamental problem is to *give a meaning to a sum of an infinite number of terms*, imposing, of course, reasonable conditions; we do not propose to explain what the sum $1 - 2 + 3 - 4 + 5 - \dots$ might signify, let alone the sum of all the real numbers[15], for it is prudent to restrict oneself to sums having no more than a *countable* infinity of terms. The terms of such a sum may, by hypothesis, be put in the form of a sequence $u_1, u_2, \dots$. Most often, they are given in advance in this form, but there are also cases where it would be quite artificial to order the terms of the sum as a sequence. Consider for example the "lattice" $\mathbb{Z}^2$ of points with integer coordinates in the Cartesian

---

[15] An ingenious innocent might observe that, each real number being cancelled by its opposite, the sum in question is "obviously" 0 . . .

plane $\mathbb{R}^2$, and suppose that you are interested in the sum of inverses of the $k^{\text{th}}$ powers of the distances from the origin to the points of the lattice, i.e. you wish to assign a meaning to the sum of numbers $1/(m^2 + n^2)^{k/2}$, where $m$ and $n$ are rational integers, not both zero. We know that $\mathbb{Z}^2$ is countable, but there is no privileged, natural or obvious bijection of $\mathbb{N}$ onto $\mathbb{Z}^2$. This poses, in this case, the problem of defining the "unordered" sum of the terms of the series. We shall study this later (n° 11), but for the moment we confine ourselves, maybe wrongly, to the traditional situation of a sum whose terms are given in the form of an ordered sequence.

To obtain a *simple series*, or series for short, without explicit mention to the contrary, one starts with a sequence $(u_n)$ of complex numbers. Since the total sum of the $u_n$ can reasonably be defined only through an approximation procedure involving only the sums of a finite number of terms – the only ones that we know at this stage of the exposition –, it is natural to consider the numbers

$$s_1 = u_1$$
$$s_2 = u_1 + u_2$$
$$s_3 = u_1 + u_2 + u_3$$

etc. One calls them the *ordered partial sums*, or partial sums for short when there is no fear of confusion, of the series with general term $u_n$, and one says that this is *convergent* when $\lim s_n = s$, the *sum* of the series considered, exists. Then one writes

$$s = \sum_{n=1}^{\infty} u_n, \qquad \text{or } s = u_1 + u_2 + \dots,$$

or simply $s = \sum u_n$. So, by definition,

(6.1) $$s = \lim(u_1 + \dots + u_n).$$

The notation $u_1 + u_2 + \dots$, as used by all the Founding Fathers, is now in total desuetude, but the reader will maybe find it, at the beginning, easier than the other.

As the convergence of a series reduces to that of a sequence, conversely the relation

$$u_n = u_1 + (u_2 - u_1) + \dots + (u_n - u_{n-1})$$

reduces the convergence of a sequence to that of a series.

Some people seem to believe that it is contrary to the principles of sane pedagogy to introduce series at the beginning of teaching analysis. At the risk of traumatising the reader, let us observe that the nonterminating decimal expansion

$$x = x_0.x_1x_2 \dots$$

of a real number means that it is the limit of the sequence whose terms are

$$
\begin{aligned}
s_0 &= x_0 \\
s_1 &= x_0 + x_1/10 \\
s_2 &= x_0 + x_1/10 + x_2/100
\end{aligned}
$$

etc., in other words that

(6.2) $$ x = x_0 + x_1/10 + x_2/100 + \ldots = \sum_{n=0}^{\infty} x_n/10^n. $$

Consider for example the number

$$ 2/15 = 0.13333333333333\ldots, $$

according to commercial arithmetic. By the above, the right hand side is the sum of the series $1/10 + 3/100 + 3/1000 + \ldots$, plausibly equal to

(6.3) $$ 1/10 + 3.10^{-2}(1 + 1/10 + 1/100 + \ldots), $$

so that it comes down to calculating the sum of the *geometric series*

$$ 1 + q + q^2 + \ldots $$

for $q = 1/10$. Now the partial sums are, for $q \neq 1$, the numbers

$$ 1 + q + q^2 + \ldots + q^n = \frac{1 - q^{n+1}}{1 - q} = \frac{1}{1 - q} - \frac{q^{n+1}}{1 - q}. $$

When $n$ increases indefinitely, $q^{n+1}$ tends to 0 if $|q| < 1$ (n° 5, example 5), so also does $q^{n+1}/(1 - q)$ by the more elementary rules that one will find in n° 8. One deduces that

(6.4) $$ 1 + q + q^2 + \ldots = \sum_{n \in \mathbb{N}} q^n = \frac{1}{1 - q} \quad \text{if } |q| < 1. $$

In particular,

$$ 1 + 1/10 + 1/100 + \ldots = \frac{1}{1 - 1/10} = 10/9. $$

Thus one finds the value $1/10 + 3.10^{-2}.10/9$ for the series (3) and it remains to verify that this result is in fact the fraction $2/15$ from which we started.

On replacing $q$ by $-q$ in (4), one finds the formula

(6.5) $$ 1 - q + q^2 - q^3 + \ldots = \sum_{n=0}^{\infty} (-1)^n q^n = \frac{1}{1 + q} $$

already known to Viète, Newton and Mercator, the last two having used it around 1665 to calculate the area of a segment of a hyperbola. The mathematicians of the XVII$^{th}$ century, Newton in the first place, obtained these series by a very different procedure, the division of 1 by $1 + q$ according to the *increasing* powers of $q$; one proceeds as one would in commercial arithmetic if $q$ were equal to $1/10$:

$$
\begin{array}{r|l}
1 & 1+q \\
\phantom{1} & \overline{\phantom{1-q+q^2-q^3}} \\
-q & 1 - q + q^2 - q^3 \\
+q^2 & \\
-q^3 &
\end{array}
$$

etc., with successive "remainders" equal to $-q$, $q^2$, $-q^3$, $\ldots$ A more economical procedure consists of writing

$$(1+q)(1-q+q^2-q^3+\ldots) =$$
$$= (1-q+q^2-q^3+\ldots) + (q - q^2 + q^3 - \ldots) = 1.$$

One should pay attention to the fact that these formulae assume $|q| < 1$ since otherwise the term $q^{n+1}$ appearing in the partial sum does not tend to any limit (n° 5, example 5), so that the geometric series is divergent. Otherwise, one could also suppose that $q = 1$ in (5) and thus obtain the relation

$$1 - 1 + 1 - 1 + 1 - 1 + \ldots = 1/2$$

which, fascinating though it is - Jakob Bernoulli "discovered" it in 1696 and others got trapped by it before or after this date -, has no meaning: the partial sums of the series of the left hand side being alternately $1, 0, 1, 0, \ldots$, one cannot see how they could converge! Absurdity would reach even more extravagant heights if one put $q = 2$ in (4); one would thus "discover" that $1 + 2 + 4 + 8 + 16 + 32 + \ldots = -1$, an example which Nikolaus I Bernoulli produced in 1743 in a letter to Euler to warn him away from divergent series[16].

Series lead to much stranger formulae, such as

$$
\begin{aligned}
1 + 1/2^2 + 1/3^2 + 1/4^2 + \ldots &= \pi^2/6, \\
1 + 1/2^4 + 1/3^4 + 1/4^4 + \ldots &= \pi^4/90, \\
1 + 1/2^6 + 1/3^6 + 1/4^6 + \ldots &= \pi^6/945,
\end{aligned}
$$

$$\cot x = \frac{1}{x} + 2x \sum_{n=1}^{\infty} \frac{1}{x^2 - n^2\pi^2},$$

---

[16] Moritz Cantor, *Vorlesungen über Geschichte der Mathematik* (Teubner, Vol. III, 1901), p. 691. Euler was not convinced; he believed that every series, even divergent, had a hidden meaning, and, in fact, was the first to calculate with the "formal series" of which we will speak in n° 22. But these are not series of *numbers*.

a formula valid for all $x$ not a multiple of $\pi$; they are all due to Euler (1707–1783), as is the expansion

$$\sin x = x \prod_{n=1}^{\infty} \left(1 - x^2/n^2\pi^2\right),$$

valid for all $x \in \mathbb{R}$, of the function sine as an *infinite product*[17].

We said above that many of the elementary functions (and infinitely many others) can be represented conveniently by series, and particularly by *power series* of the form

$$a_0 + a_1 x + a_2 x^2 + \ldots = \sum a_n x^n,$$

where the $a_n$ are numerical coefficients; it was Newton who, first, made systematic use of them to resolve all sorts of problems; he explains that they play a rôle in analysis analogous to the decimal notation in arithmetic, as the relation (2) above confirms, and that the two techniques can even be used in the same way, which is a little optimistic. Not yet being able to justify them at this stage of the exposition, we shall give examples of remarkable power series which, in this chapter, we shall use frequently as experimental material to illustrate the interest of theorems of which, otherwise, the reader might not see the necessity:

$$
\begin{aligned}
\sin x &= x - x^3/3! + x^5/5! - x^7/7! + \ldots & \text{for any } x, \\
\cos x &= 1 - x^2/2! + x^4/4! - x^6/6! + \ldots & \text{for any } x, \\
e^x &= 1 + x/1! + x^2/2! + x^3/3! + \ldots & \text{for any } x, \\
\log(1+x) &= x - x^2/2 + x^3/3 - x^4/4 + \ldots & \text{for } -1 < x \leq 1, \\
(1+x)^s &= 1 + sx + s(s-1)x^{[2]} + s(s-1)(s-2)x^{[3]} + \ldots \\
& & \text{for } |x| < 1
\end{aligned}
$$

for all real exponents $s$, for example $s = 1/2$, the first case treated by Newton; this is the famous "binomial formula of Newton" which, for $s \in \mathbb{N}$, reduces to the known algebraic formula since the coefficient of $x^n$ is then clearly zero for all $n > s$; see Chap. IV, n° 11. These formulae were discovered by Newton (1642–1727) when in 1665–67 the "Great Plague" which ravaged the region of London, see Samuel Pepys and Daniel Defoe, forced him to return to the countryside of his adolescence where, among other occupations, he discovered the composition of white light and the first idea of the law of universal gravitation, work in the fields not attracting him particularly.

---

[17] Given a sequence of numbers $(u_n)$ all $\neq 0$, one says that the infinite product of the $u_n$ converges if the partial products $p_n = u_1 \ldots u_n$ tend to a *nonzero* limit; it is necessary for this that $\lim u_n = 1$. The theory reduces easily to that of series thanks to the function log, which transforms a product into a sum (Chap. IV, n° 17). We shall show in n° 21 how Euler discovered his formula.

In practice, all the functions that one meets in *classical* analysis are representable as power series or, if necessary, by series with a further finite number of terms of negative degree like

$$1/x^2(1-x) = x^{-2} + x^{-1} + 1 + x + x^2 + \ldots,$$

or by series of this type where the variable is a fractional power of $x$, as in the relation

$$(x - x^2)^{1/2} = x^{1/2} - x^{3/2}/2 - x^{5/2}/8 - x^{7/2}/16 - 5x^{9/2}/128 - \ldots$$

which Newton deduced from the binomial series for $s = 1/2$. This fact led the mathematicians to study systematically from the XIX$^{\text{th}}$ century – there was an attempt by Lagrange a little earlier which came to nothing because he restricted himself to functions of a real variable – the *analytic functions* of a *complex* variable, which one can define as follows. Consider a function $f$ with complex values defined on an *open* subset $G$ of $\mathbb{C}$, i.e. such that, for all $a \in G$, the set $G$ contains an open disc $d(a, z) < r$ with centre $a$ and radius $r > 0$ (depending on $a$). The function $f$ is called analytic if, for all $a \in G$, there exists a power series

$$c_0(a) + c_1(a)(z-a) + c_2(a)(z-a)^2 + \ldots = \sum c_n(a)(z-a)^n$$

whose coefficients depend on $a$ and which (i) converges for $|z - a|$ sufficiently small, (ii) has sum $f(z)$ on a neighbourhood of $a$, whence necessarily $c_0(a) = f(a)$. Take for example the function $f(z) = 1/z$, defined on the open set $z \neq 0$. For $a \neq 0$, one can write

$$\frac{1}{z} = \frac{1}{a - (a - z)} = \frac{1}{a} \frac{1}{1 - (a - z)/a} = \sum_{n \in \mathbb{N}} (a - z)^n/a^{n+1}$$

on condition that $|(a - z)/a| < 1$, i.e. $|z - a| < |a|$; the function $1/z$ is thus represented by the power series in $z - a$ that we have just written, in the largest disc with centre $a$ not containing – which is normal – the point $z = 0$. Here $c_n(a) = (-1)^n/a^{n+1}$.

Cauchy (1789–1857) was the first to observe, having consecrated thirty years of work to them before seeing it clearly, that these functions possess extraordinary properties which nearly all the mathematicians of the XIX$^{\text{th}}$ century, and a good subset of their successors, have put to work from one time to an other, and have in particular generalised to functions of several complex variables.

## 7 – The marvels of the harmonic series

Let us return to the elementary theory of convergent sequences. Remember that a sequence $(v_n)$ is a *subsequence* of a sequence $(u_n)$ if there exists a strictly increasing sequence of integers $p_1, p_2, \ldots$ such that[18]

$$v(n) = u(p_n) \text{ for all } n.$$

For example, the sequence $(1/n^2)$ is a subsequence of the sequence $(1/n)$.

*If a sequence $(u_n)$ converges to a limit $u$, then every subsequence of $(u_n)$ converges to $u$.* With the notation above, one has $p_n \geq n$ for all $n$ since $p_n > \ldots > p_1 \geq 1$; the relation

$$d(u, u_n) < r \text{ for all } n > N$$

then implies that $d(u, v_n) < r$ for all $n > N$, qed.

This trivial result translates usefully into the language of series. To extract a subsequence from the sequence of partial sums $s_n$ of a series $u(n)$ one chooses as above a sequence of integers $p_n$ and considers the sequence whose terms are

$$s(p_1) = u(1) + \ldots + u(p_1), \qquad s(p_2) = u(1) + \ldots + u(p_2),$$

etc. These are manifestly the partial sums of the series whose successive terms are

$$v_1 = u(1) + \ldots + u(p_1), \qquad v_2 = u(p_1 + 1) + \ldots + u(p_2),$$

$$v_3 = u(p_2 + 1) + \ldots + u(p_3), \ldots$$

in other words, of the series obtained by grouping the terms of the initial series into blocks of $p_1$, $p_2 - p_1$, $p_3 - p_2$, $\ldots$ terms as if dealing with a finite sum. Theorem 1 shows that, if the initial series converges, so does the new, the two series having the same sum. This is an extension of the associativity of addition: one has

$$u(1) + u(2) + u(3) + \ldots = [u(1) + \ldots + u(p_1)] + [u(p_1 + 1) + \ldots + u(p_2)] + \ldots$$

as for finite sums so long as the *left hand side* converges. One should be aware of the fact that though a series may become convergent *after* grouping its terms, it does not follow that it was already so *before* this operation: the series $(1 - 1) + (1 - 1) + \ldots$ has no merit in being convergent, and the series $1 - 1 + 1 - 1 + \ldots$ is divergent. This difficulty does not arise with series *of positive terms* as we shall see in n° 12.

The preceding artifice can serve to prove the divergence of a series, for example of the *harmonic series*

$$1 + 1/2 + 1/3 + \ldots .$$

If it were indeed convergent, so would be the series

---

[18] The functional expression $u(n)$ shows that a subsequence is only a particular case of the general concept of composition of maps: compose $n \mapsto p_n$ and $n \mapsto u_n$.

$$1/2 + (1/3 + 1/4) + (1/5 + 1/6 + 1/7 + 1/8) + \ldots$$

obtained by grouping of 1, 2, 4, 8, 16, ... terms in the initial series (one omits the first term which clearly plays no rôle in the question). Now the first group of terms has a value $\geq 1/2$, the second, being the sum of two terms greater than $1/4$, is the same, the third again, since it is the sum of four terms greater than $1/8$, etc. One thus finds, for the new series, partial sums successively greater than $1/2$, 1, $3/2$, etc.; whence the divergence of the new series and so of the harmonic series. This kind of argument will be generalised in n° 12 ("Cauchy's condensation criterion").

There are many variants of the preceding proof; they have sometimes given rise, historically, to spectacular errors[19] of a nature to commend prudence to readers who are starting on the subject. A little after 1650, the Italian Pietro Mengoli observed that one always has

$$\frac{1}{n-1} + \frac{1}{n} + \frac{1}{n+1} > \frac{3}{n};$$

so, by grouping the terms of the harmonic series, one obtains

$$1 + (1/2 + 1/3 + 1/4) + (1/5 + 1/6 + 1/7) + \ldots > 1 + 3/3 + 3/6 + \ldots$$
$$= 1 + 1 + (1/2 + 1/3 + 1/4) + (1/5 + 1/6 + 1/7) + (1/8 + \ldots$$
$$> 1 + 1 + 3/3 + 3/6 + \ldots = 1 + 1 + 1 + (1/2 + 1/3 + 1/4) + \ldots$$

etc., which shows that the proposed sum $s$ of the series is greater than any integer. Instead of advertising his ingenuousness, Mengoli should have confined himself to observing that the first of his inequalities already provides the rather fishy relation $s > s + 1$.

Forty years later, Johann Bernoulli used an analogous idea. He started from the relation

$$1/1.2 + 1/2.3 + 1/3.4 + \ldots = 1,$$

obvious if one writes[20] it as

$$(1 - 1/2) + (1/2 - 1/3) + \ldots = 1,$$

and then remarked that

$$1/2 + 1/3 + 1/4 + \ldots = 1/1.2 + 2/2.3 + 3/3.4 + \ldots$$
$$= (1/1.2 + 1/2.3 + 1/3.4 + \ldots) + (1/2.3 + 1/3.4 + \ldots) + (1/3.4 + \ldots)$$
$$= 1 + (1 - 1/2) + (1 - 1/2 - 1/6) + (1 - 1/2 - 1/6 - 1/12) + \ldots$$
$$= 1 + 1/2 + 1/3 + 1/4 + \ldots,$$

---

[19] I find them in Cantor, *Vorlesungen* ..., Vol. III, particularly pp. 94–96.

[20] The relation in question is obvious if one calculates as with a finite sum since all the terms apart from the first "visibly" cancel in pairs. But the correct proof consists of remarking that the sum of the first terms $n$, namely $1 - 1/n$, tends to 1 as $1/n$ tends to 0.

whence the relation $s = 1 + s$ for the "sum" $s$ of the series. Immediately Jakob Bernoulli, his elder brother, observed that the partial sum

$$1/(a+1) + \ldots + 1/a^2$$

has $a^2 - a$ terms all greater than $1/a^2$, so has a value greater than $1 - 1/a$, whence it follows that $1/a + \ldots + 1/a^2 > 1$. From here it is easy to extract from the harmonic series groupings of terms greater than an arbitrary integer. Jakob Bernoulli observed at that point that a series whose terms tend to 0 – a condition clearly *necessary* for convergence since $u_n = s_n - s_{n-1}$ is the difference of two sequences which tend to the same limit – can still be divergent.

The fact that he had revealed the absurd marvels of the harmonic series did not prevent Jakob, several years later, from baldly putting

$$A = 1/1 + 1/2 + 1/3 + \ldots,$$

and then deducing that

$$A - 1 - 1/2 = 1/3 + 1/4 + \ldots$$

and then, by subtraction, that

$$3/2 = 2/1.3 + 2/2.4 + 2/3.5 + \ldots,$$

which provided him a "new proof", this time correct, of a formula obtained by Leibniz in 1682:

$$1/1.3 + 1/2.4 + 1/3.5 + \ldots = 3/4.$$

Similarly, putting $E = 1/1 + 1/3 + 1/5 + \ldots$ (the series diverges), whence of course $E - 1 = 1/3 + 1/5 + \ldots$, he obtains by difference and division by 2 another correct formula of Leibniz':

$$1/1.3 + 1/3.5 + 1/5.7 + \ldots = 1/2.$$

These calculations are meaningless. The usual rules of algebra were developed for calculating *finite* sums, i.e. consisting of only a finite number of terms; it is sometimes legitimate to apply them to convergent series, and almost always, as we shall see later, to the *absolutely* convergent series which we shall introduce in n° 15, not because they are obvious, but because the mathematicians of the XIX[th] century have proved the indispensable general theorems.

The same Jakob Bernoulli would give another precarious example in 1692. He starts from the relations

$$
\begin{aligned}
1/1 + 1/2 + 1/4 + 1/8 + \ldots &= 2/1, \\
1/3 + 1/6 + 1/12 + 1/24 + \ldots &= 2/3 \\
1/5 + 1/10 + 1/20 + 1/40 + \ldots &= 2/5
\end{aligned}
$$

obtained by dividing the relation

$$1 + 1/2 + 1/2^2 + 1/2^3 + \ldots = 2,$$

itself obtained on putting $q = 1/2$ in the sum (6.4) of the geometric series,
by 1, 3, 5, ... Having done this, Bernoulli adds these relations side-by-side as
one might do with a finite number of finite sums. One obtains the harmonic
series $\sum 1/n$ for the left hand side for the reason that any integer $n$ can be
written in one and only one way as the product of a power of 2 and of an odd
number, and so appears once and only once among the left hand side terms.
By this ingenious procedure one finds the formula

$$1 + 1/2 + 1/3 + \ldots = 2(1 + 1/3 + 1/5 + \ldots)$$

whence "evidently"

$$1/2 + 1/4 + 1/6 + \ldots = 1 + 1/3 + 1/5 + \ldots$$

despite the fact that each term of the left hand side is strictly less than the
corresponding term of the right hand side. One might push the paradox even
further, substituting the right hand side into the left hand side; one would
thus obtain the formula

$$1/1.2 + 1/3.4 + 1/5.6 + \ldots = 0,$$

particularly miraculous since the terms of the left hand side are all $> 0$. We
will see in n° 18, Corollary of Theorem 13, how one can justify this type of
operation, subject to hypotheses not satisfied in the preceding case.

One would be wrong to laugh at the Bernoullis. Even if, like the majority
of their contemporaries, they evinced an excessive penchant for virtuosity,
they did not have behind them three centuries of mathematicians who had
totally eliminated the difficulties inherent in the conception and use of series;
they were in process *of inventing the subject starting from nothing* or nearly
so. This protestant family, which left Anvers for Frankfort and then Bâle
when the henchmen of the supercatholic Philippe II put down the revolt of
the Low Countries at the end of the XVI[th] century, produced eight mathematicians – the two brothers Jakob (1654–1705) and Johann (1667–1748),
the son Nikolaus (1687–1759) of a brother of these two, three sons of Johann,
Nikolaus II (1695–1726), Daniel (1700–1782) and Johann II (1710–1790) and
two sons of his, Johann III (1744–1807) and Jakob II (1759–1789) – well
known or famous, without speaking of their activities as physicists, jurists,
doctors, hellenists, etc.; see their notices in the DSB. The whole XVIII[th]
century calculated like them, notably Euler, the Bach of mathematics, student of Johann, and we shall yet see Fourier obtain prodigious results around
1807, by using series much more outrageously divergent than those of the
Bernoullis.

After these examples, and those we shall present later in this chapter, one will understand perhaps a little better why the mathematicians of the $\text{XIX}^{\text{th}}$ century and yet more of the $\text{XX}^{\text{th}}$, tired of false proofs, by great mathematicians, of generally correct theorems (there were surely also innumerable false theorems produced by lesser masters, but they have not passed to posterity), have finally propounded, at least implicitly, the following basic principles:

(i)    *every assertion which is not fully proved is potentially false and is only, at best, an interesting conjecture,*

(ii)   *using an incompletely proved assertion to prove others increases the risk of error exponentially,*

(iii)  *the duty to prove an assertion falls on the author*

even if his colleagues do not refrain, on occasion, from doing so in his place, or from demolishing it. Naturally there are conjectures which people have tried to prove for decades, even centuries: Fermat's Last Theorem, the Goldbach and Riemann Conjectures, etc. But to formulate or prove such hypotheses is not the lot of everyone ...

The observation of these principles has led to the formidable intellectual discipline that mathematicians have progressively imposed on themselves for a century. One finds it nowhere else to the same degree. In physics theoreticians often take great liberties with the mathematics, their inspired intuitions being sufficient; the experimentalists insist on the reproducibility of their experiences, which, in *Big Science*, can lead far, even though one sometimes works on hypotheses which may be revealed to be totally false. True historians may attempt to observe the mathematicians' rules, but the inevitable gaps in their information, the need to check sometimes falsified documents, and to interpret them objectively, makes it difficult. And imagine the career of a politician who applied these rules.

An example of an assertion falling directly within the scope of principles (i), (ii) and (iii): it is thanks to nuclear arms that the Third World War has been avoided. Repeated *ad nauseam* for decades without the least *proof* being provided (one invokes Munich, or the Soviet regime, imperialism and arms, or even the precedent of Pearl Harbor; whereas the Foreign Policy of the Soviets has always been radically different from that of the Nazis or the Japanese before 1939, etc.), this assertion ignores all sorts of arguments which do not lead in the same direction.

(1) Nuclear or not, the horrors of two World Wars and the almost total unpredictability of this kind of enterprise, might have been enough to dissuade amateurs less mad than Adolf Hitler; look at the enthusiasm of the French, British and Soviet leaders faced with Hitler already before September 1939; the USSR did not enter the war until attacked by the Nazis; the USA waited for Pearl Harbor and a

declaration of war by Hitler. With the exception of the USA, which came out of the war much more powerful that it was at the beginning, and without comparatively great human loss (300 000 dead – the War of Secession caused double – against the twenty millions of the USSR, seven of Germany, etc.), all the belligerents were ruined or demolished to a point heretofore unimaginable; as the American diplomat George Kennan wrote from Moscow, "one has to see it to believe it".

(2) It has been recognised in the USA since 1947 that the East-West conflict takes place on an ideological much more than territorial level. The theoretical programme of the Soviets put much more stress on helping the internal revolutions in the Third World rather than on military conquests; their occupation of Central Europe, was, from the point of view of their strategy, in the first place justified by a possible resurrection of the German peril (also as much feared by the French) and, later, to provide them a defense or advance starting point in case of war against NATO. The first priority of the Soviets has always been to preserve their regime: no adventurism, except under Khrushchev who lost his position on this account. That of the Americans since 1945 is to retain their *"preponderance of power"*, as Leffler says, their influence on the noncommunist world, and a technological superiority intended to maximise the losses of a possible adversary while minimising their own. The Pacific War cost one hundred and four thousand American dead but at least nine hundred thousand to the Japanese military (and nearly as many civilians). The wars of Korea and Vietnam each brought some tens of thousands of American dead, but two or three million Koreans and Chinese and as many in the Indo-Chinese population; they were not all due to the Americans – all the belligerents behaved like savages – but the enormous superiority of American arms explains a great deal. The first war in Iraq cost 147 or 148 American dead, and maybe 100 000 Iraqis.

(3) It was the Hiroshima bomb which instantly convinced the Soviets that America constituted a "mortal menace" to them and engaged them in a frenetic nuclear arms race; one knows now that, from the end of August 1945, the Pentagon had a list of several dozens of Soviet cities and industrial zones to atomise in case of war[21]. Addressed in particular to the head of the Manhattan Project, the plan of 1945 and those, much more apocalyptic, which have followed it, clearly indicated the desire of the military to be in a position to devastate the USSR in future if need be.

---

[21] Edward Zuckerman, *The Day After World War III* (Avon Books, 1987), pp. 181–183, who refers on p. 368 to a memorandum of General Norstad, and Richard Rhodes, *Dark Sun. The Making of the Hydrogen Bomb* (Simon & Schuster, 1995), pp. 23–24, who cites the archives of the Manhattan Project.

From the political point of view, it was clearly not a question of executing such a plan in 1945. In fact, the production of bombs, which would have reached six per month by the end of the year if the war had continued, ended in practice thus: the majority of the participants in the project returned to civil life, the much too expensive installations were suppressed, and the others improved without great urgency, though the reduced scientific teams remaining in place prepared the future improvements (miniaturisation and thermonuclear) with the help of university consultants. The result was that America had only 13 bombs in 1947, to the great stupefaction of President Truman himself, who was relying on his *winning weapon* to base his politics; the Soviets, probably better informed, had not passed onto the offensive, if not on the diplomatic plane ...

In fact, the policy of "containment" of the USSR inaugurated in 1947 on the inspiration of Kennan did not truly emphasise nuclear arms until after the first Soviet explosion of August 1949, the principal effort concerning, before, aeronautics and principally the production or the development of the strategic bombers B-36, B-47 and B-52. Foreseen since 1947 by Kennan, the final collapse or the *"gradual mellowing"* of the Soviet regime, as he called it, would have to be caused by internal problems.

(4) The creation in January 1947 of a new administrative structure, the *Atomic Energy Commission* under civil control, and the acceleration of the Cold War raised the stock to 298 in 1950, 2,280 in 1955, 12,305 in 1959 (official figures) and to 32,500 in 1967 and then to diminish progressively. The Soviet stock, estimated at 200 in 1955 and at 1,050 in 1959, widely surpassed the American maximum at the end of the 1970s (*Bulletin of the Atomic Scientists*, 12/1993). The study by Stephen I. Schwartz et al., *Atomic Audit. The Cost and Consequences of U.S. Nuclear Weapons since 1940* (Brookings Inst. Press, 1998, 680 p.), where one finds not only the figures, estimates the minimum cost of nuclear weapons for the USA alone at 5 821 billion dollars at 1996 values (French GNP for that year: 1 280), of which 409 was for nuclear arms in the strict sense, the rest covering vectors, anti-aircraft and then anti-missile defence, satellites, etc.

The multiplication of nuclear arms and of their vectors in the two camps could only powerfully contribute to accentuate their feelings of insecurity and their mutual hostility, as shown by the innumerable allusions to the "Soviet menace" in the West, and notably in France (it threatened in the first place Soviet citizens, particularly under Stalin ...), also the recent evidence of Russian atomic physicists seeking to protect their country from the "American peril", not to be neglected at the beginning of the[22] 1950s and considered, right or wrong, as very serious during the Reagan period.

(5) By far the most grave crisis of the Cold War was triggered by the clandestine installation in Cuba in 1962 of Soviet *nuclear* arms in response to the American arms in Europe, principally the missiles in Turkey, and the preparatory moves to invading Cuba. Some claim that the peaceful resolution of the crisis proved the efficacity of "deterrence". Apart from the fact that there would not have been a "Cuban crisis" of such dimensions if not for nuclear arms, new information has been available for some years. At the height of the crisis, at the point of being stopped, the famous Colonel Penkowski, who had communicated a mass of information to the Americans concerning the arms of his country, is supposed to have sent a coded message to the CIA informing them of an imminent Soviet attack; in view of his personality the two employees of the CIA who received the message decided not to transmit it to their superiors (information impossible to confirm, but published by Raymond Garthoff, author of massive studies on American-Soviet relations). One knows that the Soviets left more than 40,000 men in Cuba, and not 10,000 as the CIA believed, with tactical nuclear arms which the local Soviet commandant was authorised to use in case of an American invasion, a contingency which might very well have been realised if Khrushchev had waited a few more days before throwing in the sponge: imagine the American reaction in such a case. One knows that, in the two camps, the military chiefs protested violently against the outcome of the crisis: they were for immediate invasion by the USA – everything was ready – and against "capitulation" in USSR. At the height of the crisis, the chief of the *Strategic Air Command* took it on himself to send the order *in clear* to the B-52s loaded with bombs which flew constantly towards the USSR to go on a state of quasi-maximum alert, the American Navy harassing the Soviet submarines even in the Pacific. The then Secretary of Defense, Robert McNamara, later became the principal advocate of total abolition of nuclear arms.

---

[22] At this time, which follows the first Soviet explosion of August 1949, one sees a part of the press and numerous civilian or military personnel advocate a preventive attack while the Russians did not yet have a stock of bombs; for example, our colleague von Neumann (Logic and Set Theory, Hilbert Spaces, Game Theory, Implosion Theory for the Nagasaki Bomb, Programmable Computers, H Bomb, Missiles), then one of the most influential advisers of the Pentagon, is supposed to have said: *If you say why not bomb them tomorrow, I say why not today. If you say today at five o'clock, I say why not one o'clock*, if you believe *Life*, 25/2/1957, on the occasion of his death. But neither Truman nor Eisenhower was disposed to take this risk, as, among other historians, Marc Trachtenberg explains lucidly, *History and Strategy* (Princeton UP, 1991), in a chapter entitled *A "Wasting Asset": American Strategy and the Shifting Nuclear Balance, 1949–1954*. See also Rhodes, *Dark Sun*, pp. 562-568.

(6) The balance of power between the two camps was considered highly precarious for a long time. In clear logic, it would have been enough for both sides to have had about a hundred submarine-borne nuclear arms to be able to immobilise each other; the fact that this strategy was not adopted indicates that no one was confident in nuclear arms as a guarantee of deterrence, despite the fact that nobody has yet found a way to protect oneself from them. To mention just one significant detail, it was to limit the damage from a possible Soviet attack that Paul Baran, at the Rand Corporation, invented packet switching in 1964–5, and out of this came the Arpanet and Internet. The Rand Corporation, the foremost "think tank" working for the Pentagon, was at the time developing strategies for nuclear war (Herman Kahn, Albert Wohlstetter, etc.).

Maintaining the balance of power has, in fact, justified a race for technological innovations, almost all born in America and always replicated more or less faithfully in the USSR, a race destined to assure that neither of the two protagonists had a sufficient superiority to attempt to rub out the other without taking risks said to be "unacceptable" (at least 20% of the population and 60% of industry ...). This suggests that despite the nuclear arms intended to guarantee the peace, each of the two protagonists attributed to the other, right or wrong, the temptation to attack unannounced. Concerning possible ground-based operations in Europe, the USSR has consistently given itself the very expensive means to set off a massive offensive, while NATO has provided itself very early with the means, principally nuclear because more economical, to stop her (for the 1950s see the memoirs of General Gallois, who at the time was the French member of NATO's Nuclear Planning Group), after which the Soviets in their turn adopted "tactical" nuclear arms. The invention by the Americans of missiles with multiple heads (MIRV) and the race to precision, officially justified by the need to destroy the enemy missiles before launch, would have, in case of an acute crisis, obliged each camp to shoot first in order not to destroy empty silos. This was pointed out at the time by American physicists very competent in the matter, and opposed to this strategy of which the existence of almost invulnerable submarines on both sides reinforced the absurdity.

(7) An explanation which does not exclude the preceding one consists of thinking that in reality we are dealing with a race to bankruptcy in which the winner will be the loser. The Western economy, or even only the American, has always represented, at the very least, three times the Soviet (ten times in 1945 according to Kennan). In terms of GNP the arms race has necessarily weighed more heavily on the Soviet economy rather than on the American throughout this period

(Soutou, p. 666, and Russian economists speak of 20 - 40% of the Soviet GNP during the 1980s, others of 15-20% during the whole period). The arms race thus contributed to prove Kennan right (he, opposed to nuclear madness, retreated rapidly and for life to the Institute for Advanced Study at Princeton ... ). It also confirmed the true or assumed ineffectiveness of the socialist system, the besieged the besieger since its birth. Further, it dispensed the United States from maintaining a vast ground army which would have been much more expensive.

While raising, thanks to the Korean War, the military budget to over 10% of the American GNP – an evolution advocated, three months before the outbreak of the war, by a *National Security Council* well aware of the immense superiority of American industry –, President Eisenhower had, to protect the American economy, always refused to go further because, as he said at the time, the confrontation risked lasting *until the end of the century*; America never spent more than 7% after 1970. About fifteen years ago it was again claimed seriously that it was economically impossible, for the West, to produce 3,000 heavy tanks per year as, we have been told, the Soviets were doing. It is true that at this period the NATO countries were only just capable of producing more than twenty million cars each year and some hundreds of thousands of heavy vehicles, agricultural machines and machinery of all sorts ...

Let us add that the documents, mainly Soviet and French, which would be indispensable to a true comprehension of the situation, remain largely inaccessible. The only certitudes are that the ideological hostility between the USA and the USSR dates from 1917 (no diplomatic relations before 1933) and not from 1945; that after 1945 the two camps, like the French and the Germans before 1914, steadily perfected their arms and their war plans; that nuclear arms have, in the two camps, contributed to *reinforce* the instinctive recoil from the catastrophe; but that their existence would have, in the case of an acute crisis, precipitated it because of the drastic reduction in the delay of possible reaction and of the fact that, compared to these, *"the concentration camps and the gas chambers are just the work of artisans"* (Pierre Sudreau, *L'enchaînement*, Plon, 1967, p. 209, by a non-orthodox Gaullist). As for imagining a realistic version of what history would have been without nuclear arms, the exercise is meaningless: history is not an experimental science.

In a century, the dominant theory will perhaps be that it was *despite* nuclear arms that the Third World War was avoided before 1997 (wait before pronouncing on the future). This is what some politologues or historians have begun to suggest, for example, Michael MccGwire, *Deterrence: the problem – not the solution* (International

Affairs, 1986, pp. 55–70), Soutou, too, being rather sceptical. John Mueller, *Retreat from Doomsday. The Obsolescence of Major War* (Basic Books, 1989) and John Keegan, *The Second World War* (Penguin Books, 1990, pp. 594–595) invoke the experience of two world wars. Some others think that nuclear arms have not served for anything from this point of view, which is surely the case of the French under de Gaulle, that had principally for their military mission to transform a possible classical conflict into a nuclear war. If they are right, the success of "nuclear deterrence" would be due to the fact that there was no one to deter, or, as Mueller says, because *the true deterrent was Detroit*: the enormous American industrial superiority, nuclear or not.

One can, in French, assess the extent of the subject in Pierre Grosser, *Les temps de la guerre froide* (Bruxelles, Ed. Complexe, 1996) who cites almost all that is in the public domain, but confines himself to very brief generalities on what concerns the arms race and its connections with scientific and technological progress. The same is true of George-Henri Soutou, *La Guerre de Cinquante Ans* (Paris, Fayard, 2001), an otherwise superb book by a top historian. The majority of other French authors, particularly the experts in strategy, have restrained themselves for decades to *sound the tocsin for a conflagration* [the invasion of Europe] *which never happened* (Samuel Huntington) while frequently throwing oil on the flames. We can cite, from the immense American literature, several serious books: Daniel Yergin, *Shattered Peace. The Origins of the Cold War and the National Security State* (Houghton Mifflin, 1977), George F. Kennan, *The Nuclear Delusion* (Pantheon Books, 1982), Fred Kaplan, *The Wizards of Armageddon* (Simon & Schuster, 1983) on the history of American nuclear strategy, McGeorge Bundy, *Danger and Survival. Choices About the Bomb in the First Fifty Years* (Vintage, 1990), Samuel R. Williamson, Jr and Steven L. Rearden, *The Origins of U.S. Nuclear Strategy, 1945-1953* (St. Martin's Press), Melvin Leffler, *A Preponderance of Power. National Security, the Truman Administration, and the Cold War* (Stanford UP, 1992), and *The Specter of Communism. The United States and the Origins of the Cold War, 1917-1953* (Hill and Wang, 1994), John Lewis Gaddis, *We Know Now. Rethinking Cold War History* (OUP, 1997). For a balanced and easy to read exposition see Martin Walker, *The Cold War and the Making of the Modern World* (Vintage, 1994), by a British journalist who has covered the subject but, like many other authors, does not seem to have observed that the marvels of *high tech* are one of the principal contributions of the Cold War to the construction of the "Modern World", and that, in this domain too, the children inherit their parents' genes.

Let us return to the mathematicians. As in every very hierarchical community obeying strict rules, for about fifty years discipline has been imposed by a sort of policing. Before publishing an article, and of course not only in mathematics, every serious journal now submits it to specialists, the "referees" or "gate keepers of Science" as a sociologist has called them, or, if one prefers a French term, the sentinels of Science. They do not fail to detect the weaknesses, particularly when the results announced are so very important that some of the referees are angry at not having found them themselves before the author; this system has contributed to improving the level of publications considerably. Moreover, all the authors of scientific articles now send "preprints" ("hard" or "virtual") to their colleagues, to arrive six or twelve months before the printed article. This is useful to one's colleagues and frees them from following the same path if they were on it – thus doubly assuring one's priority –, but can also lead these to criticise more or less severely the articles that they receive. Not everyone reacts like Legendre, the great expert on the theory of elliptic functions under the Revolution and the Empire, who, in his late days, about 1825, on learning the results of Abel and of Jacobi which totally revolutionised the subject, was extremely delighted at such a "great step forward" and only complained that these young people born with the century advanced too fast for him at seventy still to be able to hope to follow them ...

This is what happened a few years ago to Andrew Wiles with his proof, seven years of work, of "Fermat's Last Theorem" which generations of mathematicians have ogled for three centuries, as generations of alpinists ogled Everest before Edmund Hillary. One might well think that the referees set to on Wiles' preprints, breaking off all other activities, fell onto the object in the hope or fear, according to the degree of generosity of the person considered, of finding gaps or errors as had always been the case in the past; there was indeed an error in the calculation which, apparently, vitiated it all. But Wiles corrected it a year later with the help of one of the referees; those of the gate keepers who were angry at having lost their "first" when they were so close had their hopes dashed. It is better not to think of what would have happened to Wiles if his proof had been irremediable, as happened recently à propos another Everest. Some of the more prestigious authorities decided that the author – although brilliant and, it is the least one can say, not lacking in courage – of this false proof that used mathematics other than their own "was not of the required level" to succeed. Neither were they, until proof to the contrary ...

One can also meditate on the recent affair of the "memory of water" when a French biologist in other respects well-reputed, having published a revolutionary theory depending on difficult or impossible to reproduce experiments, was expelled with the ultimate brutality from the "community" by dozens of pontiffs of Physics and of Biology, agitating against the spectres of irrationality, of homeopathy and fraud, and claiming that his publications were

capable of "harming the image of French science". As if that depended on
the work of only one man, and as if "American science" was discredited when
a biologist made up his mice in black and white to prove his theory correct!
On this topic let us cite the opinion of Robert Hutchins[23] in 1963:

> There have been very few scientific frauds. This is because a scientist would
> be a fool to commit a scientific fraud when he can commit frauds every
> day on his wife, his associates, the president of the university, and the
> grocer ... A scientist has a limited education. He labours on the topic of
> his dissertation, wins the Nobel Prize by the time he is 35, and suddenly
> has nothing to do ... He has no alternative but to spend the rest of his
> life making a nuisance of himself.

The author of this declaration has clearly let himself be led by his taste
for paradox; one should however note that he presided over the University of
Chicago before, during and after the war and so has had some observations
to deliver ...

And Euler? His Complete Works – more precisely, those that he prepared
himself with a view to possible publication –, in course of being edited in
Germany, and then in Switzerland since 1911, must comprise about 80 volumes in quarto, of which 29 are on mathematics not to speak of mechanics,
of hydraulics, of astronomy, of mathematical physics, of optics, of shipbuild-
ing theory and navigation, of geodesy, of artillery, of financial mathematics
and of the widely circulated *Lettres à une princesse d'Allemagne sur divers
sujets de physique et de philosophie*, intended for the education of a niece
of Frederick II[24]. He would have done better to confide this heavy task to
people such as Diderot, d'Alembert or Voltaire who inspired him with considerably more sympathy than Euler and his Calvinism (his father was a pastor
in Bâle[25], and it was mathematics which without changing his ideas eventually dissuaded the young Euler from following him). There is also, as well
as these 80 volumes, an enormous correspondence of which the essential part
remains to be published. He lost an eye in 1738 and was blind from 1771, but
continued to work to the same rhythm.

Euler did not publish all that he wrote, far from it, (560 books and articles
during his life, but there about 300 others), principally because he wrote

---

[23] Cited in Daniel S. Greenberg, *The Politics of American Science* (Penguin Books,
1969, original American title: *The Politics of Pure Science*, American Library,
1967). Greenberg collaborated in the great journal *Science*, which he had to leave
because of his disrespectful points of view. His book is still worth reading.

[24] "The King calls me 'my professor', and and I am the happiest of men" (Hairer
and Wanner, p. 159). However this did not stay so for very long, Frederick II
appreciating much more his capabilities as administrator of the Berlin Academy
than his "useless" mathematics.

[25] In the XVII[th] and XVIII[th] centuries, and even more in the XIX[th], many scientists
and more generally intellectuals were, in the protestant countries, sons of pastors,
no doubt because these were educated people. The corresponding phenomenon is
found more rarely in the catholic countries, but since the instruction there was
almost totally monopolised by the religious schools, the results were not very
different until the Revolution.

too much for the capacities of the time; a legend, maybe apocryphal, but significant, tells that, when an editor came to ask him for a paper, he just handed him the topmost of the pile of his latest productions. With such habits and inspired proofs, though a little or very false as we shall see on various occasions, he would have had a lot of trouble nowadays. Of course, in our time, he would have been educated by mathematicians more "serious" than Johann Bernoulli and would have conformed to the rules of the corporation as, in his time, he conformed in his private life, to those he had absorbed in the puritan society of Bâle. Happily there were, mainly in France, people to advance other things than mathematics, hydraulics and artillery.

## 8 – Algebraic operations on limits

It is indispensable to know what happens when one performs simple algebraic operations on sequences which tend to limits. The theory rests on a very few results.

**Theorem 1.** *Let $(u_n)$ and $(v_n)$ be two convergent sequences, with limits $u$ and $v$. Then the sequences $(u_n + v_n)$ and $(u_n v_n)$ converge to $u + v$ and $uv$. If $v \neq 0$, then $v_n \neq 0$ for large $n$, and the sequence $(u_n/v_n)$, defined for large $n$, converges to $u/v$.*

In other words,

$$\lim(u_n + v_n) = \lim u_n + \lim v_n,$$
$$\lim(u_n v_n) = (\lim u_n).(\lim v_n),$$
$$\lim(u_n/v_n) = (\lim u_n)/(\lim v_n) \text{ if } \lim v_n \neq 0.$$

One can add the sums of series too – consider the partial sums:

$$\sum(u_n + v_n) = \sum u_n + \sum v_n.$$

Let us move on to the proof of Theorem 1.

**Case of a sum.** We need only write that

$$|(u_n + v_n) - (u + v)| \leq |u_n - u| + |v_n - v|;$$

for $n$ sufficiently large, each of two last differences is $< r/2$; the left hand side is thus $< r$, qed.

**Case of a product.** This relies on the following lemma (the continuity of the map $(x, y) \mapsto xy$ of $\mathbb{C}^2$ into $\mathbb{C}$):

**Lemma.** *Let $u$ and $v$ be complex numbers. For every $r > 0$ there exists a number $r' > 0$ such that the relations*

(8.1)        $|u' - u| < r'$ & $|v' - v| < r'$ *imply* $|u'v' - uv| < r.$

Note that
$$u'v' - uv = (u' - u)(v' - v) + v(u' - u) + u(v' - v).$$
The two first inequalities (1) thus imply
$$|u'v' - uv| < r'^2 + ar' \text{ where } a = |u| + |v|.$$
For $r' < 1$, the right hand side is $< (1 + a)r'$. The first will thus be $< r$ so long as one chooses $r' < \min(1, r/(1 + a))$, qed.

To deduce from this the case of Theorem 1 which interests us here, it is enough to replace $u'$ and $v'$ by $u_n$ and $v_n$; the two conditions of (1) are then satisfied for all sufficiently large $n$, and the third relation provides the result. (Here as always, one exploits the fact that if two relations are *separately* valid for $n$ large, then they are also valid *simultaneously*.)

**Case of a quotient.** Given that $u_n/v_n = u_n \times 1/v_n$, it is enough to examine $1/v_n$ and then to apply the result for a product.

The fact that $v_n \neq 0$ for large $n$ is clear: for $n$ large, one has for example $|v_n - v| < |v|/2$ since the right hand side is $> 0$; it follows that $|v_n| > |v|/2 > 0$.

To imitate the preceding argument let us put $v_n = v'$. We need to evaluate
$$|1/v' - 1/v| = |v' - v|/|vv'|.$$
For $n$ large, the numerator of the right hand side is $< r'$. Now we have just seen that the denominator is *greater* than $|v|^2/2$. The right hand side is thus $< 2r'/|v|^2$, so $< r$ provided that $r' < r|v|^2/2$, qed.

Theorem 1 allows one to calculate a large number of limits easily, if only very simple ones. For example, the sequence with general term
$$w_n = \frac{n^2 - 1}{3n^2 + n + 1} = \frac{1 - 1/n^2}{3 + 1/n + 1/n^2}$$
tends to $1/3$ since $1/n$ and $1/n^2$ tend to 0, so that the two parts of the fraction tend to 1 and 3 respectively.

More generally, let
$$f(x) = a_p x^p + a_{p-1} x^{p-1} + \ldots + a_0,$$
$$g(x) = b_p x^p + b_{p-1} x^{p-1} + \ldots + b_0$$
be two polynomials of the same degree $p$ [when one says that $f$ is of degree $p$, this means that $a_p \neq 0$]. Then[26]

---

[26] The relation below persists, with the same proof, if, in $f(x)/g(x)$, one lets $x$ tend to infinity through not necessarily integer values. We mainly confine ourselves in this chapter to limits in which the independent variable takes "discrete" values. The next chapter will show that many of these results extend to the case of "continuous" variables (in the sense where, in physics, one speaks of the "discrete spectrum" and of the "continuous spectrum" of a luminous source: the first is composed of isolated "rays" of zero width, the second of luminous "bands" of nonzero width).

$$\lim_{n \to \infty} f(n)/g(n) = a_p/b_p.$$

The proof is the same as above: one divides the two members of the fraction by $n^p$ and remarks that $1/n$ and all its powers tend to 0 as $n$ increases indefinitely.

Beyond these rules of algebraic calculation, there is another simple operation which transforms one convergent sequence into another. Let $(u_n)$ be a sequence which converges to $u$ and let $f$ be a scalar function defined on a neighbourhood of $u$, except perhaps at $u$, and such that $f(x)$ tends to a limit $v$ when $x$ tends to $u$. Then $f(u_n)$ tends to $v$. For all $r > 0$, there is indeed an $r' > 0$ such that $|x - u| < r'$ implies $|f(x) - v| < r$; then one has $|u_n - u| < r'$ for $n$ large, whence $|f(u_n) - v| < r$ for $n$ large.

*If in particular a function $f$ is continuous at a point $a$, then*

(8.2) $$\lim u_n = a \implies \lim f(u_n) = f(a),$$

a fundamental result even though almost trivial (i.e. following directly from the definitions).

## §2. Absolutely convergent series

### 9 – Increasing sequences. Upper bound of a set of real numbers

For a start let us make some remarks on passing to the limit in inequalities, essential for understanding axiom (IV) of n° 1.

First, it is obvious that if a sequence of numbers, positive for $n$ large, converges, then its limit is again positive; the reader may provide the $\varepsilon$ and the $N$ necessary for a textbook proof ... It follows from this that

$$(9.1) \qquad a \leq u_n \leq b \text{ for all } n \text{ large} \implies a \leq \lim u_n \leq b,$$

as one sees on considering the sequences $u_n - a$ and $b - u_n$. Similarly

$$(9.2) \qquad u_n \leq v_n \text{ for all } n \text{ large} \implies \lim u_n \leq \lim v_n$$

since the terms of the sequence $v_n - u_n$ are positive for $n$ large.

In other words, weak inequalities are preserved under passage to the limit.

Not so for *strict* inequalities: one has $1/n > 0$ for all $n$, but $\lim 1/n = 0$. Without explicit evidence to the contrary, passing to the limit transforms strict inequalities into weak inequalities: the inequality $1/n > -2$ is preserved in the limit since there exists a number $r > 0$ such that $1/n > -2 + r$ for all $n$, so that the limit is $\geq -2 + r > -2$.

These results, though trivial, bring us back to axiom (IV) for $\mathbb{R}$ mentioned in n° 1 of this chapter. Consider an *increasing* sequence

$$(9.3) \qquad u_1 \leq u_2 \leq \ldots \leq u_n \leq u_{n+1} \leq \ldots$$

of real numbers. For a sequence to converge, it is clearly necessary that, increasing or not, there exists a positive number $M$ such that $|u_n| \leq M$ for all $n$, i.e. that the sequence should be *bounded*. In the case of interest this means the existence of numbers $M$ which *majorise* the sequence, i.e. satisfy $M \geq u_n$ for all $n$, the weak inequality being essential in what follows. One also says that $M$ is a *majorant* of the given sequence and that the latter is *majorised* by $M$, or majorised for short if one does not want to specify $M$ exactly, or also *bounded above*.

Suppose now that the sequence (3) converges to a limit $u$. By property (1) above, the relation

$$(9.4) \qquad u_p \leq M \text{ (for all } p) \text{ implies } u \leq M.$$

Since on the other hand $u_p \leq u_n$ for all $n \geq p$, one sees similarly, on passing to the limit over $n$, that also

$$(9.5) \qquad u_p \leq u \text{ for all } p.$$

The relation (4) shows that every majorant is $\geq u$, and (5) that $u$ is one of these majorants; conclusion:

(9.6) *If an increasing sequence $(u_n)$ converges, then the set of its majorants possesses a least element, namely the limit of the given sequence.*

Conversely:

(9.7) *Let $(u_n)$ be an increasing sequence, and bounded. Suppose that the set of its majorants possesses a least element $u$. Then $u_n$ converges to $u$.*

Take a number $r > 0$. Since $u$ is the least number which majorises the sequence, the number $u - r$ does not majorise it. There is therefore an index $p$ such that $u - r < u_p$. Since the sequence is increasing, one again has $u - r < u_n$ for all $n \geq p$. But since $u$ majorises the sequence, one has finally

(9.8) $$u - r < u_n \leq u \text{ for all } n \geq p,$$

which establishes (7).

We have no alternative. When a "magnitude", as one used to call it, increases *constantly* but not *indefinitely*, i.e. remains below a certain finite value, then common sense – the most widely spread out thing in the world according to a philosopher and mathematician who believed only what he could prove or verify himself –, common sense, then, indicates that this magnitude must necessarily accumulate towards a *limit*. In mathematical terms: every increasing bounded-above sequence converges.

For example, common sense indicates that the sequence

$$3 \quad 3.1 \quad 3.14 \quad 3.141 \quad 3.1415 \quad 3.14159 \ \ldots$$

must converge to something. Alas, this something is not rational, so common sense will not help us at all if we know only about $\mathbb{Q}$. This will not wipe out any of the banalities which we have already established in this chapter, since they rely only on axioms (I), (II) and (III) common to $\mathbb{Q}$ and to $\mathbb{R}$, including (6) and (7); but to go further one clearly needs an axiom specific to $\mathbb{R}$ so as not to have false theorems such as: "every bounded increasing sequence of rational numbers converges to a rational limit" or, what would hardly be better, under penalty of being unable to attribute a limit to almost all the convergent sequences that one meets in analysis.

It is clearly the axiom (IV) of n° 1 that we lack. This affirms that if a nonempty set $E \subset \mathbb{R}$, for example the set of numbers of the form $u_n$ in what precedes, is bounded above, then the set of numbers which majorise it, its majorants, possesses a least element, its *least upper bound*. This makes the following theorem obvious, by (7):

**Theorem 2.** *For an* increasing *sequence of real numbers to converge it is necessary and sufficient that it be bounded above. Its limit is then the least number which majorises it, i.e. the least upper bound of the set of its terms.*

The crucial point masked by this statement is the postulate that, among all the majorants of the given sequence, *there exists* a number smaller than all the others.

There is a similar statement for decreasing sequences: such a sequence converges if and only if it is *minorised*, i.e. if there exist numbers $m$ less than all its terms. Its limit is then the *largest* of these *minorants* i.e. the *greatest lower bound* of the set of its terms.

To clarify the reader's thoughts, it is indispensable to introduce, or to revise more systematically than we have done so far, some very easy definitions in constant use.

One says that a set $E \subset \mathbb{R}$ is *bounded above* , or *majorised*, if there exists a number $M$ such that $x \leq M$ for all $x \in E$; one then says that $M$ *majorises* $E$, or is a *majorant* of $E$, or that $E$ is *majorised by* $M$. There are analogous definitions for sets *bounded below*, or *minorised*, and numbers which *minorise* the set, etc. Finally, one says that a set $E \subset \mathbb{C}$ is *bounded* if there exists a number $M \geq 0$ such that $|x| \leq M$ for all $x \in E$.

Let $E \subset \mathbb{R}$ be a set bounded above and let $M$ and $M'$ be two majorants of $E$. If $M < M'$, the relation $\{x \in E \implies x \leq M\}$ is clearly stronger than the similar relation with $M'$. For example, it is not without interest to know that, of the human species, everyone dies before attaining the age of 500 years, but to avoid surprises it is better to know that everyone dies before 250 years. This information again not being, it would seem, the best possible, one might try to determine as small an age as possible before which everyone dies. This would be the precise *least upper bound* of human life. Whence the concept of the least upper bound of a set $E \subset \mathbb{R}$ bounded above: it is a number $u = \sup(E)$ satisfying the two following conditions:

(SUP 1)   $x \leq u$ *for all* $x \in E$, i.e. $u$ majorises $E$;

(SUP 2)   $u \leq M$ *for any other majorant* $M$ *of* $E$.

In other words, $\sup(E)$ is *the least majorant of* $E$.

One could replace (SUP 2) by

(SUP 2')   *for all* $r > 0$ *there exists an* $x \in E$ *such that* $u - r < x$.

If indeed $u$ is the least possible majorant, then the number $u - r$ does not majorise $E$, whence the existence of $x$. If conversely (SUP 2') holds, then every $M$ majorising $E$ majorises, for any $r > 0$, an $x > u - r$, so majorises $u - r$ for any $r > 0$, so majorises $u$ (modified Archimedes' axiom), whence (SUP 2).

(SUP 2') also implies that $u$ *is the limit of a sequence of elements of* $E$: choose $x_n \in E$ such that $u - 1/n < x_n$.

As we have seen above, the difficulty in proving that an increasing sequence tends to a limit disappears the moment the *existence* of the least upper bound is established. The problem is solved by axiom (IV) of n° 1 which the majority of authors call Bolzano's "Theorem" because he was the first, it seems, to formulate it more or less clearly, in 1817. He did not prove it, no doubt because it seemed too obvious to him, and, in reality, because no one of his time yet had a sufficiently clear idea of the concept of real number to be able to provide a correct proof. Hardly a surprising situation: if you want to *prove* such a "theorem" you have to rely on other previous results; now the axioms (I), (II) and (III) of n° 1 clearly have not the least chance of sufficing, since, if such were the case, they would prove that Bolzano's Theorem is valid in $\mathbb{Q}$; the invention of real numbers would then be totally superfluous. If one really wants to make the existence of least upper bounds a *theorem*, one needs to have either an *axiom* valid in $\mathbb{R}$ but not in $\mathbb{Q}$, as do those who prefer the "nested intervals" axiom (for us, a theorem of Chap. III), or a rigorous construction of $\mathbb{R}$, for example that provided by Dedekind sections of which we have spoken in the introduction and at the end of n° 1.

It may all the same be interesting to show that Bolzano's Theorem, as it is called, which trivially implies Theorem 2, is, conversely, a consequence of it, so that Theorem 2 might have also have been taken as axiom (IV).

**Bolzano's "Theorem" (1817).** *Every nonempty bounded-above subset $E$ of $\mathbb{R}$ possesses one and only one least upper bound.*

Uniqueness is clear: one cannot see how a set of numbers, the majorants of $E$, which possesses a least element could possess two different ones, for each has to be smaller than the other. Paul Klee once depicted two naked bureaucrats bowing to each other *with their backs at the horizontal*, each thinking the other of a higher rank than his own. This has not prevented some authors, including the present one formerly, from providing a textbook proof of the uniqueness of the least upper bound. Now let us prove its existence.

Let $M$ be a majorant of $E$. There are integers $n \in \mathbb{Z}$ which majorise $E$, for example those larger than $M$. There are also integers which do not majorise $E$ since $E$ is nonempty; they are all $< M$, so one can consider the largest of them, say $u_0$. It does not majorise $E$, but $u_0 + 1$ majorises $E$, for otherwise $u_0$ would not be the largest possible.

Among the numbers $u_0 + n/10$, where $n$ is an integer $\geq 0$, let $u_1$ be the largest of those which do not majorise $E$; one has $n \leq 9$ for this number since $u_0 + 10/10$ majorises $E$. So $u_0 \leq u_1$, $u_1$ does not majorise $E$, but $u_1 + 1/10$ does.

Similarly, let $u_2$ be the largest of the numbers of the form $u_1 + n/100$ which do not majorise $E$. One has $n \leq 9$ since $u_1 + 10/100$ majorises $E$, $u_1 \leq u_2$, $u_2$ does not majorise $E$, but $u_2 + 1/100$ does.

On repeating this construction indefinitely, one obtains an increasing sequence of numbers $u_n$ possessing the following properties:

(i) $u_n$ does not majorise $E$, (ii) $u_n + 1/10^n$ majorises $E$.

Since $u_n \leq M$ for all $n$, the sequence $u_n$ tends to a limit $u$, *by virtue of Theorem 2*. Let us show that $u$ satisfies the conditions (SUP 1) and (SUP 2') which characterise the least upper bound of $E$.

For all $x \in E$ we have $x \leq u_n + 1/10^n$ for any $n$, since the right hand side majorises $E$. On the other hand, it converges to $u$. The remarks at the beginning of this n° then show that $x \leq u$, whence (SUP 1).

Since $u_n$ does not majorise $E$, for each $n$ there is an $x_n \in E$ such that $u_n < x_n$. One has also $x_n \leq u_n + 1/10^n$ since the right hand side majorises $E$. In consequence, $\lim x_n = \lim u_n = u$, whence (SUP 2'), and this completes the proof.

We have quite deliberately used decimal numbers in the preceding proof, to construct successive terminating decimal expansions $u_0, u_1, u_2, \ldots$ for $u$. To be precise, if one writes the *default* decimal expansion of each $x \in E$ in the form

$$x = x_0.x_1x_2\ldots$$

with an integer part $x_0$ and decimals $x_1, x_2, \ldots$ between 0 and 9, then one obtains $u_0$, $u_1$, etc. by the following procedure: $u_0$ is the maximum value taken by $x_0$ when $x$ varies in $E$; $u_1$ has integer part $u_0$ and its first decimal (the following are zeros) is the maximum value of $x_1$ when $x$ runs through the set $E_0$ of $x \in E$ such that $x_0 = u_0$; the decimal expansion of $u_2$ starts like that of $u_1$, but has one more decimal, namely the maximum value of $x_2$ when $x$ runs through the set $E_1 \subset E_0 \subset E$ of $x \in E$ such that $x_0, x_1 = u_1$, and so on. In other words, one considers the $x \in E$ whose integer part is a maximum, then, *among them*, those whose first decimal is a maximum, then, among these, those whose second decimal is a maximum, etc. On pursuing this construction indefinitely we find the successive decimals of the least upper bound of $E$ which we seek.

This construction allows one to prove the theorem "without knowing anything" subject to accepting that *any* nonterminating expansion decimal corresponds to a real number; but this comes back to accepting either axiom (IV) of n° 1, or Theorem 2 which, as we have seen, is equivalent to it. More ingenious arguments will never let you escape this. On the contrary, it is axiom (IV) which justifies the decimal representation of the real numbers.

The concept of least upper bound also applies to the case of a *family* $(u_i)$, $i \in I$, of real numbers – in other words, up to notation, of a map of the set $I$ into $\mathbb{R}$; the least upper bound of the set $E$ of $u_i$ ($x \in E \Longleftrightarrow$ there exists an $i$ such that $x = u_i$) is denoted by

$$\sup_{i \in I} u_i$$

or simply $\sup(u_i)$ if there is no fear of ambiguity. This is the least number which majorises all the $u_i$ or again, among the numbers which majorise it, that which, for all $r > 0$, can be approximated to within $r$ by at least one of the $u_i$.

This concept is useful in formulating the *associativity of least upper bounds*: suppose that the set of indices $I$ is the union of a family $(I_j)$ of sets indexed by a set $J$ and not necessarily disjoint; for every $j \in J$ let $M_j$ be the least upper bound of the partial family $(u_i)$, $i \in I_j$; then the least upper bound $M$ of the full family $(u_i)$, $i \in I$, is equal to that of the family $(M_j)$, $j \in J$, of partial least upper bounds: this can be expressed as

$$(9.9) \qquad \sup_{i \in I} u_i = \sup_{j \in J} \left( \sup_{i \in I_j} u_i \right).$$

The proof is easy. First $M$, majorising all the terms of the full family, majorises all those of each partial family and thus also all the $M_j$. But there exists, for any $r > 0$, an $u_i > M - r$; one has $i \in I_j$ for some $j \in J$, whence $M_j \geq u_i$ for this index $j$ and *a fortiori* $M_j \geq M - r$, qed.

## 10 – The function $\log x$. Roots of a positive number

Theorem 2 allows us to prove immediately some results which we will need in any case; this and the following n° will be devoted to them.

*Example 1.* Consider, for $x \geq 0$, the sequence

$$(10.1) \qquad u_n = 1 + x/1! + x^2/2! + \ldots + x^n/n!.$$

It is clearly increasing. To show that it is bounded, one remarks that, for any number $z \in \mathbb{C}$, the sequence $(z^n/n!)$ is bounded (n° 5, example 6). For $z = 2x$, one thus has $(2x)^n/n! \leq M$, whence $x^n/n! \leq M/2^n$ and thus

$$u_n \leq M(1 + 1/2 + \ldots + 1/2^n) \leq 2M$$

by formula (6.5) for the sum of a geometric progression, qed.

This argument shows that the series

$$(10.2) \qquad \exp x = \sum x^n/n! = \sum x^{[n]} = \lim u_n$$

converges for $x \geq 0$. In fact, it converges for all $x \in \mathbb{C}$ as we shall see in n° 14.

*Example 2.* Consider, again for $x \geq 0$, the sequence

$$(10.3) \qquad y_n = (1 + x/n)^n.$$

The binomial formula

$$y_n = 1 + n(x/n)/1! + n(n-1)(x/n)^2/2! + n(n-1)(n-2)(x/n)^3/3! + \ldots$$
$$+ n(n-1)\ldots(n-n+2)(n-n+1)(x/n)^n/n!$$

shows that

$$(10.4) \quad y_n = 1 + x + \left(1 - \frac{1}{n}\right) x^{[2]} + \left(1 - \frac{1}{n}\right)\left(1 - \frac{2}{n}\right) x^{[3]} + \dots$$
$$+ \left(1 - \frac{1}{n}\right) \dots \left(1 - \frac{n-1}{n}\right) x^{[n]}.$$

The coefficient of $x^{[p]} = x^p/p!$ lies between 0 and 1 for any $p$, whence

$$(10.5) \quad (1 + x/n)^n \leq 1 + x + x^{[2]} + \dots + x^{[n]} \leq \exp x \text{ for } x \geq 0,$$

which reduces to the preceding example once we show that the sequence is increasing.

Now $y_n$ and $y_{n+1}$ are polynomials in $x$ of degrees $n$ and $n+1$, with all coefficients positive. Since we are assuming $x > 0$, all their terms are positive. The second possesses one more term than the first. In the coefficients of the powers of $x$, the factor $1 - p/n$ increases as one passes from $n$ to $n+1$, and so, therefore, do the products of such factors since they are all positive. The coefficient of $x^{[p]}$ in $y_{n+1}$ is thus greater than its coefficient in $y_n$, qed.

In fact one has

$$(10.6) \quad \lim(1 + x/n)^n = \exp x$$

for any $x \geq 0$ (and even, we shall see later, for any $x \in \mathbb{C}$). The relation (4) makes this formula *plausible* and, for Euler and his contemporaries, made it obvious. As $n$ increases, the coefficient of $x^{[p]}$ in (4) tends to 1, the $p^{\text{th}}$ term of (4) thus tends to the $p^{\text{th}}$ term of the series for $\exp x$ and since one can pass to the limit in a sum of convergent sequences (n° 8, Theorem 1), the result follows ...

Alas, the petty Theorem 1 of n° 8 on a sum of limits assumes that one has a *finite and fixed number* of convergent sequences. The right hand side of (4) does have a finite number of terms, but this number increases indefinitely with $n$ and this "detail" thwarts the conclusion[27]. Consider the following example:

$$1 = 1$$
$$1/2 + 1/2 = 1$$
$$1/3 + 1/3 + 1/3 = 1$$

etc. All the terms of the $10^{100\text{th}}$ sum are equal to 0 to within $10^{-100}$, but $10^{100}$ errors of the order of magnitude of $10^{-100}$ do not necessarily provide a better approximation than 10 errors of the order of $1/10$. In this example, the $p^{\text{th}}$ term of the $n^{\text{th}}$ sum tends to 0 for any $p$ when $n$ increases indefinitely,

---

[27] One can escape from this "without knowing anything", as does Houzel, *Analyse mathématique* (Belin, 1996), pp. 72–75, with the aid of three pages of astute calculations with the binomial coefficients. Too difficult to type out.

so that, arguing like Euler, one might deduce that the total of the $n^{\text{th}}$ sum also tends to to 0. But it is equal to 1 for any $n$.

Later we shall meet other cases where Euler passed to the limit automatically in a sequence of convergent series as if "it went without saying". Now it is quite difficult to believe that he did not know the particularly trivial counterexample which we have just presented, and which spices up the "marvels of the harmonic series". Conclusion? The most charitable is that he employed numerical methods to confirm his intuitions[28] and that, on the contrary, not having available the methods invented in the XIX[th] century to justify his calculations, he preferred (rightly) to advance rather than to abstain from publishing such beautiful (and correct) formulae. All his contemporaries were content to admire them, without asking questions of themselves or of him: if "Euler says that ...", it is surely true. One meets the same reaction nowadays, no longer *à propos* theorems or formulae, but conjectures.

We can now justify what we have said in n° 5, example 7, *à propos* logarithms:

**Theorem 3.** *For all $x > 0$, the sequence with general term $n(x^{1/n} - 1)$ tends to a limit $f(x)$. The function $f$ is $C^\infty$, strictly increasing, and satisfies*

$$(10.7) \qquad f(xy) = f(x) + f(y), \qquad f'(x) = 1/x.$$

Suppose first that $x > 1$ and put

$$(10.8) \qquad u_n = n(x^{1/n} - 1),$$

whence $u_n > 0$. One has

$$(10.9) \qquad x = (1 + u_n/n)^n \quad \text{for all } n.$$

Now we saw in Example 2 that, for all $u > 0$, the sequence $(1 + u/n)^n$ is increasing. For $u = u_n$ one thus has

$$x = (1 + u_n/n)^n < [1 + u_n/(n+1)]^{n+1},$$

whence, using (9) for $n$ and $n + 1$,

---

[28] See in Hairer and Wanner, *Analysis by Its History* (Springer, 1995), p. 25, a table of values of $(1 + 1/n)^n$ and $1 + 1/1! + \ldots + 1/n!$ for $n \leq 28$. The second sequence provides, for $n = 28$, the 29 first decimals of the limit $e = \exp 1 = 2.718\ldots$, though the first, for $n = 28$, is equal to 2.671. Recourse to brute numerical calculation to verify formula (14) thus presupposes rather a lot of work. A more practical method, at that time, would have been to use a table of logarithms – they had been available to 20 decimals for some time. Now $\log[(1 + x/n)^n] = n.\log(1 + x/n)$, so that, for $x < 1$ for example, it would be possible to calculate without difficulty the left hand side exact to 10 decimal places for $n = 10^6$ and to deduce a much better approximation to $(1 + x/n)^n$ than by brute calculation.

$$x = [1 + u_{n+1}/(n+1)]^{n+1} < [1 + u_n/(n+1)]^{n+1}.$$

Since, for $u, v > 0$ and $m \geq 1$, the relation $u^m < v^m$ implies $u < v$, one deduces that $u_{n+1} < u_n$. The sequence (8) is therefore decreasing, and, since all its terms are $> 0$, it tends to a limit $f(x)$ as stated.

For $0 < x < 1$, one puts $x = 1/y$, whence

$$n(x^{1/n} - 1) = n(1 - y^{1/n})/y^{1/n}.$$

The factor $y^{1/n}$ (Example 7, n° 5) tends to 1 and $n(1 - y^{1/n})$ converges to $-f(y)$ by what we have just established, whence convergence again.

To establish the first relation (7) one observes that

$$\begin{aligned} f(xy) - f(x) - f(y) &= \lim n \left[ (xy)^{1/n} - x^{1/n} - y^{1/n} + 1 \right] = \\ &= \lim n(x^{1/n} - 1)(y^{1/n} - 1) \end{aligned}$$

since $(xy)^{1/n} = x^{1/n}y^{1/n}$. Now $n(x^{1/n} - 1)$ tends to $f(x)$ and $y^{1/n} - 1$ to 0. The right hand side thus tends to 0, whence (7). Note that (7) implies $f(1) = 0$ and $f(1/x) = -f(x)$.

To establish the differentiability of $f$ and the formula $f'(a) = 1/a$ one observes first that

$$f(a + h) - f(a) = f(1 + h/a)$$

by the first formula (7), whence

$$\frac{f(a+h) - f(a)}{h} = \frac{1}{a} \frac{f(1 + h/a)}{h/a} = \frac{1}{a} \frac{f(x) - f(1)}{x - 1}$$

on putting $x = 1 + h/a$. Since $h/a = k$ tends to 0 with $h$ it thus suffices to establish that $f(x)/(x-1)$ tends to 1 when $x$ tends to 1, i.e. that $f$ possesses a derivative equal to 1 at the point 1, in order to show that $f'(a) = 1/a$,

Consider first the ratio $(x - 1)/u_n$. On putting $x^{1/n} = y$ one finds

$$(x - 1)/u_n = (y^n - 1)/n(y - 1) = (1 + y + \ldots + y^{n-1})/n.$$

Whether $y$ is $< 1$ or $> 1$, the $n$ numbers $y^k$ whose arithmetic mean is being calculated all belong to the closed interval with end points 1 and $y^n = x$, so their mean $(x - 1)/u_n$ does too. The quotient $u_n/(x - 1)$ thus belongs, for all $n$, to the closed interval with end points 1 and $1/x$. It is thus the same for its limit $f(x)/(x - 1)$. Since $1/x$ tends to 1 when $x$ tends to 1, the ratio $f(x)/(x - 1)$ tends to 1. Hence the existence of the derivative and the second relation (7). Moreover, $f$ is $C^\infty$ (i.e. has derivatives of all orders) because $f'$ is.

This little calculation in fact shows that

(10.10)     $1 - 1/x = (x - 1)/x \leq f(x) \leq x - 1$  for all $x > 0$

since $f(x)$ and $x - 1$ always have the same sign.

It remains to verify that the function $f$ is strictly increasing, i.e. that $0 < x < x'$ implies $f(x) < f(x')$. But $x' = xy$ with $y > 1$ and so $f(x') = f(x) + f(y)$. It is therefore enough to show that

$$(10.11) \qquad y > 1 \Longleftrightarrow f(y) > 0,$$

which follows from (10), qed.

These results, which will be revisited in detail and by other methods in Chap. IV, are to be compared with the remarks following Example 7 of n° 5. We can now *define* the *Napierian logarithm* by the formula

$$(10.12) \qquad \log x = \lim n(x^{1/n} - 1) = \lim u_n.$$

The formulae (7) are then written

$$(10.13) \qquad \log(xy) = \log x + \log y, \qquad \log' x = 1/x$$

where $\log'$ denotes the derivative of the function log. The relation (10) applied to $y/x$ shows that

$$\frac{1}{y} < \frac{\log y - \log x}{y - x} < \frac{1}{x} \qquad \text{for } 0 < x < y.$$

As we shall see later, the functions log and exp are inverses of each other. In other words,

$$(10.14) \qquad \exp(\log x) = x, \qquad \log\left(\exp y\right) = y$$

for all $x > 0$ and $y \in \mathbb{R}$. We can now show that these relations are *plausible*. For $n$ large, $y = \log x$ is "almost " equal to $n(x^{1/n} - 1)$ and $\exp(y)$ "almost" equal to $(1 + y/n)^n$, by (6), "so" $\exp(\log x)$ is "almost" equal to

$$\left[1 + n(x^{1/n} - 1)/n\right]^n = x,$$

"qed". Similarly, $x = \exp(y)$ is "almost" equal to $(1 + y/n)^n$ by (6), "so" $x^{1/n}$ is "almost" equal to $1 + y/n$, "so" $n(x^{1/n} - 1)$ "is almost" equal to $y$, again "qed" in quotes.

These arguments appear obvious only so long as one has not understood the problems that they pose: there is no *general* theorem in analysis that will allow us to legitimise them in the form, due to Halley the astronomer, and to Euler, in which we have just presented them. A correct argument would be, for example, to write

$$\begin{aligned} \log\left(\exp x\right) &= \lim_{n \to \infty} n\left\{\exp(x)^{1/n} - 1\right\} = \\ &= \lim_{n \to \infty} n\left\{\left[\lim_{m \to \infty}(1 + x/m)^m\right]^{1/n} - 1\right\} \end{aligned}$$

where the dummy variables $m$ and $n$ are *independent of each other*. From this form, the sophism – it is one – stands out clearly: it is to assume that given a double sequence $u(m,n)$ whose terms depend on two integers, in this case

$$u(m,n) = n\left\{[(1+x/m)^m]^{1/n} - 1\right\},$$

one is entitled to confuse

$$\lim_{n\to\infty}\left[\lim_{m\to\infty} u(m,n)\right] \qquad \text{with} \qquad \lim_{n\to\infty} u(n,n).$$

But if one chooses $u(m,n) = m/(m+n)$, one has $\lim_m u(m,n) = 1$ for all $n$, so that $\lim_n [\lim_m u(m,n)] = 1$, while $\lim_n u(n,n) = 1/2$.

A more reasonable method of proving (14) uses the fact – which we have not yet proved – that

$$\exp(x) = \lim(1+x/n)^n.$$

Since the function $\log x$ is differentiable and so continuous, one has, by the relation (8.2) at the end of n° 8,

$$\begin{aligned}
\log\exp(x) &= \lim\log[(1+x/n)^n] = \lim n.\log(1+x/n) = \\
&= x.\lim\frac{\log(1+x/n) - \log 1}{x/n};
\end{aligned}$$

since $x/n$ tends to 0, the quotient tends to the derivative of the function $\log$ at $x = 1$, i.e. to 1, whence one concludes that $\log\exp(x) = x$. We shall replace these wishy-washy proofs by correct arguments in Chap. IV and even, on occasion, earlier (Chap. III, n° 2).

However it may be, these calculations show that it will not be without value to establish the following result, another example of an application of Theorem 2:

*Example 3. Let $p$ a nonzero integer. Every real number $a > 0$ possesses one and only one positive $p^{\text{th}}$ root.*

In other words, the equation $x^p = a$ has a unique root $x > 0$. One may assume $a > 0$ and $p > 0$, since the case where $p < 0$ reduces to this, by the identity $x^{-p} = 1/x^p$. First, there exists an $x_1 > 0$ such that $x_1^p > a$ since, for $x$ large, one has $x^p > x > a$. Next, we define a sequence of numbers $x_n > 0$ by

$$px_2 = (p-1)x_1 + a/x_1^{p-1}, \qquad px_3 = (p-1)x_2 + a/x_2^{p-1},$$

and, generally,

(10.15)     $$x_{n+1} = \frac{1}{p}\left[(p-1)x_n + a/x_n^{p-1}\right].$$

We shall see that this sequence of positive numbers is decreasing, so tends to a limit which will be the sought-for $p^{\text{th}}$ root of $a$.

Since one passes from $x_n = x$ to $x_{n+1} = y$ by the formula

$$(10.16) \qquad y = \frac{1}{p}\left[(p-1)x + a/x^{p-1}\right] = \left[(p-1)\,x^p + a\right]/px^{p-1},$$

it suffices to establish that, for $x > 0$, the relation $x^p > a$ implies $y < x$ and also $y^p > a$: then we can apply this result repeatedly and get $x_1 < x_0$, then $x_2 < x_1$, etc. Now

$$(10.17) \qquad x - y = \frac{x^p - a}{px^{p-1}} > 0,$$

whence $y < x$. Moreover

$$x^p - y^p = (x - y)\left(x^{p-1} + \ldots + y^{p-1}\right) < px^{p-1}(x - y)$$

since $x^k y^h < x^{k+h}$ and $x - y > 0$. Using (17), one thus finds $x^p - y^p < x^p - a$, whence $y^p > a$ as stated.

We have seen that the $x_n$ decrease to a limit $x$. The relation (15) then shows that

$$x = \frac{1}{p}\left[(p-1)x + a/x^{p-1}\right],$$

which is equivalent to $x^p = a$. As for the uniqueness of the root, this follows from the fact that the function $x \mapsto x^p$ is strictly increasing, qed.

In the case where $p = 2$, the relation (15) takes the form

$$2x_{n+1} = x_n + a/x_n.$$

If one takes $a = 2$ and chooses $x_1 = 3/2$, one finds $x_2 = 17/12$ and then

$$\begin{aligned} x_3 &= (17/12 + 24/17)/2 = (289 + 288)/24.17 = \\ &= 1 + 24/60 + 51/60^2 + 10/60^3 + \ldots \end{aligned}$$

in the sexagesimal numeration. This value was once found on a Babylonian tablet of the $18^{\text{th}}$ century before our era. Moreover, an Indian text[29] which may date from the sixth century before our era, presents without explanation

---

[29] Tropfke, *Geschichte der Elementar-Mathematik*, Vol. III, p. 172, refers to an article by L. F. Rodet in the Bulletin de la Société Mathématique de France (vol. 7, 1879) for the Indian calculus. See also A. P. Juschkevitsch, *Geschichte der Mathematik im Mittelalter* (Moscow, 1961, trad. Teubner, 1964), p. 100, who systematically expounds the mathematical activities in the Indian and Arab lands in the Middle Ages. The Babylonians, who used the sexagesimal numeration, have been the subject of impressive works by Otto Neugebauer; the problem is not only to discover and decipher the tablets that deal with Mathematics (or, as often, of Astronomy), it is also to interpret them. We note finally that the western historians of Mathematics or of Astronomy – Montucla in the $XVIII^{\text{th}}$ century, Delambre at the beginning and Moritz Cantor at the end of the $XIX^{\text{th}}$ – did not wait for decolonisation to study the Arabs and Indians with the means available

the value $1 + 1/3 + 1/3.4 - 1/3.4.34$, or $1.4142157$ for the square root of 2, equal to $1.4142136\ldots$ By a coincidence as surprising as unexpected, one has

$$\begin{aligned} x_3 &= \frac{17.17 + 12.24}{24.17} = \frac{2.17.17 - 1}{2.3.4.17} = 17/3.4 - 1/3.4.34 = \\ &= (12 + 4 + 1)/3.4 - 1/3.4.34 = 1 + 1/3 + 1/3.4 - 1/3.4.34. \end{aligned}$$

## 11 – What is an integral?

The concept of an *integral* is a particularly striking and fundamental application of axiom (IV); it is not difficult to introduce it here, without waiting for Chap. V.

First we note that one can state axiom (IV) as follows:

(IV bis) *Let $E$ and $F$ be two nonempty sets of real numbers. Suppose that every $u \in E$ is smaller than every $v \in F$. Then there exists a number $w$ such that $u \le w \le v$ for all $u \in E$ and $v \in F$. For $w$ to be unique it is necessary and sufficient that for every number $r > 0$ there exist a $u \in E$ and a $v \in F$ such that $v - u < r$.*

It is clear that $E$ is bounded above, so possesses a least upper bound $m$, and that $F$ is bounded below, so possesses a greatest lower bound $M$. Since every $v \in F$ majorises $E$, one has $m \le v$. In consequence, $m$ minorises $F$, whence $m \le M$ (which justifies the notation $\ldots$). Every $w \in [m, M]$ will then do. Finally, for $w$ to be unique, it is necessary and sufficient that $m = M$; and then one can always find an $u \in E$ and an $v \in F$ arbitrarily close to $m$ and $M$, whence the second point.

This done, let us return to integrals. The problem is the following: given a function $f(x)$ with real values (for simplicity) on a compact interval $I = [a, b]$, one wants to give a meaning to the measure

$$m(f) = \int_a^b f(x)dx$$

of the area between the $x$-axis, the graph of $f$ and the verticals $x = a$ and $x = b$, it being understood that one counts as positive the areas lying above the $x$-axis and the others as negative, so as to obtain the linearity relation $m(\alpha f + \beta g) = \alpha m(f) + \beta m(g)$ for any constants $\alpha$ and $\beta$.

---

in their time. It has also been known for a long time that in Newton's time, the Japanese, for their part, were in process of making very similar discoveries to his. The reasons why all these advances were transformed into an enormous delay behind the Westerners might form the subject of all sorts of theories and are surely not the same in Japan, which isolated herself voluntarily in order to escape possible colonisation, as in the Arab countries of the Middle East or of Central Asia; see Bernard Lewis, *The Muslim Discovery of Europe* (Norton, 1982). In both cases, the interest in Western Civilisation first came from Western weapons $\ldots$

The simplest case is that where $f$ is a *step function*, i.e. where one can divide $I$ into a finite number of pairwise disjoint intervals $I_1, \ldots, I_p$ of any sort – some of the $I_k$ may reduce to a single point – on each of which the function $f$ is constant. The graph of $f$ thus has an appearance as in figure 2, and if, for all $k$, one denotes the length of the interval $I_k$ by $m(I_k)$ and the value, constant, of $f$ on $I_k$ by $c_k$, "it is clear" that

$$(11.1) \qquad m(f) = \sum c_k . m(I_k)$$

since "everyone knows" that the area of a union of pairwise disjoint rectangles is the sum of their areas. To avoid recourse to folklore, it is better to take (1) as the *definition* of the area considered in the present case.

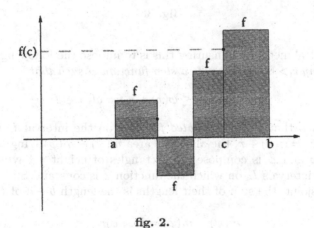

**fig. 2.**

Now consider the general case of a real function that is not a step-function. Assume that it is *bounded* on the interval $I$, i.e. that there exists a number $M > 0$ such that $|f(x)| \leq M$ for all $x \in I$. We can now consider on the one hand the step functions $\varphi$ such that $\varphi(x) \leq f(x)$ for all $x$, and on the other hand the step functions $\psi$ such that $\psi(x) \geq f(x)$. Common sense indicates that the integrals of functions $\varphi$, $f$ and $\psi$ must satisfy the relation

$$m(\varphi) \leq m(f) \leq m(\psi).$$

We are thus prompted to define two sets $E$ and $F$ of real numbers: the set $E$ of numbers $u = m(\varphi)$, where $\varphi$ is any step function $\leq f$, and the set $F$ of numbers $v = m(\psi)$, where $\psi$ is any step function $\geq f$. To assign a meaning to the sought-for integral $m(f)$ it *suffices* that the sets $E$ and $F$ should satisfy the conditions of axiom (IV bis).

It is obvious geometrically (and one can prove it easily) that $u \leq v$ for all $u \in E$ and $v \in F$. The crucial point is thus the existence, for all $\varepsilon > 0$, of two step functions $\varphi$ and $\psi$ that frame $f$ and such that $m(\psi) - m(\varphi) < \varepsilon$.

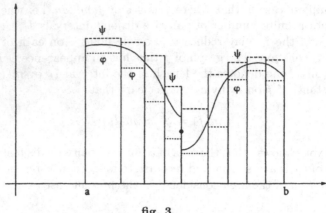

fig. 3.

An economical method of ensuring this is to impose the following hypothesis on $f$: *for any $r > 0$, there exists a step function $\varphi$ such that*

(11.2)     $$\varphi(x) \leq f(x) \leq \varphi(x) + r \text{ for all } x \in I;$$

one says then that $f$ is a *regulated function* on the interval $I$. If one then chooses $\psi(x) = \varphi(x) + r$, the algebraic area $m(\psi) - m(\varphi)$ lying between the graphs of $\varphi$ and $\psi$ is composed of rectangles of height $r$ having for bases the various intervals $I_k$ on which the function $\varphi$ is constant. Since the $I_k$ are pairwise disjoint, the sum of their lengths is the length $b - a$ of $I$, so that

(11.3)     $$m(\psi) - m(\varphi) = (b - a)r < \varepsilon$$

if $r < \varepsilon/(b-a)$. The second condition of the axiom (IV bis) is thus satisfied.

The linearity of $m(f)$ reduces to the two following propositions: (i) $m(cf) = cm(f)$ for every constant $c \in \mathbb{R}$; (ii) $m(f + g) = m(f) + m(g)$ for any regulated $f$ and $g$. (i) is obvious for a step function and one proves the general case thanks to (3). Additivity is shown similarly, observing that, if one has two step functions $\varphi'$ and $\varphi''$ on $I$, one can always divide $I$ into intervals on each of which $\varphi'$ and $\varphi''$ are constant: whence (ii) trivially for step functions $f$ and $g$, the general case following in the limit.

*Example 1.* Suppose that you wish to calculate the area $L(a, b)$ lying between the $x$-axis, the curve $y = 1/x = f(x)$ and the verticals with abscissae $a > 0$ and $b > a$. Let us choose an integer $n$, put $q = (b/a)^{1/n}$, whence $b = aq^n$, and divide the interval $[a, b]$ by the points $a = aq^0, aq, aq^2, \ldots, aq^n = b$. Consider the step function $\varphi$ which, between $aq^k$ and $aq^{k+1}$, is equal to $1/aq^{k+1}$; it is $\leq f$ since $f$ is decreasing and

$$\begin{aligned}
m(\varphi) &= (aq - a)/aq + (aq^2 - aq)/aq^2 + \ldots + (aq^n - aq^{n-1})/aq^n = \\
&= n(q - 1)/q = n(c^{1/n} - 1)/c^{1/n} \text{ where } c = b/a.
\end{aligned}$$

If you replace $\varphi$ by the step function $\psi \geq f$ which, between $aq^k$ and $aq^{k+1}$, is equal to $1/aq^k$, so that $\psi(x) = q\varphi(x)$ for all $x$, you find

$$m(\psi) = qm(\varphi) = n(c^{1/n} - 1).$$

So

$$n(c^{1/n} - 1)/c^{1/n} \leq L(a, b) \leq n(c^{1/n} - 1).$$

As $n$ increases indefinitely, the left hand side tends to $\log c$, and the third does so too, since the denominator $c^{1/n}$ tends to 1. We have thus proved that the function $1/x$ is integrable and that

(11.4)
$$\int_a^b dx/x = \log b - \log a \qquad (0 < a < b).$$

*Example 2.* The same method, a little less simply, can be applied to calculate the integral of the function $f(x) = x^s$ between $x = a > 0$ and $x = b > a$, where $s \in \mathbb{Z}$, $s \neq -1$ (the case where $s = -1$ has just been treated and is quite different). One uses the same subdivision of $[a, b]$ and the same definitions of $\varphi$ and $\psi$ as above, though of course the values of $\varphi$ and $\psi$ on $(aq^k, aq^{k+1})$ must now be those which $f(x)$ takes at the end points[30], i.e. $(aq^k)^s = a^s q^{sk}$ and $(aq^{k+1})^s$. Now

$$
\begin{aligned}
m(\varphi) &= \sum_{0 \leq k < n} (aq^{k+1} - aq^k)a^s q^{sk} = (q - 1)a^{s+1} \sum_{0 \leq k < n} q^{(s+1)k} = \\
&= (q - 1)a^{s+1}\left[1 + q^{s+1} + \ldots + q^{(s+1)(n-1)}\right] = (q - 1)a^{s+1}\frac{q^{n(s+1)} - 1}{q^{s+1} - 1} \\
&= \frac{q - 1}{q^{s+1} - 1}\left[(aq^n)^{s+1} - a^{s+1}\right] = \frac{q - 1}{q^{s+1} - 1}\left(b^{s+1} - a^{s+1}\right)
\end{aligned}
$$

since $aq^n = b$. The area sought is, on the other hand, smaller than $m(\psi)$, the number one obtains on replacing $f(aq^k)$ by $f(aq^{k+1}) = q^s f(aq^k)$ in the preceding calculations, so $m(\psi) = q^s m(\varphi)$.

As $n$ increases indefinitely, $q = (b/a)^{1/n}$ tends to 1, so that the ratio $(q^{s+1}-1)/(q-1) = (q^{s+1}-1^{s+1})/(q-1)$ tends, by definition, to the derivative of the function $x \mapsto x^{s+1}$ at $x = 1$, i.e. to[31] $s + 1 \neq 0$; its reciprocal thus tends to $1/(s + 1)$, so that

$$\lim m(\varphi) = (b^{s+1} - a^{s+1})/(s + 1).$$

Since $m(\psi) = q^s m(\varphi)$ tends to the same limit as $q$ tends to 1, one finally has

---

[30] This choice assumes that $f$ is increasing, i.e. $s > 0$. We leave to the reader the trouble of switching the letters $\varphi$ and $\psi$ when $s < 0$.

[31] We have seen this for $s \in \mathbb{N}$ in n° 4, but the result persists for $s \in \mathbb{Z}$ and even for $s \in \mathbb{R}$, so the calculations and result apply in this last case too.

$$(11.5) \qquad \int_a^b x^s\, dx = \frac{b^{s+1} - a^{s+1}}{s+1}$$

for $s \in \mathbb{Z}$ (in fact, $s \in \mathbb{R}$) different from $-1$ and $0 < a < b$ (and in fact for any $a$ and $b$ if $s \in \mathbb{N}$). The method above is due to Fermat who, curiously, did not apply it to the case $s = -1$.

The concepts of integral and derivative are, in classical analysis, linked by the "fundamental theorem of the integral calculus": *if $f$ is continuous*, then (i) every function $F$ such that $F' = f$ satisfies

$$(11.6) \qquad \int_a^x f(t)\, dt = F(x) - F(a)$$

for any $x$, (ii) if, conversely, one uses (6) to define a function $F$ – no matter what value one chooses for $a$ –, then $F' = f$. Example 2 confirms point (ii); Example 1 and point (ii) imply the formula $\log' x = 1/x$ of n° 10, Theorem 3. (To follow.)

## 12 – Series with positive terms

After this anticipation of Chap. V, we note several immediate applications of Theorem 2 to series.

**Theorem 4.** *For a series with positive terms to converge, it is necessary and sufficient that its partial sums be bounded above. The sum of the series is then the the least upper bound of the partial sums.*

This is clear since they form an increasing sequence.

If for example one has two series $\sum u_n$ and $\sum v_n$ with positive terms, if the second converges, and if $u_n \leq v_n$ for all $n$ (or only for all $n$ sufficiently large), then the first converges too, since its partial sums are smaller than those of the second.

Theorem 4 shows also that *if, in a series $\sum u(n)$ with positive terms, one regroups the terms*

$$[u(1) + \ldots + u(p_1)] + [u(p_1 + 1) + \ldots + u(p_2)] + \ldots$$

*then the initial series and the new series $\sum v(n)$ are simultaneously convergent or divergent.* We already know that the convergence of the first series implies that of the second without any hypothesis of positivity. To establish the converse, one observes that the partial sum $u(1) + \ldots + u(n)$ of the first is majorised by the partial sum

$$v(1) + \ldots + v(n) = u(1) + \ldots + u(p_n)$$

of the second, since it contains more terms, all positive, than the first. So if the partial sums of the second series are majorised, so similarly are those of the initial series, qed. We will show further that one can even perform regroupings of infinitely many terms and permute the terms of the series arbitrarily without changing the result, which is "obvious" only so long as one has not understood the difference between algebra (finite sums) and analysis (infinite sums).

We said at the beginning of n° 6 that the fundamental problem of the theory of series would be to give a meaning to the sum of an infinite family $(u(i))_{i\in I}$ of real or complex numbers indexed by any *countable* set $I$, for example $I = \mathbb{N} \times \mathbb{N} = \mathbb{N}^2$, or $\mathbb{N} \times \mathbb{N} \times \mathbb{N} = \mathbb{N}^3$, etc. The preceding arguments already allow us to treat the case of a sum of positive numbers.

An "obvious" method to give a meaning to the expression

$$(12.1) \qquad \sum_{i\in I} u(i)$$

would be to choose a bijection $f : \mathbb{N} \longrightarrow I$, to put

$$v(n) = u(f(n))$$

and to declare that the sum (1) is, by definition, equal to the sum of the series $v(n)$. This would presume that we had established that the result is independent of the choice of $f$, and since two bijections $f$ and $g$ of $I$ onto $\mathbb{N}$ differ from one another by a bijection of $\mathbb{N}$ onto $\mathbb{N}$, i.e. by a *permutation* of $\mathbb{N}$, this reduces to showing that the sum of a convergent series with positive terms is independent of the order of its terms. This is indeed the case, but one can in fact proceed directly, i.e. without choosing a particular bijection of $\mathbb{N}$ onto $I$.

Let us take inspiration from the case where $I = \mathbb{N}$. The sum of the series is then the least upper bound of the set of its partial sums, i.e. the *least* number which majorises them. But these are the *ordered* partial sums and not the *unordered partial sums*

$$(12.2) \qquad s(F) = \sum_{i\in F} u(i)$$

obtained by adding those terms of the series whose index belongs to an arbitrary finite subset $F$ of the set of indices $I = \mathbb{N}$. If nevertheless you add the terms of index 1492, 1776, 1812, 1861, 1898, 1917 and 1941 the fact that the series has positive terms shows that the result is $\leq u(1) + u(2) + \ldots + u(1940) + u(1941)$. A number which majorises all the ordered partial sums thus also majorises all the unordered partial sums, and conversely, since the ordered partial sums appear among the unordered partial sums.

We thus might also define the sum of a series with positive terms as the least upper bound of the set of its unordered partial sums.

It is now easy to define directly the "unconditional" sum (hint: total) (1) in the general case where the set $I$, though identifiable with $\mathbb{N}$ – choose a bijection –, is not so identified: this will be *the least upper bound*[32] *of the set of its unordered partial sums* (2). If, as suggested above, one chooses a bijection $f$ of $\mathbb{N}$ onto $I$ arbitrarily, which transforms the sum (1) into the classical series $u(f(n))$ and, consequently, the unordered partial sums (2) of the series (1) into those of this series, it is clear that (i) the family of $u(i)$ has a finite unordered sum if and only if the series $u(f(n))$ converges, (ii) the unordered sum (1) is equal to the sum $\sum u(f(n))$ in the sense of n° 6. There is thus no difference between unordered convergence and convergence in the classical sense of the series obtained by ordering the terms of the family $(u(i))_{i \in I}$ arbitrarily.

This result immediately leads to another: subjecting the terms of a series with positive terms to an arbitrary permutation, i.e. replacing $u(i)$ by $u(f(i))$ where $f : I \longrightarrow I$ is bijective, does not change its sum, for the reason that this operation clearly does not change *the set* of unordered partial sums[33] of the series; it merely permutes them. In other words, *the rule of commutativity of addition applies unrestrictedly to convergent series with positive terms*, whether indexed by $\mathbb{N}$ or by an arbitrary countable set $I$. We shall see soon that it also applies, more generally, to "absolutely convergent" series, and to them alone.

Let us return to the classical series indexed by $\mathbb{N}$. On generalising the argument used by Cauchy to establish the divergence of the harmonic series (n° 7), one obtains the following result:

**Cauchy's condensation criterion.** *Let* $\sum u(n)$ *be a series whose terms tend to 0 while* decreasing *and put* $v(n) = 2^n . u(2^n)$. *Then the given series is of the same nature as the series* $\sum v(n)$.

The argument of n° 7 shows that

$$(12.3) \quad \begin{aligned} v(n)/2 &= 2^{n-1} u(2^n) \\ &\leq u(2^{n-1}) + u(2^{n-1} + 1) + \ldots + u(2^n - 1) = \\ &= w(n) \leq 2^{n-1} u(2^{n-1}) = v(n-1) \end{aligned}$$

since the block considered has $2^n - 2^{n-1} = 2^{n-1}$ terms whose values all lie between $u(2^{n-1})$ and $u(2^n)$. The relation $v(n) \leq 2w(n) \leq 2v(n-1)$ shows that the series $\sum v(n)$ has the same nature as the series $\sum w(n)$ that one obtains by grouping the terms of the series $\sum u(n)$ in blocks of $2^n$ terms; now

---

[32] Assuming this least upper bound finite. In the opposite case, it is natural to consider that it is the symbol $+\infty$, as we shall explain in n° 17.

[33] The set $E$ of unordered partial sums is defined as follows: $x \in E$ if and only if there exists an $F$ such that $x = s(F)$. The definition is analogous for the set of elements of a sequence, of an arbitrary family $(u_i)$, of values of a function, etc. In the case of a family $(u_i)_{i \in I}$ for example, it is thus the image of $I$ under the map $i \mapsto u_i$. The concept of least upper bound, applied to the terms of a sequence or to the partial sums of a series, really involves only the *set* of these terms.

we have just seen that we do not change the convergence or the divergence of a series with positive terms by performing arbitrary regroupings of its terms.

**Theorem 5.** *The series* $\sum 1/n^k$ *converges for* $k > 1$ *and diverges for* $k \leq 1$.

For $k = 1$, this is the harmonic series, already investigated. For $k < 1$, one has $n^k < n$, thus $1/n^k > 1/n$, whence all the more divergent. For $k > 1$, the condensation criterion leads to the series with general term

$$v(n) = 2^n . 1/(2^n)^k = 1/2^{n(k-1)} = q^n$$

where $q = 1/2^{k-1}$ is $< 1$ since

$$a^b > 1 \text{ for } a > 1 \text{ and } b > 0.$$

The convergence of the geometric series, see (6.4), thus entails that of the series considered[34], qed.

Consider now the series with general term

$$u(n) = 1/n.(\log n)^k,$$

where $n \geq 2$ since $\log 1 = 0$. All we need to know is that $\log x > 0$ for $x > 1$, that $\log xy = \log x + \log y$ for $x, y > 0$, so that the function log is increasing[35], then that $\log(x^n) = n. \log x$ by the preceding formula. Now

$$v(n) = 2^n/2^n(\log 2^n)^k = c/n^k$$

where the constant $c = 1/(\log 2)^k$ is of little importance. Conclusion: the series converges if and only if $k > 1$. The reader may go on to amuse himself by treating the series

$$u(n) = 1/n. \log n.(\log \log n)^k,$$

then the series

$$u(n) = 1/n. \log n. \log \log n.(\log \log \log n)^k$$

and so on indefinitely. We write $\log \log x$ though we ought to write $\log(\log(x))$, etc.

As for the series $\sum 1/\log n$, this is clearly divergent since it notorious that $\log n$ is smaller than $n$ [see furthermore (10.10)] and even, as we shall see, is $o(n)$ when $n$ increases indefinitely.

Maybe some readers will ask: why $2^n$ rather than $3^n$? Because Cauchy chose $2^n$, of course; he might also have, in the statement of his criterion,

---

[34] The proof is valid not only for $k$ an integer, but also for any real $k$ once one accepts that the rules of calculation for integer exponents extend without formal modification to real exponents. We shall show this in Chapter IV.

[35] Every $x' > x$ is of the form $xy$ with $y > 1$, whence $\log x' = \log x + \log y > \log x$.

chosen $p^n$ for any integer $p > 1$, considering that there are $(p-1)p^n$ integers between $p^n$ and $p^{n+1}$, so that the inequalities (3) remain valid in this case after trivial modifications. In fact, Cauchy must surely have observed that the criterion with $p^n$ applies precisely to the same series as the criterion with $2^n$ so he must have thought it better to leave to others, if by chance there were such people, the glory of a futile generalisation.

Theorem 5 admits generalisations to multiple series, which will allow us to understand the utility of "unconditional convergence" introduced above. Consider for example the sum

$$(12.4) \qquad \sum_{(m,n)\in\mathbb{Z}^2-\{0\}} 1/(m^2+n^2)^{k/2} = \sum u(m,n)$$

mentioned in n° 6. Its resemblance to the series

$$(12.5) \qquad \sum_{n=1}^{\infty} 1/n^k = \frac{1}{2} \sum_{n\in\mathbb{Z}-\{0\}} 1/(n^2)^{k/2}$$

of Theorem 5 is sufficiently striking for one to hope for a similar result. More to explain the general method than to press on further, let us consider for every integer $p \geq 1$ the set $F_p \subset I = \mathbb{Z}^2 - \{0\}$ of pairs $(m,n)$ such that $|m| + |n| = p$. There are $4p$, apart from error or omission (diagram!). For such a pair one has

$$p^2/4 \leq m^2 + n^2 \leq 2p^2$$

since each of the two integers $|m|$, $|n|$ is $\leq p$ and one of the two is $\geq p/2$; the partial sum corresponding to $F_p$ thus satisfies the relation

$$(12.6) \qquad 4p/2^{k/2}p^k = m/p^{k-1} \leq s(F_p) \leq 4p2^k/p^k = M/p^{k-1}$$

with constants $m, M > 0$ independent of $p$ and whose exact values matter little. Since the set of indices $I$ is the union of the pairwise disjoint sets $F_p$ one can assume – associativity ... – that the convergence of (4) is governed by that of the classical series $\sum s(F_p)$. Let us first justify this point. Let $F$ be any finite subset of $I$. It is the union of the sets $F \cap F_p$, which are pairwise disjoint, and only a finite number of them are not empty, since $F$ is finite. One thus has $s(F) = \sum s(F \cap F_p) \leq \sum s(F_p)$ since all the $u(i)$ are positive. The convergence of the series $s(F_p)$ then shows the existence of a number which majorises all the partial sums $s(F)$, namely

$$(12.7) \qquad s' = \sum s(F_p);$$

so the sum (4) converges unconditionally, with a total sum $s \leq s'$. Conversely, the ordered partial sums

$$s(F_1) + \ldots + s(F_n) = s(F_1 \cup \ldots \cup F_n)$$

of the series $\sum s(F_p)$ are particular unordered partial sums of $s(F)$; thus if (4) converges unconditionally and has sum $s$, then the series $s(F_p)$ converges and has sum $s' \leq s$, whence finally $s = s'$.

So it all reduces to deciding the convergence of the classical series (7). This follows from the comparison (6) with the series $1/p^{k-1}$ since the $s(F_p)$ [resp. the partial sums of (7)] are, up to factors independent of $p$, minorised and majorised by the terms (resp. the partial sums) of the series $\sum 1/p^{k-1}$. Theorem 5 then provides us the condition for convergence, namely $k > 2$.

The reader can easily treat the case of the sum[36]

$$\sum 1/\left(m^2 + n^2 + p^2\right)^{k/2}$$

where, this time, the summation is extended over $\mathbb{Z}^3 - \{0\}$, or to the sum

$$\sum 1/\left(3m^2 + 5n^2\right)^{k/2},$$

etc. The principle always remains the same, and consists of effecting regroupings of the terms to suit the problem. These sums, where the general term depends on two, three, ... integers, are called *double, triple series*, etc. They make beginners tremble with their "flood of indices". But there is no more in the general form (1) than in the classical theory, and when one is obliged, as we shall be, to make it all explicit and to work in $\mathbb{Z}^2$ or $\mathbb{N}^7$, one has the opportunity to familiarise oneself with the principles of Elementary Set Theory as expounded in Chap. I.

## 13 – Alternating series

Theorem 4 allows us to elucidate a class of series which, though convergent, do not generally possess the commutativity property that we have just established for series with positive terms.

Consider for example the *alternating harmonic series*

(13.1)                     $1 - 1/2 + 1/3 - 1/4 + \ldots$

and, first, its partial sums of odd order

$$\begin{aligned}
s_1 &= 1, \\
s_3 &= 1 - (1/2 - 1/3), \\
s_5 &= 1 - (1/2 - 1/3) - (1/4 - 1/5),
\end{aligned}$$

etc. Since $1 \geq 1/2 \geq 1/3 \geq \ldots$, it is clear that they decrease while remaining positive, since one has also, for example,

---

[36] A particularly simple method for treating the series in two variables is to remark, with Weierstrass, that $m^2 + n^2 \geq 2|mn|$, whence a majorisation by the product of the series $\sum 1/|m|^{s/2}$ and $\sum 1/|n|^{s/2}$ and thus the condition $s > 2$. But this method does not adapt to series in more than two variables.

$$s_5 = (1 - 1/2) + (1/3 - 1/4) + 1/5.$$

It thus converges to a limit $s \geq 0$.

The sums of even order

$$s_{2n} = s_{2n-1} - 1/2n$$

also converge to $s$ since $1/2n$ tends to 0. It is now clear that the series (1) converges and has sum $s$.

These arguments, like the following theorem due to Leibniz, extend immediately to *alternating series*, those of the form

$$u_1 - u_2 + u_3 - \ldots,$$

where the $u_n$ are all positive.

**Theorem 6.** *Every alternating series whose terms tend to* 0 *while decreasing in absolute value, is convergent.*

Now, the partial sums

$$s_{2n+1} = u_1 - (u_2 - u_3) - \ldots - (u_{2n} - u_{2n+1})$$

decrease and are positive since

$$s_{2n+1} = s_{2n} + u_{2n+1} \geq s_{2n} = (u_1 - u_2) + \ldots + (u_{2n-1} - u_{2n}) \geq 0.$$

In consequence $s_{2n+1}$ tends to a limit $s \geq 0$, which is also the limit of $s_{2n}$ since $u_{2n+1}$ tends to 0, whence the convergence of the series.

Note that while the $s_{2n+1}$ decrease, the $s_{2n}$, on the contrary, increase; so

(13.2)
$$s_{2p} \leq s \leq s_{2q+1}$$

for all $p$ and $q$. For $p = q = n$, this relation may be written

$$s_{2n+1} - u_{2n} \leq s \leq s_{2n+1},$$

or

$$s_{2n} \leq s \leq s_{2n} + u_{2n+1}.$$

We deduce that

(13.3)
$$|s - s_r| \leq u_{r+1}$$

for any $r$. In other words, *the error committed in replacing the total sum by a partial sum is, in absolute value, smaller than the first term neglected.* Nor is it much smaller: the relation

$$s - s_r = (u_{r+1} - u_{r+2}) + (u_{r+3} - u_{r+4}) + \ldots$$

shows that also

(13.4)                           $|s - s_r| \geq u_{r+1} - u_{r+2}.$

   If one considers for example the series

$$1 - 1/3 + 1/5 - 1/7\ldots,$$

for which $u_n = 1/(2n - 1)$, whence $u_{r+1} - u_{r+2} = 2/(2r+1)(2r+3)$, one has

$$2/(2r + 1)(2r + 3) \leq |s - s_r| \leq 1/(2r + 1).$$

In order to calculate $s$ exact to 20 decimal places, one must therefore calculate $s_r$ to the same precision for an integer $r$ such that

$$2/(2r + 1)(2r + 3) < 10^{-20},$$

i.e. such that $(2r+1)(2r+3) > 2.10^{20}$, which requires $(2r+3)^2 > 2.10^{20}$, so $2r + 3 > 1.4 \times 10^{10}$; one must therefore calculate about seven billion terms of the series and their sum, all exact to 20 decimal places. Since errors have a strong tendency to reinforce, and since $10^{-20}$ is almost equal to $7.10^9.10^{-31}$, the calculations must be performed to 31 exact decimals. *Exercise:* calculate $1/1234567898$ by hand to 31 exact decimals (in fact 21, since the result obviously starts with ten 0s).

   One will therefore understand the hilarity – a very rare event, it seems – which must have seized Newton when in 1676, on the occasion of a brief exchange of correspondence with Leibniz (1646–1716) – an exchange in the course of which each very obviously strained to show that he knew more than the other, particularly Newton who was several years in advance but had *published* nothing –, Leibniz informed him of a sort of mechanical procedure for calculating an arbitrarily high number of decimals of $\pi$, namely the formula (not obvious at this stage of the exposition)

$$\pi/4 = 1 - 1/3 + 1/5 - 1/7 + \ldots.$$

Newton who a dozen years earlier, had amused himself – one amuses oneself as one can, and he adduced his youth in excusing himself to Leibniz – in calculating $\log(1 + x)$ to about 50 places, with the aid of the analogous series $x - x^2/2 + x^3/3 - \ldots$, *but only for values of x very close to 0* and not, for example, for $x = 1$, a desperate case, replied to Leibniz that it would require a million years of work to calculate the first twenty decimals of $\pi$ by his method, and further that his formula was already known in 1671 to his compatriot James Gregory. One notes, a curious coincidence, that in 1673 Leibniz had presented his calculating machine to the Royal Society of London and to the Académie des sciences of Paris. It was inspired by that of Pascal and capable of performing not only additions and subtractions like that one, but also multiplications and divisions; Leibniz developed it over decades, at great cost, but it was certainly not yet the central computer of the American *Strategic Air Command* capable of organising mathematically (?) an attack leaving 400 millions of corpses[37] on the ground, as well as others uncounted, including, according to an American admiral, the sailors of the Pacific Fleet who would be too busy cleaning themselves from radioactive fallout to have time to chase the Soviet submarines.

After this incursion into the domain of psychopathological additions – a French humorist remarked in the 1930s that the kepis of the military of high rank are surrounded by several braids to prevent their skulls from exploding –, let us return to mathematical additions to show that the commutativity property established above for series with positive terms no longer applies when, in a convergent series of real terms, the series of its positive terms (and so also that of its negative terms) diverges, as is the case for the alternating

---

[37] See in Desmond Ball and Jeffrey Richelson, eds., *Strategic Nuclear Targeting* (Cornell University Press, 1986) the articles of David Alan Rosenberg and Desmond Ball on the (one and only) American plan of war of 1960, particularly pp. 35–36 and 62 for the extent of the attack: at least 1,400 bombs with a total power of 2,100 megatonnes onto 650 multiple objectives, to a maximum of 3,500 weapons onto 1,050 multiple objectives of which 151 were industrial-urban centres. The plan estimates that the maximum attack would cause between 360 and 425 million deaths from the Iron Curtain to China inclusive. The operation would take twenty four hours and, in theory, would be launched in the event that preparations for a Soviet attack were detected. In *Secrets. A Memoir of Vietnam and the Pentagon Papers* (Viking, 2002), pp. 58-59, Daniel Ellsberg, who is in the best position to know, confirms, with in addition up to 100 million deaths in neutral countries and as many in allied ones, as told in 1964 to President Johnson; "A hundred Holocausts", as Ellsberg puts it. The Soviets *at that period* had much less means of reaching the USA, but as a former advisor of President Kennedy, McGeorge Bundy, who had lived through the Cuban Crisis in 1962, remarked much later, a single bomb on a single large town, let alone ten, would already be an unimaginable catastrophe.

The Soviet plans, analysed in the first book by a member of the American intelligence services, put much more emphasis on the destruction of military forces and of a certain number of industries and of vital means of transport, avoiding uselessly devastating Europe whose resources would be indispensable to the reconstruction of the USSR after the "victory"; the prevailing Westerly winds would anyway demand a certain moderation, also needed so that one might not exterminate the populations that one desired convert to socialism ... The objective assigned by General de Gaulle in the 1960s to the French nuclear arms was the power, in the end, to eliminate about 60 million citizens and 50% of the Soviet industrial potential. The Soviet reaction to this program is not known.

See also Roger Godement, "Aux sources du modèle scientifique américain" (*La Pensée*, n° 201, 203 and 204, 1978/79) where one can see, for example, that in 1958, an IBM 704 "scientific" computer – no matter that a present day PC is more powerful – had the task of managing about 3,000 planes spread over 70 bases. These uses contributed powerfully to launching the computer industry, the Internet included, and the rôle that it now plays in the civil sector has by no means caused its other face to disappear, quite the contrary since the military sector now profits greatly from the civil sector. We do not know what Pascal and Leibniz would have thought of this.

harmonic series; this clearly amounts to saying that the series $\sum |u_n|$ diverges. In fact, one can even, by reordering the terms of such a series, transform it into a *divergent* series.

Take for example the case of an alternating series. The series

(13.5) $$u_1 + u_3 + u_5 + \dots$$

of its positive terms being divergent, there is an index $2p - 1$ for which the sum of the terms of index $\leq 2p - 1$ of (5) exceeds $10 + u_2$. To construct the new series, one writes first all the terms of index $\leq 2p - 1$ of (5), then the term $-u_2$, which has a new partial sum again $> 10$, then the terms of index $2p + 1, \dots$ of the series (5) until one obtains, at a rank $2q - 1$, a partial sum greater than $100 + u_4$; one then writes the term $-u_4$, whence a partial sum again $> 100$, then the terms of index $2q + 1, \dots$ of (5) until one obtains a partial sum $> 1000 + u_6$, then the term $-u_6$, whence a new partial sum $> 1000$, and so on indefinitely. It is clear that, in this way, one reorders the initial series so that the new partial sums take arbitrarily large values, which rules out convergence.

Historically the first known example (Dirichlet, 1837), is that of the alternating series where $u_n = 1/n^{1/2}$; it becomes divergent when written in the form

$$u_1 + u_3 - u_2 + u_5 + u_7 - u_4 + u_9 + u_{11} - u_6 + \dots;$$

if it were indeed convergent, it would remain so if one grouped the terms in threes; now the sum

$$1/(4n - 3)^{1/2} + 1/(4n - 1)^{1/2} - 1/(2n)^{1/2}$$

is positive for $n$ large (exercise!) and of the same order of magnitude as $1/n^{1/2}$; whence divergence.

The reason for this phenomenon is simple: it is not because the terms of the given series tend to 0 sufficiently fast that it converges – if such were the case, the series $\sum |u_n|$ would converge –, it is because the terms, when one adds them *in the prescribed order*, change sign often enough for the decrease in the negative sums to compensate miraculously for the increase in the positive sums. These compensations may evanesce when the order of terms is upset. A construction analogous to the preceding would show (Riemann) that, for any $s \in \mathbb{R}$, one can reorder the terms so as to obtain a convergent series with sum $s$. In other words, if the series $\sum u_n$ converges while the series $\sum |u_n|$ does not, its sum can be defined only through its ordered partial sums $u_1 + \dots + u_n$. For this reason, some authors prefer to speak of *semi-convergent* series.

## 14 – Classical absolutely convergent series

When a series $\sum u(n)$ is *absolutely convergent*, i.e. when the series $\sum |u(n)|$ converges, the commutativity problem does not arise: the series is a combination of convergent series with positive terms.

The best way of seeing this is to associate to any $x \in \mathbb{R}$ the positive numbers $x^+$ and $x^-$ defined as follows:

$$
\begin{aligned}
x^+ &= \sup(x,0) = x \text{ (if } x \geq 0) \text{ or } 0 \text{ (if } x \leq 0), \\
x^- &= \sup(-x,0) = 0 \text{ (if } x \geq 0) \text{ or } -x \text{ (if } x \leq 0),
\end{aligned}
$$

whence
$$
x = x^+ - x^-, \qquad |x| = x^+ + x^-.
$$

On applying this stratagem to the terms of a real series $\sum u(n)$ one exhibits it as the difference of two series with positive terms whose sum is the series $\sum |u(n)|$; since $u(n)^+$ and $u(n)^-$ are $\leq |u(n)|$, it is clear that the $|u(n)|$ series converges if and only if these two series with positive terms do so too. For a series of complex terms, one can apply the method to the real and imaginary parts $v(n)$ and $w(n)$ of its terms; one then has

(14.1) $$ u(n) = v(n)^+ - v(n)^- + iw(n)^+ - iw(n)^-, $$

four series with positive terms, which all converge if and only if the series of the $u(n)$ is absolutely convergent. It is clear that, more generally, any linear combination of absolutely convergent series is absolutely convergent, since

$$
|a + b + c + \ldots| \leq |a| + |b| + |c| + \ldots.
$$

An absolutely convergent series is thus convergent, and since we have commutativity for each of series $v(n)^+$, etc. appearing in (1), we have it for the given series. Furthermore

(14.2) $$ \left| \sum u(n) \right| \leq \sum |u(n)| $$

as one sees in passing to the limit in the analogous inequality for the partial sums of the two sides.

*Example 1. The exponential series*

(14.3) $$ \exp(z) = \sum_{n=0}^{\infty} z^n / n! = \sum z^{[n]} $$

*converges absolutely for all $z \in \mathbb{C}$ and its sum is a continuous function of $z$.*

We showed in n° 10, example 1, that it converges for $z > 0$, whence absolute convergence in the general case since

$$
|z^n / n!| = |z|^n / n!.
$$

The continuity of the function exp is an immediate result of the general theorems of n° 19 on power series, but we shall need this earlier, so we shall

have recourse to a particularly simple workmanlike method, which, generalised, will also serve in n° 19. To do this, we observe that

$$|\exp(z+h) - \exp z| \le \sum \left|(z+h)^{[n]} - z^{[n]}\right|;$$

and we have shown, see (4.6), that

$$\left|(z+h)^{[n]} - z^{[n]}\right| \le |h|\,(|z| + |h|)^{[n-1]}$$

for any $h, z \in \mathbb{C}$. Thus

$$|\exp(z+h) - \exp z| \le |h| \sum (|z| + |h|)^{[n-1]} = |h|\exp(|z| + |h|),$$

whence continuity, since, for $|h| < 1$ for example, the right hand side is
$< M|h|$ where $M = \exp(|z| + 1)$, so is $< r$ once $|h| < r/M$.

*Example 2.* We know from n° 6, or shall see later, that

$$\sin x = x - x^3/3! + x^5/5! - \dots,$$
$$\cos x = 1 - x^2/2! + x^4/4! - \dots.$$

These two series are absolutely convergent for any $x \in \mathbb{C}$. Indeed if one
replaces their terms by their absolute values one obtains the series of terms of
odd or even degree of the exponential series for $z = |x|$. Since that converges,
so do the two series considered.

The resemblance to the exponential series has doubtless struck the reader.
It so impressed Euler that about 1740 he had the idea of writing

$$\begin{aligned}
\exp(ix) &= 1 + ix + (ix)^{[2]} + (ix)^{[3]} + (ix)^{[4]} + (ix)^{[5]} + \dots \\
&= (1 - x^{[2]} + x^{[4]} - \dots) + i(x - x^{[3]} + x^{[5]} - \dots)
\end{aligned}$$

and concluding that

(14.4) $$\exp(ix) = \cos x + i \sin x$$

for $x \in \mathbb{R}$. This discovery can surely not have cost him more effort than it
would have us – it "suffices" to conceive the imaginary exponential ... –,
he knew the three series in question, and, as we have already quoted, "he
calculates as one breathes". Nevertheless the preceding relation plays a rôle
in analysis disproportionate to its apparent triviality.

The formula (4) of course relies on the expansions of the trigonometric
functions as power series. It was unfortunately almost impossible to justify
this rigorously for as long as one had only the traditional geometric definition
of sine and cosine: one had to understand what an angle is, or what the length
of a arc of circle (since the definition of an angle depends on this); it was also
necessary to know that the total length of a circumference of radius 1 is

measured by the mysterious number $2\pi$, and other things too, particularly from the theory of the integration. Clearly the founders of analysis did not bother themselves with such rigour, quite unachievable in their time, and this allowed them to progress. But we are no longer in the XVII[th] century, and the best reply to these questions will be, as we shall do in n° 14 of Chap. IV, to *define* the functions $\sin x$ and $\cos x$ by their power series and then to *deduce* the elementary properties which we all expect: the addition formulae, derivatives, relation $\cos^2 x + \sin^2 x = 1$, number $\pi$, etc.

*Example 3.* Consider the series $\sum z^n/n^2$. Since $|u_n| \leq 1/n^2$ if $|z| \leq 1$ and since the series $\sum 1/n^2$ converges, the series $\sum |u_n|$ satisfies the hypothesis of Theorem 4 at least as well as the second does. It is thus absolutely convergent for $|z| \leq 1$. We shall show later, with the help of certain simple criteria, that it diverges for $|z| > 1$.

*Example 4.* Consider a power series $\sum a_n z^n$ and suppose that it converges absolutely for $z = u$. Since $|a_n z^n| \leq |a_n u^n|$ for $|z| \leq |u|$, one can conclude that the series again converges absolutely for $|z| \leq |u|$. The fundamental concept of *radius of convergence* of a power series, due to Cauchy, is obtained by a similar argument. Consider the set $E$ of numbers $r \geq 0$ for which the sequence with general term $|a_n|r^n$ is *bounded*; denote by $R$ either the least upper bound of $E$ if $E$ is bounded, or the symbol $+\infty$ if it is not (n° 17). Then *the series $\sum a_n z^n$ converges absolutely for $|z| < R$ and diverges for $|z| > R$* (one cannot say anything a priori as to what happens at the points of the circumference $|z| = R$: anything is possible).

The second point is obvious since then the terms of the series are not bounded, so do not tend to 0. To establish the first, one notes that, by definition of a least upper bound, there exists an $r \in E$ such that $|z| < r < R$, whence $|z| = qr$ with $q < 1$; putting $M = \sup(|a_n|r^n)$, one then has $|a_n z^n| = q^n|a_n r^n| \leq Mq^n$, whence absolute convergence, since $q < 1$.

These small calculations show a little more: the coefficients of a power series with radius of convergence $R > 0$ cannot increase faster than a geometric progression. In other words, there are always positive constants $M$ and $q$ such that

(14.5) $$|a_n| \leq Mq^n \text{ for all } n$$

or, in the language of n° 3,

$$a_n = O(q^n), \qquad n \to +\infty.$$

This is obvious since, for all $r < R$, the sequence $(a_n r^n)$ is bounded (and even tends to zero), so that it suffices to take $q = 1/r$ to obtain (5). Conversely, if the $a_n$ satisfy the relation (5), one has $R > 0$ since then the series converges absolutely for $q|z| < 1$. The existence of a relation (5) thus *characterises* power series with radius of convergence $> 0$ or, as one calls them dangerously, the *convergent power series*.

"Dangerously" since there are those who believe that only series which converge for any $z \in \mathbb{C}$ should be called convergent. This is not the case; one demands only that they should not diverge for every $z \neq 0$, which is quite different. These series appear everywhere in classical analysis. The others never, because one can do nothing with them; all the same, see the "formal series" of n° 22.

It is obvious that in the interior of its circle of convergence the sum $f(z) = \sum a_n z^n$ of a power series is the limit of a sequence of polynomials in $z$, namely the partial sums of the series. One can make this result more specific by working within a disc of radius $r$ *strictly* smaller than the radius of convergence $R$ of the series. Now $|z^{n+1+p}| \leq |z|^{n+1} r^p$ for any $p \geq 0$, whence

$$|f(z) - a_0 - a_1 z - \ldots - a_n z^n| \leq |z|^{n+1} \left( |a_{n+1}| + |a_{n+2}| r + \ldots \right);$$

since the sum $\sum_p |a_{n+p}| r^p = M$ is convergent – it differs from the series $\sum |a_p r^p|$ only by some terms at the beginning and a factor $r^n$ –, one finds the inequality

(14.6)     $|f(z) - a_0 - a_1 z - \ldots - a_n z^n| \leq M|z|^{n+1}$ for $|z| \leq r$.

In the notation of n° 3, this implies

(14.7)     $f(z) = a_0 + a_1 z + \ldots + a_n z^n + O(z^{n+1})$ when $z \to 0$,

but (6) is a little more specific because it is valid not only for $r$ "small", but even for all $r < R$ (and so for all finite $r$ if $R = +\infty$, as in the case of the series for exp for example). The constant $M$ of (6) depends on $n$ and on $r$; it is generally impossible to find an $M$ for which (6) will be valid in the whole disc of convergence: this is already false for the series $\sum z^n$. See n° 8 of Chap. III.

## 15 – Unconditional convergence: general case

After this incursion into the classical – the concept of absolute convergence does not however antedate Cauchy and above all it was Weierstrass who exploited it systematically – we come to that of unconditional convergence for a sum of complex numbers $\sum u(i)$ where the index $i$ varies in an arbitrary countable set $I$. It allows us to elucidate the concept of absolutely convergent series and, indeed, reduces to this; but it is indispensable for other reasons. First we show how one can define it in a way similar to that of n° 12.

Suppose first that the $u(i) \geq 0$. The sum[38] $s = s(I)$ is, by definition, the least upper bound of the unordered partial sums

---

[38] We use the notation $s(F)$ for any subset $F$ of $I$, finite or not, under the clear condition that the sum has a meaning or, in the case of a sum of *positive* terms, agreeing that $s(F) = +\infty$ if the sum does not converge unconditionally.

(15.1)
$$s(F) = \sum_{i \in F} u(i)$$

extended over all, arbitrary, finite subsets $F$ of $I$. For every number $r > 0$ one can then choose $F$ so that

(15.2)
$$s(I) - r \le s(F) \le s(I).$$

But since the $u(i)$ are positive, it is clear that

(15.3)
$$G \supset F \Longrightarrow s(F) \le s(G) \le s(I);$$

one thus one sees that, for every finite subset $G$ of $I$,

(15.4)
$$G \supset F \Longrightarrow |s(G) - s(I)| \le r.$$

The analogy with the classical definition is clear: instead of being indexed by an integer $n$, the unordered partial sums are indexed by arbitrary finite subsets $F$ of the set $I$, the order relation $q > p$ is replaced by the inclusion $G \supset F$ and the relation (4) replaces the classical definition

$$q \ge p \Longrightarrow |s_q - s| \le r.$$

The generalisation to sums of complex numbers is now obvious. The sum of the $u(i)$ will be said to be *unconditionally convergent* if there is a number $s = s(I)$ possessing the following property: for every $r > 0$ there exists a finite subset $F$ of $I$ satisfying (4). In the case where all the $u(i)$ are positive, this means, as we have seen, that there exists a real number $M$ which majorises all the unordered partial sums. We shall show that, in the general case, unconditional convergence of the sum of the $u(i)$ *is equivalent* to that of the sum of the $|u(i)|$, i.e. to the fact that the partial sums (unordered – there are no others) of the second series are bounded above.

First of all we remark that since unconditional convergence does not presuppose writing the elements of $I$ as a sequence in any particular way, a permutation of $u(i)$ does not change the situation at all: here again we have *total commutativity of addition*.

As in the classical case, a sum of complex terms converges unconditionally if and only if the sums obtained on replacing the terms by their real and imaginary parts do so too: replace $s(F)$, $s(G)$ and $s(I)$ by their real or imaginary parts in (4).

It is also almost obvious that if two series $\sum u(i)$ and $\sum v(i)$ indexed by the same set $I$ converge unconditionally and have sums $s$ and $t$, then the series of $w(i) = u(i) + v(i)$ converges unconditionally to $s + t$. If, indeed, for every $r > 0$ there are finite subsets $F'$ and $F''$ of $I$ such that the unordered sum extended over $G$ of the first (resp. second) series is equal to $s$ (resp. $t$) to within $r$ once $G \supset F'$ (resp. $G \supset F''$), it is clear that if $G \supset F = F' \cup F''$, then the sum of $w(i)$, $i \in G$, will be equal to $s + t$ to within $2r$, qed[39].

It follows immediately from this that series with complex terms reduce to series with real terms as in the classical case. Let us examine these, in order to show that one can even, in fact, reduce to unordered families or sums with *positive terms*.

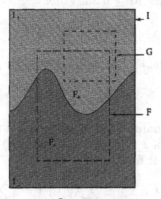

**fig. 4.**

If all the $u(i)$ are real (and, one may assume, $\neq 0$), one can partition $I$ into two disjoint sets: the set $I_+$ of $i$ such that $u(i) > 0$ and the set $I_-$ of $i$ such that $u(i) < 0$. Let us choose a finite subset $F$ of $I$ satisfying (4), for example for $r = 1$. $F$ is the union of the disjoint sets $F \cap I_+ = F_+$ and $F \cap I_- = F_-$. Now let $G$ be an arbitrary finite subset of $I_+$. Since $G \cup F_+$ contains more terms $u(i)$, all positive, than $G$, one has

$$s(G) \leq s(G \cup F_+) = s(G \cup F) - s(F_-)$$

since $G \cup F$ is the union of the disjoint sets $G \cup F_+$ and $F_-$. But the inequality (4) for $r = 1$ applies to $G \cup F \supset F$ and implies that

$$s(G \cup F) \leq s(I) + 1.$$

Substituting in the preceding result, one deduces that

$$s(G) \leq s(I) + 1 - s(F_-)$$

for *every* finite subset $G$ of $I_+$. The right hand side being independent of $G$, it follows that the series of the $u(i)$, $i \in I_+$, converges unconditionally (n° 12).

---

[39] This is the analogue of properties valid "for $n$ sufficiently large" of n° 3: in the present case we are dealing with an assertion about a finite subset $G$ of $I$ which is valid so long as it *contains* a suitably chosen subset $F$. Do not confuse with: "so long as $G$ contains sufficiently many elements of $I$". If for example you wish to calculate the sum of the series $1 + 1/2! + 1/3! + \ldots$ to within 0.1, and if you add ten billions of terms chosen arbitrarily, forgetting to include the first term of the series, you will not obtain the result $\ldots$

Since we can reason in this way about $I_-$ as about $I_+$, we see that each of these sums taken over $I_+$ and $I_-$ converges unconditionally. The converse being obvious, one thus obtains the following result:

**Theorem 7.** *For a series* $\sum u(i)$, $i \in I$, *to converge unconditionally it is necessary and sufficient that the series* $\sum |u(i)|$ *converges unconditionally, i.e. that there is a number* $M \geq 0$ *such that*

$$(15.6) \qquad \sum_{i \in F} |u(i)| \leq M$$

*for every finite subset $F$ of $I$. One then has*

$$(15.7) \qquad \left| \sum u(i) \right| \leq \sum |u(i)|.$$

To establish the inequality (7), one uses the fact that for every $r > 0$ there exists a finite subset $F$ of $I$ such that the sum $s(F)$ is equal to the total sum $s(I)$ of $u(i)$ to within $r$; it follows that

$$|s(I)| \leq |s(F)| + r \leq \sum_{i \in F} |u(i)| + r \leq \sum_{i \in I} |u(i)| + r,$$

whence the result.

This theorem shows that, for *unconditional* convergence, there is no difference between convergence (in short) and absolute convergence: this is the great difference from the classical concept. For this reason, one often speaks of *absolutely summable* instead of unconditionally convergent families, and, in practice, one says only that the series $u(i)$ is *absolutely* or *commutatively convergent* ; the terminology depends on the author. Although there are important series which do not fit into this scheme – one meets them for example in the theory of Fourier series of a single variable, without speaking of alternating series *à la* Leibniz –, unconditional convergence suffices in the great majority of cases because it is the "discrete" analogue of, and much simpler than, the modern theory of integration, which one might define as a theory of unconditional convergence for "continuous" sums, i.e. indexed by the points of an interval of $\mathbb{R}$ or something analogous, a cube in $\mathbb{R}^3$ for example. These two theories allow one to calculate in a quasi-algebraic way, and one uses hardly any other nowadays in analysis for the reason that simple *ordered* series, and even less, semi-convergent series, appear only very rarely in dimensions greater than 1. The Riemann series in two variables $m$ and $n$ studied in n° 12 is typical in this respect.

**Corollary (Cauchy-Schwarz inequality for series).** *Let* $(u_i)$ *and* $(v_i)$ *be two families of complex numbers such that the series* $\sum |u_i|^2$ *and* $\sum |v_i|^2$ *converge unconditionally. Then so does the series* $\sum u_i \bar{v}_i$, *and*

$$\left| \sum u_i \bar{v}_i \right|^2 \leq \sum |u_i|^2 . \sum |v_i|^2.$$

The inequality, where one replaces the $u$ and the $v$ by their moduli, is certainly valid if one restricts the sums to a finite subset $F$ of the set $I$ of indices (Chap. III, Appendix). It is *a fortiori* valid if one sums the left hand side over $F$ and the right hand side over $I$. The squares of the partial sums of the series $\sum |u_i \bar{v}_i|$ are thus bounded above by the right hand side of the relation to be established, qed.

To finish this study of unconditional convergence for the time being – the law of associativity remains to be proved, n° 18 –, let us return to the absolutely convergent classical series of the preceding n°. By definition, the series $\sum |u(n)|$ converges; it therefore satisfies (6), and therefore the given series $\sum u(n)$ converges unconditionally, and vice-versa.

Thus, for the classical ordered series, there is no difference between unconditional convergence and absolute convergence as in n° 14.

We have, however, to show that the classical definition of the sum

$$s = \lim \left( u(0) + \ldots + u(n) \right)$$

provides the same result as definition (4) of the unordered sum $s(\mathbb{N})$, supposing that the series starts with the term $u(0)$. To do this let us choose a number $r > 0$ and a finite subset $F$ of $\mathbb{N}$ satisfying (4). Since $F$ is finite, the set $F_n = \{0, 1, \ldots, n\}$ contains $F$ for $n$ sufficiently large. So $|s(F_n) - s(\mathbb{N})| \leq r$ for $n$ large. But $s(F_n)$, the sum of $n$ first terms of the series, is equal to $s$ to within $r$ for $n$ large; whence $|s - s(\mathbb{N})| \leq 2r$, qed.

Finally, consider an absolutely summable family $(u(i))$, $i \in I$, and choose a bijection $f$ of $\mathbb{N}$ onto $I$. The unordered sums of the given family and those of the series $\sum u(f(n))$ are clearly the same, and similarly for those obtained by replacing the $u(i)$ by their absolute values. It follows that the series $\sum u(f(n))$ is absolutely convergent and has the same sum as the given family. If, conversely, a bijection of $\mathbb{N}$ onto $I$ transforms the family $(u(i))$ into an absolutely convergent series, it is clear that the given family is absolutely summable. This shows that *there is in fact no difference in nature between general unconditional convergence and classical absolute convergence*.

The interest of unconditional convergence will appear later, when after establishing associativity, we will apply it to multiple series for which the choice of a bijection $f$ does not provide any result, except by a miracle[40]. For the moment, we shall return to more classical, not to say more hackneyed, aspects of the traditional theory of series.

---

[40] Hardly surprising. If you subject the terms of a geometric series, for example, to an arbitrary permutation, you will have severe difficulties in demonstrating the convergence of the new series using only its ordered partial sums – unless you reconstitute an *ad hoc* proof of the general theorem.

## 16 – Comparison relations. Criteria of Cauchy and d'Alembert

Given two scalar sequences $(u_n)$ and $(v_n)$, one says that the first is *dominated* by the second if there exists a number $M \geq 0$ such that $|u_n| \leq M.|v_n|$ for $n$ large, which one can express by writing (end of n° 3)

$$(16.1) \qquad u_n = O(v_n) \qquad \text{when } n \to +\infty.$$

On the other hand one says (n° 3, (ii)) that the sequences $(u_n)$ and $(v_n)$ are *of the same order of magnitude* at infinity if both $u_n = O(v_n)$ and $v_n = O(u_n)$, which one writes

$$(16.2) \qquad u_n \asymp v_n \qquad \text{when } n \to +\infty.$$

This means that there exist numbers $m > 0$ and $M > 0$ such that

$$(16.3) \qquad m.|v_n| \leq |u_n| \leq M.|v_n| \qquad \text{for } n \text{ large.}$$

For $n$ large, the two ratios $u_n/v_n$ and $v_n/u_n$ thus should stay off 0, as sailors would say.

These definitions will provide us with comparison principles between series, quasi-trivial it is true, but very useful all the same:

**Theorem 8.** *Let $\sum u_n$ and $\sum v_n$ be two series with complex terms.*

*(CP 1) If $u_n = O(v_n)$ and if the series $\sum v_n$ is absolutely convergent, then so is the series $\sum u_n$.*

*(CP 2) If $u_n \asymp v_n$ and if one of the two series is absolutely convergent, then so is the other.*

If indeed $|u_n| \leq M.|v_n|$ for $n$ large, and if the partial sums of the series $\sum |v_n|$ are bounded, then so clearly are those of the series $\sum |u_n|$, whence (CP 1). The second part follows from the first on exchanging the rôles of two series.

**Corollary.** *For a series $\sum u_n$ to be absolutely convergent, it suffices that there exists a number $q$ such that*

$$u_n = O(q^n), \qquad 0 \leq q < 1,$$

*or that there exists a number $s$ such that*

$$u_n = O(1/n^s), \qquad s > 1.$$

In the case of series *with positive terms* one can go a little further:

$(16.4) \qquad$ if $u_n \asymp q^n$, $\quad$ the series converges if $q < 1$ and diverges if $q \geq 1$;

$(16.5) \qquad$ if $u_n \asymp 1/n^s$, $\quad$ the series converges if $s > 1$ and diverges if $s \leq 1$.

This follows from (CP 2).

Theorem 8 has no application outside the domain of absolutely convergent series: the series $\sum 1/n$ diverges and the alternating series $\sum (-1)^n/n$ converges even though (CP 2) holds in this case. In a case of this kind, one clearly has to take account of the signs of the terms, or, in the complex case, of their arguments. The relation

$$u_n \sim v_n$$

allows one to go a little further; by definition this means (n° 4) that

(16.6) $\lim u_n/v_n = 1, \quad \lim v_n/u_n = 1, \quad u_n = v_n w_n$ with $\lim w_n = 1$,

these three relations clearly being equivalent[41].

Since a sequence which tends to 1 is bounded and remains bounded away from 0, this relation implies $u_n \asymp v_n$.

**Theorem 9.** *Let $\sum u_n$ be a series with terms $> 0$ and $\sum v_n$ a series with complex terms such that $u_n \sim v_n$ for $n$ large.*

*(i) If the series $\sum u_n$ converges then the series $\sum v_n$ converges absolutely;*
*(ii) If the series $\sum u_n$ diverges, so does the other.*

Case (i) follows from (CP 1). In case (ii), consider the series $\sum w_n$ where $w_n = \mathrm{Re}(v_n)$. Since $\mathrm{Re}(v_n/u_n) = w_n/u_n$ tends to 1 for the same reason as $v_n/u_n$, and since $u_n > 0$, one also has $w_n > 0$ for $n$ large, so the series $\sum u_n$ and $\sum w_n$ are simultaneously convergent or divergent. But if $\sum w_n$ diverges, so a fortiori does $\sum v_n$, qed.

For example, for a series with complex terms,

(16.7) $\qquad u_n \sim c/n^s \Longrightarrow \begin{cases} \text{absolute convergence if } s > 1, \\ \text{divergence if } s \leq 1. \end{cases}$

Suppose for example that $u_n = f(n)/g(n)$ where $f$ and $g$ are polynomials of degrees $p$ and $q$. One has

$$u_n = \frac{a_p n^p (1 + ?/n + \ldots + ?/n^p)}{b_q n^q (1 + ?/n + \ldots + ?/n^q)} = c n^{p-q} \frac{1 + \ldots}{1 + \ldots}$$

with $c = a_p/b_q \neq 0$ and numerical coefficients ? whose values do not matter. Since the fraction on the right hand side tends to 1, it is clear that $u_n \sim c n^{p-q}$ for $n$ large; hence

$$\sum f(n)/g(n) \quad \begin{cases} \text{converges absolutely if } d°(g) \geq d°(f) + 2 \\ \text{diverges if } d°(g) \leq d°(f) + 1. \end{cases}$$

Thus the series $\sum (n^2 + 3in - 5)/(n^4 - 2i)$ is absolutely convergent, while the series $\sum (n^2 + 3in - 5)/(n^3 - 2i)$ is divergent.

---

[41] This obviously assumes that, for $n$ large, $u_n$ and $v_n$ are never zero. Vain pedantry: one never meets any other case in practice.

We have seen above that a series $\sum u_n$ whose terms satisfy a relation of the form $u_n = O(q^n)$ with $0 < q < 1$ is absolutely convergent. There are two classical criteria, due respectively to d'Alembert, the man of the Encyclopaedia and of the Enlightenment, militant atheist, and to Cauchy, the ultra-legitimist of the Polytechnique School, and ultra-catholic of the Restoration, who developed this quasimechanistic method of comparison. Since, in practice, they frequently apply to the same series, readers of the left may prefer d'Alembert, those of the right Cauchy:

**Theorem 10.** *Let $\sum u_n$ be a series with complex terms. Suppose either that the ratio $|u_{n+1}/u_n|$ (d'Alembert), or that $|u_n|^{1/n}$ (Cauchy) tends to a limit $q$ as $n \to \infty$. Then the series is absolutely convergent if $q < 1$ and divergent if $q > 1$.*

If indeed $\lim |u_{n+1}/u_n| = q < 1$ and if you choose a number $q'$ such that $q < q' < 1$, then $|u_{n+1}/u_n| < q'$ for $n$ large, say for $n > p$. Then $|u_{p+r}| < q'|u_{p+r-1}| < q'^2|u_{p+r-2}| < \ldots < q'^r|u_p|$ for all $r > 1$ and thus $|u_n| < Mq'^n$ for $n$ large, with a constant $M = |u_p|/q'^p$ (put $n = p+r$). Since $q' < 1$, the geometric series $\sum q'^r$ converges, whence the absolute convergence of the given series. The argument is even simpler in the Cauchy case: one has $|u_n|^{1/n} < q'$ for $n$ large, i.e. $|u_n| < q'^n$, and concludes as before.

If $q > 1$, this time one chooses $q'$ so that $1 < q' < q$. One obtains the same inequalities, but with $>$ signs instead of $<$ signs and then, for $n$ large, the expressions of d'Alembert and of Cauchy are $> q'$. But then the inequality $|u_n| > Mq'^n$ proves not only that the series diverges, but that $|u_n|$ increases indefinitely, qed.

As examples, consider the following power series:

(a) $\sum z^n/n!$; here $|u_{n+1}/u_n| = |z|/(n+1)$ as one line of calculation shows; the ratio thus tends to 0 for any $z$, whence the absolute convergence of the exponential series; the radius of convergence (n° 14, example 4) is $+\infty$.

(b) $\sum z^n/n^s$; the d'Alembert ratio, equal to $|z|/(1 + 1/n)^s$, tends to $|z|$; therefore absolute convergence for $|z| < 1$ and divergence for $|z| > 1$; the radius of convergence is $R = 1$. Note that the Cauchy expression, equal to $|z|.n^{1/n}$, also tends to $|z|$ (n° 5, example 8). Note also that, for $|z| = 1$, one can say nothing, so that the Riemann series $\sum 1/n^s$ does not come within the scope of the very weak Theorem 10.

(c) $\sum n^2 z^n$; the Cauchy expression, namely $|z|(n^{1/n})^2$, again tends to $|z|$; same conclusions.

(d) $\sum n! z^n$; the d'Alembert ratio, namely $(n + 1)|z|$, increases indefinitely for $z \neq 0$. The series is always divergent, except of course for $z = 0$. The radius of convergence is zero.

These traditional criteria are so useful that one is often tempted to believe that a "small" modification of the hypotheses of Theorem 10 will be of little importance, as if a theorem obeyed the same laws of stability as the pendulum of a clock. In reality, a well constructed theorem resembles a pendulum in upwards vertical equilibrium: a small impulse and it runs away. *Examples*:

(a) the series $1/n$: the d'Alembert ratio, $n/(n+1)$, tends to 1 and the series diverges;

(b) the series $1/n^2$: the ratio, $n^2/(n+1)^2$, tends to 1 and the series converges;

(c) the series

$$1/2^2 + 1/1^2 + 1/4^2 + 1/3^2 + \ldots$$

obtained from the preceding by interchanging the terms in pairs: the d'Alembert ratio again tends to 1 but is alternately $> 1$ and $< 1$, and the series converges (compare its partial sums to those of the series $1/n^2$, or apply n° 11 bluntly).

(d) the series $\sin n/2^n$: this is absolutely convergent since dominated by the series $1/2^n$, but the d'Alembert ratio, namely $|\sin(n+1)/2\sin n|$, does not tend to any limit and, in fact, oscillates randomly between 0 and $+\infty$.

There are, of course, lots of more subtle criteria, applicable to the case where the d'Alembert ratio tends to 1 through values $< 1$ (the series is *evidently* divergent if it is $> 1$ for $n$ large). The most famous, due to Gauss, assumes a relation of the form

$$u_{n+1}/u_n = 1 - s/n + O(1/n^2) \qquad \text{when } n \to +\infty;$$

if the $u_n$ are positive, the series converges for $s > 1$ and diverges for $s \leq 1$. We shall return to this in Chap. VI.

For the moment we content ourselves with illustrating the theorem for *Newton's binomial series*

$$(16.8) \qquad N_s(z) \quad = \quad 1 + sz + s(s-1)z^2/2! +$$
$$+ s(s-1)(s-2)z^3/3! + \ldots = \sum_{n=0}^{\infty} \binom{s}{n} z^n,$$

where $z$ and $s$ are complex and where the notation used for the coefficient of $z^n$ is self-explanatory. It reduces to the polynomial expansion of $(1+z)^s$ for $s \in \mathbb{N}$ and it was in extrapolating this case and that where $2s$ is integer that Newton was led to his series, by a process more akin to divination than to standard mathematics. Here $u_3/u_2 = (s-2)z/3$, and, more generally,

$$u_{n+1}/u_n = (s-n)z/(n+1).$$

The ratio thus tends to $-z$, whence it follows that Newton's series *converges absolutely for $|z| < 1$ and diverges for $|z| > 1$*. And for $|z| = 1$? More difficult; see Chap. VI, hypergeometric series.

The convergence of a *sequence* $(u_n)$ of complex numbers being equivalent to that of the *series* $\sum(u_n - u_{n+1})$, any convergence criterion applicable to series will provide a criterion for sequences. The most obvious follows from Theorem 4 for series with positive terms:

**Theorem 11.** *For a sequence* $(u_n)$ *of complex numbers to converge it* suffices *that* $\sum |u_n - u_{n+1}| < +\infty$.

This theorem applies principally in the case where one has an estimate of the form $|u_n - u_{n+1}| < Mq^n$ with $0 < q < 1$, or else $< M/n^k$ with $k > 1$, where $M$ is a positive constant.

As an illustration, consider an interval $I \subset \mathbb{R}$, a map $f : I \longrightarrow I$, and let us set ourselves to solve the equation

$$(16.9) \qquad\qquad f(x) = x$$

in $I$. The *method of iteration* consists of choosing an $x_0 \in I$ arbitrarily, and then examining the points

$$(16.10) \qquad\qquad x_1 = f(x_0), \qquad x_2 = f(x_1), \ldots$$

The figures below demonstrate what can happen.

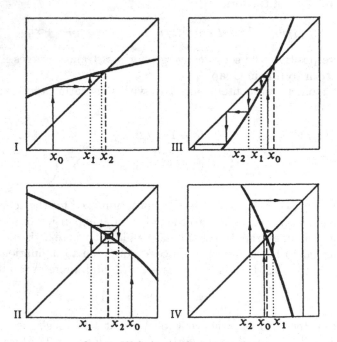

**fig. 5.** Walter, *Analysis* 1, p. 314

**Theorem 12.** *Let $I$ be a closed interval, $f$ a map from $I$ to $I$, and assume that there exists a positive number $q < 1$ such that*

$$(16.11) \qquad |f(x) - f(y)| \le q|x - y|$$

*for all $x, y \in I$. Then the equation $f(x) = x$ has one and only one solution in $I$ and the sequence (10) converges to it for any $x_0$.*

Indeed, put $M = |x_1 - x_0|$ and apply (11) repeatedly, replacing $x$ and $y$ by $x_p$ and $x_{p-1}$; we obtain

$$|x_n - x_{n-1}| \le q|x_{n-1} - x_{n-2}| \le \ldots \le q^{n-2}|x_2 - x_1| \le Mq^{n-1}.$$

Since $q < 1$, the sequence $(x_n)$ converges to a limit $x$, by Theorem 11. Since $|f(x) - f(x_n)| \le q|x - x_n|$, the sequence with general term $f(x_n)$ converges to $f(x)$. But $f(x_n) = x_{n+1}$ converges to $x$. Therefore $f(x) = x$.

If $y$ also satisfies $f(y) = y$, one has

$$|x - y| = |f(x) - f(y)| \le q|x - y|$$

and thus $x = y$ since $q < 1$, qed.

In practice, the theorem is applied to functions possessing a derivative $f'(x)$ everywhere, and such that $|f'(x)| \le q$ for any $x$; the mean value theorem (Chap. III, n° 16) shows that for all $x$ and $y$, there exists a $z$ between $x$ and $y$ such that

$$f(x) - f(y) = (x - y)f'(z),$$

whence (11). There are many other methods of approximating roots of a equation; the first truly effective one was discovered by Newton around 1665 and is in widespread use.

The idea of the method of iteration seems first to have appeared among the Arabs in the IX$^{\text{th}}$ century[42] *à propos* the equation

$$u - e.\sin u = \omega t \qquad (0 < e < 1)$$

which appears in the elliptic motion of the planets and is attributed to Kepler. The method consists of choosing the function

$$f(u) = e.\sin u + \omega t$$

and $I = \mathbb{R}$; since one has always $|\sin x - \sin y| \le |x - y|$ – this is obvious from the graph of the function sine – even without the mean value theorem, the relation $0 < e < 1$ allows one to apply this method.

---

[42] Unless some day it is discovered, for example, that the Indians knew it before the Arabs. The questions of priority being sometimes difficult to unravel even when it concerns the XX$^{\text{th}}$ century, it is strongly advisable to show prudence in what concerns times for which we have only very fragmentary information, for lack of which a militant in a cause courts the risk of one day finding himself the biter bit.

## 17 – Infinite limits

When a sequence of numbers $u_n$ does not converge, it may happen that as $n$ increases indefinitely, the $u_n$ vary in an irregular way (the case of the sequence $\sin n$ for example), or that they increase, or decrease, indefinitely, in other words "tend to $+\infty$ or to $-\infty$". The aim of this n° is to study this case summarily – experience is more helpful than theorems in situations of this type – showing which algebraic operations on sequences lead to predictable results. We shall consider only sequences of real numbers in what follows.

First, we shall write

$$\lim u_n = +\infty$$

when, for any number $A$, one has $u_n > A$ for $n$ sufficiently large; the relation

$$\lim u_n = -\infty$$

has a similar meaning, except that one has $u_n < A$ for $n$ sufficiently large. Obviously one tends to choose "very large" positive numbers for $A$ in the first case (or very large negative numbers in the second) since if one can check that $u_n > 10^{100\,000}$ for $n$ sufficiently large, there is no point in taking the trouble to check that $u_n > -10^{123}$ too ... But, strictly speaking, the definition imposes no hypothesis on $A$.

Consider for example an increasing sequence $(u_n)$; to say that, for any $A$, one has $u_n > A$ for $n$ large is clearly equivalent to saying that the sequence is not majorised, in other words that it diverges. One might therefore say that *every increasing sequence converges, possibly to $+\infty$*.

This remark also allows one to write

$$\sum u_n < +\infty$$

to express the fact that a series with *positive* terms converges and to attribute to it the sum $+\infty$ in the opposite case; one can similarly express absolute convergence by writing that

$$\sum |u_n| < +\infty.$$

These are pure conventions of language or of expression and are not to be confused with theorems; reread Hardy, end of n° 2. One can apply them to unordered series $\sum_{i\in I} u(i)$.

The rules of calculus that apply to finite limits extend to a certain extent to infinite limits too.

(I)    If $\lim u_n = +\infty$ and $v_n \geq u_n$ for $n$ large, then $\lim v_n = +\infty$. Evident.
(II)   If $\lim u_n = +\infty$, then $\lim c.u_n = +\infty$ or $-\infty$ as $c$ is $> 0$ or $< 0$. Evident.
(III)  The relation $\lim u_n = +\infty$ implies $\lim 1/u_n = 0$. Evident. The converse is true if $u_n > 0$ for $n$ large. This comes from the fact that, if $u_n$ is $> 0$ for $n$ large, the relation $1/u_n < r$ is equivalent to $u_n > 1/r$.

(IV)  If $\lim u_n = +\infty$ and if $v_n$ tends to a finite limit or to $+\infty$, then $\lim(u_n + v_n) = +\infty$. It is even enough that the sequence $(v_n)$ be bounded below: if indeed $v_n > B$ for all $n$, the relation $u_n + v_n > A$ is satisfied once $u_n > A - B$, so for $n$ large.

On the other hand one can say nothing when $u_n$ and $v_n$ tend respectively to $+\infty$ and $-\infty$, as trivial examples show: $n^2$ tends to $+\infty$, $-n$ tends to $-\infty$, but $n^2 - n$ tends to $+\infty$; $n^2$ tends to $+\infty$, $-2n^2$ tends to $-\infty$, but $n^2 - 2n^2$ tends to $-\infty$; $n$ tends to $+\infty$, $-n+\sin n$ tends to $-\infty$, but $\sin n$ has no limit. Whence a great mystery which has mystified a number of mystics meditating on the infinite: *the difference* $(+\infty) - (+\infty)$, which some write $\infty - \infty$, is *meaningless* even though one can always claim that the relations

$$(+\infty) - 10^{100\,000\,000} = +\infty, \quad (+\infty) + (+\infty) = \infty,$$
$$(-\infty) + (-\infty) = -\infty, \quad (+\infty) - (-\infty) = +\infty$$

have a meaning and are even correct if one states precisely what they mean: back to Hardy again . . .

(V)  If $\lim u_n = +\infty$ and if $v_n$ tends to $+\infty$ or to a *strictly* positive limit, then $\lim u_n v_n = +\infty$. Indeed, there is a number $m > 0$ such that one has $v_n > m$ for $n$ large, so $u_n v_n > m.u_n$ since all are $> 0$ for $n$ large, and it remains to apply the rules (I) and (II) – or we may write that $u_n > A/m$ implies $u_n v_n > A$.

Part (V) remains valid, with a change of sign in the limit, if $v_n$ tends to $-\infty$ or to a finite strictly negative limit.

If on the other hand $v_n$ tends to 0, anything may happen: $n^2.1/n$ tends to $+\infty$; $n^2.1/3n^2$ tends to $1/3$; $n^2.1/n^3$ tends to 0; and $n^2.\sin n/n^2$ has no limit. Whence a new mystery: *the product* $0.\infty$ *is meaningless*, as clearly is the so-called quotient $\infty/\infty$, as shown by the preceding examples. One cannot deduce anything without imposing much more precise hypotheses on the behaviour of the sequences $u_n$ and $v_n$ as $n$ increases indefinitely. This is one of the aims of the theory of asymptotic expansions of Chap. VI.

## 18 – Unconditional convergence: associativity

As we saw in n° 12 *à propos* the sum $\sum 1/(m^2+n^2)^{k/2}$, it can be helpful, when determining whether such an expression converges, to regroup the terms, though not by a simplistic procedure, as described in n° 12, where one groups in their natural order the terms of a series indexed by $\mathbb{N}$. We shall now show that this type of operation is allowable unrestrictedly when applied to sums which converge unconditionally.

Let us start from a family $(u(i))$, $i \in I$, where $I$ is countable. To group the terms of the sum $\sum u(i)$ is to construct a *partition*

(18.1)
$$I = \bigcup_{j \in J} I_j$$

of $I$ into pairwise disjoint sets $I_j$ (so as to avoid repeating the terms of the given family several times), indexed by a set $J$, which must be finite or countable like $I$ and the $I_j$, and to calculate the sum $s(I)$ of all the $u(i)$ by adding the partial sums $s(I_j)$ corresponding to the various sets $I_j$. Associativity can then be expressed as follows:

**Theorem 13.** *Let $(u(i))$, $i \in I$, be a family of complex numbers indexed by a countable set $I$ and let $I = \bigcup I_j$, $j \in J$, be a partition of $I$. For the given family to converge unconditionally it is necessary and sufficient that the following conditions be satisfied:*

*(i) each of the partial families $(u(i))$, $i \in I_j$, converges unconditionally;*
*(ii) the family of sums*

(18.2)
$$S(I_j) = \sum_{i \in I_j} |u(i)|$$

*converges unconditionally.*
    *If these conditions are satisfied each of the sums*

(18.3)
$$s(I_j) = \sum_{i \in I_j} u(i)$$

*converges unconditionally, the sum of the $s(I_j)$ converges unconditionally, and*

(18.4)
$$s(I) = \sum_{i \in I} u(i) = \sum_{j \in J} s(I_j) = \sum_{j \in J} \left( \sum_{i \in I_j} u(i) \right).$$

In what follows we shall use the letter $s$ to denote the partial sums of the family of the $u(i)$ and the letter $S$ to denote those of the family of the $|u(i)|$.
    To prove the necessity of (i), one remarks that the partial sums $S(F)$ corresponding to all the finite subsets $F$ of $I$ are bounded above, by the definition of unconditional convergence of a sum with positive terms. Similarly for $S(F)$ for those $F$ contained in a given $I_j$ or more generally in any subset $E$ of $I$; but this is precisely the condition for the sum $u(i)$, $i \in E$, to converge unconditionally (n° 15), whence (i).
    To prove that the family of partial sums (2) or (3) is summable, it suffices to do this for (2) since

$$|s(I_j)| \leq S(I_j)$$

(n° 15, Theorem 7). Let $G$ be a finite subset of $J$, and for each element $j$ of $G$ choose a finite set $F_j \subset I_j$ and let $F$ be the union, finite, of these $F_j$. Since the $F_j$ are, like the $I_j$, pairwise disjoint, we have

$$(18.5) \qquad \sum_{j \in G} S(F_j) = S(F) \le S(I);$$

on the other hand one can, for all $r > 0$, choose each of $n = \text{Card}(G)$ subsets $F_j$ so that $S(F_j) \ge S(I_j) - r/n$; for such a choice one clearly has

$$(18.6) \qquad \sum_{j \in G} S(I_j) \le \sum_{j \in G} S(F_j) + r \le S(I) + r;$$

the right hand side being independent of the finite set $G \subset J$, so the sum of $S(I_j)$ and *a fortiori* that of $s(I_j)$ converges unconditionally, which proves the necessity of (ii).

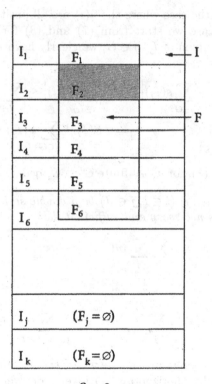

fig. 6.

It remains to prove the identity (4), i.e. the associativity formula. Let us do this, first assuming that the $u(i)$ are positive, in which case the sums denoted by $s$ and $S$ are identical.

The relation (6) already shows that

$$(18.7) \qquad \sum_{i \in J} s(I_j) \le s(I).$$

To establish the opposite inequality it suffices to show that, for all $r > 0$, one can find a finite subset $G$ of $J$ such that

(18.8) $$\sum_{j \in G} s(I_j) \geq s(I) - r.$$

To do this we choose a finite subset $F$ of $I$ such that $s(F) \geq s(I) - r$ and let $G$ be the set, clearly finite, of $j \in J$ such that $F_j = F \cap I_j$ is nonempty. Since the $I_j$, and so the $F_j$, are pairwise disjoint, we have

$$s(I) - r \leq s(F) = \sum_{j \in G} s(F_j) \leq \sum_{j \in G} s(I_j) \leq \sum_{j \in J} s(I_j),$$

which proves (4) in the case where the $u(i)$ are all positive.

In the general case we start from (7) and (8) for the series $\sum |u(i)|$; denoting by $I_G$ the union of $I_j$, $j \in G$, we clearly have

$$\left| s(I) - \sum_{j \in G} s(I_j) \right| = \left| \sum_{i \notin I_G} u(i) \right| \leq \sum_{i \notin I_G} |u(i)|$$

$$= S(I) - \sum_{j \in G} S(I_j) \leq r,$$

and this is valid, as (8), for every finite $G' \supset G$, qed.

**Corollary.** *Let $\sum u(i,j)$ $(i \in I, j \in J)$ be a double series. For it to converge unconditionally it is necessary and sufficient that*

(18.9) $$\sum_i \sum_j |u(i,j)| < +\infty;$$

*then*

(18.10) $$\sum_i \sum_j u(i,j) = \sum_j \sum_i u(i,j) = \sum_{(i,j) \in I \times J} u(i,j) = \text{etc.}$$

One applies the theorem in the case where the set of indices is $I \times J$, using the partition either by "horizontals" ($j$ given), or "verticals" ($i$ given), the etc. at the end of the preceding relation signifying any other partition of $I \times J$ one might use, reasonable or not. We leave it to the reader to generalise this himself to triple, quadruple series, etc. The most important case is clearly that where $I = J = \mathbb{N}$, but this is not the only one that arises in practice.

*Example 1.* Let us consider the sum

(18.11) $$\sum_{(m,n) \in \mathbb{Z}^2 - \{0\}} 1/(m^2 + n^2)^{k/2} = \sum u(m,n)$$

already studied in n° 12 again. A possible way of partitioning $I = \mathbb{Z}^2 - \{0\}$, where 0 denotes the pair $(0,0)$, consists of grouping all the terms for which $m$ has a given value, i.e. all the points of the lattice $\mathbb{Z}^2$ situated on the same vertical in the plane. Unconditional convergence then reduces to the relation

$$(18.12) \qquad \sum_{m \in \mathbb{Z}} \sum_{n \in \mathbb{Z}} 1/(m^2 + n^2)^{k/2} < +\infty$$

since all the terms are positive; it goes without saying that one excludes the pair $(0,0)$. Convergence as a series in $n$ is clear for $k > 1$ since, for $m$ given, one manifestly has $(m^2 + n^2)^{k/2} \asymp |n|^k$; but the sum, which we do not know, depends on $m$ in a way too little obvious for (12) to be usable here. One might also interchange the rôles of $m$ and $n$, i.e. reverse the order of summations in (12), but one would not advance further for all that . . .

A more natural partition consists of grouping all the pairs $(m, n)$ for which $m^2 + n^2$ has a given value $p \geq 1$. There are only finitely many such, since then $|m|, |n| \leq p$. Let $N(p)$ this number; the corresponding partial sum is clearly equal to $N(p)/p^{k/2}$, so instead of verifying (12) one might instead attempt to verify that

$$(18.13) \qquad \sum N(p)/p^{k/2} < +\infty,$$

which brings us back to the case of a simple series; for this method to be usable one would need to know, if not the exact value, at least an order of magnitude of $N(p)$ for $p$ large. This is unfortunately not obvious[43]. Instead of considering the pairs $(m, n)$ lying on a given circumference with centre 0, one could also group those for which $|m| + |n|$ has a given value $p$, whence another translation of unconditional convergence:

$$(18.14) \qquad \sum_{p=1}^{\infty} \sum_{|m|+|n|=p} u(m, n) = \lim_{p \to \infty} \sum_{|m|+|n| \leq p} u(m, n) < +\infty.$$

One might imagine all sorts of other more or less bizarre groupings of terms, for example grouping together the terms situated on the same equilateral hyperbola $m^2 - n^2 = p$, but such a grouping would hardly be very well adapted to the situation.

As we saw in n° 12 by using *ad hoc* arguments, version (14) is the best adapted to the problem. The terms for which $|m| + |n|$ has a given value $p$ all have the same order of magnitude, namely $1/p^k$, and there are $4p$ of them. One is thus led to the Riemann series $\sum 1/p^{k-1}$, whence the condition $k > 2$ for convergence.

---

[43] The calculation of $N(p)$ is a very interesting problem in number theory. The first remark to make is that $N(p)$ is frequently zero. In fact, $N(p) = 4(n'_p - n''_p)$ where $n'_p$ (resp. $n''_p$) is the number of divisors of $p$ of the form $4k + 1$ (resp. $4k + 3$).

*Example 2 (multiplication of absolutely convergent series).* Consider two absolutely summable families $(u_i)$, $i \in J$ and $(v_j)$, $j \in J$ and the family $(u_i v_j)$ of products, indexed by the Cartesian product $I \times J$; this is again absolutely summable and

$$\sum_{(i,j) \in I \times J} u_i v_j = \sum_{i \in I} u_i . \sum_{j \in J} v_j$$

as in ordinary algebra. This follows immediately from the corollary above, since on putting $S = \sum |v_j|$, the sum of the $|u_i v_j|$ for given $i$ is equal to $S|u_i|$, the general term of a series which is absolutely convergent by hypothesis. We shall return to this example in n° 22 *à propos* power series, but the following example will immediately give us another application.

*Example 3 (convolution product on a discrete group).* Consider the set $L^1(\mathbb{Z})$ of functions $f : \mathbb{Z} \longrightarrow \mathbb{C}$ such that

$$(18.15) \qquad \|f\|_1 = \sum |f(n)| < +\infty$$

and, for two such functions, let us define their *convolution product* $f \star g = h$ by the formula

$$(18.16) \qquad h(n) = \sum f(n - p)g(p) = f \star g(n)$$

where $f \star g(n)$ denotes the value at $n$ of the function $f \star g$. The series converges unconditionally since (15) shows that the numbers $|f(n)|$ are bounded above – in fact, they tend to 0 at infinity –, so up to a constant factor the general term of (16) is, in modulus, smaller than that of the series (15) for the function $g$.

If one performs the permutation $p \mapsto n - p$ on the terms of (16), which transforms $n - p$ to $p$ and does not change the sum of the series, one finds $h(n) = \sum f(p)g(n - p)$, whence commutativity of the product:

$$(18.17) \qquad f \star g = g \star f.$$

We now show that $f \star g \in L^1(\mathbb{Z})$ and even that

$$(18.18) \qquad \|f \star g\|_1 \leq \|f\|_1 . \|g\|_1.$$

For this, consider the double series with general term $f(p)g(q)$, indexed by $\mathbb{Z} \times \mathbb{Z}$. By Example 2, it converges unconditionally and

$$(18.19) \qquad \sum |f(p)g(q)| = \sum |f(p)| \sum |g(q)| = \|f\|_1 . \|g\|_1.$$

Now one can regroup the terms of the series $\sum |f(p)g(q)|$ as one wishes, and, for example, group them according to the value of the integer $p + q = n$. One finds then that

$$\sum |f(p)g(q)| = \sum_n \sum_p |f(p)g(n - p)|;$$

and since
$$|f \star g(n)| \leq \sum_p |f(p)g(n-p)|,$$

one deduces that the series $\sum |f \star g(n)|$ converges unconditionally and that

$$\sum |f \star g(n)| \leq \sum |f(p)g(q)|,$$

whence (18) in view of (19).

Since the convolution product does not take one out of the set $L^1(\mathbb{Z})$ where it is defined, one can (and one must ...) ask oneself whether it is *associative*, i.e. whether

(18.20) $$f \star (g \star h) = (f \star g) \star h$$

for any $f, g, h \in L^1(\mathbb{Z})$. To show this, remark first that (16) can be rewritten as

(18.21) $$f \star g(s) = \sum_{q+r=s} f(q)g(r).$$

Thus

$$
\begin{aligned}
f \star (g \star h)(n) &= \sum_{p+s=n} f(p)g \star h(s) = \sum_{p+s=n} f(p) \sum_{q+r=s} g(q)h(r) = \\
&= \sum_{p+s=n} \sum_{q+r=s} f(p)g(q)h(r);
\end{aligned}
$$

but the triple series $\sum f(p)g(q)h(r)$ converges unconditionally, and the expression we have just obtained is, precisely, by the associativity theorem, its partial sum over the system of all integers $(p, q, r)$ such that $p + q + r = n$, a partial sum in which we have grouped together the terms for which $q + r$ has a given value $s$. In other words,

(18.22) $$f \star (g \star h)(n) = \sum_{p+q+r=n} f(p)g(q)h(r).$$

A similar calculation provides the same result for $(f \star g) \star h(n)$, but one can also argue in the reverse direction, and, in (22), group together all the terms for which $p + q$ has a value given $s$; one finds

$$\sum_{r+s=n} h(r) \sum_{p+q=s} f(p)g(q) = \sum_{r+s=n} h(r)f \star g(s) = (f \star g) \star h(n),$$

whence associativity.

It is obvious that the sum of two $L^1(\mathbb{Z})$ functions is again in $L^1(\mathbb{Z})$ and that the convolution product is distributive with respect to addition:

$$(f + g) \star h = f \star h + g \star h.$$

In other words, $L^1(\mathbb{Z})$ is a *commutative ring*, to use the term from algebra. This ring possesses a unit element, namely the function

$$e(n) = 1 \text{ if } n = 0, \quad = 0 \text{ if } n \neq 0$$

as a trivial calculation shows.

The ring $L^1(\mathbb{Z})$ arises in the theory of absolutely convergent Fourier series (Chap. VII). There one considers series of the form

$$(18.23) \qquad \sum f(n)u^n = \hat{f}(u)$$

where $f \in L^1(\mathbb{Z})$ and where $u$ is a complex variable such that $|u| = 1$, so that the series converges absolutely. If $f, g \in L^1(\mathbb{Z})$, then

$$\hat{f}(u)\hat{g}(u) = \sum_{p,q \in \mathbb{Z}} f(p)g(q)u^{p+q}$$

and, by grouping the terms for which $p + q$ has a given value $n$,

$$\hat{f}(u)\hat{g}(u) = \sum_{n} u^n \sum_{p+q=n} f(p)g(q);$$

whence the formula

$$(18.24) \qquad \hat{f}(u)\hat{g}(u) = \widehat{f \star g}(u).$$

A famous theorem of Norbert Wiener states that, if the function (23) is $\neq 0$ for every $u$, then $1/\hat{f}(u) = \hat{g}(u)$ for some $g \in L^1(\mathbb{Z})$.

These calculations can be generalised considerably. To do this one needs to know what a *group* is, namely a set $G$ in which one has defined a "multi-plication" $G \times G \longrightarrow G$, generally denoted $(x, y) \mapsto xy$, which has to satisfy the three following conditions: (i) associativity, i.e. $(xy)z = x(yz)$, (ii) the existence of a "neutral element" $e$ such that $ex = xe = x$ for all $x$, (iii) the existence, for all $x$, of an "inverse" $y$ such that $xy = yx = e$ (one writes $x^{-1}$ for this). One does not assume commutativity $xy = yx$; when this is satisfied, one often writes $x + y$ for the product $xy$ ("sum") and $-x$ for the inverse ("opposite") of $x$; such is the case in $\mathbb{Z}$, in the additive groups $\mathbb{Q}$, $\mathbb{R}$, $\mathbb{C}$, of the Cartesian spaces $\mathbb{R}^p$, etc. The sets $\mathbb{Q}^*$, $\mathbb{R}^*$ and $\mathbb{C}^*$, endowed with their usual multiplication, are also commutative groups, similarly the set $\mathbb{T}$ of complex numbers $u$ such that $|u| = 1$ (the unit circle in $\mathbb{C}$). The set $\mathbb{Z}$, with 0 removed, and endowed with the usual multiplication, is not a group: mathematics would be appreciably simplified (no more of rational numbers nor of real numbers, no more of convergence, etc.) if one could find an $x \in \mathbb{Z}$ such that $2x = 1$.

The simplest example of a noncommutative group is obtained by considering a set $X$ and the set $\mathfrak{S}(X)$ of bijective maps $X \longrightarrow X$, i.e. of permutations of $X$, the "product" of two such permutations being the composed map (Chap. I); we know that if $X$ is finite with $n$ elements then $\mathfrak{S}(X)$ possesses $n!$ elements. Linear algebra provides other examples of noncommutative groups: the group $GL_n(K)$ of *invertible* matrices $n \times n$ with coefficients in a field $K$, for example $\mathbb{Q}$, $\mathbb{R}$ or $\mathbb{C}$, or even with coefficients in a ring such as $\mathbb{Z}$, for example the sets of matrices

$$\begin{pmatrix} a & b \\ c & d \end{pmatrix} \text{ with } a, b, c, d \in \mathbb{Z}, \quad ad - bc = \pm 1,$$

the "orthogonal" matrices with coefficients in $K$, the "unitary" matrices with coefficients in $\mathbb{C}$, etc. There are countless interesting examples.

That said, consider an arbitrary group $G$ (not endowed with a "topology" or, as one says, "discrete") and denote by $L^1(G)$ the set of functions $f : G \longrightarrow \mathbb{C}$ such that

$$(18.25) \qquad \|f\|_1 = \sum_{x \in G} |f(x)| < +\infty.$$

For two such functions, put

$$(18.26) \quad f \star g(x) = \sum_{yz=x} f(y)g(z) = \sum_{z \in G} f(xz^{-1})g(z) = \sum_{y \in G} f(y)g(y^{-1}x)$$

as in (21). *Everything* we have said about the convolution product in $\mathbb{Z}$ extends to this case – except the formula $f \star g = g \star f$ if $G$ is not commutative – and with exactly the same proofs, since, for $a \in G$ given, the maps $x \mapsto ax$ and $x \mapsto xa$ are *permutations* of the set $G$. For example, to prove the associativity of the convolution product, one starts from the series

$$\sum_{xyz=u} f(x)g(y)h(z)$$

extended over all the systems of elements of $G$ such that $xyz$ has a given value $u \in G$; it converges unconditionally since it is a partial sum of a product of three absolutely convergent series; one can therefore group the terms *ad libitum*. If you group them according to the value of the product $yz$ you find the value at the point $u \in G$ of $f \star (g \star h)$; if you group them according to the value of the product $xy$ you find the value at $u$ of $(f \star g) \star h$, qed.

Moral: *unconditional convergence allows one to apply the rules of algebra valid for finite sums to infinite sums also*. There are no problems of convergence at all, since one knows in advance that everything converges.

# §3. First concepts of analytic functions

### 19 – The Taylor series

First recall that given an *open* set[44] $G \subset \mathbb{C}$ and a scalar function $f$ defined on $G$ one says that $f$ is *analytic* in $G$ if, for all $a \in G$, there is a series expansion

$$f(z) = \sum f_n(a)(z - a)^n$$

valid for $|z - a|$ sufficiently small, with coefficients $f_n(a)$ which necessarily depend on $a$ and on $f$. The first example of an analytic function is provided by the following theorem:

**Theorem 14.** *The sum of a power series $f(z) = \sum a_n z^n$ of radius of convergence $R > 0$ is analytic in the disc $D : |z| < R$.*

We need to show that, for all $a \in D$, the sum $f(z)$ can be represented by a power series in $z - a = h$ on a neighbourhood of $a$ and even, as we shall see, for $|h| < R - |a|$, i.e. in *the largest disc with centre $a$ contained in $D$.*

Let us first calculate formally, using, to simplify the calculations, the divided powers $z^{[n]} = z^n/n!$ as in n° 1. Now every power series can be written in the form

$$(19.1) \qquad f(z) = \sum c_n z^{[n]} = c_0 + c_1 z/1! + c_2 z^2/2! + \ldots$$

with scalar coefficients $c_n$. Using the binomial formula of n° 1, one has, if $|a|$ and $|a + h|$ are $< R$,

$$
\begin{aligned}
f(a + h) &= \sum c_n (a + h)^{[n]} = \sum_{n \in \mathbb{N}} c_n \sum_{0 \leq p \leq n} a^{[n-p]} h^{[p]} = \\
&= \sum_{(p,n) \in I} c_n a^{[n-p]} h^{[p]}
\end{aligned}
$$

where $I \subset \mathbb{N} \times \mathbb{N}$ is the set of pairs $(p, n)$ such that $0 \leq p \leq n$. Theorem 13 shows that, *if* this sum converges unconditionally, then one can group its terms arbitrarily. It is natural to consider the partition $(I_p)$ of $I$, where, for each $p \in \mathbb{N}$, $I_p$ denotes the set of pairs $(n, p)$, $n \geq p$: the partial sum will be the product of $h^{[p]}$ by the power series $\sum_{n \geq p} c_n a^{[n-p]} = f_p(a)$ and $f(z) = f(a + h)$ will thus, for given $a$, be a power series in $h = z - a$.

To prove unconditional convergence of the sum that we have just obtained we use another partition $(J_n)$ of $I$, that obtained by grouping together all the pairs $(n, p)$ for which $n$ (and no longer $p$) has a given value. To be able to apply Theorem 13 we need to verify that if one replaces the general term

---

[44] Recall the definition: for all $a \in G$ there is an open disc $B(a, r)$ such that $B(a, r) \subset G$.

in the preceding sum by its absolute value then (i) for fixed $n$, the sum over $p$ converges (obvious: it is a finite sum), (ii) the sum over $n$ of the sums over $p$ converges. But the sum over $p$ is obviously equal to $|c_n| (|a| + |h|)^{[n]}$ – it suffices to retrace the calculation –, so it all reduces to proving the convergence of the series

$$\sum |c_n| (|a| + |h|)^{[n]} = \sum |c_n| u^{[n]}$$

where $u = |a| + |h|$. By definition of the radius of convergence $R$ this is the case for $|u| < R$, i.e. for $|h| < R - |a|$.

Since the sum considered converges unconditionally, one can calculate it according to the partition $(I_p)$, in other words permute the order of summations relative to $n$ and $p$, qed.

Writing $z$ for what we have just denoted by $a$, we see that, subject to the conditions $|z| < R$ and $|h| < R - |z|$, we have a relation of the form

$$(19.2) \quad f(z + h) = \sum_{p \geq 0} f_p(z) h^{[p]} = f_0(z) + f_1(z)h/1! + f_2(z)h^2/2! + \dots$$

with coefficients $f_0(z) = f(z)$ and, for $p > 0$,

$$(19.3) \qquad f_p(z) = \sum_{n \geq p} c_n z^{[n-p]} = c_p + c_{p+1}z/1! + c_{p+2}z^2/2! + \dots.$$

Replacing $n$ by $n + p$, this relation can be written as

$$(19.3') \qquad\qquad f_p(z) = \sum_{n \geq 0} c_{n+p} z^{[n]},$$

which exhibits how these series are formed starting from the given series $\sum c_n z^{[n]}$: one shifts the coefficients $c_n$ to the left.

These formulae can be interpreted in a much more striking way. We showed in n° 4, equation (4.8), that, in $\mathbb{C}$, the function $z^{[n]}$ possesses a derivative (in the complex sense) equal to $z^{[n-1]}$, so has a $p^{\text{th}}$ derivative equal to $z^{[n-p]}$. The series (3) may thus be obtained from the initial power series by replacing each of its monomials by its $p^{\text{th}}$ derivative; in other words, one passes from $f$ to $f_p$ by applying the traditional rule of calculus for the derivatives of a *polynomial* to the *power series* $f(z)$. Another dangerous bend: do not generalise this to arbitrary series of differentiable functions on $\mathbb{R}$ ...

To clarify the situation, let us agree to call the series defined by the formula

$$(19.4) \qquad\quad f'(z) = \sum c_{n+1} z^{[n]} = c_1 + c_2 z + c_3 z^{[2]} + \dots$$
$$\text{if } f(z) = \sum c_n z^{[n]} = c_0 + c_1 z + c_2 z^{[2]} + \dots$$

the *derived series* of the power series, or, in more traditional notation,

$$(19.4') \qquad f'(z) \;=\; \sum na_n z^{n-1} = a_1 + 2a_2 z + 3a_3 z^2 + \cdots$$

$$\text{if } f(z) \;=\; \sum a_n z^n = a_0 + a_1 z + a_2 z^2 + a_3 z^3 + \cdots;$$

formally, one obtains it by replacing each monomial $z^n$ by its "derivative" $nz^{n-1}$, which makes the constant term $c_0$ disappear and replaces $z^{[n]} = z^n/n!$ by $z^{[n-1]} = z^{n-1}/(n-1)!$. The series $f'(z)$ is thus the term $f_1(z)$ in (3). Now the unconditional convergence argument used above shows that the series (3) converges unconditionally, i.e. absolutely, for all $z$ such that $|z| < R$ [replace $a$ by 0 and $h$ by $z$ in (2)], so we see that the derived series converges for $|z| < R$; it diverges for $|z| > R$ as does the given series by reason of the integer factors in (4'), which can only diminish its chances of converging. In other words, *the series $f$ and its derivative $f'$ have the same radius of convergence*[45]. Since, on the other hand, the function $f'(z)$ is, like $f(z)$, the sum of a power series, $f'$ is, like $f$, *analytic*, i.e. expandable as a power series in $z - a$ on a neighbourhood of any point $a$ of the disc of convergence $|z| < R$.

On iterating the law (4) of passage from $f$ to $f'$, one finds the successive derivatives defined or given by

$$f''(z) \;=\; \sum c_{n+2} z^{[n]} = c_2 + c_3 z/1! + c_4 z^2/2! + \cdots,$$

$$f'''(z) \;=\; \sum c_{n+3} z^{[n]} = c_3 + c_4 z/1! + c_5 z^2/2! + \cdots,$$

and, generally,

$$(19.5) \qquad f^{(p)}(z) = \sum_{n \geq 0} c_{n+p} z^{[n]} \text{ if } f(z) = \sum c_n z^{[n]}$$

or, if you prefer,

$$(19.5') \qquad f^{(p)}(z) \;=\; \sum_{n \geq p} n(n-1)\ldots(n-p+1)a_n z^{n-p}$$

$$\text{if } f(z) = \sum a_n z^n;$$

the coefficients which we denoted above by $f_p(z)$ are just the derived series $f^{(p)}(z)$ of $f$, so, returning to (2), one thus finally finds

$$(19.6) \quad f(z+h) = \sum_{p \geq 0} f^{(p)}(z) h^{[p]} = f(z) + f'(z)h/1! + f''(z)h^2/2! + \cdots$$

---

[45] Direct proof: if $|z| < R$, the radius of convergence, there exists an $q < 1$ such that $|a_n z^n| = O(q^n)$, whence $na_n z^{n-1} = O(nq^n)$, and it remains to verify that $\sum nq^n < +\infty$ for $q < 1$, which is obvious from the Cauchy and d'Alembert criteria.

for $|h| < R - |z|$, agreeing to put $f^{(0)} = f$. This is the famous formula for power series named after (Brook) *Taylor* (1715) even though it was more or less known to Newton by 1691, who did not exploit it further than his exponential series, as well as to Johann Bernoulli who published it, replacing $f$ by its primitive, in 1694 (M. Cantor, III, pp. 229 and 383); one can verify it very easily algebraically when $f$ reduces to a polynomial in $z$. If, in particular, in (6) one replaces $z$ by 0, $h$ by $z$ and $p$ by $n$, one finds *MacLaurin's formula*

$$(19.6')\quad f(z) = \sum f^{(n)}(0)z^{[n]} = f(0) + f'(0)z/1! + f'(0)z^2/2! + \ldots$$

We shall see later (Chap. V, n° 18) that there is a similar result, but with a finite number of terms and a "remainder" that one can evaluate, valid for functions of a *real* variable possessing derivatives (in the usual sense) up to a certain order, much more general than those which can be represented by power series.

Finally, the algorithm for passing from the series $f$ to the series $f'$ can be inverted; one thus obtains the *primitive series*

$$F(z) = c + a_0 z + a_1 z^2/2 + a_2 z^3/3 + \ldots$$

of $f$, with an arbitrary constant $c$, the same radius of convergence and the relation $F' = f$ which "justifies" the terminology to those who have learnt that the primitive, in the elementary sense, of the function $x^n$ is $x^{n+1}/(n+1)$; see n° 11.

As we have already said, Newton was the first to make systematic use of power series to solve all sorts of problems. The greatest mathematician of the following century, Euler, also used them freely. But neither Newton nor Euler elaborated any general theory of power series, and it was Lagrange (1736–1813) who, in his *Théorie des fonctions analytiques* (1797), founded it, as he said, to liberate analysis from infinitely small quantities, from limits, from "fluxions" (derivatives) *à la* Newton, etc. and to transform "infinitesimal" analysis into an "algebraic" analysis. He it was who introduced the notation $f'(x)$ and the word "derivative" for analytic functions; he proved the Taylor and MacLaurin formulae as we have done above, though without worrying too much about questions of convergence, nor working in $\mathbb{C}$. Lagrange's idea that all the useful or interesting functions are analytic, except at isolated points[46], was doomed to failure, but his theory prepared the ground for Cauchy and his

---

[46] This is false in general for functions of a real variable, no matter how differentiable they may be, because, as Cauchy was first to remark, the function $f(x)$ equal to $\exp(-1/x^2)$ for $x \neq 0$ and to 0 for $x = 0$ possesses at $x = 0$ successive derivatives of every order, all of which are zero, as we will show later; MacLaurin's formula would then show that $f$ is identically zero. In fact, on $\mathbb{R}$ one can construct indefinitely differentiable functions having arbitrarily given successive derivatives at $x = 0$ (Chap. V, n° 29).

theory of analytic functions of a *complex* variable, which would revolutionise all of classical analysis in the $XIX^{th}$ century.

We defined the "derived series" of a power series $f$ above by an algebraic algorithm. The terminology is justified by reason of the fact that $f'(z)$ is also the limit, in the sense of n° 4, of the ratio

$$\frac{f(z+h) - f(z)}{h} = f'(z) + h\left[f''(z)/2! + f'''(z)h/3! + \ldots\right]$$

when $h \in \mathbb{C}$ tends to 0. The expression between the brackets [ ] is indeed a convergent power series; its sum is thus, in modulus, majorised by a fixed number $M$ for $|h|$ sufficiently small (see the end of n° 14). The preceding relation thus shows that

(19.7)    $$\left|\frac{f(z+h) - f(z)}{h} - f'(z)\right| \leq M|h| \quad \text{or} \quad = O(h)$$

for $|h|$ sufficiently small, so $< r$ once $|h| < r' = r/M$; this is precisely the definition of a limit for functions defined on a subset of $\mathbb{C}$.

One can prove this result directly without the double series which we employed above. To do this one starts from the relation (4.9)

$$\left|(z+h)^{[n]} - z^{[n]} - hz^{[n-1]}\right| \leq |h|^{[2]}\left(|z| + |h|\right)^{[n-2]}.$$

On replacing $f(z)$ by $\sum c_n z^{[n]}$ in (7), $f(z+h)$ by the analogous expression in $z + h$, and $f'(z)$ by its definition (4), it is clear, since $h^{[2]}/h = h/2$, that the left hand side of (7) is majorised by

$$\frac{1}{2}|h| \sum |c_n| \left(|z| + |h|\right)^{[n-2]};$$

if one has proved directly that the series $f'$ and $f''$ have the same radius of convergence $R$ as $f$, then the series we have just obtained will converge for $|z| + |h| < R$ and one obtains the estimate (7) again.

This argument shows in passing that, on a neighbourhood of a given point $a$, a function $f$ cannot be represented by two different power series in $z - a$. If indeed $f(a+h) = \sum c_n(a)h^{[n]}$ for $|h|$ sufficiently small, one has first $f(a) = c_0(a)$; then the function $f'(z) = \lim[f(z+h) - f(z)]/h$ is necessarily given by the derived series $c_n(a)h^{[n-1]}$ as the preceding argument shows, whence $c_1(a) = f'(a)$. Iterating, one finds successively that $c_2(a) = f''(a)$, $c_3(a) = f'''(a)$, etc., which determines the coefficients of the series: these are the values at $a$ of the *functions* $f, f', f'', \ldots$ defined as the limits of quotients and no longer as the sums of the series obtained from the derivation algorithm, even though the results are the same.

This apparently banal property of analytic functions, the existence of derivatives in the sense of n° 4, has far-reaching consequences, as Cauchy was the first to see. Just for the moment let us note that it implies a simple relation between the derivatives of $f(x + iy) = f(x, y)$ with respect to the *real* variables $x$ and $y$. These are defined, as for every function of several real variables, by the formulae[47]

$$f'_x(x, y) = \lim \frac{f(x + u, y) - f(x, y)}{u},$$

(19.9)

$$f'_y(x, y) = \lim \frac{f(x, y + v) - f(x, y)}{v}$$

when $u$ and $v$ tend to 0 through *real* values: one fixes all the variables except the one with respect to which one differentiates; there is no need for a long discussion. But if, in (7), one lets $h$, *a priori* complex, tend to 0 through *real* values $h = u$, one finds the first limit in (9) since

$$f(z + h) = f(x + u, y),$$

whence $f'_x = f'(z)$. If on the other hand one lets $h$ tends to 0 through *purely imaginary* values $h = iv$ with $v$ real, the ratio (7) again tends to $f'(z)$, and in it one now has

$$f(z + h) = f(x, y + v),$$

so the ratio tends to $f'_y/i$ because of the factor $i$ in the denominator $iv$. On comparing these results one finds the relation

(19.10)                    $i f'_x = f'_y$      or    $i D_1 f = D_2 f,$

so

(19.10')                            $f' = D_1 f = -i D_2 f.$

Cauchy showed that this relation $D_2 f = i D_1 f$ *characterises* those functions $f(x, y)$ which are analytic as functions of $z$. Since we do not know this yet, and to avoid confusion, we shall use exclusively, up to Chap. VII, the word *holomorphic* to denote functions defined on *open* sets $U \subset \mathbb{C}$ and possessing *continuous*[48] partial derivatives $D_1 f$ and $D_2 f$ satisfying (10) there; every analytic function $f$ is thus holomorphic (the partial derivatives of $f$ are continuous because proportional to the analytic function $f'$), but we do not yet know that the converse is valid. Should the opportunity arise we shall prove

---

[47] The notation $f'_x$, or $\partial f/\partial x$, even though (or because ...) traditional, has the great drawback of using the same symbol for both the variable $x$ and for the symbol indicating that one is differentiating with respect to it. It would be much preferable to use the notation $f'_1$ and $f'_2$, or $D_1 f$ and $D_2 f$, as in fact is becoming more and more frequent, and as we shall do, in order to indicate that one differentiates with respect to the *first* and *second* variable, no matter what letters are used to denote them.

[48] A theoretically superfluous hypothesis, but no harm in adopting it.

theorems that apply in certain cases to holomorphic functions and, in others, to analytic functions. *After* Chap. VII, the adjectives "holomorphic" and "analytic" will become strictly synonymous for us, as they are for everyone, and the theorems already obtained will apply to both cases, for in reality there is only one case.

If one puts $f = u + iv$ where $u$ and $v$ are functions with real values, whence $D_1 f = D_1 u + i D_1 v$, etc., then (10) translates as the two formulae

$$(19.10") \qquad\qquad D_1 u = D_2 v, \qquad D_2 u = -D_1 v.$$

For example, the function $f(z) = \mathrm{Re}(z) = x$ is not holomorphic: here $u'_x = 1$ and $v'_y = 0$. The function $f(x,y) = x^2 y^3 + i x^4 y^2$ neither: one has $D_1 f(x,y) = 2xy^3 + 4ix^3 y^2$, $D_2 f'(x,y) = 3x^2 y^2 + 2ix^4 y$ and Cauchy's relation is clearly not satisfied. To speak of the derivative $f'(z)$ of such a function *is meaningless*, quite simply because the ratio $[f(z+h) - f(z)]/h$ which would define it has no limit when $h$ tends to 0 in $\mathbb{C}$; it converges when $h$ tends to 0 through real, or purely imaginary values, or even along a line with origin 0, etc., but these partial limits are generally different one from the other, while they would be identical for a holomorphic function[49]. The fact that a function of two real variables $x$ and $y$ can be considered as a function of $z = x + iy$ is a pure triviality: a complex number is a pair of real numbers. No matter how "regular" $f(x,y)$ may be as a function of the two real variables $x$ and $y$ it need not be an analytic function of $z$. Set theoretic sleight of hand has never proved any true theorem whatever: principle of conservation of intellectual energy in mathematics.

One can understand this difficulty by considering a polynomial function in $x$ and $y$. By definition this can be written as a finite sum

$$f(x,y) = \sum a(p,q) x^{[p]} y^{[q]}$$

with scalar coefficients $a(p,q)$. Clearly

$$D_1 f(x,y) = \sum a(p,q) x^{[p-1]} y^{[q]} = \sum a(p+1,q) x^{[p]} y^{[q]},$$
$$D_2 f(x,y) = \sum a(p,q) x^{[p]} y^{[q-1]} = \sum a(p,q+1) x^{[p]} y^{[q]}.$$

---

[49] Let $g(z)$ be a function defined for $z \neq 0$ and tending to a limit $u$ when $z \in \mathbb{C}$ tends to 0 to the sense of n° 4. Let $h(t)$ be a function defined on a neighbourhood of 0 in $\mathbb{R}$, with complex values, continuous at $t = 0$ and such that $h(0) = 0$, $h(t) \neq 0$ for $t \neq 0$. Then the composed function $g(h(t))$, defined for $t \neq 0$ sufficiently small, tends to $u$ when $t$ tends to 0. This follows from the general theorems of Chap. III, but can be verified immediately: for every $r > 0$, there is an $r' > 0$ such that $|z| < r'$ implies $|f(z) - u| < r$, then an $r' > 0$ such that $|t| < r''$ implies $|h(t)| < r'$, and so $|g(h(t)) - u| < r$, qed. This said, you can, in the ratio $[f(a+z) - f(a)]/z$ which defines the derivative at $a$ of an analytic function, replace $z$ by $h(t)$ and let $t$ tend to 0 in order to calculate the derivative $f'(a)$. For example, $h(t) = ct$ where $c \in \mathbb{C}$ is a constant (a line issuing from the origin), $h(t) = t(\cos t + i \sin t)$ (spiral), $h(t) = t(\cos 1/t + i \sin 1/t)$ (another spiral, which makes increasingly tight rotations around 0 when $t$ tends to 0), etc.

Cauchy's relation then forces

$$ia(p+1,q) = a(p,q+1)$$

since the coefficients of $x^p y^q$ must be the same in $if'_x$ and $f'_y$. But this relation implies

$$a(p,q) = ia(p+1,q-1) = i^2 a(p+2,q-2) = \ldots = i^q a(p+q,0).$$

If now one puts $a(n,0) = a_n$, one has $a(p,q) = i^q a_{p+q}$ and consequently

$$f(x,y) = \sum a_{p+q} x^{[p]}(iy)^{[q]} = \sum_n a_n \sum_{p+q=n} x^{[p]}(iy)^{[q]} = \sum_n a_n (x+iy)^{[n]}$$

by the algebraic binomial formula; consequently, $f(z) = \sum a_n z^{[n]}$. In conclusion, *a polynomial in $x$ and $y$ satisfies Cauchy's equation if and only if it is a polynomial in $z = x + iy$*, which proves Cauchy's Theorem in a case so particular as to be almost trivial. Even though any polynomial in $x$ and $y$ is a polynomial in $z$ and $\bar{z}$, since $x = (z+\bar{z})/2$, $y = (z-\bar{z})/2i$, in general it will not be a polynomial in $z$.

To obtain the Baccalaureat in mathematics, one learned in 1939 that the successive derivatives of the function $1/(1-x) = (1-x)^{-1}$ are

$$(1-x)^{-2}, \ 2(1-x)^{-3}, \ 2.3(1-x)^{-4}, \ldots, p!(1-x)^{-p-1}, \quad \text{etc.}$$

This is for functions of a real variable, but the Lord, though subtle, is not malicious, as Albert Einstein said in another context, and surely would not have decided that these formulae should no longer be valid in the complex domain; besides, the derivative $f'(z)$ of an analytic function of $z \in \mathbb{C}$ coincides, for $z \in \mathbb{R}$, with its derivative in the elementary sense of the term, principally because we have seen above that $f'_x = f'$; this allows one to calculate the complex derivatives of $1/(1-z)$ by differentiating the function $1/(1-x-iy)$ with respect to $x$, whence $(1-z)^{-2}$. We can then differentiate the formula

$$(19.11) \qquad (1-z)^{-1} = 1 + z + z^2 + z^3 + \ldots + z^n + \ldots (|z| < 1)$$

term-by-term *ad libitum* to obtain successively

$$(19.12) \quad (1-z)^{-2} \ = \ 1 + 2z + 3z^2 + 4z^3 + \ldots + (n+1)z^n + \ldots,$$
$$(19.13) \ \ 2(1-z)^{-3} \ = \ 2 + 2.3z + 3.4z^2 + \ldots + (n+1)(n+2)z^n + \ldots$$

and generally

$$(p-1)!(1-z)^{-p} = \sum (n+1)(n+2)\ldots(n+p-1)z^n = \sum (n+p-1)! \, z^{[n]},$$

whence, since $(n+p-1)! = (p-1)!p(p+1)\ldots(p+n-1)$, the formula

$$(19.14) \qquad (1-z)^{-p} = \sum p(p+1)\ldots(p+n-1)z^{[n]}.$$

On putting $-p = s$ and replacing $z$ by $-z$, we get

$$(19.15) \qquad (1+z)^s = \sum s(s-1)\ldots(s-n+1)z^{[n]}.$$

If $s$ were a positive integer the coefficient of $z^n$ would be zero for $n \geq s+1$ since it would contain $s - s = 0$ as a factor; the preceding formula would reduce to the algebraic binomial formula recalled in n° 1. The result obtained shows that, for $s$ a negative integer, it remains valid on condition:

(i) one writes its coefficients in the form above, and not in the form $s!/n!(s-n)!$: this is meaningless apart from the case where $s \in \mathbb{N}$ and Newton, who very happily did not know this – it would have led him down a blind alley –, never did so, contenting himself with the form (15) which has a meaning for any $s \in \mathbb{C}$,

(ii) one replaces the finite sum of algebra by an infinite series,

(iii) one does not forget that the formula obtained is valid only for $|z| < 1$, since otherwise the series is divergent as we saw in n° 16 *à propos* Newton's binomial series, which is identical to (15), except that, for Newton, the exponent $s$ must be rational.

The formula (15) is thus only Newton's series for $s \in \mathbb{Z}$. The real problem, more difficult, which we shall resolve in Chap. IV, is to prove (15) for all exponents $s \in \mathbb{R}$ or even $\mathbb{C}$, at least for real $z$ between $-1$ and $1$; the result which we have just obtained is a small first step in this direction.

Another method for establishing (15) would be to check that on multiplying the series (15) with exponent $s$ by $1 + z$ one obtains the series for exponent $s + 1$. This manifestly reduces to verifying the famous relation

$$(19.16) \qquad \binom{s}{p} + \binom{s}{p-1} = \binom{s+1}{p}$$

for the binomial coefficients, which, for $s \in \mathbb{N}$, explains the no less famous "Pascal triangle". One lends only to the rich: it was known in the West fully a century before Pascal, for example to the German Stifel (1486–1567) and the Italian Tartaglia (1500–1557), and to the Arabs, Indians and Chinese one or two centuries earlier; but of course they did not write such edifying *Pensées*. In fact, the contribution of Pascal – and of Fermat – is to have exhibited the binomial coefficients in relation to the calculus of combinations and permutations (whence the birth of the calculus of probabilities) and to have provided a correct proof, by induction, of the relation (16) for $s \in \mathbb{N}$.

To show the usefulness of MacLaurin's formula, let us *suppose* that the function $\sin x$ is, for $x \in \mathbb{R}$, the sum of a power series in $x$. It is then easy to calculate what the series must be by Maclaurin's formula: the successive derivatives of $\sin x$ are $\cos x$, $-\sin x$, $-\cos x$, etc. .... and their values at $x = 0$ are $1, 0, -1, 0, 1, 0$, etc.

Since $\sin 0 = 0$, the only power series that can possibly represent $\sin x$ is thus

$$(19.17) \qquad s(x) = x - x^3/3! + x^5/5! - \ldots = x - x^{[3]} + x^{[5]} - \ldots,$$

i.e. the one we stated in n° 6. All this, as one sees, is very coherent, which is generally a good sign in mathematics. But we have still not *proved* that this series really represents $\sin x$, for which reason we shall call its sum $s(x)$. And anyhow, what is $\sin x$? Until further notice, a sketch on a sheet of paper!

A similar calculation shows that the only power series that can represent $\cos x$ is

$$(19.18) \qquad c(x) = 1 - x^2/2! + x^4/4! - \ldots = 1 - x^{[2]} + x^{[4]} - \ldots$$

The two series that we have just obtained being manifestly convergent for any $x$, like the exponential series $\sum x^{[n]}$, one can calculate their first derived series by applying the general rule (4). The simplest of calculations then shows that

$$(19.19) \qquad\qquad s'(x) = c(x), \qquad c'(x) = -s(x),$$

which reinforces "coherence". We shall show in Chap. IV how to deduce all the properties of the trigonometric functions either from (17) and (18), or (19), or the addition formulae.

In 1693 Leibniz, who had begun to be interested in using series to integrate differential equations, i.e. to discover functions whose derivatives satisfy a given relation – a vast programme –, applied the method to the function $\sin x$ by using the fact that $\sin'' x + \sin x = 0$. If one supposes that $\sin x = \sum a_n x^n$, whence $\sin'' x = \sum n(n-1)a_n x^{n-2}$, one obtains the relation $a_n = a_{n-2}/n(1-n)$ which, bearing in mind the obvious formulae $a_0 = 0$ ($\sin 0 = 0$) and $a_1 = 1$ ($\sin x \sim x$ for $x$ small), immediately gives the result already found, but not made public, by Newton (M. Cantor, III, p. 198); Leibniz published it himself, which of course annoyed the Englishmen enormously.

Theorem 14 has an important consequence *à propos* the *zeros of an analytic function* , i.e. points $a$ where $f(a) = 0$; in algebra, one calls these the roots of the equation $f(z) = 0$. Let $f$ be an analytic function in an open set $G$ in $\mathbb{C}$ and, for an $a \in G$, return to the Taylor formula

$$f(a + h) = f'(a)h + f''(a)h^2/2! + \ldots$$

where one omits the term $f(a)$ since it is zero. Two cases are possible.

(i) *All* the derivatives $f^{(p)}(a)$ are zero. Clearly then $f(z) = 0$ for all $z$ such that $|z - a| < R$, where $R$ is the radius of the disc in which the preceding formula is valid.

(ii) There is an integer $p$ such that

$$f(a) = f'(a) = \ldots = f^{p-1}(a) = 0, \quad f^{(p)}(a) \neq 0.$$

One then says that $a$ is a *zero of order $p$* of $f$. The situation is the opposite to that of the case (i): there exists an open disc with centre $a$ in which $a$ is the only zero of $f$ *(principle of isolated zeros)*. Indeed

(19.20)    $$f(a+h) = f^{(p)}(a)h^{[p]}[1+?h+?h^2 + \ldots]$$

with scalar coefficients denoted by ? since their values are of little importance. For $|h|$ small, the sum $?h+\ldots$ is, in modulus, $\leq M|h|$ with a constant $M > 0$, since it is the product of $h$ by a convergent power series. Since there exists a number $r > 0$ such that one has $M|h| < 1/2$ for $|h| < r$, the sum between the brackets [ ] is in modulus $> 1/2$ for $|h| < r$, so that $f(a+h)$ cannot vanish in the disc $D(a,r)$, except at the point $a$, qed.

An immediate consequence of this result is that *if mutually distinct zeros of $f$ tend to a point $a \in G$, then $f$ vanishes on a neighbourhood of $a$*, since any disc with centre $a$ then contains infinitely many zeros of $f$.

## 20 – The principle of analytic continuation[50]

Case (i) above leads to an *a priori* curious result: if all the derivatives of $f$ vanish at $a$, then the function $f$ is *identically zero on $G$* – and not only on a neighbourhood of $a$ – subject to a "connectedness" hypothesis on $G$. This "rigidity" property of analytic functions is quite special to them; in $\mathbb{R}$ (or in any Cartesian space), a function as "smooth" as one wants, i.e. possessing derivatives of arbitrarily high order with respect to the *real* variables on which it depends, can very well vanish on an open set without being identically zero: this is the case for $f(x) = \exp(-1/x^2)$ for $x > 0$, $= 0$ for $x \leq 0$ (the existence of all the successive derivatives at 0 is not obvious; see Chap. IV, n° 5). Equivalent formulation: if two analytic functions $f$ and $g$ on an open connected $G$ coincide on the neighbourhood of a point $a$ of $G$ or, it comes to the same, have equal successive derivatives at $a$, then they are equal on all of $G$.

For let $f$ be an analytic function on $G$ and suppose that $f$ and all its derivatives vanish at $a \in G$; then, by Taylor's formula, $f$ vanishes on a neighbourhood of $a$. Let $b \in G$ and first assume that the line segment $[a,b]$ lies entirely in $G$. We shall show that $f(z) = 0$ on a neighbourhood of $b$ also. The segment $[a,b]$ is the set of complex numbers of the form

$$z(t) = a + t(b-a) = (1-t)a + tb, \qquad 0 \leq t \leq 1,$$

the points $a$ and $b$ corresponding to $t = 0$ and $t = 1$. Let $E \subset [0,1] = I$ be the set of $t$ possessing the following property: the function $f$ vanishes on a

---

[50] This n° will be called on only sporadically before Chap. VII and is not indispensable for the moment.

disc with centre $z(t)$. The set $E$ is not empty: it contains 0 since $z(0) = a$. Let $u \le 1$ be the least upper bound of $E$; we need to show that $u \in E$ and that $u = 1$.

First note that if $E$ contains a $t \in I$, it also contains all $t' \in I$ sufficiently close to $t$: if indeed $f$ vanishes on an open disc $D$ with centre $z(t)$, one has clearly (continuity) $z(t') \in D$ for $t'$ close to $t$ and consequently $f$ vanishes on every open disc with centre $z(t')$ contained in $D$.

If on the other hand a $t \in I$ is the limit of a sequence of points $t_n \in E$, then again one has $t \in E$, since in the opposite case the $t_n$ would be distinct from $t$ and this would contradict the principle of isolated zeros established at the end of the preceding n°. [We have just shown that $E$ is simultaneously open and closed in $I$.]

Let us now return to the least upper bound $u$ of $E$. This is the limit of points of $E$, so $u \in E$. Every $t \in I$ sufficiently close to $u$ therefore also belongs to $E$. If we had $u < 1$, we would then have a $t > u$ in $E$, impossible since $u = \sup(E)$. Thus $u = 1$, and since $z(u) = b$ the function $f$ vanishes on a neighbourhood of $b$ in $\mathbb{C}$, qed.

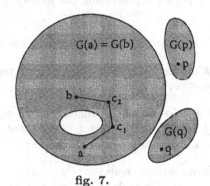

fig. 7.

This argument presumes that one can join $a$ to $b$ by a line segment entirely contained in $G$, which clearly is not always the case. But it remains valid if one can join $a$ to $b$ by a broken line entirely in $G$ and formed of a finite number of line segments $[a, c_1], [c_1, c_2], \ldots, [c_n, b]$: since $f$ vanishes on a neighbourhood of $a$, it does so on a neighbourhood of $c_1$ too; since it vanishes on a neighbourhood of $c_1$, it does so on a neighbourhood of $c_2$ too; etc.

The set $G(a)$ of $b \in G$ that can so be joined to $a$ in $G$ is *open*. $G$ contains a disc $D(b)$ with centre $b$, so that if one can join $a$ to $b$ in $G$ one can also join $a$ to all $c \in D(b)$ by adjoining the segment $[b, c]$ to the path going from $a$ to $b$. The set $G'(a)$ of $b \in G$ which one cannot join to $a$ in this manner is open too. For let $D(b)$ be a disc with centre $b$ contained in $G$; if one could join the point $a$ to a $c \in D(b)$ by a path in $G$, it would be enough to complete it with the segment $[c, b]$ to join $a$ to $b$, absurd. So $G'(a)$ contains $D(b)$.

To sum up, we have partitioned the open set $G$ as two disjoint open sets $G(a)$ and $G'(a)$, the first being nonempty since $a \in G(a)$. When $G(a) = G$, one says that the open set $G$ is *connected*. One can then, without leaving $G$, join any two points $b$ and $c$ of $G$ by a broken line: one goes first from $b$ to the point $a$ then from this to the point $c$. It is clear that this property characterises the open connected sets: one has $G(a) = G$ for any $a \in G$. In the general case, $G(a)$, which is clearly connected, is called *the connectedness component* of $a$ in $G$ or, if the point $a$ is not explicit, a connectedness component of $G$. Figure 7 shows an open set $G$ possessing three connectedness components. One can also define an open connected set as follows: it is impossible to partition $G$ as two open (disjoint, as in every partition) *nonempty* sets.

In order to formulate these results conveniently, it is useful to introduce the concept of an *isolated point* of a subset of $\mathbb{C}$: this means that there exists an open disc having this point as its centre and not containing any other point of the set than the point considered; we shall meet this concept again in Chap. III *à propos* limits. In particular this allows us to define the *isolated zeros* of an analytic function $f$: these are the isolated points of the set $Z(f)$ of zeros of $f$, i.e. the points $a$ possessing the following property: there exists an $r > 0$ such that

$$|z - a| < r \ \& \ f(z) = 0 \iff z = a.$$

It is now clear that at the end of the preceding n° we proved the following result:

**Theorem 15.** *Let $f$ be an analytic function on an open subset $G$ of $\mathbb{C}$ and let $a$ be a zero of $f$ in $G$. Then either $a$ is an isolated zero of $f$, or else $f$ vanishes on the connectedness component of $a$ in $G$.*

An equivalent formulation *assuming $G$ connected*: either all the points of $Z(f)$ are isolated, or else $f = 0$. If the function $f(z) = \operatorname{Re}(z)$ were analytic in $\mathbb{C}$, it would be identically zero since its zeros are manifestly not isolated. Note that the zeros of an analytic function in an open connected $G$ subset of $\mathbb{C}$ may all the same accumulate at the frontier of $G$; example: $G = \mathbb{C} - \{0\}$, $f(z) = \sin(1/z)$, which vanishes at all the points $1/n\pi$, which converge to $0 \notin G$. There are functions $f$ defined and analytic for $|z| < 1$ and such that there are infinitely many zeros of $f$ in the neighbourhood of every point of the boundary circle $|z| = 1$ (the "automorphic functions" of Henri Poincaré for example).

**Corollary (Principle of analytic continuation).** *Let $f$ and $g$ be two analytic functions on an open connected subset $G$ of $\mathbb{C}$. Suppose that there are mutually distinct points $a_n$ of $G$ converging to a limit $a \in G$ and such that*

$$f(a_n) = g(a_n)$$

*for all $n$. Then $f(z) = g(z)$ for any $z \in G$.*

It is clear that then the point $a$ is a non-isolated zero of the function $f - g$. It is important that the limit must belong to $G$.

Theorem 15 is trivially false for open nonconnected sets: take for $G$ the union of two disjoint open discs and for $f$ the function equal to 0 on the first and to 1 on the second.

Although the preceding results concern functions defined on an open subset of $\mathbb{C}$, they also apply to a class of functions defined on an interval $I$ of $\mathbb{R}$. Such a function $f$, with possibly complex values, is said to be *real-analytic* (or analytic for short if no confusion with the *complex-analytic* functions of n° 19 is possible) if, for all $a \in I$, it can be represented on an interval with centre $a$ by a power series in $x - a$; this is the same definition as in the case of $\mathbb{C}$, except that the variable takes only real values.

In truth, there is hardly any difference between the two concepts: for a function $f(x)$ defined in an interval $I \subset \mathbb{R}$ to be real-analytic it is necessary and sufficient that there exists an open subset $G$ in $\mathbb{C}$ containing $I$ and a function $g(z)$ defined and complex-analytic in $G$ such that

$$f(x) = g(x) \text{ for all } x \in I.$$

This condition is clearly sufficient. To show the necessity, one starts from the fact that, for all $a \in I$, there exists a convergent power power series $g_a(z) = \sum c_n(a)(z - a)^n$ and an open disc $D(a) \subset \mathbb{C}$ with centre $a$ such that (i) $g_a$ converges in $D(a)$, and possibly elsewhere, (ii) $g_a(x) = f(x)$ for all $x \in I \cap D(a)$. Now let $G$ be the union of all the open discs $D(a)$. This is clearly an open subset of $\mathbb{C}$. It is connected: if $u \in D(a)$ and $v \in D(b)$, one can join $u$ to $v$ in $G$ by following the ray $[ua]$, the segment $[ab]$ of $I$ and the ray $[bv]$ of $D(b)$. If, moreover, $D(a)$ and $D(b)$ intersect, then we have

$$(20.1) \qquad\qquad g_a(z) = g_b(z)$$

in $I \cap D(a) \cap D(b)$, an interval of $\mathbb{R}$ not reducing to a point. Since $D(a) \cap D(b)$ is connected the principle of analytic continuation then shows that (1) is true in all of $D(a) \cap D(b)$.

Having done this one can define the function $g$ on $G$ by putting $g(z) = g_a(z)$ for all $a \in I$ such that $z \in D(a)$. Even though the point $a$ can be chosen in many different ways for a given $z$, this definition is not ambiguous because of[51] (1). It is clear that $g$ is complex-analytic in each $D(a)$, so on $G$, and that $g = f$ on $I$, qed.

---

[51] More generally: let $E$ be a set, and $(E_i)$ any family of subsets of $E$ having union $E$, and, for each $i$, let $g_i$ be a function with arbitrary values defined on $E_i$. For there to exist a function $g$ on $E$ such that $g = g_i$ on $E_i$ for all $i$, it is necessary and sufficient that $g_i = g_j$ in $E_i \cap E_j$ for all $i$ and $j$. Chap. I, §1.5.

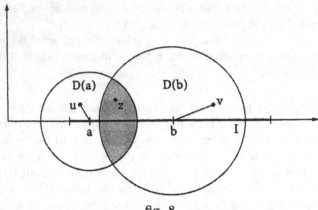

**fig. 8.**

## 21 – The function $\cot x$ and the series $\sum 1/n^{2k}$

Let us *accept* Euler's relation

$$(21.1) \qquad \pi \cot \pi x = \frac{1}{x} + 2x \sum_{n=1}^{\infty} \frac{1}{x^2 - n^2} = \sum_{n \in \mathbb{Z}} \frac{x}{x^2 - n^2}$$

mentioned at the end of n° 6 (though we have now replaced $x$ by $\pi x$ to eliminate the awkward $\pi^2$ factors), valid for all non-integer $x \in \mathbb{R}$, *though we have not yet established it*. Suppose that $|x| < 1$, whence $x^2/n^2 < 1$ for any $n$. One can then write, using the geometric series for $1/(1 - u)$,

$$(21.2) \qquad \frac{1}{x^2 - n^2} = \frac{-1}{n^2} \cdot \frac{1}{1 - x^2/n^2} =$$

$$= \frac{-1}{n^2} \sum_{p \geq 0} x^{2p}/n^{2p} = -\sum_{p \geq 0} x^{2p}/n^{2p+2}.$$

On substituting this result in the series (1) and calculating formally, one obtains

$$(21.3) \quad \pi \cot \pi x - 1/x = -2x \sum_{n \geq 1} \sum_{p \geq 0} x^{2p}/n^{2p+2} =$$

$$= -2x \sum_{p \geq 0} \sum_{n \geq 1} x^{2p}/n^{2p+2} = -\sum_{p \geq 0} a_{2p+1} x^{2p+1}$$

with

$$(21.4) \qquad a_{2p+1} = 2 \sum 1/n^{2p+2} = 2\zeta(2p+2)$$

on putting, after Riemann,

$$\zeta(s) = \sum_{1}^{\infty} 1/n^s.$$

If one knew (Chap. VI) a way to calculate these coefficients directly, i.e. to represent the function $\cot x - 1/x$ as a power series in $x$, namely

(21.5) $$\cot x - 1/x = -x/3 - x^3/45 - \ldots$$

as we shall see at the end of the next n°, it would follow that

$$\pi \cot \pi x - 1/x = -\pi^2 x/3 - \pi^4 x^3/45 - \ldots = -2 \sum \zeta(2p+2) x^{2p+1}$$

and thus that

$$a_1 = \pi^2/3, \quad a_3 = \pi^4/45, \ldots .$$

One would thus have calculated the sums

$$\zeta(2) = \sum 1/n^2 = a_1/2 = \pi^2/6, \qquad \zeta(4) = \sum 1/n^4 = a_3/2 = \pi^4/90,$$

etc. whose bizarre values we have already mentioned at the end of n° 6.

It all reduces to verifying that interchanging the summations with respect to $n$ and $p$ in (3) is justified and, for this, that the double series in $n$ and $p$ appearing there converge unconditionally. Since all the terms have the same sign there is no need to pass to absolute values. By Theorem 13 or its Corollary, it suffices to show that (i) the sum over $p$ converges for all $n$, (ii) the sum over $n$ of the sums over $p$ converges. Point (i) is clear since the sum over $p$ is just the geometric series (2), which converges since $|x| < 1$. On replacing the series relative to $p$ by its sum, one finds the left hand side of (2) – we are "retracing" the calculations – so that (ii) reduces to the absolute convergence of the series appearing in (1); this is clear since its general term is of the same order of magnitude as $1/n^2$. So all is justified, and it only remains to calculate the $a_p$ directly, i.e. the formula (5), which we shall do at the end of the following n° by using the power series representing $\sin x$ and $\cos x$. We must not forget to prove the formula (1), which can be done by an elementary argument, over-ingenious for some tastes, or by an almost direct argument using Fourier series (Chap. VII), or by the general theory of analytic functions[52].

The reader may have noticed that

$$2x/(x^2 - n^2) = 1/(x - n) + 1/(x + n),$$

in consequence of which (1) may be written

---

[52] See Walter, *Analysis I*, pp. 181–182 for the elementary proof and Remmert, *Funktionentheorie 1* (Springer, 1995), pp. 258–270 for a proof using a little general theory and some clever manipulations, due to Eisenstein, with series of the form $\sum 1/(z+n)^k$.

$$(21.6) \qquad \pi \cot \pi x = \sum_{n \in \mathbb{Z}} \frac{1}{x - n},$$

since the term $n = 0$ gives $1/x$ and one recovers the $\sum$ of (1) by grouping the terms for $n$ and $-n$. This new formula is even more beautiful than (1). It has only one drawback: the series on the right hand side, considered as an unordered sum, *diverges*, otherwise the partial series $\sum_{n>0} 1/(x - n)$ would converge, which clearly is not the case, since for $x$ real the general term is of constant sign for $n$ large and, up to sign, equivalent to $1/n$. A pity, since (6) exhibits the periodicity of the left hand side: replacing $x$ by $x + 1$ is the same as replacing $n$ by $n - 1$, i.e. to permuting the terms of the right hand side, and, as everyone knows, this does not change the value of a sum (except perhaps, alas, when it does not converge absolutely ...). But one can add $1/n$ to the general term for $n \neq 0$; this does not change the sum since these additional terms disappear when one groups the terms for $n$ and in $-n$, and the new series is absolutely convergent since its general term

$$u_n(x) = 1/(x - n) + 1/n = x/n(x - n)$$

is of the same order of magnitude as $1/n^2$. Then periodicity, no longer so obvious for this new general term, can be demonstrated as follows. Denote by $f(x)$ the sum of the series of $u_n(x)$. When one replaces $x$ by $x + 1$, $u_n(x)$ is replaced by

$$1/(x + 1 - n) + 1/n = 1/[x - (n - 1)] + 1/(n - 1) + [1/n - 1/(n - 1)],$$

so that

$$u_n(x + 1) = u_{n-1}(x) - 1/n(n - 1);$$

this calculation assumes $n$ different from 0 and from 1; if $n = 0$ or 1, one observes that $u_0(x + 1) = 1/(x + 1) = u_{-1}(x) + 1$, and that $u_1(x + 1) = 1/x + 1 = u_0(x) + 1$. Finally, one sees that

$$f(x + 1) = u_{-1}(x) + 1 + u_0(x) + 1 + \sum [u_{n-1}(x) - 1/n(n - 1)]$$

where the $\sum$ is extended over all $n \in \mathbb{Z}$ except 0 and 1. Since the series with general term $1/n(n - 1)$ converges, the preceding expression can be written

$$\sum u_{n-1}(x) + 2 - \sum 1/n(n - 1)$$

where the first $\sum$ is taken over *all* $n \in \mathbb{Z}$, so has sum $f(x)$ since one has performed the permutation $n \longrightarrow n - 1$ on the set of indices $\mathbb{Z}$, and where the second $\sum$ omits the values $n = 0$ and $n = 1$. To show that $f(x+1) = f(x)$, it therefore remains to verify that

$$
\begin{aligned}
2 &= \sum 1/n(n - 1) \\
&= \ldots + 1/3.4 + 1/2.3 + 1/1.2 + 1/2.1 + 1/3.2 + 1/4.3 + \ldots \\
&= 2(1/1.2 + 1/2.3 + 1/3.4 + \ldots),
\end{aligned}
$$

which brings us back to Johann Bernoulli and the marvels of the harmonic series. So we have, changing the notation,

$$(21.7) \qquad \pi \cot \pi x = \frac{1}{x} + \sum_{n \in \mathbb{Z}, n \neq 0} \left( \frac{1}{x-n} + \frac{1}{n} \right).$$

Since the reader is probably familiar with the elementary properties of derivatives, he may also have had the no less ingenious idea of differentiating the sum (6) or (7) term-by-term; the derivative of the function $\cot x$ being $-1/\sin^2 x$ and that of $1/(x-a)$ being $-1/(x-a)^2$, one obtains another remarkable formula:

$$(21.8) \qquad \frac{1}{\sin^2 x} = \sum_{x \in \mathbb{Z}} \frac{1}{(x-n\pi)^2}.$$

This is perfectly correct for all $x \in \mathbb{C}$ not a multiple of $\pi$ (so long as we define $\sin z$ for $z \in \mathbb{C}$ by its power series) and makes manifest the periodicity of the function sine as well as the points where it vanishes. But the "proof" which we have just given is (again) not one: it is obvious that the derivative of a *finite* sum of differentiable functions is the sum of their derivatives, but here we have to deal with an *infinite* sum. Weierstrass has shown in a general way that this type of operation is justified when dealing with *analytic* functions of a complex variable (Chap. VII) or even, in the real case, when the series of *derivatives*, and not only that of the given functions, converges "uniformly" (Chap. III, n° 17).

*Exercise*: deduce the "power" series of $1/\sin^2 x$ from (8) by imitating the passage from (1) to (3).

Finally we remark that Euler uses a very different method to calculate the sums $\sum 1/n^2$, etc. He starts from the expansion of the function $\sin x$ as an infinite product mentioned at the end of n° 6 or, what comes to the same, from the formula

$$(21.9) \qquad \prod \left( 1 - x^2/n^2 \right) = \sin(\pi x)/\pi x = 1 - \pi^2 x^2/3! + \pi^4 x^4/5! - \ldots$$

and calculates the left hand side as if dealing with a finite product, i.e. by choosing a term in each factor arbitrarily, finding the product of these terms, grouping the terms corresponding to the same power of $x$ and then adding the lot. [One can justify this argument by noting that the infinite product (9) is the limit of its partial products, and these are polynomials whose coefficients may be calculated by the method just indicated, and then passing to the limit]. One finds first the product of all the numbers 1. If, in the product, one chooses the number 1 everywhere except in the $n^{\text{th}}$ factor one finds a contribution equal to $-x^2/n^2$, and every other way of forming a term of the product yields a term of degree $> 2$; whence $\sum 1/n^2 = \pi^2/3!$. The coefficient of $x^4$ is obtained by choosing the factor 1 everywhere except in the factors $p$ and $q$, which yields a contribution $x^4/p^2q^2$; one thus has

$$\pi^4/5! = \sum_{p<q} 1/p^2q^2,$$

the condition $p < q$ assuring that one does not count the same term twice. The identity

$$\left(\sum x_i\right)^2 = \sum x_i^2 + 2\sum_{i<j} x_i x_j$$

then shows that

$$\sum 1/n^4 = (\pi^2/6)^2 - 2\pi^4/5! = \pi^4/90,$$

etc. Of course, Euler did not worry over such justifications; such extraordinary results were enough to make him happy. He might have checked them numerically, having discovered very effective and ingenious methods for calculating series such that those we are dealing with here.

His argument for "proving" the relation (9) in 1734 was to observe that if an algebraic equation of degree $n$

(21.10)     $$P(x) = a_0 + a_1 x + \ldots + a_n x^n = 0$$

has $n$ distinct roots $x_1, \ldots, x_n$, real or complex, then the left hand side of (10) is identical to the polynomial $a_n(x - x_1) \ldots (x - x_n)$, a perfectly precise statement, and easy to prove[53] and that Baccalaureat candidates are even supposed to know for $n = 2$. It follows that

$$a_0 = (-1)^n a_n x_1 \ldots x_n$$

(for $n = 2$, the product $c/a$ of the roots of the trinomial $ax^2 + bx + c$), thus that

$$\begin{aligned} P(x) &= (-1)^n a_0(x - x_1) \ldots (x - x_n)/x_1 \ldots x_n = \\ (21.11) &= a_0(1 - x/x_1) \ldots (1 - x/x_n) \end{aligned}$$

if $a_0 \neq 0$. Now "the algebraic equation of infinite degree"

(21.12)     $$\sin x/x = 1 - x^2/3! + x^4/5! - \ldots = 0,$$

---

[53] For $u \in \mathbb{C}$ given, $P(y+u)$ is a polynomial in $y$ whose term independent of $y$, the value for $y = 0$, is $P(u)$. If now $P(u) = 0$, then $P(y+u)$ is divisible by $y$, so that $P(x) = (x - u)Q(x)$ where $Q$ is a polynomial, with $d^\circ(Q) = d^\circ(P) - 1$. If $v \neq u$ is another root of $P$, one has $Q(v) = 0$, whence $P(x) = (x - u)(x - v)R(x)$, etc. This argument assumes that the $n$ roots of $P$ are distinct, for if not the factors $x - u$, $x - v$, etc. might be repeated ("order of multiplicity" of a root).

as Euler called it with characteristic aplomb – no one before or after him has ever met this kind of object in algebra –, has for its roots the points where the function $\sin(x)/x$ vanishes, i.e. $n\pi$, $n \neq 0$. [The function $\sin z$ happily has no complex nonreal roots ...] Since the first term $a_0$ is equal to 1, the general formula (11) shows "obviously" that the "polynomial of infinite degree" (12) is identical to the product of all the expressions of the form $1 - x/n\pi$; on replacing $x$ by $\pi x$ and grouping the terms for $n$ and $-n$ you obtain (9) with the greatest of ease[54].

Providence was manifestly on Euler's side – Providence, and a formidable intuition for correct formulae – since his argument could just as well prove that any other everywhere convergent power series has an expansion as an infinite product featuring the points where it vanishes; this is flagrantly not possible, particularly, but not only, when the series never vanishes: the exponential is an example of an "algebraic equation of infinite degree" having no root. The theory of analytic functions here again provides general theorems (Weierstrass, Chap. VII, n° 20), but much less simple than Euler's ideas; he later gave, in his *Introductio in Analysin Infinitorum* of 1748, a proof which, without being perfectly correct, is at least salvageable (Chap. IV, n° 18) since it is based on a correct (and very ingenious) idea.

## 22 – Multiplication of series. Composition of analytic functions. Formal series

Let $(u_i)$, $i \in I$ and $(v_j)$, $j \in J$ be two absolutely summable families and let us consider the family of products $u_i v_j$, indexed by the Cartesian product $I \times J$. As we have seen in n° 18, example 2 the product family converges unconditionally and its sum is the product of those of the two given series:

$$(22.1) \qquad \sum_{(i,j)\in I \times J} u_i v_j = \left( \sum_{i \in I} u_i \right) \cdot \left( \sum_{j \in J} v_j \right).$$

As an application, consider two series $\sum_{n \geq 0} u_n$ and $\sum_{n \geq 0} v_n$, absolutely convergent in the sense of n° 14, so converging unconditionally as we have seen in n° 15. Writing $U$ and $V$ for their sums we then have

$$(22.2) \qquad UV = \sum u_p v_q$$

where the right hand side is the sum of a double series which also converges unconditionally. Let us group the terms according to the value of $p + q = n$ (associativity! see also the much more general example 3 of n° 14). We obtain

---

[54] Moritz Cantor, *Vorlesungen ...*, Vol. III, pp. 658–659. One can judge Euler's influence by the fact that, more than seventy years later, Fourier reproduced his proof in his works on trigonometric series without voicing the slightest doubt as to its validity. It is true that, with his grossly divergent series, he was ill-placed to argue with Euler ...

the algebraically obvious formula (apart from the fact that we are not doing algebra[55] ...)

$$(22.3) \quad \sum_0^\infty u_n \cdot \sum_0^\infty v_n = \sum_0^\infty w_n \quad \text{where} \quad w_n = u_0 v_n + u_1 v_{n-1} + \ldots + u_n v_0.$$

The principal application of this formula is

**Theorem 16.** *Let $f(z) = \sum a_n z^n$ and $g(z) = \sum b_n z^n$ be two power series which converge absolutely for $|z| < R$. Then*

$$(22.4) \qquad f(z)g(z) = \sum c_n z^n \quad \text{with} \quad c_n = a_n b_0 + \ldots + a_0 b_n$$

*and the series $\sum c_n z^n$ converges absolutely for $|z| < R$.*

In other words, *the rule for multiplication of polynomials applies to power series*, of course under the restriction that one is at a point $z$ where the two series are absolutely convergent, e.g. in the interior of the smaller of the discs of convergence of the two given series. It applies also more generally to the *Laurent series*

$$f(z) = \sum a_n z^n$$

where this time one sums over all $n \in \mathbb{Z}$; being the sum of a power series in $z$ and of a power series in $1/z$, such a series converges, in general, only in an annulus (possibly empty: the case of the series $\sum z^n$ where one sums over $\mathbb{Z}$) defined by inequalities $0 \leq R' < |z| < R''$; there it converges absolutely as do power series in their disc of convergence, and for the same reason: comparison with geometric series. We shall return to this point in Chap. VII. So, if one has two Laurent series converging in the same nonempty annulus, one can form the unordered product and then group the terms:

$$\sum a_n z^n \sum b_n z^n = \sum c_n z^n \quad \text{with} \quad c_n = \sum a_p b_q$$

where the series is extended over all pairs $(p, q)$ such that $p + q = n$: one recovers the convolution product over $\mathbb{Z}$ of n° 18, though the series $\sum a_n$ and $\sum b_n$ need not be absolutely convergent even if the series defining their convolution product is, as we have just seen. These series arise in the study of analytic functions on the neighbourhood of an isolated singular point; example: the function $\exp(z + 1/z)$, whose Laurent series expansion uses the following result (the fact that this is presented as a "Corollary" should not lead one to forget how fundamental it is):

---

[55] This is so little algebraic that, if one applies the rule to non-absolutely convergent series, one may obtain a divergent series, as Cauchy showed by choosing $u_n = v_n = (-1)^{n+1}/n$. Then $|w_n| > 2n/(n-1)$, an expression which tends to 2 and not to 0. Paul Dugac, *Sur les fondements de l'Analyse de Cauchy à Baire* (doctoral thesis, Université Pierre et Marie Curie, 1978) p. 19.

**Corollary 1 (addition formula for the exponential series).**

$$(22.5) \qquad\qquad \exp(x).\exp(y) = \exp(x+y),$$

*for all $x, y \in \mathbb{C}$.*

Since in general $\exp(z) = \sum z^{[n]}$, the term $w_n$ of (3) is then indeed equal to the sum of the products $x^{[p]}y^{[q]}$ for all pairs of integers $p, q \geq 0$ such that $p + q = n$; but this is just $(x+y)^{[n]}$ by the binomial formula of n° 1; since there is no problem of convergence, the result follows. In fact, the corollary consists of passing to the limit over $n$ in the formula

$$\sum_{0 \leq m \leq n} (x+y)^{[m]} = \sum_{p+q \leq n} x^{[p]}y^{[q]}$$

which follows from the binomial formula.

**Corollary 2.** *The product of two functions $f$ and $g$ defined and analytic in an open subset $G$ of $\mathbb{C}$ is analytic in $G$, and*

$$(fg)' = f'g + fg'.$$

The given functions $f$ and $g$ are, by hypothesis, represented on a neighbourhood of any $a \in G$ by power series in $z - a$; so their product is too.

To show that the derivative of $fg$ is given by the "obvious" formula – one could establish this directly, and also for the holomorphic functions, on defining the derivatives by passage to the limit –, one remarks that on a neighbourhood of a $z \in G$, $f$ and $g$ are given by Taylor's formula (19.6)

$$\begin{aligned} f(z+h) &= f(z) + f'(z)h + f''(z)h^2/2! + \ldots, \\ g(z+h) &= g(z) + g'(z)h + g''(z)h^2/2! + \ldots. \end{aligned}$$

Since these power series in $h$ converge and represent the left hand sides for $|h|$ sufficiently small, one has

$$f(z+h)g(z+h) = c_0(z) + c_1(z)h + c_2(z)h^2/2! + \ldots$$

where the coefficients $c_n(z)$ are given by Theorem 16. In particular,

$$c_1(z) = f(z)g'(z) + f'(z)g(z),$$

and since $c_1(z)$ must be the derivative at the point $z$ of the function $fg$ by Taylor's formula for the latter, the result follows.

Of course the reader would like to know if the *quotient* of two analytic functions is again analytic or, what comes to the same by the preceding theorem, if the recipocal $1/f(z)$ of an analytic function is again analytic. This is so, provided, as always, that one does not divide by 0.

But there is much more. Put $g(z) = 1/z$, an analytic function on the open subset $z \neq 0$ of $\mathbb{C}$ as we have known since n° 6. One has $1/f(z) = g[f(z)]$. We are thus led to ask much more generally if one again obtains an analytic function on composing two analytic functions, for example $\cos(\sin z)$ where, on $\mathbb{C}$, one defines $\sin z$ and $\cos z$ by the power series which have already appeared several times in this chapter. Here again the response is affirmative and the proof, once again, uses the associativity theorem, but in a case where the set of indices is essentially more complicated than $\mathbb{N}$ or $\mathbb{N} \times \mathbb{N}$. This is really the only difficulty, the calculations being, as one says in English, pedestrian.

**Theorem 17.** *Let $U$ and $V$ be two open subsets of $\mathbb{C}$, let $f$ be an analytic function defined on $U$ with values in $V$, and $g$ an analytic function defined on $V$. Then the composed function $h(z) = g[f(z)]$ is analytic in $U$, and $h'(z) = g'[f(z)].f'(z)$.*

Consider an $a \in U$ and the point $b = f(a)$ of $V$. On a neighbourhood of $b$ the function $g(z)$ is a power series in $z - b = Y$, so that $h(z)$ is obtained by substituting $f(z)$ for $z$ in this; one thus obtains a power series in $f(z) - b = f(z) - f(a)$. Since $f(z)$ is, in its turn, a power series in $z - a = X$ whose constant term is $f(a)$, the difference $f(z) - f(a)$ is a power series *without constant term* in $X$.

The situation is thus the following: we have two convergent power series (i.e. converging at points other than the origin) in variables $X$ and $Y$, the first in $X$ without constant term, and we wish to show that on substituting the first for the variable $Y$ in the second, manipulating crudely, and then grouping together the terms containing the same powers of $X$, we will again find a *convergent* power series. We may also assume that the second series has zero constant term, since this poses no problem.

So – we return to the traditional notation $z$ for the variable – let

$$(22.6) \qquad f(z) = a_1 z + a_2 z^2 + \ldots, \qquad g(z) = b_1 z + b_2 z^2 + \ldots$$

be the given series, and let us first calculate formally.

We have

$$(22.7) \qquad h(z) = g[f(z)] = \sum b_p \left( \sum a_n z^n \right)^p.$$

Now the formula for multiplication of series, which clearly generalises to more than two factors, shows that

$$(22.8) \qquad \left( \sum_{n>0} a_n z^n \right)^p = \sum a_{n_1} a_{n_2} \ldots a_{n_p} z^{n_1 + n_2 + \ldots + n_p}$$

where one sums over all families $(n_1, \ldots, n_p)$ of $p$ integers $> 0$ chosen arbitrarily (for $p = 2$, these are ordered pairs in the sense of set theory): to multiply $p$ sums one by the other, one chooses arbitrarily a term in each sum, multiplies them, and adds all the products so obtained; here we have chosen the term n° $n_1$ from the first sum, the term n° $n_2$ from the second, etc. Writing $I$ for the set of integers $n \geq 1$, the preceding series, the product of $p$ absolutely convergent series for $|z|$ sufficiently small – but this is not very important for the moment –, is thus an unordered sum where the index varies over the Cartesian product $I \times \ldots \times I = I^p$, with $p$ factors.

Substituting this result in (7) and continuing to calculate formally, one then finds that

$$(22.9) \qquad h(z) = \sum_{p,n_1,\ldots,n_p > 0} b_p a_{n_1} \ldots a_{n_p} z^{n_1 + \ldots + n_p};$$

the general term depends simultaneously on the integer $p$ and on an element of $I^p$, so we are dealing with an unordered sum extended over the set

$$J = I \cup I^2 \cup \ldots \cup I^p \cup \ldots \,,$$

the union of all the Cartesian products $I^p$. These are pairwise disjoint, since one does not see how a sequence of 3 integers could also be a sequence of 7 integers. $J$ is thus the set of all the sequences $(n_1, \ldots, n_p)$ of any number of integers $> 0$. To clarify the notation, we ought to put

$$(22.10) \qquad u_j = b_p a_{n_1} \ldots a_{n_p} z^{n_1 + \ldots + n_p} \text{ if } j = (n_1, \ldots, n_p) \in J.$$

*If the sum* (9) *converges unconditionally* we can group the terms *ad libitum*, in particular group together all the terms containing the same power of $z$, say $z^k$; the result is clearly the power series $\sum c_k z^k$ with

$$(22.11) \qquad c_k = \sum_{\substack{p,n_1,\ldots,n_p > 0 \\ n_1 + \ldots + n_p = k}} b_p a_{n_1} \ldots a_{n_p},$$

the sum being extended over the set, clearly *finite*, of systems $(n_1, \ldots, n_p)$ of any number of integers $> 0$ satisfying $n_1 + \ldots + n_p = k$.

[Note in passing that the general term of (11) does not change if one permutes the indices $n_1, \ldots, n_p$, since each term actually features several times in the sum if the $n_i$ are not all equal to each other. Likewise, in the identity $(\sum x_i)^2 = \sum x_i x_j$, the terms for which $i \neq j$ appear twice, since the pairs $(i,j)$ and $(j,i)$ are distinct. But no matter; here we are calculating crudely; to introduce factorials or binomial coefficients in (11) would be the best way of not seeing anything.]

It remains to prove the unconditional convergence of (9). The convergence hypothesis on the given series ensures that their coefficients are dominated by geometric progressions (n° 14, example 4) and this will save the situation. One may even assume that there exist an $r > 0$ and a constant $M > 0$ such that, simultaneously[56],

---

[56] Choose $r$ so that the two series converge for $|z| = 1/r$; one then has bounds $|a_n| r^{-n} < M'$ and $|b_n| r^{-n} < M''$, so that $M = \max(M', M'')$ is as required.

(22.12)    $$|a_n| < Mr^n \quad \text{and} \quad |b_n| < Mr^n$$

for all $n$. The terms of the unordered sum (9) are then, in modulus, majorised by those of the analogous sum obtained by replacing $a_n$ and $b_n$ everywhere by $Mr^n$ – this is the *method of majorants* of Weierstrass, applicable in many other situations – and $z$ by $|z|$. All now reduces to establishing the unconditional convergence of the new sum, which has positive terms. But this is just

(22.13)    $$\sum M^{p+1} r^{p+n_1+\cdots+n_p} |z|^{n_1+\cdots+n_p} = \sum M^{p+1} r^p u^{n_1+\cdots+n_p}$$

where we have put $u = r|z|$ and where the sum is extended over the same set $J$ of indices as in (9). To prove that (13) converges for $|z|$ sufficiently small, we shall apply the associativity theorem, grouping together the terms corresponding to the same value of $p$, in other words according to the partition of $J$ by the Cartesian products $I^p$ introduced above. So now it reduces to verifying that (i) the sum of the terms of (13) for which $p$ is fixed converges, (ii) the sum over $p$ of these partial sums converges.

For $p$ given, one obtains, up to a factor $M^{p+1} r^p$, the series

(22.14)    $$\sum u^{n_1+\cdots+n_p} = \sum u^{n_1} \cdots u^{n_p} = \left(\sum u^{n_1}\right) \cdots \left(\sum u^{n_p}\right)$$

from the general multiplication formula

$$\sum a_i \sum b_j \sum c_k \cdots = \sum a_i b_j c_k \cdots$$

for unconditional convergence; as we saw at the beginning of this n°, this formal calculation will be justified if each of simple series appearing on the right of (14) is convergent (remember that here all the terms are positive). Since we are dealing with the same geometric progression multiplied $p$ times by itself the condition we seek is therefore $|u| < 1$, i.e.

(22.15)    $$|z| < 1/r.$$

Point (i) above is therefore established, modulo the condition (15).

It remains to establish that the sum over all $p$ converges. By (14), this is just

$$\sum M^{p+1} r^p \left(\sum u^n\right)^p = M \sum \left(Mr \sum u^n\right)^p.$$

It thus converges so long as one has $Mr \sum u^n < 1$. But since, in (14), all the $n_i$ are $> 0$, the series $\sum u^n$ is a geometric progression without constant term; its sum is thus equal to $u/(1-u)$. The series above converges then if $Mru/(1-u) < 1$, i.e. if $u < 1/(1+Mr)$. As we wrote above, $u = r|z|$, so all is justified provided that $z$ simultaneously satisfies (15) and $|z| < 1/r(1+Mr)$, in other words for $z$ sufficiently small.

It remains to calculate the derivative of the composed function.

Newton would have explained to you that, to do this without wasting energy, you write

$$f(z) = a_1 z + \ldots, \qquad g(z) = b_1 z + \ldots$$

where the unwritten terms are of degree $> 1$ in $z$. Then

$$h(z) = a_1(b_1 z + \ldots) + \ldots$$

with the same commentary. Thus $h(z) = a_1 b_1 z + \ldots$, and since $h'(0)$ must be the coefficient of $z$, one obtains

$$h'(0) = f'(0)g'(0) = f'(0)g'[f(0)].$$

The general case where one works at a point $a$ of $U$ and the point $b = f(a)$ of $V$ reduces immediately to this: on a neighbourhood of $a$, $f(z) - f(a)$ is a series without constant term in $X = z - a$, the coefficient of $X$ being $f'(a)$ by Taylor's formula; on a neighbourhood of $b$, $g(z)$ is a series in $Y = z - b$, the coefficient of $Y$ being $g'(b) = g'[f(a)]$. In consequence, in the series in $X$ obtained in substituting the series in $X$ which represents $f(z) - f(a)$ for $Y$, the coefficient of $X = z - a$ is the number $f'(a)g'[f(a)]$, qed.

The inventor of power series, Newton, clearly knew all these results, more precisely he used them constantly in his calculations as if they were self evident, and as if it was not necessary to worry about convergence and even less so to formulate general theorems. He is, for example, capable of calculating the quotient of two power series by the method of division applicable to decimal fractions[57], $x$ playing the rôle of $1/10$ as we mentioned above in the quotient $1/(1 + q)$. He is capable of inverting a power series[58], i.e. deducing from a relation

$$y = x + a_2 x^2 + a_3 x^3 + \ldots$$

a relation

$$x = y + b_2 y^2 + b_3 y^3 + \ldots$$

and, for example, of working out the exponential series by inverting the logarithmic series, or the series for the sine by inverting that for arcsin.

---

[57] *I am amazed that it has occurred to no one (if you except N. Mercator with his quadrature of the hyperbola) to fit the doctrine recently established for decimal numbers in similar fashion to variables, especially since the way is then open to more striking consequences.* First page of the English translation of the Newton's manuscript mentioned below. For division, one has to widen the definition of power series a little and allow a finite number of negative powers of $z$:

$$1/(z^2 - z^3) = z^{-2} + z^{-1} + 1 + z + \ldots$$

After all, conventional arithmetic does not confine itself to considering numbers with zero integer part.

[58] There is of course a general theorem, but it appeared two centuries later.

In studying algebraic plane curves, i.e. defined by a polynomial relation between $x$ and $y$, he is capable, for any point $(a, b)$ of the curve, of calculating with power series in $x - a$ (or, if necessary, in a fractional power of $x - a$) starting with $b$ and which, substituted for $y$, satisfy the equation of the curve; they allow one to study the curve in the neighbourhood of the point $(a, b)$ and to determine its singular points (multiple points where several branches of the curve cross, cusps, etc.) or the asymptotes.

He explains this with the example $y^3 + y + xy - 2 - x^3 = 0$, which represents a plane curve of third degree. For $x = 0$, the equation possesses, among others, the solution $y = 1$. Newton then sets $y = 1 + p$, where $p$ is a power series starting with a term in $x$; substituting in the initial equation, the constant terms must cancel, and one finds an equation $4p + x + \ldots = 0$, where the unwritten terms contain, taking account of $p$, only powers $\geq 2$ of $x$; in consequence, $p = -x/4 + q$ where the series $q$ starts with a term in $x^2$; one then substitutes in the equation in $p$ and one finds an equation $4q - x^2/16 + \ldots = 0$ where the unwritten terms all contain at least $x^3$. One has in consequence $q = x^2/64 + r$ where $r$ starts with a term in $x^3$, and so on indefinitely. Newton thus finds, in the example considered,

$$y = 1 - x/4 + x^2/64 + 131x^3/512 + 509x^4/16384 + \ldots$$

setting out the calculations fully in a very elegant way[59].

The only problem is that these formal calculations prove nothing as to the convergence of the series obtained, of which Newton never exhibits more than the first terms. Here as always, it was the XIX[th] century, and quite particularly Weierstrass, who invented *a priori* methods for majorising the coefficients of a series so obtained *without calculating them explicitly*, and showed that it converged otherwise than for $x = 0$; this in fact is what Theorem 17 proves. One can see no other way, in the example just presented *à la* Newton, of how, "without knowing anything", one might obtain an idea of its convergence when its coefficients are of such complexity and do not obey any obvious general law of formation.

More beautiful *in this context* is that the problem of convergence is of hardly any interest. The theory of *algebraic* curves (or surfaces, or ...), i.e. defined by polynomial equations between the coordinates, lies in the domain of algebra for the excellent reason that it still has a meaning if one replaces $\mathbb{R}$ or $\mathbb{C}$ by any commutative field, as discovered in the XX[th] century. Even if the series obtained by Newton were completely divergent, in other words reduced to formal series, his calculations retain an *algebraic* meaning, so that his method is in fact applicable (and applied) to an arbitrary field: the four

---

[59] See, in Vol. III, pp. 32–225, of the complete edition of the *Mathematical Papers* by D. T. Whiteside, the "tract" *De Methodis Serierum et Fluxionum* of 1670–1671 where Newton systematically expounds his discoveries, especially pp. 55–57 for the example in question. Whiteside reproduces the Latin text and an English translation from 1710.

operations of algebra are enough to get down to work. After all, this is algebra for computers, apart from the fact that machines do not know where to go if one does not tell them where to start; this is one of the numerous differences between what they call artificial intelligence and what one used to call intelligence in short and what must now be called natural intelligence to avoid confusion, assuming that there is a risk of confusion...

We have just spoken of "formal series"; what is a *formal series* with coefficients in an arbitrary field $K$? It is an expression of the form $\sum a_n X^n$ with coefficients $a_n \in K$ depending on an index $n \in \mathbb{Z}$, but *zero for $n$ negative and sufficiently large*; the so-called "series" contains only a finite number of negative powers of the so-called "variable" $X$, which it must contain if one wants to be able to write something like

$$1/X^3(1 - X) = X^{-3} + X^{-2} + \ldots.$$

Such a "series" has no *a priori* meaning: it is possible to give one to certain infinite sums in $\mathbb{R}$ or $\mathbb{C}$, but quite impossible in a field where one has no concept of a limit. One extricates oneself by considering – this was Hamilton's idea for defining the complex numbers – that a formal series is simply a family $(a_n)$ of elements of the field $K$, and *defining* the sum and the product of two such series by the natural formulae:

$$(a_n) + (b_n) = (a_n + b_n),$$

$$(a_n).(b_n) = (c_n) \quad \text{where} \quad c_n = \sum_{p+q=n} a_p b_q;$$

again a convolution product ... The fact that the $a_n$ and $b_n$ are zero for $n < 0$ large ensures that the definition of $c_n$ involves only a finite sum, so has a meaning in an arbitrary field. With these definitions of the sum and of the product, the formal series form a new *commutative field*, i.e. satisfy the axioms (I) of n° 1: an elementary exercise in algebra, asking only a little patience. If the field $K$ appears too abstract to you, abstract yourself from $K$ and calculate without trying to understand the concrete significance of the letters: they have none, so here again we have mechanics for computers.

The notation $\sum a_n X^n$ can be justified – or rather explained – as follows. First, one identifies each $a \in K$ with the formal series all of whose coefficients are zero except $a_0 = a$. The letter $X$ denotes the formal series all of whose coefficients are zero except $a_1 = 1$. On applying the rules of calculus above, one finds that, for all $n \in \mathbb{Z}$, the series $X^n$ has all its coefficients zero except $a_n = 1$, so that the product $a_n X^n$ is the series all of whose the terms are zero except the $n^{\text{th}}$, which equals $a_n$. The addition law then shows that a formal series having no more than a *finite* number of nonzero coefficients is just the sum of the series $a_n X^n$ for those $n$ such that $a_n \neq 0$; one generally calls this a polynomial in $X$ and $X^{-1}$ ... One then extends this expression to the general case, remaining aware of the fact that this is a pure convention to ease the calculations – one really can calculate with these "series" as if they were polynomials in $X$ and $X^{-1}$ – and not a theorem in the genre "limit of partial sums", an expression with no meaning in algebra. The first mathematician deliberately to use formal series (with coefficients in $\mathbb{C}$) was, naturally, Euler who, of course, provided neither justifications nor explanations: he calculated.

One can consider a power series [or even a series with possibly a finite number of terms of negative degree – cf. $\cot x$] as a formal series with coefficients in $\mathbb{C}$, but the difference between algebra and analysis lies in the fact that the formal series which converge only for $x = 0$, for example $\sum n! x^n$, have no interest *in analysis* as we have already noted in n° 15, example 4. The miracle lies in the fact that all the reasonable operations – and not only the algebraic, consider derivation for a start – that one performs, in $\mathbb{C}$, on *convergent* formal series, those having a radius of convergence $> 0$, lead again to *convergent* series.

All the same one must be prudent. If for example you calculate the expression $1/\cos z$ *à la* Newton, replacing $\cos z$ by its power series, you will again find a convergent power series by Theorem 17. But while that for $\cos z$ converges for any $z$, the new series will converge in no more than (and, in fact, exactly in) the largest disc with centre 0 not containing any zero of the function $\cos z$, in other words for $|z| < \pi/2$. Such a result could never be established if one were restricted to examining the coefficients of this series: they are quite complicated and $\pi$ does not appear in them; one needs the general theory of analytic functions: if a function $f$ is analytic in an open subset $G$ of $\mathbb{C}$, then for all $a \in G$ the Taylor series of $f$ about $a$ converges and represents $f$ on the *largest* open disc with centre $a$ contained in $G$; we will show this in Chap. VII. For the function $1/\cos z$, $G$ is the set of $z \in \mathbb{C}$ where $\cos z \neq 0$, whence one finds the radius of convergence, so long as one knows that, in $\mathbb{C}$ as in $\mathbb{R}$, the series vanishes only at the points $z = (2k+1)\pi/2$.

Newton's method, in its simplest form, allows one to calculate the coefficients in the formula (21.4), i.e. to justify the series (21.5) for the function $\cot x$, subject to the assumption of the existence of such a formula – this detail would have held back neither Newton nor his successors for 150 years – and to accepting the series expansions of the functions sine and cosine already mentioned.

The first point raises no difficulty. Indeed

$$\cot x = \cos x/\sin x = (1 - x^2/2! + \ldots)/x(1 - x^2/3! + \ldots),$$

so that $x \cdot \cot x$ is the quotient of two power series starting with 1. Theorem 17 was introduced precisely to show that this quotient is itself a power series converging for $|x|$ sufficiently small, and its constant term is clearly equal to 1.

Now put

$$(22.16) \qquad \cot x = 1/x - c_1 x - c_3 x^3 - c_5 x^5 - \ldots;$$

the term $1/x$ is indispensable from what we have just remarked; the exponents are necessarily odd for the obvious reason. With the notation (21.4) one thus has

$$(22.17) \qquad \pi^{2p+2} c_{2p+1} = a_{2p+1} = 2 \sum 1/n^{2p+2} = 2\zeta(2p+2).$$

The problem is now to calculate the $c_i$. On transferring the factor $x$ from $\sin x$ to the series for $\cot x$ we get

$$(22.18) \quad \left(1 - x^2/3! + x^4/5! - x^6/7! + \ldots\right)\left(1 - c_1 x^2 - c_3 x^4 - c_5 x^6 - \ldots\right) =$$
$$= 1 - x^2/2! + x^4/4! - x^6/6! + \ldots.$$

Since the left hand side of (18) is equal to

$$(22.19) \qquad 1 - (c_1 + 1/3!)\, x^2 + (-c_3 + c_1/3! + 1/5!)\, x^4 -$$
$$- (c_5 - c_3/3! + c_1/5! + 1/7!)\, x^6 - \ldots$$

one clearly finds[60] the relations

$$
\begin{aligned}
c_1 + 1/6 &= 1/2, &\text{whence } c_1 &= 1/3, \\
-c_3 + 1/3.3! + 1/5! &= 1/4!, &\text{whence } c_3 &= 1/45, \\
c_5 - 1/45.3! + 1/3.5! + 1/7! &= 1/6!, &\text{whence } c_5 &= 2/945,
\end{aligned}
$$

so that $\sum 1/n^6 = a_5/2 = \pi^6 c_5/2 = \pi^6/945$ by (17). The reader can pursue the calculations *ad libitum*; this is what Newton would have done to distract himself from the rainbows in his room. In fact, there is a recurrence formula for calculating the $c_p$ one-by-one: it suffices simply to calculate the term in $x^{2p}$ in the product (18). This would take us too far for the moment, and especially to the strange numbers of Bernoulli (Jakob) to be discussed in Chap. VI and which, to general surprise, resurfaced a few decades ago in problems of algebraic topology having no near or distant relation to the mathematics of Euler.

---

[60] So long as one knows that two power series having the same sum have the same coefficients. By subtracting, it suffices to show that if $\sum a_n z^n = 0$ for all sufficiently small $z$, then $a_n = 0$ for all $n$. This is the principle of "isolated zeros" for analytic functions, expounded in n° 20.

## 23 – The elliptic functions of Weierstrass

To end this chapter with a "bonus to the brave reader", we shall return to the general associativity formula for unconditional or absolute convergence, and illustrate it by a example which will not profit us for the moment but which, in contrast to the traditional gymnastic exercises, has considerable mathematical interest.

Series like $\sum 1/\left(m^2 + n^2\right)^{k/2}$ arise in the theory of numbers, and also appear in a closely related form in the theory of elliptic functions of a complex variable, one of the great inventions of the XIX[th] century, more than ever the order of the day in "pure" mathematics after its fusion with the theory of algebraic numbers, algebraic geometry, etc. Its "users" have for a long time exploited the more manipulatory parts of the theory, principally to calculate numerically the integrals of apparently very elementary functions (square roots of a polynomial of degree 3 or more), yet whose primitives are not known, more precisely are not expressible in terms of elementary functions. Besides, the historical origin of the elliptic functions[61] in the XVIII[th] century was as much in mathematics as in mechanics (oscillations of a simple pendulum), and since many interesting applications were found for them, to mechanics, to physics, etc., they were honoured in the courses of analysis at the Ecole polytechnique during the second half of the XIX[th] century, a little beyond the level of the majority of the future artillery men[62] ... The current developments, much closer to algebraic geometry and to number theory than to analysis, are difficult of access for the majority of professional mathematicians who are not specialists. They are very likely not within the reach of users and, in any case, have long since passed very far beyond the stage of numerical calculations in the usual sense of incorporating arithmetic calculations that machines can perform.

---

[61] See for example C. Houzel, *Analyse mathématique* (Foris, Belin, 1997), pp. 290–302, or, more difficult, Henry McKean and Victor Moll, *Elliptic Curves* (CUP, 1997), Chap. 2.

[62] Jacobi also taught them at Koenigsberg about 1830, Liouville at the Collège de France in 1850, Weierstrass in Berlin before and after 1870, etc. But admission to these courses was free and there was no final examination. Adolf Kneser, who followed Weierstrass' course in Berlin in the 1880s, spoke in 1925 of "two hundred young people" who followed Weierstrass' course on the elliptic functions from beginning to end "in full awareness of the fact that they would not appear at this time in any state examination, resounding testimony to the scientific mind of this time"; Remmert, *Funktionentheorie 1*, p. 336. There were no internal examinations in the German Universities at this period; one prepared for the State examinations leading to the "professions".

The elliptic functions are *analytic* functions defined on $\mathbb{C}$, except for isolated points where they have *poles*[63], and they are *doubly periodic*[64]: there are two numbers $\omega_1, \omega_2 \in \mathbb{C}$ such that (the use of the letter $u$ instead of $z$ is traditional)

$$f(u + n_1\omega_1 + n_2\omega_2) = f(u)$$

for all $u$ and $n_1, n_2 \in \mathbb{Z}$; the theory is of no interest if the ratio $\omega_1/\omega_2$ is real. One of the methods for constructing such functions explicitly consists of considering for all $k \in \mathbb{N}$ the series

$$(23.1) \qquad f_k(u) = \sum 1/(u - \omega)^k$$

extended over the set $L$ of periods $\omega = n_1\omega_1 + n_2\omega_2$. Suppose that it converges unconditionally and let us calculate $f_k(u + \omega')$, where $\omega'$ is a period. This is the same as replacing $\omega$ by $\omega - \omega'$ in the general term. But since the sum or the difference of two periods is again a period, the map $\omega \mapsto \omega - \omega'$ is, for $\omega'$ given, a permutation of the set L. Whence $f_k(u + \omega') = f_k(u)$ to the general satisfaction.

The problem of unconditional or absolute convergence remains; it alone allows us to justify this too easy formal calculation. One must clearly eliminate the term $(u - \omega)^{-k}$ if $u = \omega$ is a period. So let us work in a disc $|u| < R$ with $R > 0$. There can be only a finite number of periods such that $|\omega| \leq 2R$ (see figure 9) and the corresponding terms of the series influence neither its convergence nor its analyticity. For the rest, one has $|\omega| > 2R > 2|u|$; the relation

$$|\omega| - |u| \leq |u - \omega| \leq |\omega| + |u|$$

then shows that

$$|\omega|/2 \leq |u - \omega| \leq 3|\omega|/2.$$

It all reduces to deciding on the unconditional convergence of the *Eisenstein series* (1823–1852: tuberculosis like Abel and Riemann)

---

[63] We shall prove (Chap. VII) that a function $f(z)$ defined and analytic in a neighbourhood of a point $a$ except at the point $a$ itself is representable on a neighbourhood of $a$ by a Laurent series, i.e. of the form $f(z) = \sum a_n(z - a)^n$ with, in general, infinitely many nonzero terms of negative degree [example: $\cos z . \sin(1/z)$, analytic for $z \neq k\pi$]. When they are only finite in number, one says that $f$ possesses a pole at $a$ and, more precisely, a *pole of order $p$* if

$$f(z) = a_{-p}(z - a)^{-p} + a_{-p+1}(z - a)^{-p+1} + \ldots$$

with $a_{-p} \neq 0$; an obvious connection with the formal series of n° 22. For example, the function $\cot \pi z$ has a pole of order 1 at each of the points $n \in \mathbb{Z}$. Note that then the function $g$ given by $g(z) = (z - a)^p f(z)$ for $z \neq a$, $g(a) = a_{-p}$, is analytic in all of a neighbourhood of $a$, including and even particularly at the point $a$. In other words, $f(z) = g(z)/(z - a)^p$ where $g$ is a power series in $z - a$.

[64] The construction of such functions is immediate if one omits the hypothesis of analyticity: the function $f(z) = \cos x + \sin y$ has the periods $2\pi$ and $2\pi i$. But it is not analytic.

(23.2)     $$G_k(L) = \sum 1/\omega^k = \sum 1/\left(n_1\omega_1 + n_2\omega_2\right)^k$$

of 1847, where one sums over all the periods $\omega \in L$, with 0 obviously excluded. These vanish for $k$ odd, but they are the series $\sum 1/|\omega|^k$ which interest us here.

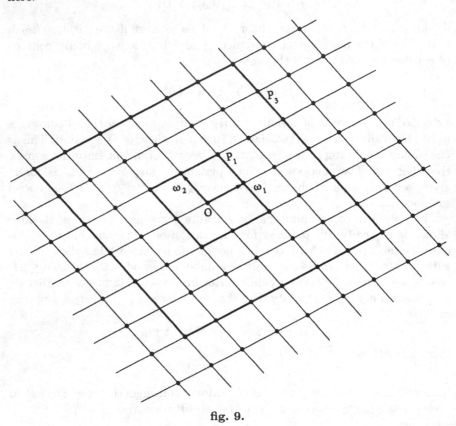

fig. 9.

So, for each $n \in \mathbb{N}$ consider in the complex plane the parallelogram $P_n$ with centre 0 formed by the points of the form $t_1\omega_1 + t_2\omega_2$ with real "coordinates" $t_1$ and $t_2$ satisfying either $|t_1| = n$ and $|t_2| \leq n$, or $|t_1| \leq n$ and $|t_2| = n$ or, to make ourselves understood to a physical or biological computer,

$$\left(\left(\left(|t_1| = n\right) \text{ and } \left(|t_2| \leq n\right)\right) \text{ or } \left(\left(|t_1| \leq n\right) \text{ and } \left(|t_2| = n\right)\right)\right).$$

Let $L_n \subset L$ be the set, finite, of periods such that $\omega \in P_n$. The "coordinates" $t_1$ and $t_2$ of $\omega$ being integers, it is clear that $L_n$ contains $8n$ elements. What is their order of magnitude?

As the figure above shows, the parallelogram $P_n$ is derived from $P_1$ by the homothety with centre 0 and ratio $n$. It is clear that the interior of $P_1$

contains a disc with centre 0 and radius $r > 0$ and that $P_1$ is contained in a disc with centre 0 and radius $R > 0$. The distance to the origin of any point $u$ of $P_1$ is thus between $r$ and $R$. By homothety, the distance to the origin of any point of $P_n$ lies between $nr$ and $nR$, whence $nr < |\omega| < nR$ for every period $\omega \in L_n$. In (2), the partial sum $S(L_n)$ of the moduli thus lies between $8n/(nR)^k$ and $8n/(nr)^k$, i.e. is of the same order of magnitude ($\asymp$) as $1/n^{k-1}$. The condition of convergence that we seek is thus $k > 2$, i.e. $k \geq 3$ since $k$ is an integer.

It happens that the elliptic function which "controls" all the others corresponds to $k = 2$: the $f_k$ are, up to numerical factors, its derivatives in the sense of n° 19, as is "obvious" from the formulae, though one does not differentiate an infinite sum without taking precautions (but see Chap. VII for series of analytic functions). One proves this by modifying (1) so as to make the series converge, as was done in n° 21 to make the series with general term $1/(x - n)$ converge.

For this, one remarks that, for $|\omega| > |u|$,

$$(23.3) \qquad \frac{1}{(u - \omega)^2} = \frac{1}{\omega^2} \cdot \frac{1}{(1 - u/\omega)^2}$$

$$= \frac{1}{\omega^2} \left( 1 + 2u/\omega + 3u^2/\omega^2 + 4u^3/\omega^3 + \dots \right)$$

by (19.12). The quasi-geometric series (3) shows that, for $|\omega|$ large, the general term of (1) is approximately equal to $1/\omega^k$, the reason why (1) does not converge for $k = 2$. The solution now lies in diminishing the order of magnitude of the general term by subtracting $1/\omega^2$ from it, for $\omega \neq 0$, in other words we introduce the famous function

$$(23.4) \qquad \wp(u) = 1/u^2 + \sum_{\omega \neq 0} \left[ 1/(u - \omega)^2 - 1/\omega^2 \right]$$

of Weierstrass (it already appeared in Eisenstein), with a $p$ which smacks of the gothic, of the italic and of the cursive, chosen by the inventor[65] and retained by posterity. Formula (3) above shows that, for $|u/\omega| < 1/2$ for example, i.e. for "almost all" the $\omega \in L$, the set of periods, the general term of (4) is of the same order of magnitude as $1/\omega^3$, whence convergence.

More precisely, let us work as above in a disc $|u| < R$ and, to decide on the convergence of the series, eliminate from the series the finite number of terms for which $|\omega| \leq R$. For all $u$ such that $|u| < R$, there then exists a number $q < 1$ such that one has $|u/\omega| < q$ in the retained terms; the difference

$$(23.5) \qquad 1/(u - \omega)^2 - 1/\omega^2 = \omega^{-2} \left[ 2u/\omega + 3(u/\omega)^2 + \dots \right] =$$

$$= \sum_n \omega^{-2}(n + 1)(u/\omega)^n$$

---

[65] His biography in the DSB tells us that in the course of his fourteen years of high-school teaching he had to teach mathematics, physics, German, botany, geography, history, gymnastics "and even calligraphy".

is thus majorised in modulus by

$$|\omega|^{-2}|u/\omega|(2 + 3q + 4q^2 + \ldots) = M|\omega|^{-3}$$

where $M < +\infty$ does not depend on $\omega$. It remains to remark that the series $\sum |\omega|^{-3}$ converges, as we established earlier. The periodicity of the function is less obvious than in the case of the functions (1); we shall obtain it below, although a direct argument would be possible.

One may note that in view of the general associativity theorem (sum over $n$ then over $\omega$), this argument actually proves the unconditional convergence of the sum

$$(23.6) \qquad \sum_{|\omega|>R} \sum_{n>0}(n + 1)\omega^{-2}(u/\omega)^n.$$

We can then permute the $\sum$ that appear in (6); in the disc $|u| < R$, the function $\wp(u)$ is then the sum of the function $1/u^2$, of the functions $1/(u - \omega)^2 - 1/\omega^2$ for the finite number of $\omega$ such that $|\omega| \leq R$, and finally of a power series in $u$ converging in $|u| < R$. The function $\wp$ is thus, in the disc considered, the sum of a power series and of a finite number of analytic functions each having a double pole in a period of modulus $< R$. Since $R$ is arbitrary, the function $\wp$ is analytic apart from at the points of the lattice of periods.

One can in particular, in the above, choose

$$R = \inf |\omega|,$$

the inf being taken over the nonzero periods, so that $R$ is the radius of the largest open disc with centre 0 not containing any nonzero period. The sum (6) is then extended over all the nonzero periods, so that, in the disc considered, one has

$$(23.7) \qquad \wp(u) = 1/u^2 + \sum_{\omega\neq0,n>0}(n + 1)\omega^{-2}(u/\omega)^n.$$

It follows that

$$\wp(u) = 1/u^2 + a_1 u + a_2 u^2 + \ldots$$

with

$$(23.8) \qquad a_n = (n + 1)\sum 1/\omega^{n+2} = (n + 1)G_{n+2}(L),$$

the series being extended over the nonzero periods. Its sum in fact vanishes for $n$ odd since then the terms $\omega$ and $-\omega$ cancel each other, so that one has a series expansion for the function $\wp$

$$(23.9) \qquad \wp(u) = 1/u^2 + a_2 u^2 + a_4 u^4 + a_6 u^6 + \ldots$$

whose coefficients are the sums $3\sum 1/w^4$, $5\sum 1/w^6$,... This formula is valid for $|u| < R$, i.e. in the largest open disc with centre 0 not containing any nonzero period, but clearly not outside it because of the presence on the circumference $|u| = R$ of at least one period, i.e. of a pole of $\wp$.

One could expand the series (1) similarly in the same disc for $k \geq 3$. For example

$$f_3(u) = \sum (u-w)^{-3} = u^{-3} - \sum_{w \neq 0} w^{-3}(1 - u/w)^{-3} =$$

$$= u^{-3} - \frac{1}{2}w^{-3}\sum (n+1)(n+2)(u/w)^n$$

$$= 1/u^3 + \sum b_n u^n$$

with

$$b_n = -\frac{1}{2}(n+1)(n+2)\sum 1/w^{n+3} = -(n+1)a_{n+1}/2$$

by (8). This time $b_n = 0$ for $n$ even, in other words,

$$-2f_3(u) = -2/u^3 + 2a_2 u + 4a_4 u^3 + 6a_6 u^5 + \ldots$$

Comparing with (9), one sees that, for $|u| < R$, one has

(23.10) $$-2f_3(u) = \wp'(u),$$

the derived function of $\wp(u)$ in the sense of the theory of analytic functions. Since the derivative of the general term of (4) is $-2/(u-w)^3$ by the traditional rules of calculus, this result confirms that one can deduce (10) from (4) by differentiating the series[66] which defines $\wp(u)$ term-by-term with respect to $u$. And since $f$ and $\wp'$ are analytic in the *connected* open set $G = \mathbb{C} - L$, (10) remains valid on all of $G$ (n° 20).

However it may be, (10) allows us to show the periodicity of $\wp(u)$; this is not obvious from its definition, in contrast to the case of $f_3(u)$. (10) shows that $\wp'(u+w) = \wp'(u)$, so that the function

$$g(u) = \wp(u+w) - \wp(u),$$

analytic in $G$, has zero derivative. Taylor's formula then shows that it is constant on a neighbourhood of each point of $G$, so on $G$ (principle of analytic continuation). One therefore has a relation

$$\wp(u+w) = \wp(u) + c(w)$$

---

[66] We have shown in n° 19 that one can differentiate a *power* series term-by-term, but here we are dealing with a series whose general term, though analytic, is not a power of $u$.

with constants $c(\omega)$. If one takes for $\omega$ one of the periods of base $\omega_1$, $\omega_2$, it is clear that $\omega/2$ is not a period, which legitimates the formula

$$\wp(\omega/2) = \wp(-\omega/2) + c(\omega).$$

Now the function $\wp$ is clearly even. So $c(\omega) = 0$ in this case, qed.

Although we cannot justify it entirely here, we would not hide from the reader the fundamental property of the function $\wp$: it satisfies the differential equation

$$(23.11) \qquad \wp'(u)^2 = 4\wp(u)^3 - 20a_2\wp(u) - 28a_4,$$

where $a_2$ and $a_4$ are the coefficients (8). The proof is very simple, up to a "detail". By using the power series found above for $\wp$ and $\wp' = -2f_3$, one finds, by a small calculation *à la* Newton that, in the difference between the two sides of (11), the negative powers of $u$ as well as the constant terms cancel. This means that this difference, which *a priori* is analytic in $\mathbb{C}$ with the periods removed, does not have a pole at $u = 0$ and, in fact, is zero for $u = 0$. But this difference is clearly a function that is also doubly periodic like $\wp$ and $\wp'$ and it can have no more poles than those of $\wp$ and $\wp'$, i.e. the periods; if then 0 is not a pole, the other periods will not be either. In other words, the difference is an elliptic function which is holomorphic or analytic *in all* $\mathbb{C}$ *without any exception*; further, it is *bounded on* $\mathbb{C}$, since (by periodicity) it takes the same values on $\mathbb{C}$ as on the *compact* parallelogram constructed on $\omega_1$ and $\omega_2$ (Chap. III, n° 9, Theorem 11). It then remains to invoke – this is the "detail" – a general theorem of Joseph Liouville (Chap. VII) which says that such a function is necessarily constant and so zero if its power series around $u = 0$ has zero constant term.

*Exercise.* Apart from several negative powers of $u$, each of the two sides of (11) is a power series in $u^2$ whose coefficients can in principle be calculated from the coefficients $a_n$ of $\wp(u)$. On writing that these two power series are identical, one obtains strange algebraic relations between the $a_n$, i.e. between the Eisenstein series $G_{2k}(L)$. Perform, *à la* Newton, this calculation for the coefficients of $u^2$ and of $u^4$.

The equation (11) allows one to integrate the square roots of polynomials of third degree *if* one knows that one can always choose the lattice of periods so that the coefficients $a_2$ and $a_4$ take values given in advance[67]. Correct, but very far from obvious (Chap. XII, n° 18).

Traditionally one writes (11) in the form

---

[67] One has to exclude the case, which can be treated elementarily, where the polynomial $4X^3 - 20a_2X - 28a_4$ has a double root. For a polynomial of the form $X^3 + pX + q$, this means $4p^3 + 27q^2 = 0$ (write that it and its derivative have a common root).

(23.12) $$\wp'(u)^2 = 4\wp(u)^3 - g_2\wp(u) - g_3$$

where the coefficients

(23.13) $$g_2 = 60 \sum_{w \in L} 1/w^4 = 60G_4(L), \quad g_3 = 140 \sum_{w \in L} 1/w^6 = 140G_6(L)$$

depend on the lattice $L$ of periods and where one omits $w = 0$ from the summation.

Given that in general

$$\text{Im}(z) > 0 \Longleftrightarrow \text{Im}(1/z) < 0$$

for all nonreal $z \in \mathbb{C}$, one can always assume that the base $w_1, w_2$ of $L$ satisfies $\text{Im}(w_2/w_1) > 0$, by swapping the two periods if need be; on putting

(23.14) $$z = w_2/w_1, \quad \text{whence} \quad \text{Im}(z) > 0,$$

it is clear that in general

(23.15) $$G_k(L) = \sum 1/w^k = w_1^{-k} \sum 1/(cz + d)^k$$

where the summation is – change of notation – extended over all pairs $(c, d) \in \mathbb{Z}^2$ apart from $(0, 0)$. On putting

(23.16) $$G_k(z) = \sum 1/(cz + d)^k$$

for $\text{Im}(z) > 0$ and $k$ even $> 2$, one thus has

(23.17) $$g_2 = 60w_1^{-4}G_4(z), \qquad g_3 = 140w_1^{-6}G_6(z).$$

It is no less clear from the definitions that the Weierstrass function $\wp$ and its derivative $\wp'$ depend on the lattice $L$ in a way analogous to the relation (15), after replacing the variable $u$ by $w_1 u$. It is therefore unprofitable to study lattices of general period; it suffices to study the elliptic functions having periods 1 and $z$, with $\text{Im}(z) > 0$.

The study of (11) thus reduces to that of the differential equation

$$\wp'(u)^2 = 4\wp^3(u) - 60G_4(z)\wp(u) - 140G_6(z)$$

for $z$ given, $\text{Im}(z) > 0$, where the function $\wp$ is given by

$$\wp(u) = 1/u^2 + \sum \left[ \frac{1}{(u - cz - d)^2} - \frac{1}{(cz + d)^2} \right],$$

the summation being extended over all pairs $(c, d)$ of rational integers not simultaneously zero.

Thus it is all governed by the properties of the Eisenstein series (16). This is the theory of *modular functions* which, for a century and a half, has spurred exciting research (for those enthused by it) and great generalisations which chase one another incessantly. Although it now uses the most "modern" methods and results, it is one of the more spectacular achievements of classical analysis. We shall give an idea of it in Chap. XII of Vol. IV.

# III – Convergence: Continuous variables

§1. The intermediate value theorem – §2. Uniform convergence –
§3. Bolzano-Weierstrass and Cauchy's criterion –
§4. Differentiable functions – §5. Differentiable functions of several variables

## §1. The intermediate value theorem

### 1 – Limit values of a function. Open and closed sets

The concept of a limit value of a function defined on a set $X \subset \mathbb{C}$ whose elements are not necessarily integers has already been introduced in n° 4 of Chap. II, but we have hardly used it up to now. In this chapter we shall develop it much further.

(a) *Limit as $x \to +\infty$*. Here we suppose that $X \subset \mathbb{R}$ is not bounded above. We then say that $f(x)$ tends to a limit $u$ as $x$ tends to $+\infty$ if, for all $r > 0$,

(1.1) $$d[u, f(x)] < r \quad \text{for all sufficiently large } x \in X,$$

i.e. for $x > N$ where in general $N$ depends on $r$. We use the notation

$$u = \lim_{\substack{x \to +\infty \\ x \in X}} f(x)$$

omitting "$x \in X$" when there is no likelihood of ambiguity.

For example, for any integer $n > 1$,

$$\lim_{x \to +\infty} 1/x^n = 0$$

since $1/x^n < 1/x$ for $x > 1$, whence $|1/x^n| < r$ for $x > \max(1, 1/r)$.

(b) *Limit as $x \to -\infty$*. Almost identical definition. In decimal language: for every $n \in \mathbb{N}$ there exists a $p \in \mathbb{N}$ such that

(1.2)           $(x \in X)$ & $(x < -10^p) \implies d[u, f(x)] < 10^{-n}$.

(c) *Limit as* $x \to a \in \mathbb{C}$; this is by far the most important case. Here one demands that for all $r > 0$

(1.3)           $|u - f(x)| < r$ for all $x \in X$ sufficiently close to $a$,

i.e. that there exists an $r' > 0$ such that, for all $x \in X$,

(1.4)                      $d(a, x) < r' \implies d[u, f(x)] < r$.

Then one writes

$$u = \lim_{\substack{x \to a \\ x \in X}} f(x).$$

This is the type of situation which we met in the complex case in Chap. II, n° 19, in showing that the sum $f(z)$ of a power series is differentiable at all points $a$ in the disc of convergence $D$: in this case $X = D - \{a\}$, the function to consider is the quotient of $f(z) - f(a)$ by $z - a$ (not defined for $z = a$), and the limit value when $z$ tends to $a$ is the sum $f'(a)$ of the derived power series.

Before proceeding, some fundamental remarks.

In the first place, one neither insists on the point $a$ belonging to the set $X$ on which the function is defined, nor does one forbid this. It can happen, in the case where the function $f$ is defined on a set $E$ containing $a$, that one studies the behaviour of $f$ on the set $X = E - \{a\}$ obtained by deleting $a$ from $X$, or on the set $X$ of $x \in E$ such that $x > a$, etc. This then has to be specified through the notation, for example in the following way:

$$\lim_{\substack{x \to a, x \neq a \\ x \in E}} \quad \text{or} \quad \lim_{\substack{x \to a, x > a \\ x \in E}};$$

the second case, which arises mainly in the theories of integration and of Fourier series, assumes that one is working in $\mathbb{R}$ since inequalities are not defined between non-real complex numbers.

In the second place, we note that the concept of convergence when $x \in X$ tends to a point $a$ imposes an hypothesis on $a$. Relative to $X$, the points of $\mathbb{R}$ (or of $\mathbb{C}$ according to the case under consideration) decompose into two disjoint sets:

(i) First, it may happen that there exists a ball $B$ with centre $a$ and of radius $R > 0$ such that $B \cap X$ is *empty*, in which case one says that $a$ is *exterior* to $X$ in $\mathbb{R}$ (or $\mathbb{C}$); in this case, the relation (4) is trivially satisfied for all $u$ and $r$ once $r' < R$ since then there is nothing to verify. This shows that if one takes the definition literally, a function defined on $X = (0, 1)$ may, as $x \in X$ tends to 2, converge indifferently to 1815, to $\pi$ or to $-10^{123}$; absurd. So we have to exclude those points exterior to $X$ in the definition.

It is important to understand that though points exterior to $X$ are clearly "outside" $X$ the converse does not hold: if $X$ is the interval $[0, 1[$ the point 1 is outside $X$ but is not exterior to it, neither in $\mathbb{R}$ nor in $\mathbb{C}$; in this example, the points of $\mathbb{R}$ (resp. $\mathbb{C}$) exterior to $I$ are those which satisfy $x < 0$, or $x > 1$, with strict inequalities (resp. these same points *together with* the non-real points). For an arbitrary interval $I$ the points of $\mathbb{R}$ (or of $\mathbb{C}$) exterior to $I$ are those which do not belong to the *closed* interval having the same end points as $I$. In $\mathbb{C}$, the points exterior to a ball with centre $a$ and of radius $R$ are those which satisfy the strict inequality $|z - a| > R$.

(ii) On the contrary it can happen that *every* ball $B$ with centre $a$ meets $X$; one then says that $a$ is *adherent* to $X$, or is a *cluster point* of $X$. In this case, the limit $u$, if it exists, is unique: if $f(x)$ tends simultaneously to $u$ and to $v$, then for any $r > 0$ there exists an $x \in X$ satisfying the two inequalities $d[u, f(x)] < r$ and $d[v, f(x)] < r$, since each is separately satisfied on a neighbourhood of $a$; whence $|u - v| < 2r$ and finally $u = v$. See n° 3 of Chapter II.

Every $x \in X$ is clearly adherent to $X$, but the converse is false: the adherent points of an arbitrary ball are those of the corresponding *closed* ball. In the general case, for any $n$ there is an $x_n \in X$ such that $d(a, x_n) < 1/n$, which shows that a point adherent to $X$, though not necessarily belonging to $X$, is *the limit of a sequence of points of $X$*. The converse is obvious since every ball $B(a, r)$ contains $x_n \in X$ for $n$ large. A set $X$ containing all its adherent points is said to be *closed*, which justifies the terminology adopted in Chap. II, n° 2 for intervals and balls; in the general case the set of adherent points to $X$ is called the *adherence* (or the *closure*) of $X$ and is often denoted by $\bar{X}$. For subsets of $\mathbb{R}$ one does not need to distinguish between "closed in $\mathbb{R}$" and "closed in $\mathbb{C}$": the adherent points are the same. The empty set is both open and closed.

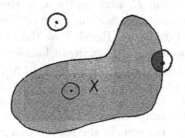

fig. 1.

If a point $a$ is exterior to a set $X \subset \mathbb{R}$ (resp. $\mathbb{C}$) then the complement $Y = \mathbb{R} - X$ (resp. $\mathbb{C} - X$) of $X$ contains an open ball of $\mathbb{R}$ (resp. $\mathbb{C}$) with centre $a$; one says then that $a$ is *interior* to $Y$ in $\mathbb{R}$ (resp. $\mathbb{C}$), or interior for short if no confusion is possible. The interior points in $\mathbb{R}$ of an interval $(a, b)$ are those which satisfy $a < x < b$, with strict inequalities; but they are clearly not interior, in $\mathbb{C}$, to the interval in question. In $\mathbb{C}$, the interior points of a ball $B$ with centre $a$ and of radius $R$ are those of the *open* ball with the same centre and radius. Indeed, if a point $z$ lies on the circumference then every open ball with centre $z$ meets both $B$ and $\mathbb{C} - B$; so it is neither interior nor exterior to them, though it is adherent to both $B$ and $\mathbb{C} - B$. This argument extends to any set $X$: to the splitting of points of $\mathbb{R}$ (resp. $\mathbb{C}$) into the exterior and adherent points of $X$ there corresponds a splitting of the adherent points to $X$ into the interior and *boundary points* of $X$ (or of the complement of $X$: these are *clearly* the same). If one chooses for $X$ a circumference $d(a, x) = R$, then all the points of $X$ are boundary points and all those $z \notin X$ are exterior; no point is interior to $X$, and $\bar{X} = X$.

If all the points of a set are interior to it one says that the set is *open*; this definition, here again, tallies with that of Chap. II when applied to intervals or balls, and with that which we have given *à propos* analytic functions in n° 6 and in n° 19: an open subset $G$ of $\mathbb{C}$ must, for any $a \in G$, contain an open ball with centre $a$. One should pay attention to the fact that, contrary to the concept of a closed set, that of an open set is not the same in $\mathbb{R}$ as in $\mathbb{C}$: a "ball" in $\mathbb{R}$ is an interval, which is not a ball in $\mathbb{C}$. The interval $]0, 1[$ is open in $\mathbb{R}$ but not in $\mathbb{C}$.

By definition, a set $X \subset \mathbb{R}$ (resp. $\mathbb{C}$) is closed if and only if every point of the complement $Y$ of $X$ in $\mathbb{R}$ (resp. $\mathbb{C}$) is exterior to $X$, i.e. interior to $Y$ in $\mathbb{R}$ (resp. $\mathbb{C}$); in other words, *the complement in $\mathbb{R}$ (resp. $\mathbb{C}$) of a set closed in $\mathbb{R}$ (resp. $\mathbb{C}$) is open in $\mathbb{R}$ (resp. $\mathbb{C}$), and conversely*.

These sets possess other important properties that are easy to establish. First, *the union of any family $(U_i)$ of open sets is an open set*: for a point $a \in \bigcup U_i$ belongs to some $U_i$ which contains a ball with centre $a$ necessarily contained in the union. On the other hand, *the intersection of a finite family of open sets is an open set*; for a point $a \in \bigcap U_i$ belongs to each $U_i$, a set which contains an open ball $B_i$ with centre $a$, so that $\bigcap U_i \supset \bigcap B_i$; now it is clear that the intersection of a *finite* number of open balls with centre $a$ is again an open ball with centre $a$. Recall that the radius of an open ball is, by definition, strictly positive. The argument fails for an infinite intersection: that of the intervals $] - 1/n, 1/n[$ reduces to $\{0\}$ (Archimedes' axiom).

The corresponding properties of closed sets: *every intersection of closed sets $(F_i)$ is a closed set; the union of a finite number of closed sets is a closed set*. In the first case, the complement of the intersection is the union of complements, open, of $F_i$, so is open. In the second case, the complement of the union is the intersection of the complements, open.

After these set-theoretic gymnastic exercises let us return to the limit values of a function. A first property is that *if $f(x)$ tends to a limit $u$ when $x \in X$ tends to $a$, then, for every sequence of points $x_n \in X$, the relation*

$$(1.5) \qquad \lim x_n = a \quad \text{implies} \quad \lim f(x_n) = u.$$

For any $r > 0$ there is an $r' > 0$ such that, for $x \in X$, the inequality $d(x, a) < r'$ implies $d[f(x), u] < r$; but for $n$ large one has $d(x_n, a) < r'$ and so $d[f(x_n), u] < r$, qed.

Conversely one can show that if, for every sequence $x_n \in X$ converging to $a$, the sequence of values $f(x_n)$ converges to a limit which, *a priori*, depends on the sequence considered, then that limit is in reality independent of it, and $f(x)$ tends to the latter when $x \in X$ tends to $a$. This converse is rarely used, but since it may oblige the reader to revise his ideas of Socratic logic, always useful, let us give the proof.

The fact that the limit is always the same is immediate: if $f(x_n)$ tends to $u$, if $f(y_n)$ tends to $v$ and if one denotes by $(z_n)$ the sequence

$$x_1, y_1, x_2, y_2, \ldots$$

obtained by interlacing the two given sequences, it tends to $a$ as do the two chosen sequences, so the sequence $f(z_n)$ can converge only if $u = v$. It remains to show that $f(x)$ tends to this limit $u$ common to all the sequences considered. We argue by contradiction, assuming this assertion false. Now the relation $\lim f(x) = u$ means that "for all $r > 0$ *there exists* an $r' > 0$" having a certain property. If such is not the case, *there exists* an $r > 0$ such that, *for all* $r' > 0$, the property in question is false. This property is that the relations $x \in X$ and $|x - a| < r'$ imply $|f(x) - u| < r$. The negation of this property is that *there exists* an $x \in X$ satisfying $|x - a| < r'$ while *not* satisfying $|f(x) - u| < r$, so satisfying $|f(x) - u| \geq r$.

Applying this argument to $r' = 1$, $1/2$, $1/3$, etc., one finds a sequence of points $x_n \in X$ satisfying

$$|x_n - a| < 1/n \quad \text{and} \quad |f(x_n) - u| \geq r \text{ for all } n.$$

In other words, contrary to the hypothesis, there exists a sequence $x_n \in X$ which converges to $a$ but for which $f(x_n)$ does not converge to $u$, qed.

A second type of property of limits concerns what happens when one performs simple algebraic operations on functions which tend to limits. We have to distinguish several cases; we restrict to the case of a real variable which increases indefinitely, the statements and proofs being practically identical in the others, both in $\mathbb{R}$ and in $\mathbb{C}$. The theory is really nothing new, since the essential ideas, if one can speak of ideas, have already been explained for sequences.

**Theorem 1.** *Let $f$ and $g$ be scalar functions defined on a set $X \subset \mathbb{R}$ which is not bounded above, and suppose that $f(x)$ and $g(x)$ tend to $u$ and $v$ as $x \in X$ increases indefinitely. Then $f(x) + g(x)$ and $f(x)g(x)$ tend respectively to $u + v$ and $uv$. If $v \neq 0$ then $g(x) \neq 0$ for $x$ large and the function $f(x)/g(x)$ tends to $u/v$.*

The proofs are those of Theorem 1 of Chap. II, n° 8: one replaces $u_n$ and $v_n$ by $f(x)$ and $g(x)$ and "for $n$ large" by "for $x$ large", or, in the case where $x$ tends to a finite $a$, by "for $x$ sufficiently close to $a$". In particular, and this is important, we note that if the limit $v$ of $g$ is $\neq 0$ then $g(x) \neq 0$ for all $x \in X$ sufficiently close to $a$, for

$$d[v, g(x)] < |v|/2$$

on a neighbourhood of $a$, so $|g(x)| \geq |v|/2 > 0$.

## 2 – Continuous functions

Let $f$ be a scalar function defined on a set $X \subset \mathbb{C}$ (it is unnecessary, when considering continuity, to specify whether in $\mathbb{R}$ or in $\mathbb{C}$). Under what conditions does $f(x)$ tend to a limit $u$ when $x \in X$ tends to an $a \in X$?

If it does, one can, for all $r > 0$, find an $r' > 0$ such that, for $x \in X$, the relation $d(a, x) < r'$ implies $d[u, f(x)] < r$. Since $a \in X$ and since $d(a, a) = 0 < r'$, we will have $d[u, f(a)] < r$ for all $r > 0$. *The limit must be $f(a)$.*

It remains to express the relation

$$\lim_{x \to a} f(x) = f(a),$$

which, by the definition of a limit, means that *for all $r > 0$ one has* $d[f(a), f(x)] < r$ *for all $x \in X$ sufficiently close to $a$*, or again that there exists an $r' > 0$ such that, in fairly correct logical language – omitting those jarring "quantifiers" $\forall$ and $\exists$ to spare the reader –, the following assertion is true:

(2.1)        $\{(x \in X) \ \& \ (|x - a| < r')\} \Longrightarrow \{|f(x) - f(a)| < r\}.$

If so, one says that $f$ is *continuous at the point $a$*. If $f$ is continuous at every point $a \in X$, one says simply that $f$ is *continuous on $X$*.

While the concept of continuity is a simple particular case of that of limit value defined in the preceding n°, conversely the latter reduces immediately to continuity: for $f$, defined on a set $X$, to tend to $u$ as $x \in X$ tends to an $a \notin X$ it is necessary and sufficient that the function $g$ defined on the set $X' = X \cup \{a\}$ by $g(x) = f(x)$ for all $x \in X$ and $g(a) = u$ be continuous at $a$; this follows directly from the definitions.

We note that while the concept of a limit value when $x \in X$ tends to $a$ is absurd for $a$ exterior to $X$, that of continuity is of no interest when $a$ is an *isolated point* of $X$, i.e. if there exists a ball $B(a, R) = B$ such that $B \cap X = \{a\}$; for in this case the one and only point of $X$ arbitrarily close to $a$ is $a$ itself; $f(x)$ is automatically arbitrarily close to $f(a)$ when $x \in X$ tends to $a$ (i.e. is equal to $a$).

Continuity can be expressed in decimal language: to calculate $f(a)$ to within $10^{-n}$ it suffices to know a sufficiently large number of decimal places of $a$. This is why "users" have a marked weakness for continuous functions (and, in fact, for functions more special than just continuous). Dirichlet's function, equal to 1 for $x$ rational and to 0 for $x$ irrational, does not lend itself to numerical analysis even though it was very interesting to the founders of the set theory, who invented far more bizarre functions[1]; one finds good approximations of them today in the business of the computer science, under the name of "fractals".

One can translate this fundamental definition into yet another language. Let us say that a function defined onto a set $M$ (and maybe elsewhere) is *constant to within $r$ on $M$* if

$$d[f(x'), f(x'')] \leq r \ \text{ for any } x', x'' \in M.$$

Continuity at $a$ can then be expressed by saying that *for all $n \in \mathbb{N}$, there exists an open ball $B$ with centre $a$ such that $f$ is constant to within $10^{-n}$ on* $X \cap B$. This condition is clearly sufficient since it shows in particular that, for all $n$, one has $|f(x) - f(a)| \leq 10^{-n}$ for all $x \in X$ sufficiently close to $a$. If, conversely, $f$ is continuous at $a$, then $|f(x) - f(a)| \leq r/2$ for all $x \in X \cap B$, where $B$ is a suitably chosen ball with centre $a$, whence $|f(x') - f(x'')| \leq r$ for all $x', x'' \in X \cap B$. We have seen, for example, in Chap. II (n° 4, 10, 14), that the functions $x^n$ ($n \in \mathbb{N}$) and $\exp x$ are continuous on $\mathbb{C}$ and that the function $\log x$ is continuous (and even differentiable) on $\mathbb{R}_+^*$. Theorem 1 provides the following result immediately:

**Theorem 2.** *Let $f$ and $g$ be scalar functions defined on a set $X \subset \mathbb{C}$ and let $a$ be a point of $X$ where $f$ and $g$ are continuous. Then the functions*

$$f + g : x \longmapsto f(x) + g(x) \ \text{ and } \ fg : x \longmapsto f(x)g(x)$$

*are continuous at $a$. If $g(a) \neq 0$ then $g(x) \neq 0$ on a neighbourhood of $a$ and the function $f/g : x \mapsto f(x)/g(x)$, defined for $g(x) \neq 0$, is continuous at $a$.*

First consequence: on $\mathbb{R}$, *every rational function $h(x) = p(x)/q(x)$, where $p$ and $q$ are polynomials, is continuous at any $x$* [of course, one excludes the points where $q(x) = 0$]. Indeed, the constant functions and the function $x \mapsto x$ are trivially continuous; hence all monomials $ax^n$, then every polynomial, then every quotient of polynomials.

There is a similar statement for $\mathbb{C}$. It is of course clear that the functions $\mathrm{Re}(z)$ and $\mathrm{Im}(z)$ are continuous on $\mathbb{C}$ since, for example,

---

[1] Let $f$ be a function defined on an interval $I$, with complex values. If $f$ is continuous, one can consider the point $f(t)$ of $\mathbb{C}$ as a point moving in the plane as a function of time, which suggests the naïve idea that the "trajectory" is a highly regular curve "of one dimension". In 1890, a clever Italian, Giuseppe Peano, also one of the creators of mathematical logic, demonstrated a quite simple construction of a trajectory which passes through *all* the points of a square, in other words, a *continuous and surjective* map $I \longrightarrow I \times I$; "bijective" is impossible. The effect on the mathematicians was no less sensational than that of Weierstrass' non-differentiable continuous function mentioned later. See Hairer and Wanner, *Analysis by Its History*, pp. 289 and 296 (pictures).

$$|\text{Re}(z) - \text{Re}(a)| = |\text{Re}(z-a)| \le |z-a|.$$

On putting $z = x + iy$, the functions $z \mapsto x^p y^q$ are similarly continuous for all $p, q \in \mathbb{N}$, hence so is every function of $z$ that can be expressed in terms of polynomials with coefficients in $\mathbb{C}$ in the two real variables $x$ and $y$, hence also every function of the form $h(z) = p(x,y)/q(x,y)$ where $p$ and $q$ are polynomials. But here the points to exclude, defined by the equation $q(x,y) = 0$, are not necessarily finite in number as in the case of $\mathbb{R}$: the function $x^2/(x^2 - y^2 - 1)$ is defined only off the equilateral hyperbola $x^2 - y^2 = 1$.

This example is interesting in that it shows us that the quotient $p/q$ is defined on an *open* subset of $\mathbb{C}$. This is a perfectly general fact: *if $f$ is a function defined and continuous on $\mathbb{R}$ (resp. $\mathbb{C}$) then the relation $f(x) \ne 0$ defines an open subset of $\mathbb{R}$ (resp. $\mathbb{C}$)*. If indeed $f(a) \ne 0$, we again have $f(x) \ne 0$ on a neighbourhood of $a$, as we saw in Theorem 2; the point $a$ is thus interior to the set $f \ne 0$.

More generally: *for any open $V$ in $\mathbb{R}$ (resp. $\mathbb{C}$), the inverse image $f^{-1}(V) = U$ is open in $\mathbb{R}$ (resp. $\mathbb{C}$)*. For, given $a \in U$ and $b = f(a) \in V$, there exists an open ball $B(b, r) \subset V$; since $f$ is continuous, there exists an open ball $B(a, r')$ which $f$ maps into $B(b, r)$; thus $B(a, r') \subset U$, qed. This result remains valid for a function defined on an arbitrary set $X \subset \mathbb{R}$ (for example) on condition that one calls a subset $U$ of $X$ *open in $X$* if it possesses the following property: for all $a \in U$, $U$ contains all the $x \in X$ (not $\mathbb{R}$ or $\mathbb{C}$) that are sufficiently close to $a$. All this generalises immediately (and becomes clearer) in the framework of "metric spaces", described in the Appendix.

The *analytic functions* form another class of continuous functions on $\mathbb{C}$. This is immediate. From the definition of analytic functions (Chap. II, n° 19), $f(a + h)$ is, for $|h|$ small, a power series in $h$, with constant term $f(a)$; by Chap. II, (14.7), we thus have

$$f(a+h) - f(a) = O(|h|), \qquad |h| \to 0,$$

from which continuity is clear.

Every reasonable algebraic operation (i.e. excluding division by 0) performed on continuous functions yields a continuous function again (Theorem 2). Another rather important operation, though not algebraic, possesses the same property:

**Theorem 3 (continuity of composite functions).** *Let $X$ and $Y$ be two subsets of $\mathbb{C}$, let $a$ be a point of $X$, and $b$ a point of $Y$; let $f$ be a map of $X$ into $Y$ such that $f(a) = b$ and $g$ a map of $Y$ into $\mathbb{C}$. Suppose that $f$ is continuous at the point $a$ and $g$ is continuous at the point $b$. Then the composite function*

$$h = g \circ f : x \longmapsto g[f(x)]$$

*is continuous at $a$.*

Direct proof: for any $r > 0$ there is an $r' > 0$ such that $|g(y) - g(b)| < r$ for $|y - b| < r'$, then an $r'' > 0$ such that $|f(x) - f(a)| < r'$ for $|x - a| < r''$. It is now clear that $|x - a| < r''$ implies $|h(x) - h(a)| < r$, qed.

If for example one knows that the functions $\sin x$ and $\cos x$ are continuous on $\mathbb{R}$ one can deduce that the function $\sin(\cos x)$ is so too.

*Example 1.* The function $f(x) = \log[\exp(x)]$ is continuous on $\mathbb{R}$. This result, though trivial, nevertheless has an important consequence. For

$$f(x + y) = \log[\exp(x + y)] = \log[\exp(x)\exp(y)] = \log[\exp(x)] + \log[\exp(y)],$$

in other words $f(x + y) = f(x) + f(y)$. It follows immediately that

$$f(x + 0) = f(x) + f(0) \quad \text{whence} \quad f(0) = 0,$$

then that $0 = f(x - x) = f(x) + f(-x)$, whence $f(-x) = -f(x)$. Moreover, $f(nx) = f(x + \ldots + x) = nf(x)$ for $n \in \mathbb{N}$, then for $n \in \mathbb{Z}$ since $f(-nx) = -f(nx) = -nf(x)$. This also shows that $f(x/n) = f(x)/n$ for $n \neq 0$, whence $f(px/q) = pf(x)/q$ for $p, q \in \mathbb{Z}$, $q \neq 0$. On putting $f(1) = a$, one thus has $f(x) = ax$ for all $x \in \mathbb{Q}$. But since $f$ is continuous, this relation persists when $x \in \mathbb{Q}$ tends to any limit in $\mathbb{R}$ [2]. Conclusion: there exists a constant $a = \log[\exp(1)]$ such that

$$\log[\exp(x)] = ax \quad \text{for all } x \in \mathbb{R}.$$

This argument, alas, will not show that $a = 1$ since it applies equally well to the function $3 \log[\exp(x)]$.

But we know that, by definition, $\log t = \lim n(t^{1/n} - 1)$, from which $\log[\exp(x)] = \lim n\left[\exp(x)^{1/n} - 1\right]$. Now the addition formula for the exponential function shows that $\exp(x) = \exp(n.x/n) = \exp(x/n)^n$ for any $x \in \mathbb{R}$ and so

$$\log[\exp(x)] = \lim n[\exp(x/n) - 1].$$

The power series $\exp(t) = 1 + t + t^2/2! + \ldots$ shows on the other hand that $\exp(t) - 1 \sim t$ when $t$ tends to 0. Consequently $\exp(x/n) - 1 \sim x/n$; since the ratio of the two sides of this relation is equal to $n[\exp(x/n) - 1]/x$, and tends to 1, we deduce that

$$\lim n[\exp(x/n) - 1] = x,$$

whence the relation we seek:

---

[2] More generally, let $f$ and $g$ be two functions defined and continuous on the closure $\bar{X}$ of a set $X$ and suppose that $f = g$ on $X$; then $f = g$ on $\bar{X}$. For if $x_n \in X$ tends to $x \in \bar{X}$ then $f(x_n)$ and $g(x_n)$, which are equal, tend to $f(x)$ and $g(x)$ respectively.

(2.3)                    $\log[\exp(x)] = x$  for all $x \in \mathbb{R}$.

Since the functions log and exp are injective we can also write (3) in the form

(2.3')                   $\exp(\log y) = y$  for all $y \in \mathbb{R}$, $y > 0$.

We shall go over all this in detail in Chap. IV.

Most of the functions that one meets – or that the Founders met – at the start of analysis are continuous and even much more, because they are given by "formulae" which are quasi-algebraic or originate from geometry, mechanics or physics, or from power series, etc. This explains why the very concept of continuity was not isolated explicitly before Bolzano and Cauchy, around 1820. Furthermore, one has no desire to traumatise the innocent neophyte with horrors he has never yet seen.

But it is not necessary to go very far to find one. Electricians test their apparatus by setting them to reproduce *square waves*, i.e. functions represented by a graph of the type below:

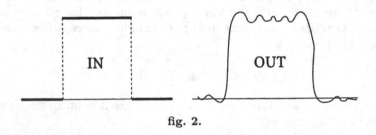

fig. 2.

As we shall see *à propos* Fourier series, the "square waves" arose for the first time in a very different problem and, for the age, even more theoretical: the propagation of heat, more precisely the evolution of the temperature of a circular metallic ring whose two halves are initially at different temperatures, say $-273°$ and $+3000°$ Celsius for realism. The first series which Fourier exhibits in his *Mémoire sur la propagation de la chaleur* of 1809 is

$$(2.4) \quad \cos x - \cos 3x/3 + \cos 5x/5 - \ldots = \begin{cases} +\pi/4 & \text{for} \quad |x| < \pi/2 \\ 0 & \text{for} \quad |x| = \pi/2 \\ -\pi/4 & \text{for} \quad \pi/2 < |x| < \pi \end{cases}$$

whose complete graph the reader will trace easily, observing that the left hand side has period $2\pi$; these are precisely the square waves. From the electronic point of view, the signal is obtained by superimposing a fundamental frequency and all its odd harmonics, with phase shifts and intensities specified by the signs and coefficients in the formula.

We observe in passing an apparently strange consequence of the preceding Fourier series: *a series whose terms are everywhere continuous functions of a real variable $x$ and which converges everywhere may have a discontinuous function for its sum.* We shall return to this very important problem in n° 5 and 6. For the moment we only remark that Fourier's formula assumes that his series converges; the proof, unfortunately, is not entirely obvious. We shall give it at the end of n° 11 since it provoked one of the first efforts to ground analysis on a solid base, and applies to many similar series.

Most of the engineers and technicians who use these series find no problem: the result is part of the folklore of the profession to the same extent as the dangers of alternating current. The "high fidelity" journals sometimes speak without explanation of the unpropitious influence of the "odd harmonics" on the fidelity of the hi-fi: this impresses the punters to the same extent as the mysterious "RMS power" (root mean square) defined by an integral which we shall meet again *à propos* Fourier series.

## 3 – Right and left limits of a monotone function

Let $f$ be a real function defined on a subset $X$ of $\mathbb{R}$. We say that $f$ is *increasing* (in the wide sense) if, for $x, y \in X$, the relation $x \leq y$ implies $f(x) \leq f(y)$; on substituting strict for wide inequalities we obtain the *strictly increasing* functions; this is the case of $\log x$ as we saw in Chap. II, n° 10, and also of the function exp since $\exp(x + h) = \exp(x)\exp(h) > \exp x$ if $h > 0$ (use the power series in $h$). We define decreasing and strictly decreasing functions similarly. When a function is increasing or decreasing one may say that it is *monotone*, not specifying more precisely which. When $X$ is an interval, it may happen that one can decompose $X$ into subintervals so that the function $f$, while not monotone on all of $X$, is monotone on each of these subintervals; one says then that $f$ is *piecewise monotone* in $X$. This is the case for all the elementary functions whose graph one requires the future graduates to trace on millimeter graph paper; but contemplating these drawings, all the less realistic as they are more artistic, would never give an idea of the level of complexity a monotone function can attain; you will find an example in Chap. V, n° 32 of an increasing function which is discontinuous at all the points of $\mathbb{Q}$ and continuous elsewhere.

Now consider a real function, defined and increasing on a subset $X$ of $\mathbb{R}$, and let $c$ be an adherent point of $X$; let us write $E$ for the set of $x \in X$ such that $x < c$ (strict inequality) and confine ourselves to considering $f$ on $E$, supposing that $E$ is nonempty and that $c$ is still adherent to $E$. When $x \in E$ tends to $c$ while remaining $< c$, the values taken by $f(x)$ increase (in the wide sense); the arguments of Chap. II, n° 9 now suggest that $f(x)$ tends to a limit value, maybe $+\infty$. Arguing as in Chap. II, we are led to consider all the numbers $M$ which majorise $f(x)$ for all $x < c$ and to show that $f(x)$ converges to the least of these numbers (or to $+\infty$ if no such exists), namely $u = \sup(f(E)) \leq +\infty$, the least upper bound of the set $f(E)$ of values taken by $f$ on $E$; and this is so. If $u$ is finite, then for any $r > 0$ there exists a $c' \in E$ such that $f(c') > u - r$; we have $c' < c$ whence, as in Chap. II,

$$u - r < f(x) \leq u \text{ for all } x \in X \text{ such that } c' < x < c,$$

i.e. for all $x \in E$ sufficiently close to $c$. So clearly $f$ tends to $u$. The case where $u = +\infty$ is even more obvious: for any $M \in \mathbb{R}$, there is a $c' < c$ in $E$ such that $f(c') > M$, whence $f(x) > M$ for $c' < x < c$. In conclusion:

**Theorem 4.** *Let $X$ be a subset of $\mathbb{R}$, let $f$ be a real function defined and increasing on $X$, and let $c$ be an adherent point of the set $E$ of $x \in X$ such that $x < c$. Then $f(x)$ tends to a limit when $x \in X$ tends to $c$ remaining $< c$, and*

$$(3.1) \qquad \lim_{x \to c, x < c} f(x) = \sup(f(E)) \leq +\infty.$$

The preceding theorem applies mainly to the case where $X$ is an interval, but we shall use it to define real powers in the next chapter, in the case where $X = \mathbb{Q}$. The fact that the point $c$ has been deleted from $E$ is essential. If we choose $E$ to be the set of $x \in X$ such that $x \leq c$, then to say that $f(x)$ tends to a limit as $x \in E$ tends to $c$ would mean that the restriction of $f$ to the set $E$ was *continuous* at the point $c$ as we saw at the start of n° 1 or, if one prefers, that $f(x)$ tends to $f(c)$ as $x$ tends to $c$ remaining $< c$; one says then that $f$ is *continuous on the left* at the point $c$. But there is no reason for this to happen: consider on $X = [0,1]$ the function equal to 0 for $x < 1/2$ and to 1 for $x \geq 1/2$ and take $c = 1/2$: when $x$ tends to $1/2$ remaining in $[0, 1/2[$, it tends to $0 \neq f(c)$, but has no limit when $x$ tends to $1/2$ in the closed interval $[0, 1/2]$. Square waves are similar.

We have supposed, in the preceding theorem, that $x$ tends to $c$ remaining $< c$; there is an analogous statement for the case where $x$ remains $> c$, replacing $E$ by the set of $x \in X$ such that $x > c$ (supposing $c$ adherent to this set, i.e. a limit of points of $X$ all $> c$).

Irrespective of Theorem 4, it frequently happens that given a function, *monotone or not*, defined on a set $X \subset \mathbb{R}$ of which $a$ and $b$ are the lower and upper bounds, one is interested in what happens on a neighbourhood of a point $c \in \,]a, b[$, so that there are two cases to envisage if $c$ is adherent to the two subsets of $X$ defined by $x < c$ and $x > c$ (the case of an interval or of $\mathbb{Q}$, for example). In the first case, the limit, if it exists, of $f(x)$ when $x$ tends to $c$ remaining strictly smaller than $c$ is traditionally denoted by

$$(3.2') \qquad f(c - 0) = \lim_{x \to c, x < c} f(x),$$

while the limit when $x$ tends to $c$ remaining strictly greater than $c$ is denoted similarly by

$$(3.2") \qquad\qquad f(c+0) = \lim_{x \to c, x > c} f(x).$$

The numbers $f(c-0)$ and $f(c+0)$ are called the *limits from the left* and *from the right* of $f$ at the point $c$. These limits always exist if $f$ is monotone, even if it is not defined at the point $c$ itself (as in the case $X = \mathbb{Q}$: $c$ can be an irrational number). They exist also when $f$ is defined and continuous at the point $c$, and in fact it is clear that the equalities

$$f(c-0) = f(c) = f(c+0)$$

*characterise* the continuity of $f$ at the point $c$.

If $f$ is *monotone* one has $f(x) \le f(y)$ for $x < c < y$ if $f$ is increasing, and $f(x) \ge f(y)$ if $f$ is decreasing; so, in the first case,

$$f(c-0) \le f(c+0),$$

with the reversed inequality in the second. Now, for all $y > c$, the number $f(y)$ majorises $f(x)$ for any $x < c$; so it majorises the least upper bound $f(c-0)$ of $f(x)$; but then $f(c-0)$ minorises $f(y)$ for any $y > c$, so also minorises the greatest lower bound $f(c+0)$ of these $f(y)$. If, furthermore, $f$ is defined at the point $c$ (the case where $X$ is an interval for example), the relation $x < c < y$ implies $f(x) \le f(c) \le f(y)$, whence one concludes, by the same argument, that

$$f(c-0) \le f(c) \le f(c+0).$$

Figure 3 (or Fourier's square waves at the point $x = \pi/2$) shows that these

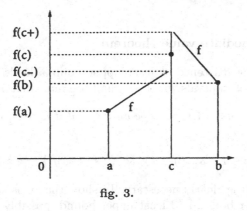

fig. 3.

three numbers can well be different.

It was Dirichlet, who, in trying to prove Fourier's formulae, introduced these concepts in 1829 for any kind of function; but clearly the limits from the left and right do not always exist beyond the case of monotone or continuous functions.

Here we must invoke the ghost of G. H. Hardy again: the symbols $c + 0$ and $c - 0$ do not denote the number $c$; if such were the case, one could indeed deduce corollaries as to the mental equilibrium of mathematicians. They have no meaning in themselves, and their sole legitimacy is to figure in the context where they have just appeared. Some people have recently invented the notation $f(c+)$ and $f(c-)$, mainly so as not to "traumatise" the little dears who, it appears, might believe that if one writes $c+0$ rather than $c$, this must be because in Transcendent Mathematics, as one sometimes called it in the romantic age, $c+0$ is not always equal to $c$. But in the presence of the notation $f(c+)$, these same will ask: "$c$ plus what?"; one will have to reply: "$c$ plus nothing", whence the following question: "then why this $+$ sign?", and the dialogue will end in general hilarity. The only advantage that one can acknowledge in the notation $f(c+)$ is that it can be typed more quickly than $f(c + 0)$. Mathematical notations are what they are: purely writing conventions[3].

Let us add that both these notations have the drawback of irresistibly suggesting that the function $f$ is defined at the point $c$. This hypothesis is not at all necessary as we have seen above, and as we shall see in the following chapter when we define real powers. In this last case, one assumes that the expression $a^x = f(x)$ is defined for $a$ real $> 0$ and $x \in \mathbb{Q}$ – it is an increasing function of $x$ if $a > 1$, decreasing if $a < 1$ – and seeks to define $a^x$ for $x \in \mathbb{R}$ so that the new function, defined on $\mathbb{R}$, will again be monotone, in the same sense as its restriction to $\mathbb{Q}$. The method consists of observing that, for $x \notin \mathbb{Q}$, the required number $a^x$ must lie between $f(x - 0)$ and $f(x + 0)$, and then to prove that these two limit values are equal, to determine $a^x$ without any ambiguity.

## 4 – The intermediate value theorem

The axiom of the existence of least upper bounds leads to a very simple characterisation of intervals which we shall use immediately:

**Theorem 5.** *For a set $I \subset \mathbb{R}$ to be an interval it is necessary and sufficient that*

$$\{(x \in I) \ \& \ (y \in I) \ \& \ (x < z < y)\} \Longrightarrow z \in I.$$

The condition is clearly necessary. To show that it is sufficient, consider the greatest lower bound and least upper bound, possibly infinite, of $I$. For any number $z$ satisfying $\inf(I) < z < \sup(I)$, with *strict* inequalities, there exist, by definition, elements $x$ and $y$ of $I$ such that $x < z$ and $z < y$. Then

---

[3] In this circle of ideas, we remark that some authors write $x \to c_-$ for what we write as $x \to c, x < c$. One might also write $x \to c - 0$.

$z \in I$ by hypothesis. And since certainly $\inf(I) \le z \le \sup(I)$ for any $z \in I$, the set $I$ can only be one of the intervals with end points $\inf(I)$ and $\sup(I)$, qed.

**Corollary.** *Every intersection of intervals is an interval.*

For if $x < z < y$ where $x$ and $y$ belong to the intersection $I$, thus to each of the given intervals, then $z$ also belongs to them, hence also to $I$; it remains to apply Theorem 5.

Of course the intersection in question can very well be empty: the intervals $(0, 1)$ and $(2, 3)$ have nothing in common.

Starting from these results, one easily obtains one of the most important properties of continuous functions:

**Theorem 6.** (Bolzano, 1817) *Let $f$ be a real function, defined and continuous in an* interval $I$. *Then the image $f(I)$ of $I$ under $f$ is an interval.*

By Theorem 5 this reduces to establishing the following result: *let $u, v, w$ be three real numbers such that $u < w < v$; suppose that there exist an $x \in I$ such that $u = f(x)$ and a $y \in I$ such that $v = f(y)$. Then there exists a $z \in I$ such that $w = f(z)$.*

Suppose, to fix our ideas, that $x < y$, and consider the set $E$ of $t \in I$ such that both

$$t \le y \qquad \text{and} \qquad f(t) \le w.$$

This set is not empty – it contains $x$ since $f(x) = u < w$ – and it is majorised by $y$. Let $z = \sup(E) \le y$.

Since $z$ is the limit of points $t \in E$ satisfying $f(t) \le w$, and since $f$ is continuous, we have $f(z) \le w$. Since $w < v$ we have $z \ne y$ and so $z < y$. For $z < t < y$, we have $f(t) > w$ since otherwise one would have $t \in E$ and so $t \le z$. Since $f$ is continuous at the point $z$ we have $f(z) > w$. Whence, finally, $f(z) = w$, qed.

The classical formulation of Theorem 6 is to say that if the function $f$ takes values $< 0$ and values $> 0$ on $I$, then the equation $f(x) = 0$ has a root in $I$. This is, for example, the case, for $I = \mathbb{R}$, if $f$ is a polynomial of odd degree. The ratio between $f(x)$ and its term of highest degree tends to 1 when $|x|$ increases indefinitely as we saw – it is quite evident – in Chap. II, n° 8, so that for $|x|$ large, $f(x)$ has the sign of its term of highest degree. "Geometrically obvious" conclusion:

**Corollary.** *Every algebraic equation of odd degree with real coefficients has a real root.*

But the most important consequence of Theorem 6 is the following:

**Theorem 7.** *Let $f$ be a real function, defined and strictly monotone on an interval $I$ and let $J = f(I)$ be the image of $I$ under $f$. The following properties are then equivalent: (i) $f$ is continuous, (ii) $J$ is an interval. The map $g : J \to I$ inverse to $f$ is then continuous and strictly monotone.*

Since $f$ is strictly monotone it is injective, so $g$ exists and, like $f$, is strictly monotone. The fact that (i) $\Longrightarrow$ (ii) has been established above. To show that (ii) $\Longrightarrow$ (i) it suffices, by (i) $\Longrightarrow$ (ii) applied to $g$, to show that $g$ is continuous. Let $b = f(a)$ be a point of $J$, with $a \in I$. Choose an $r > 0$; we need only prove that there exists an $r' > 0$ such that, for $y \in J$, the relation $|y - b| < r'$ implies $|g(y) - a| < r$. Suppose that $f$ is increasing, for definiteness.

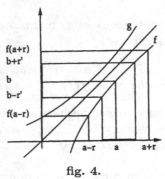

fig. 4.

Suppose first that $a$ is not an end point of $I$ (or, equivalently, since $f$ is strictly monotone, that $b$ is not an end point of $J$).

If we suppose $r$ sufficiently small, as we may, then $I$ contains the closed interval $[a - r, a + r]$, so that $J$ similarly contains $[f(a - r), f(a + r)]$. Since $f$ is strictly increasing we have $f(a - r) < f(a) = b < f(a + r)$, so that there is an $r' > 0$ such that $f(a - r) < b - r' < b + r' < f(a + r)$. It is now clear that $b - r' < y < b + r'$ implies $a - r < x = g(y) < a + r$, whence the result.

The case where $b$ is an end point of $J$ is treated similarly up to one detail: one substitutes for the interval $[a - r, a + r]$ either the interval $[a - r, a]$, or the interval $[a, a + r]$.

Theorem 7 has a useful variant:

**Theorem 7 bis.** *Let $f$ be a real function defined and continuous on an interval $I$ and let $J = f(I)$ be the image interval. The following properties are equivalent: (i) $f$ is injective, (ii) $f$ is strictly monotone. The map $g : J \to I$ inverse to $f$ is then continuous.*

It suffices to show that (i) implies (ii), since the implication (ii) $\Longrightarrow$ (i) is obvious; and if $f$ is strictly monotone and continuous, Theorem 7 shows that similarly so is $g$. The fact that $J$ is an interval is Bolzano's Theorem.

First let us show that for all $a, b \in I$ the image under $f$ of the interval with end points $a$ and $b$ is the interval with *end points $f(a)$ and $f(b)$*; in other words that for all $c$ lying between $a$ and $b$ the value $f(c)$ lies between $f(a)$ and $f(b)$. ("Between" means either $a \leq c \leq b$ or $b \leq c \leq a$.)

Suppose, for definiteness, that $a < b$ and $f(a) < f(b)$ – equality is excluded by (i); it reduces to showing that

$$a < c < b \Longrightarrow f(a) < f(c) < f(b).$$

If $f(c) < f(a) < f(b)$ then the image of $[c, b]$, which is an interval since $f$ is continuous, contains $f(a)$, whence there is a $u \in [c, b]$ such that $f(a) = f(u)$, contrary to (i). If $f(a) < f(b) < f(c)$ then $f(b)$ is in the image of $[a, c]$, a new contradiction by the same argument.

Having established this preliminary point, and with $a$ and $b$ as above, let us show that $f$ is increasing, i.e. that $x < y$ implies $f(x) < f(y)$ since $f$ is injective. It is simplest to examine the various possible positions of the pair $x, y$ relative to the pair $a, b$. If for example $x < y < a < b$, then $a$ is between $x$ and $b$, so $f(a)$ is between $f(x)$ and $f(b)$, so $f(x) < f(a)$ since $f(a) < f(b)$, and since $y$ is between $x$ and $a$, $f(y)$, which is between $f(x)$ and $f(a)$, is $> f(x)$. If, another possible case, $a < x < b < y$, one already knows that $f(a) < f(x) < f(b)$, or $f(b)$ is between $f(x)$ and $f(y)$ since $b$ is between $x$ and $y$, whence $f(b) < f(y)$ and thus $f(x) < f(y)$. Etc.

*Example 1.* Let us work on the interval $I = \mathbb{R}_+$ and consider the function $f(x) = x^n$, where $n$ is an integer $\geq 1$. It is continuous, strictly increasing, with values $\geq 0$, takes the value $0$ for $x = 0$ and tends to $+\infty$ when $x$ increases indefinitely. The interval $J = f(I)$ must therefore be $\mathbb{R}_+$. We thus find again, and much more easily, a result we have already proved (Chap. II, n° 10, example 3), and even more: *Every real positive number possesses one and only one positive $n^{\text{th}}$ root for any integer $n \geq 1$ and this determines a continuous function of $x$.* Since $x^{p/q} = (x^p)^{1/q}$, one concludes, more generally, that the function $x^{p/q}$ is continuous for $x > 0$, for all $p, q \in \mathbb{Z}$, $q \neq 0$.

*Example 2.* The same argument applies to the function $\exp(x)$. If one works on $\mathbb{R}$ the function is continuous (sum of a power series) and has values[4] $> 0$. On the other hand it takes values as close to $0$ or as large as one desires: one has $\exp(n) = \exp(1)^n > 2^n$ for $n \in \mathbb{N}$ and $\exp(-n) = 1/\exp(n) < 1/2^n$. It follows from this that the image of $\mathbb{R}$ must be the set $\mathbb{R}_+^*$ of all strictly positive real numbers. The function being strictly increasing, hence injective, the equation

$$x = \exp y$$

thus has, for all $x > 0$, one and only one solution $y \in \mathbb{R}$. As we have already said on several occasions, the inverse map is precisely the function $\log y$ (example 1 of n° 2). It is common to *define* the logarithm function in this way.

---

[4] We have seen in Chap. II, n° 22, that $\exp(x + y) = \exp(x)\exp(y)$; it follows immediately that $\exp(x) = \exp(x/2)^2 > 0$ for any $x \in \mathbb{R}$. Since it is clear that $\exp y > 1$ for $y > 0$, the function is, moreover, strictly increasing.

*Example 3.* It is "well known" that the function $\sin x$ is continuous and strictly increasing on the interval $[-\pi/2, \pi/2]$; it therefore maps this interval onto the interval $[-1, 1]$, whence arises the possibility of defining the continuous map

$$\arcsin x : [-1, 1] \longrightarrow [-\pi/2, \pi/2].$$

The same method allows one to define the functions

$$\arccos x \quad : \quad [-1, 1] \quad \longrightarrow \quad [0, \pi],$$
$$\arctan x \quad : \quad \mathbb{R} \quad \longrightarrow \quad ]-\pi/2, \pi/2[.$$

# §2. Uniform convergence

### 5 – Limits of continuous functions

When a sequence or a series of functions $f_n(x)$ converges at all points $x$ of
the set $X$ on which they are defined one says that they *converge simply*; at
first glance this is the weakest possible concept of convergence for a sequence
of functions, but others, weaker and more subtle, have been invented, mainly
for the needs of the theory of integration.

As we saw at the end of n° 2 the square waves series raises a serious
problem, one which does not arise for finite sums: that of finding conditions
to ensure that a sequence $f_n(x)$ of functions defined and *continuous* on a
set $X \in \mathbb{R}$, and converging simply, will have a limit function that is again
*continuous* on $X$. An answer is easy to find and rests on two principles better
stated naïvely:

(a) *If two functions f and g are almost equal at a point a and if they
are almost constant on a neighbourhood of a, then they are almost
equal on a neighbourhood of a;*
(b) *If f and g are almost equal on a neighbourhood of a and if g is
almost constant on a neighbourhood of a, then f is almost constant
on a neighbourhood of a.*

Let us also propose a more concrete preliminary exercise to the reader.

Two motorists in Giganes 11.2 W24 192-valve Spitzenracers drive south
along the *autoroute du Soleil*, several lengths apart, on the lane corresponding to their political convictions. Never having heard that the set of
speeds authorised (resp. tolerated) on a French motorway theoretically has
130 (resp. 180) kmh as a relatively strict upper bound, the leader drives
constantly at 220 kmh, to within 20 kmh. The second ensures that the
speeds of the two vehicles remain equal to within 10 kmh. What can one
say about the range of his speed?

Returning to the problem above, suppose that $\lim f_n(x) = f(x)$ exists for
all $x \in X$ and that $f$ is continuous at $a \in X$. Choose an $r > 0$ and consider
an $n$ sufficiently large that

$$(5.1) \qquad\qquad d\,[f(a), f_n(a)] < r.$$

Since $f$ is continuous at $a$,

$$(5.2) \qquad\qquad d[f(a), f(x)] < r \quad \text{on a neighbourhood of } a.$$

Since $f_n$ is continuous at $a$, similarly

$$(5.3) \qquad\qquad d\,[f_n(a), f_n(x)] < r \quad \text{on a neighbourhood of } a.$$

Starting from $r/3$ and using the quadrilateral inequality we obtain – principle (i) above – the following statement:

(5.4)        *For every $r > 0$ and every sufficiently large $n$ we have*

$$d\left[f(x), f_n(x)\right] < r \text{ on a neighbourhood of } a.$$

This means, more exactly, that for every large $n$ there exists an $r'_n > 0$ such that, for $x \in X$,

$$d(x, a) < r'_n \implies d\left[f(x), f_n(x)\right] < r.$$

Suppose conversely that this condition is satisfied for any $r > 0$ and let us show that then $f$ is continuous at $a$. We choose an $n$ large enough for $f(x)$ and $f_n(x)$ to be equal to within $r$ on a neighbourhood of $a$ and in particular at the point $a$. Since $f_n$ is continuous, $f_n(x)$ and $f_n(a)$ are equal to within $r$ on a neighbourhood of $a$. The quadrilateral inequality – principle (ii) above – then shows that $d[f(a), f(x)] < 3r$ on a neighbourhood of $a$ in $X$, qed.

When the condition (4) is satisfied one says that the convergence of the sequence $(f_n)$ to $f$ is *locally uniform* at the point $a$, (dangerous terminology because of the risk of confusion with the stronger concept of a sequence which converges uniformly on a fixed neighbourhood of $a$...). This is the necessary and sufficient condition for continuity at $a$ of the limit function $f$, but it is not very handy, and in practice one uses two simpler properties, more restricted than (4).

First and foremost is that of (global) *uniform convergence* on $X$. This means that for every $r > 0$ there exists an integer $N$, independent of $x$, such that

(5.5)        $\{(n > N) \ \& \ (x \in X)\} \implies d\left[f(x), f_n(x)\right] < r.$

Condition (4) is then satisfied for all $a \in X$ and for all $n > N$ without it being necessary to impose the least restriction on $d(x, a)$: $r'_n$ has no further rôle, in other words you can, for $n > N$, choose it *ad libitum*. We have met an example in Chap. II, at the end of n° 14: if a power series $\sum a_n z^n$ has radius of convergence $R > 0$, then for all $\rho < R$ there exist positive constants $M$ and $q < 1$ such that, for all $n$,

$$|f(z) - s_n(z)| < M q^n$$

on the disc $X : |z| \le \rho$, where $s_n$ denotes the $n^{\text{th}}$ partial sum of the series. Since $q^n$ tends to 0, one thus has

$$|f(z) - s_n(z)| < r \ \text{ for all } z \in X$$

provided that $n$ is sufficiently large, i.e. once $n$ exceeds an integer $N$ *independent of $z \in X$*.

Another simple case, but much less frequently met, is that where the functions $f_n$ are *equicontinuous at the point a*; this means that for all $r > 0$ there exists an $r' > 0$ such that

(5.6)          $d(x, a) < r' \Longrightarrow d\,[f_n(a), f_n(x)] < r$ for all $n$.

If this condition is satisfied, then passing to the limit, with $x$ and $a$ fixed, as $n$ increases indefinitely, shows that[5]

$$d(x, a) < r' \Longrightarrow d[f(a), f(x)] \leq r,$$

whence the continuity of $f$ at $a$ without even having to use the arguments leading to (4).

If we agree to say that a family of functions defined on $X$ is *equicontinuous on X* if it is equicontinuous at all points of $X$, we finally obtain the following result:

**Theorem 8.** *Let $(f_n)$ be a sequence of functions defined and continuous on a subset $X$ of $\mathbb{C}$ and suppose that $\lim f_n(x) = f(x)$ exists for all $x \in X$. For $f$ to be continuous in $X$ it suffices that one of the two following conditions be satisfied: (i) the given sequence converges uniformly on $X$; (ii) the functions $f_n$ are equicontinuous on $X$.*

Experience has shown that at the beginning innumerable students have very great trouble in understanding this fundamental concept of uniform convergence. Yet it is simple: supposing that the functions have real values, it means that for all $r > 0$

$$f(x) - r < f_n(x) < f(x) + r \text{ for any } x \in X$$

for all sufficiently large $n$. Or again: for all sufficiently large $n$ the graph of $f_n$ lies *entirely* in the strip in the plane of height $2r$ contained between the graphs of the functions $f(x) + r$ and $f(x) - r$ (fig. 5).

The difficulty, apparently, is with logic: one has here a proposition $P_N(x)$, namely the implication

$$(n > N) \Longrightarrow \{|f(x) - f_n(x)| < r\},$$

which depends simultaneously on $N$ and on $x$ [it does not depend on $n$, since we have very deliberately omitted the sign $(\forall n)$ which ought to figure there and to transform $n$ into a phantom variable]. Simple, nonuniform, convergence demands that

for all $x$, there exists $N$ such that $P_N(x)$;

while uniform convergence requires that

---

[5] We use the fact that if $|u_n| < r$ for all $n$ then $|\lim u_n| \leq r$ (weak inequality).

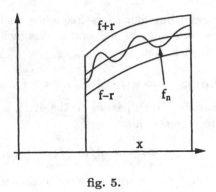

fig. 5.

there exists $N$ such that, for all $x, P_N(x)$.

There is a problem in permuting the logical operators "for all" and "there exists".

One meets this in everyday life. The two assertions

for all $x \in H$ there exists $y \in F$ such that $C(x, y)$

there exists $y \in F$ such that $C(x, y)$ for all $x \in H$

are not equivalent: even if every man had at least one partner, it would not follow that there must be a woman belonging to all the men (or the opposite, to remain politically correct).

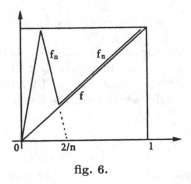

fig. 6.

As we saw in proving the preceding theorem, uniform convergence is only a *sufficient* condition for assuring the continuity of the limit function (later, Chap. V, n° 10, we will see that it is also necessary when the sequence $f_n$ is *increasing*: Dini's theorem). Figure 6 shows a sequence of continuous functions on the interval $[0, 1]$ which, while converging to the continuous function

$x \mapsto x$, does not converge uniformly. It is clear that the $f_n$ are not equicontinuous at the point $x = 0$ since, to satisfy the inequality $|f_n(x) - f_n(0)| < r$, it is necessary to suppose $|x| < r/n$ (or to confine oneself to a neighbourhood of $x = 1$, which is not an answer to the problem). For a given $r$ there is no $r' > 0$ which works for *all* the functions $f_n$ on a neighbourhood of $x = 0$.

Uniform convergence features in many very general theorems on approximation of functions by much simpler functions. Some of these results can be stated very easily, but their proofs are not accessible at this stage of the exposition.

First there is Weierstrass' famous approximation theorem: *every real function $f$ defined and continuous on a* compact *interval $I$ is the limit of a sequence of polynomials which converges to $f$ uniformly on $I$.* In other words, for every $r > 0$ there exists a polynomial $p(x)$ such that $|f(x) - p(x)| < r$ *for all $x \in I$.* Weierstrass preferred to state his theorem in terms of a series of polynomials, but this clearly comes to the same. We will establish it in Chap. V, n° 28.

To state the second result, also due to Weierstrass, let us call any function of the form

$$p(x) = c_0 + a_1 \cos x + b_1 \sin x + \ldots + a_n \cos nx + b_n \sin nx$$

(where $c_0, \ldots, b_n$ are a finite set of complex constants ) a *trigonometric polynomial* of period $2\pi$. Then *every function defined, continuous, and of period $2\pi$ on $\mathbb{R}$, is the uniform limit of a sequence of trigonometric polynomials on $\mathbb{R}$*[6].

We remark that this second result immediately provides another, analogous to the first. Consider a function $f$ defined and continuous on a compact interval $I = [a, b]$, and suppose $b < a + 2\pi$ (strict inequality). Consider a continuous function $g$, defined on the interval $[b, a + 2\pi]$, equal to $f(b)$ for $x = b$ and to $f(a)$ for $x = a + 2\pi$; we may define a periodic function $h$ on $\mathbb{R}$ by requiring it to be equal to $f$ on $I$ and to $g$ between $b$ and $a + 2\pi$ (figure 7).

The function $h$ is clearly continuous at all points of $\mathbb{R}$. If one applies the preceding theorem to $h$, and then confines oneself to examining what happens on $I$, one sees that for every $r > 0$ there exists a trigonometric polynomial $p$ such that $|f(x) - p(x)| < r$ for all $x \in I$. The hypothesis imposed on $I$ is essential since the trigonometric polynomials have period $2\pi$; if, for example, $I$ is the interval $[a, a + 3\pi]$, one will have $p(x + 2\pi) = p(x)$ for $a \leq x \leq a + \pi$, so that every function which is the uniform, or even simple, limit of trigonometric polynomials on $I$ must possess the same property, which is clearly not the case in general.

Likewise, Weierstrass' first theorem does not apply in the case of a noncompact interval $I$. If $I$ is not bounded, then every uniform limit $f$ of poly-

---

[6] This result *does not say* that every periodic continuous function is the sum of its Fourier series; this is false.

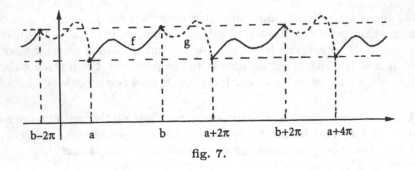

fig. 7.

*nomials* $p_n$ *is a polynomial.* For $n$ large one has $|f(x) - p_n(x)| < 1$ for all $x$, so $|p_m(x) - p_n(x)| < 2$ for $m$ and $n$ large; the polynomial $p_m - p_n$ is thus bounded at infinity, so is constant on $\mathbb{R}$ and in particular on $I$; the same must be true, for given $n$ large, for $\lim_m p_m - p_n = f - p_n$, whence $f = p_n + \text{const.}$

If $I = (a, b)$ is bounded, one can, as we shall see in Chap. V *à propos* uniform continuity, approximate by polynomials only those functions having limit values at $a$ and $b$, which brings us back to the case of continuous functions on the compact interval $[a, b]$. The function $\sin(1/x)$ is not a uniform limit of polynomials on $I = ]0, 1]$: it oscillates too rapidly between $-1$ and $1$ on a neighbourhood of $0$. If one could find a polynomial $p(x)$ such that $\sin(1/x)$ were, for all $x \in I$, equal to $p(x)$ to within $1/10$ for example, and if one observes that $p$, being continuous on the closed interval $[0, 1]$ (and even on $\mathbb{R}$), is constant to within $1/10$ on a neighbourhood of $0$, it would follow that $\sin(1/x)$ was constant to within $2/10$ on a neighbourhood of $0$; false.

In all these cases one can always obtain an approximation which is *uniform on every compact interval* contained in $I$, a mode of convergence (*compact convergence*, for short) much more widespread than uniform convergence. [For example, on its disc of convergence $D : |z| < R$, the partial sums $s_n(z)$ of a power series converge to the sum $f(z)$ uniformly on all compact $K \subset D$, since $K$ is contained in a disc $|z| \leq r$, with $r < R$, on which, as we saw above, the convergence is uniform. (The reader may confine himself to real values of $z$ while waiting to read n° 9 on general compact, i.e. closed and bounded, sets.)]

To see this, one first notes that there exists an increasing sequence of compact intervals $I_n$ with union $I$:

$$
\begin{aligned}
I_n &= [a, n] && \text{if } I = [a, +\infty[, && a \text{ finite,} \\
&= [a + 1/n, n] && \text{if } I = ]a, +\infty[, && a \text{ finite,} \\
&= [a + 1/n, b - 1/n] && \text{if } I = ]a, b[, && a \text{ and } b \text{ finite,}
\end{aligned}
$$

etc. For all $n \in \mathbb{N}$ one can then find a polynomial $p_n$ such that

$$|f(x) - p_n(x)| < 1/n \text{ for all } x \in I_n;$$

this is Weierstrass' Theorem for $I_n$. If $K \subset I$ is a compact interval there is clearly an integer $k$ such that $K \subset I_k$. Then

$$|f(x) - p_n(x)| < 1/n \ \text{ for all } x \in K \text{ and all } n \geq k,$$

so that the sequence $(p_n)$ converges uniformly to $f$ on $K$, qed.

Figure 8 shows an example of compact convergence on $\mathbb{R}$ where the $f_n$ do not converge uniformly on $\mathbb{R}$.

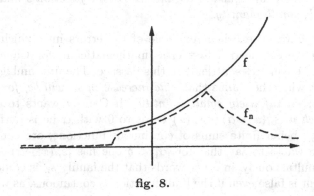

fig. 8.

## 6 – A slip up of Cauchy's

It is interesting to see[7] that Cauchy, attempting to ground analysis on a rigorous basis, nevertheless "proved" in his *Cours d'analyse* (1821) at the Ecole polytechnique, that every simple limit of continuous functions is again continuous. Figure 9 "confirms" the "theorem": the limit function equals 0 for $x = 0$ and 1 for $x > 0$. In fact, Cauchy dealt with a series of continuous functions, but this amounts to the same since he introduced the total sum $s(x)$ of the series, its partial sums $s_n(x)$ and the "remainder" $r_n(x) = s(x) - s_n(x)$. He argued as follows:

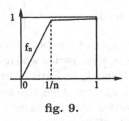

fig. 9.

---

[7] For all that follows, see Paul Dugac, *Sur les fondements de l'Analyse de Cauchy à Baire* (doctoral thesis, Université Pierre et Marie Curie, 1978).

*Consider the increases which these three functions receive when one increases x by an infinitely small quantity α [for Cauchy, this means that α tends to 0]. The increase in $s_n$ will be, for all possible values of n, an infinitely small quantity; and that in $r_n$ will become imperceptible at the same time as $r_n$, if one attributes a very considerable value to n. In consequence, the increase in the function s must be an infinitely small quantity.*

Here we have an excellent example of the errors into which the use of vague language can lead a first class mathematician, for Cauchy did not write precise inequalities, at least in this passage. The first ambiguity occurs at the point where he claims that *"the increase in $s_n$ will be, for all possible values of n, an infinitely small quantity"*. If Cauchy wants to say by this that for *each n*, $s_n(x + α) - s_n(x)$ tends to 0 with α he is perfectly right, since the $s_n$, being finite sums of continuous functions, are continuous. If, however, he wants to say that for $|α| < δ$ one has $|s_n(x+α) - s_n(x)| < ε$ for *all n* simultaneously, in other words that the family $s_n$ is equicontinuous, the argument is false even if the limit function is continuous, as we have seen above.

The second error is to claim that the increase in $r_n$ *"will become imperceptible at the same time as $r_n$ if one attributes a very considerable value to n;"* since, for x given, $|r_n(x)| < ε$ for $n > N(x)$ – this is simple convergence –, Cauchy's argument amounts to saying that for $|α| < δ$, one has $|r_n(x+α)| < ε$ for all $n > N(x)$, which is local uniform convergence as mentioned at the beginning of the preceding n°; this would *follow* from continuity of the limit function, a property that one may not use to establish continuity of itself . . .

It was only in 1853 that Cauchy corrected his "theorem" of 1821 by exploiting uniform convergence quite correctly, though without naming this concept. But his error was detected rapidly. The young Norwegian mathematician Abel, who was living in Paris, provided, in a letter of 1825 to a friend (published in 1839 in his Complete Works), as Dugac tells us, a counterexample in the series

$$(6.1) \qquad \sin x - \sin 2x/2 + \sin 3x/3 - \ldots = \begin{cases} x/2 & \text{if} \quad |x| < \pi, \\ 0 & \text{if} \quad |x| = \pi; \end{cases}$$

one of those exhibited by Fourier at the beginning of his *Théorie analytique de la chaleur*, after the square wave series which Abel would have also been well able to use, and which Cauchy, at Paris where Fourier still lived, should have known, or maybe had forgotten. One also sees Abel, in the same letter, complain about divergent series and unjustified manipulations of series of functions, such as, for example, differentiating term-by-term as if dealing with finite sums (see n° 17); Abel observes that if one differentiates the formula

$$(6.2) \qquad x/2 = \sin x - \sin 2x/2 + \sin 3x/3 - \ldots,$$

one obtains the relation

(6.3)                    $1/2 = \cos x - \cos 2x + \cos 3x - \ldots,$

*"an absolutely false result, since the series is divergent"*. This is obvious for certain values of $x$: for $x = \pi/2$ for example, one finds the "formula"

$$1/2 = 1 - 1 + 1 - 1 + \ldots,$$

which we have already met. The same is true more generally for $x$ commensurable to $\pi$ since then among the multiples of $x$ there occur infinitely many integer multiples of $\pi$ for which the corresponding terms in (3) are equal to 1 or $-1$. The case where $x$ is not commensurable to $\pi$ is even worse, the values of $\cos nx$ then being distributed at random[8] between $-1$ and 1. If he had had to, Abel could have rendered the situation even more ridiculous by differentiating the relation (2) thrice, and setting $x = 0$; then one finds that $0 = 1 - 4 + 9 - 16 + \ldots$ This is nevertheless what Fourier did: but instead of starting from the relation (2.4) that he sought to establish, he started from a series with *a priori* undetermined coefficients $a_n$, and, differentiating *ad libitum* for $x = 0$, obtained linear equations in infinitely many unknowns (!) between the $a_n$, so that he never had to write the extravagant relations which, to *verify* the correctness of his calculations, he would have obtained in giving these $a_n$ the explicit values 1, 1/3, etc. that he finally found after acrobatic calculations; this must be the absolute record in mathematical prestidigitation. His results were no less correct for all that.

In 1822, a young German, Gustav Peter Lejeune Dirichlet (1805–1859) or Dirichlet for short, already impassioned by mathematics, arrived like Abel at what was then the City of Light of mathematics and of physics[9]. A year later, he had the opportunity of finding very comfortable employment, as tutor in the household of General Fay[10], a companion in arms of Napoleon. Here he met Parisian "society" and notably Fourier whose works impressed him as much for their staggering results as by the imaginativeness of their proofs. He then tried to erect Fourier's theory on a solid base, particularly for a neighbourhood of a point where the sum is discontinuous, which dropped him right into Cauchy's "theorem" and Fourier's square wave series. His principal result (1829), now become classical, is to be found in Chap. VII; it completely justifies the relation (2) and Fourier's formulae.

---

[8] See in Hairer and Wanner, *Analysis by Its History*, p. 41, a curious graphical representation of the values of the function $n \mapsto \sin n$ in logarithmic scale for $n$.

[9] See Amy Dahan-Dalmedico, *Mathématisations: Augustin-Louis Cauchy et l'école française* (Paris, Blanchard, 1992) and I. Grattan-Guinness, *Convolutions in French Mathematics*, 1800–1840 (Birkhäuser Verlag, 3 Vol., 1990).

[10] who died in 1825. Dirichlet returned to Germany in 1826, obtained a post which did not interest him in Breslau, arrived in Berlin in 1828 where he taught banal mathematics at the military academy for a quarter of a century, obtained at the same time a post, then a chair, at the University of Berlin, which, tired by the 13 hours weekly for ill paid courses, he resigned in 1855 to succeed Gauss at Göttingen. DSB.

*Retrospectively* the most extraordinary thing about these controversies is that, concerned as they were with the necessity of relieving analysis of its artistic vagueness, neither Cauchy, nor Abel, nor Dirichlet, i.e. three of the greatest mathematicians of the XIX[th] century, apparently had the idea of using the incomparably simpler example which we have mentioned above, that one can vary *ad libitum*, and which has now acquired the status of cliché. The fact that they never traced the graph of a function is maybe to the point: these were the first, Abel and Dirichlet much more than Cauchy, who tried to *found analysis on arguments as rigorous as those of the arithmetic* and not on geometric intuition; though not dispensing their users from "arithmetic" proofs, diagrams are sometimes useful. But the more probable explanation is that, while only beginning to develop an awareness of the extreme generality of the concept of function – Dirichlet gave the first good general definition –, they were still trammelled by their predecessors. They, the functions, are those that one can represent by algebraic formulae (polynomials and rational fractions) or by power series. During the XIX[th] century, following Cauchy, the theory (of analytic functions on $\mathbb{C}$) was constructed, but, at the other end of the spectrum which spans from hypernormality to superpathology, one also introduced "bizarre" functions into analysis, not continuous, not differentiable, not representable by simple formulae nor by power series, even resistant to every theory of integration, etc. Habituated as they were to the virtuosic calculations and beautiful formulae of their predecessors, they did not see the simple things which stare our contemporaries in the face. Strange? No. New ideas do not enter the folklore of mathematicians any faster than in every other intellectual guild.

Finally, one must remark that Cauchy was neither the only nor the first to have had these ideas; but one lends only to the rich. From 1817, in rather obscure works, much in advance of their time in their spirit, and rediscovered later – some were not known before 1930 –, Bernhard Bolzano (1781–1848), priest and professor of "science of religions" at Prague[11] from 1805 to 1819, published in German (or did not publish) the majority of ideas that one attributes to Cauchy – a precise definition of convergence and the first notion of the general criterion to be discussed in n° 10, among others – with, here again, proofs which leave something to be desired; in 1835 he still believed in Cauchy's false theorem on the limits of continuous functions. Cauchy's

---

[11] Wolfgang Walter, *Analysis* I (Springer, 1992) tells us that Bolzano's chair was founded by the Emperor of Austria "to combat the ever more powerful influence of the free thinkers of the Enlightenment". Bolzano, having become the principal spokesman of the Enlightenment in Bohemia, was dismissed in 1819. See Jan Sebestik, *Logique et mathématiques chez Bernard Bolzano* (Paris, Vrin, 1992) where one will see particularly that Bolzano already had ideas on set theory, but, too much the philosopher, clearly lacked the proper language to express them mathematically, B. Bolzano, *Paradoxien des Unendlichen*, 1851 (trans. *Paradoxes of the Infinite*, Routledge and Kegan Paul, London, 1950), and François Rivenc and Philippe de Rouilhan, *Logique et fondements de mathématiques. Anthologie* (1850–1914) (Paris, Payot, 1992).

ideas and his formulations are even so close to those of Bolzano that in 1975 I. Grattann-Guinness tried, comparing the texts, to prove that Cauchy had plagiarised Bolzano, a theory refuted by Hans Freudenthal (especially in his superb article on Cauchy in the DSB) and hardly plausible in view of the obscurity in which Bolzano's activities were then surrounded: he was not, by far, a full time professional integrated into the "international mathematical community". In particular he seems to have been the first to reflect on the concept of the least upper bound of a set; he showed that one can approximate it indefinitely closely by rational numbers, but, alas, was unable to prove its existence, which continued to be impossible so long as one lacked an exact definition of the real numbers. These ideas, that one also meets at the same period with Gauss, were manifestly in the air at the time; but the problem, solved by Dedekind in the years from 1858, held up the attempts to arithmetise analysis for as long as one spoke of numbers whose nature one had not understood.

All this poses the very difficult problem of characterising by internal properties – nature of the discontinuities – the functions which are simple limits of continuous functions. It was solved by René Baire in his doctoral thesis (1899) by using extraordinarily subtle techniques of set theory in $\mathbb{R}$, which – including the plain statement of the result – go much beyond the level of the present exposition; Baire's methods paved the way for many French works (Borel, Lebesgue, Denjoy), then gave rise, between the two wars, to Russian and Polish works, more general, or analogous, and often more useful. The reader may console himself for not knowing this famous theorem of Baire's on the limits of continuous functions by reading what Dieudonné justifiably wrote to Dugac in 1976:

> with the benefit of hindsight it is quite clear today that the greater part of these works [of Baire, Borel, Lebesgue and Denjoy] came to a dead end; essentially there remain "Baire's theorem" [12] and the Lebesgue integral, two fundamental tools in all analysis; but all the rest, for the moment at least, are museum pieces; I have never seen a problem (not concocted *ad hoc*) where the famous theorem on "pointwise discontinuous" functions [i.e. the characterisation by Baire of simple limits of continuous functions] arises anywhere.

---

[12] Let us say that a set $X \subset \mathbb{R}$ is *everywhere dense* in a set $Y \subset \mathbb{R}$ if every point of $Y$ is adherent to $X$ (so a limit of points of $X$, example $\mathbb{Q}$ in $\mathbb{R}$). Then the intersection of a *countable* family of *open* everywhere dense sets in $\mathbb{R}$ is again everywhere dense in $\mathbb{R}$. On taking complements one obtains the following statement: if $F_n$ is a sequence of closed sets not containing any interior point, neither does their union. There are other formulations, some applicable to spaces much more general than $\mathbb{R}$. See Dieudonné, *Treatise on Analysis*, Vol. 2, Chap. XII, n° 16.

Yes. But before the war, in the public library at Le Havre, which was well provided in mathematics (at 17 years old the French authors were enough for me), his very beautiful *Leçons sur les théories générale de l'analyse* (first edition 1908), delivered at Dijon, determined my vocation. Try to find them in the public library of your town: this might be a better test of its cultural level that the detective stories, science fiction or the comic books for children or "adults" that one finds in quantity, in the City of Light, in the branch of the public library near my house.

## 7 – The uniform metric

For scalar sequences, the convergence of $u_n$ to $u$ can be expressed by the relation $\lim d(u_n, u) = 0$. The aim of this n° is to show that uniform convergence of a sequence of functions can be expressed in the form $\lim d(f_n, f) = 0$, after defining the "distance" between two functions suitably.

First of all, we shall recapitulate the definitions pertinent to least upper bounds (Chap. II, n° 9) so as to apply them to functions with scalar values. It is a matter of simple translation.

Let $f$ be a function defined on an arbitrary[13] set $E$, and with *real* values. The image $f(E)$, the set of numbers $f(x)$ where $x \in E$, is a subset of $\mathbb{R}$ whose least upper bound, finite or infinite, is called the *least upper bound of f over E* and is written

$$(7.1) \qquad \sup_{x \in E} f(x) = \sup(f(E));$$

this is, up to notation, the concept of the least upper bound of a family $(u_i)$ of real numbers (Chap. II, n° 9, end).

It is finite if and only if $f$ is, by definition, *bounded above*; it is then *the number $u \in \mathbb{R}$ satisfying*

> (SUP 1)    $f(x) \le u$ for any $x \in E$,
>
> (SUP 2)    $u \le M$ for any number $M$ which *majorises* $f$,
>
> i.e. such that $f(x) \le M$ for any $x \in E$.

As in Chap. II, n° 9 one can replace (SUP 2) by

> (SUP 2')  for every $r > 0$ there exists an $x$ such that $f(x) > u - r$.

It may happen that there exists an $a \in E$ where $f(a) = u$, in which case $f$ possesses an *absolute maximum*, but in general such a point does not exist[14]:

---

[13] To be "any" or "arbitrary" – some humourists even speak of "arbitrary and in other respects any" objects– is not a property, it is the total absence of any hypothesis whatever, and in particular the hypothesis $E \subset \mathbb{R}$, totally irrelevant to what follows.

[14] We will, however, show in n° 9 that if $f$ is a real function defined and *continuous* on a *compact* interval $I$, then there exists an $a \in I$ where $f(a)$ attains its maximum.

the function $x$ on $E = [0, 1[$ has least upper bound 1, but $x < 1$ for all $x \in E$. Nevertheless, there always exists a sequence $(x_n)$ of points of $E$ such that

$$(7.2) \qquad \lim f(x_n) = \sup f(x);$$

it suffices to choose $x_n$ so that $f(x_n) > u - 1/n$, as allowed by (SUP 2').

There are analogous definitions and notation for the case of greatest lower bounds. When the least upper bound is infinite, the condition (SUP 1) is empty and (SUP 2') means that for all $A \in \mathbb{R}$ there is a point $x$ in $X$ where $f(x) > A$.

If two maps $f, g : E \longrightarrow \mathbb{R}$ are bounded above or below then clearly so is the function $f + g$ and, with or without these hypotheses, we have the inequalities

$$(7.3') \qquad \sup(f(x) + g(x)) \quad \leq \quad \sup f(x) + \sup g(x),$$

$$(7.3'') \qquad \inf(f(x) + g(x)) \quad \geq \quad \inf f(x) + \inf g(x).$$

If one puts $\sup f(x) = A$ and $\sup g(x) = B$, then $f(x) \leq A$ and $g(x) \leq B$ for all $x$, whence $f(x) + g(x) \leq A + B$, and so (3').

If, further, the values of $f$ and $g$ are positive and bounded above then the map $fg$ is also bounded above, since one can multiply inequalities between *positive* numbers member-by-member.

These definitions are not meaningful for functions with complex values. In this case one can define the positive number

$$(7.4) \qquad \|f\|_E = \sup_{x \in E} |f(x)| \leq +\infty,$$

the *uniform norm of $f$ on $E$* for every set $E$ on which $f$ is defined. If $f$ is *bounded on $E$*, i.e. if there exist numbers $M \geq 0$ such that $|f(x)| \leq M$ for all $x \in E$ or, this comes to the same, if $\|f\|_E < +\infty$, this norm is then the least possible $M$. One can thus always, and generally advantageously, replace a bound of the form

$$|f(x)| \leq M \quad \text{for all } x \in E$$

by $\|f\|_E \leq M$: the two relations are strictly equivalent and the second is more concise.

It is clear that if $f$ and $g$, with complex values, are bounded, so similarly are $f + g$ and $fg$; then

$$(7.5) \qquad \|f + g\|_E \quad \leq \quad \|f\|_E + \|g\|_E,$$

$$(7.6) \qquad \|fg\|_E \quad \leq \quad \|f\|_E \cdot \|g\|_E$$

since the right hand sides majorise $|f(x) + g(x)|$ and $|f(x)g(x)|$ for any $x$. It is even more obvious that

$$\|c.f\|_E = |c|.\|f\|_E$$

for every constant $c \in \mathbb{C}$. The associativity formula for least upper bounds established at the end of n° 9 of Chap. II can be translated as follows: if $f$ is a scalar function defined on the union $E = \bigcup E_i$, $i \in I$, of any family of sets, then

$$\|f\|_E = \sup_{i \in I} \|f\|_{E_i}.$$

When $f$ and $g$ are functions with complex values defined on $E$, we define their *uniform distance* on $E$ by

$$(7.7) \qquad d_E(f,g) = \|f - g\|_E = \sup |f(x) - g(x)| = \sup d[f(x), g(x)].$$

The left hand side of (7), if it is finite (for example if $f$ and $g$ are bounded, not a necessary condition), is thus the least number $M \geq 0$ such that

$$(7.8) \qquad |f(x) - g(x)| \leq M \text{ for any } x \in X.$$

The inequality (5) shows that, if $f$, $g$ and $h$ are three bounded functions on $E$ then

$$(7.9) \qquad d_E(f,g) \leq d_E(f,h) + d_E(h,g).$$

It is moreover clear that $d_E(f,g) \geq 0$ always, and that this distance can be zero only if $f = g$: for then 0 majorises all the numbers $|f(x) - g(x)|$.

The uniform convergence on $E$ of a sequence of functions $(f_n)$ to a limit function $f$ can be translated immediately into these terms. One clearly does not change the situation if one replaces strict by weak inequality in (5.5). But to say, for a given $n$, that $d[f(x), f_n(x)] \leq r$ for any $x \in E$ is the same as saying that $d_E(f, f_n) \leq r$. Since, for all $r > 0$, this relation must be satisfied for $n$ sufficiently large, one concludes that *uniform convergence on $E$ is equivalent to the relation*

$$(7.10) \qquad \lim d_E(f, f_n) = 0$$

analogously to the convergence of a sequence of numbers. This statement, trivial though it is, will be of constant use.

From it one deduces easily, as in the case of scalar sequences, the rules of algebraic calculations under uniform convergence; using them often allows us to bypass explicit calculations.

(R 1) *If two sequences $(f_n)$ and $(g_n)$ converge uniformly to $f$ and $g$ then the sequence $(f_n + g_n)$ converges uniformly to $f + g$.*

For $d_E(f_n + g_n, f + g) \leq d_E(f_n, f) + d_E(g_n, g)$.

(R 2) *If two sequences* $(f_n)$ *and* $(g_n)$ *converge uniformly to* $f$ *and to* $g$, *and if the limit functions* $f$ *and* $g$ *are bounded, then the sequence* $(f_n g_n)$ *converges uniformly to* $fg$.

To see this we use the identity

$$f_n g_n - fg = (f_n - f)(g_n - g) + f(g_n - g) + g(f_n - f),$$

whence, by (5) and (6), omitting the subscripts $E$,

$$\|f_n g_n - fg\| \leq \|f_n - f\| \cdot \|g_n - g\| + \|f\| \cdot \|g_n - g\| + \|g\| \cdot \|f_n - f\|;$$

thus the left hand side tends to 0 so long as $\|f\|$ and $\|g\|$ are finite.

The boundedness hypothesis on $f$ and $g$ is essential, as the following counterexample shows. Take $E = [0, +\infty[$, $f_n(x) = x + 1/n$ and $g_n(x) = 1/n$, whence $f(x) = x$, $d(f_n, f) = 1/n$, $g(x) = 0$ and $d(g_n, g) = 1/n$. The functions $f_n(x) g_n(x) = x/n + 1/n^2$ converge simply to $0 = f(x)g(x)$ but not uniformly since they are not even bounded in $E$.

(R 3) *If a sequence of functions* $f_n$ *converges uniformly on* $E$ *to a limit function* $f$, *and if* $\inf |f(x)|$ *is strictly positive, then the sequence* $(1/f_n)$ *converges uniformly to* $1/f$. For

$$|1/f_n(x) - 1/f(x)| \;=\; |f_n(x) - f(x)|/|f_n(x)| \cdot |f(x)| \leq$$
$$\leq\; d_E(f_n, f)/|f_n(x)| \cdot |f(x)|,$$

so that to *majorise* the result one has to *minorise* the denominator of this fraction. If there exists a number $m > 0$ such that $|f(x)| \geq m$ for any $x \in E$, then the relation $\|f_n - f\| < m/2$, valid for $n$ large, shows that

$$|f_n(x)| \geq m/2 \text{ for all } x \in E$$

if $n$ is sufficiently large, whence $d_E(1/f_n, 1/f) \leq 2d_E(f_n, f)/m^2$ for $n$ large, qed.

Rule (R 3) requires the limit function $f$ to remain "uniformly away from zero" for $n$ large or, equivalently, *that the function* $1/f$ *be bounded*. Even if $f$ and the $f_n$ never vanish, the uniform convergence of $f_n$ to $f$ does not imply that of $1/f_n$ to $1/f$. If, for example, one takes $E = [1, +\infty[$ and $f_n(x) = 1/x + 1/nx$, then $f(x) = 1/x$ and $d(f_n, f) = \sup |1/nx| = 1/n$, so that $f_n$ converges uniformly to $f$; but

$$1/f(x) - 1/f_n(x) = x/(n+1)$$

does not converge *uniformly* to 0 and is not even bounded on $E$.

## 8 – Series of continuous functions. Normal convergence

The concept of uniform convergence applies to series of functions when we consider their partial sums: if $s(x) = \sum u_n(x)$, then uniform convergence means, by definition, that for $p \in \mathbb{N}$ given and for all $n$ sufficiently large, the partial sum $s_n(x) = u_1(x) + \ldots + u_n(x)$ is equal to $s(x)$ to within $10^{-p}$ for *all $x \in E$*, the set on which the functions $x \mapsto u_n(x)$ are defined; this also means that the "remainder" $r_n(x)$ of the series converges uniformly to 0.

A particularly simple case is that of a *normally convergent series* of functions $u_n(x)$, a concept used freely by Weierstrass, though Remmert, *Funktionentheorie 1*, p. 84, attributes the explicit definition and name to Baire, *Leçons sur les théories générale de l'analyse*, which I have not reread for more than half a century, and whose details I seem to have forgotten: one assumes that there exists a *convergent* series with positive terms $\sum v_n$ *whose terms are independent of $x$* and such that

$$(8.1) \qquad |u_n(x)| \leq v_n \text{ for all } n \in \mathbb{N} \text{ and } x \in E.$$

This definition can be expressed in a much more striking way. Using the notation $u_n(x)$, the preceding relation is equivalent, just by the definition of the uniform norm, to

$$(8.2) \qquad \|u_n\|_E \leq v_n.$$

So normal convergence implies

$$(8.3) \qquad \sum \|u_n\|_E < +\infty$$

and in fact is equivalent to it, since we have $|u_n(x)| \leq \|u_n\|_E$ for all $x \in E$, so that the series with general term $v_n = \|u_n\|_E$ satisfies the original definition. The advantage of (3) is that it does not involve any particular series $\sum v_n$, but the original definition is often more convenient in practice.

This said, it is first of all clear that the series $\sum u_n(x)$ converges absolutely for all $x$. On the other hand,

$$|s(x) - s_n(x)| = |u_{n+1}(x) + \ldots| \leq v_{n+1} + \ldots,$$

and since the difference between the total sum and the $n^{\text{th}}$ partial sum of the series $\sum v_n$ is $< r$ for $n > N$, we deduce that the same holds for $|s(x) - s_n(x)|$ *for any $x \in E$*; whence uniform convergence of $s_n(x)$ to $s(x)$, with

$$(8.4) \qquad d_E(s_n, s) \leq v_{n+1} + \ldots.$$

Note also the relation

$$(8.5) \qquad \left\| \sum u_n \right\|_E \leq \sum \|u_n\|_E$$

which follows from the fact that, for all $x$,

$$\left| \sum u_n(x) \right| \leq \sum |u_n(x)| \leq \sum \|u_n\|_E.$$

A consequence, which will be generalised in n° 13:

**Theorem 9.** *The sum of a normally convergent series of continuous functions is continuous*[15].

*Example 1.* The sum of a *trigonometric series*

$$(8.6) \qquad c_0 + \sum_{n=1}^{\infty} a_n \cdot \cos nx + \sum_{n=1}^{\infty} b_n \cdot \sin nx$$

whose coefficients satisfy

$$\sum (|a_n| + |b_n|) < +\infty$$

is a continuous function on $\mathbb{R}$ [this is by no means the case for the square wave series of n° 2 which we shall reexamine in n° 11]. We shall prove (Chap. VII) that the Fourier series of periodic functions *having an everywhere continuous derivative* are of this kind, though this condition is not necessary[16].

*Example 2.* Consider the Riemann series

$$\zeta(s) = \sum 1/n^s$$

for $s > 1$, not necessarily an integer. Since $a^s$ is an increasing function of $s$ for all $a > 1$,

$$1/n^s \leq 1/n^\sigma \text{ for } s \geq \sigma$$

and since the series $\sum 1/n^\sigma$ converges for $\sigma > 1$, one deduces that the series converges normally on every interval $[\sigma, +\infty[$ with $\sigma > 1$, strict inequality.

---

[15] The definition of normal convergence, and Theorem 9, extend in an obvious way to sums indexed by any countable set of indices $I$. See n° 18.

[16] The theory, whose more elementary aspects we shall expound in Chap. VII, associates to each regulated periodic function a trigonometric series whose coefficients satisfy a definitely weaker relation, namely $\sum(|a_n|^2 + |b_n|^2) < +\infty$, as for example in the case of the square wave series. The complete theory, valid for "square integrable" functions in the sense of the Lebesgue integral, does not go much further. To go beyond this type of series one has to use the "distributions" of L. Schwartz (Chap. V), generalised functions to which one can assign Fourier series whose coefficients increase no faster than a power of $n$, as we shall see in Chap. VII.

Let us warn the reader: beside very simple elementary aspects, the theory of trigonometric series presents very subtle aspects, far outside the framework of absolutely or uniformly convergent series; happily one does not use them in most domains of analysis, even though there are always specialists interested in them and they form the topic of much interesting work.

The functions $a^s$ being continuous, so is the zeta function for $s > 1$. It possesses more spectacular properties, but this would take us too far aside.

Note, however, that the Riemann series does not converge normally on the interval $X = ]1, +\infty[$. Indeed, for all $n$, $1/n^s < 1/n$ for all $s > 1$ and it is clear (or will become so in Chap. IV) that $1/n^s$ tends to $1/n$ when $s \to 1+0$; so here $\|u_n\|_X = 1/n$, which rules out the relation (3).

*Example 3. Every power series which converges in a disc of radius $R > 0$ converges normally on any disc of radius strictly smaller than $R$.*

Let $r < R$. We have known since Chap. II that the given series $f(z) = \sum a_n z^n$ converges absolutely for $|z| = r$; since $|a_n z^n| \leq |a_n r^n| = v_n$ for $|z| \leq r$, normal convergence follows and, incidentally, so again does the continuity of the sum; but we knew that already!

The hypothesis $r < R$ is essential. Consider for example the geometric series $\sum z^n$ which converges for $|z| < 1$ to the function $s(z) = 1/(1-z)$. We shall show that the series does not converge normally, nor even uniformly, on the disc $|z| < 1$. If it did one could find an $n$ such that

$$|s(z) - s_n(z)| \leq 1$$

on all the disc $|z| < 1$. The polynomial $s_n(z) = 1 + z + \ldots + z^n$ being, in absolute value, majorised by $n + 1$ for $|z| < 1$, one would then have $|s(z)| \leq n + 2$, i.e. $|1 - z| \geq 1/(n+2)$ for $|z| < 1$; impossible on a neighbourhood of $z = 1$. Since furthermore $|s(z) - s_n(z)| = |z|^{n+1}/|1-z|$, it is clear that, for $n$ given, the left hand side increases indefinitely as $|z|$ tends to 1, so is not even bounded on $|z| < 1$.

If one restricts oneself temporarily to real values of $z$ and if one remarks that every compact, so closed, interval contained in $X = ]-1, +1[$ is in fact contained in an interval $[-q, +q]$ with $q < 1$, one has thus obtained a series of continuous functions which, without converging uniformly on the *open* interval $X$, converges uniformly on every *compact* interval contained in $X$. A very frequent phenomenon, valid for every power series and which nevertheless suffices to assure the continuity of the sum of the series. It is in fact continuous on $[-1, 1[$ because it is equal to $1/(1-z)$; the following example will provide a better reason in a more interesting case.

*Example 4.* For $x$ real consider the series

$$(8.7) \qquad L(x) = x - x^2/2 + x^3/3 - \ldots = \sum_{n \geq 1} (-1)^{n+1} x^n / n$$

whose sum is the function $\log(1 + x)$, as we shall see (n° 16, example 1). It converges absolutely for $|x| < 1$ since its general term is, in absolute value, majorised by that of the corresponding geometric series. It converges for $x = 1$ also, since then it is the alternating harmonic series $\sum (-1)^{n+1}/n$. Finally, it diverges for $x = -1$ (harmonic series). Let us show that *its sum is continuous on the interval* $]-1, 1]$ *where it is defined.*

There is no problem for $|x| < 1$ since we are dealing with a power series whose radius of convergence is clearly 1. To examine what happens when $x$ tends to 1, let us work on $[0, 1]$. The series is then alternating with decreasing terms. Writing $L_n(x)$ for the $n^{\text{th}}$ partial sum of the series, the general result of Chap. II, n° 13 then tells us that

$$(8.8) \qquad |L(x) - L_n(x)| \leq x^{n+1}/(n+1) \leq 1/(n+1) \quad \text{for } 0 \leq x \leq 1.$$

In consequence, $L_n(x)$ converges *uniformly* to $L(x)$ on this interval, whence continuity on the *closed* interval $[0, 1]$ and so on $]-1, 1]$ as stated. *If one knows* that the function $\log(1 + x)$ is continuous for $x > -1$ and is represented by the series (7) for $|x| < 1$ (strict inequality), one can then pass to the limit when $x \to 1 - 0$, whence

$$(8.9) \qquad\qquad 1 - 1/2 + 1/3 - 1/4 + \ldots = \log 2.$$

It goes without saying that this formula is no better suited to the numerical calculation of $\log 2$ than Leibniz' formula for $\pi/4$ which evoked Newton's irony.

One should also remark that, though the series (7) converges *uniformly* on $[0, 1]$, we have not proved that it converges *normally* there, for this is patently false: a series with positive terms which majorises the moduli of the terms of (7) for any $x \in [0, 1]$ must, for $x = 1$, majorise the harmonic series and so cannot possibly converge.

*Example 5.* Consider the square wave series. Clearly it cannot converge uniformly on the interval $[0, \pi]$ since its sum $s(x)$ is discontinuous at $x = \pi/2$; Further, if you consider the subset of the plane included between the graphs of $s(x) + r$ and $s(x) - r$, with for example $r = \pi/12$, you will obtain the union of the three following sets:

(i) $\qquad\qquad 0 \leq x < \pi/2, \qquad\qquad \pi/6 < y < \pi/3,$

(ii) $\qquad\qquad\qquad x = \pi/2, \qquad\qquad |y| < \pi/12,$

(iii) $\qquad\qquad \pi/2 < x \leq \pi, \qquad\qquad -\pi/3 < y < -\pi/6.$

How can a graph lying entirely in this strip pass from the strip (i) to the strip (iii) across (ii) while remaining *continuous*? A simple sketch will let one understand the problem, which gives rise to the "Gibbs phenomenon", called *overshoot* by the electricians. While waiting to read the chapter of this book consecrated to Fourier series, the reader may practice by tracing the graphs of the first partial sums of the series; this exercise presupposes only elementary knowledge of derivatives and of the trigonometric functions, and exhibits the phenomenon very clearly[17]. Convergence of the series, which the

---

[17] One can also, of course, trace these curves with the help of computer programmes, but this is not the method to learn mathematics; mathematics is not a *black box* in whose interior phenomena are produced that one does not try to understand.

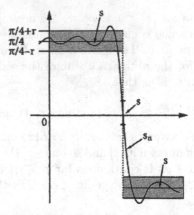

fig. 10.

simplistic criteria of Chap. II are inadequate to establish, will be established in n° 11.

*Example 6.* Consider the Weierstrass series

$$f_k(z) = \sum (z - \omega)^{-k}$$

(Chap. II, n° 23) for $k \geq 3$. Suppose that $z$ remains in a disc $D : |z| \leq R$, of finite radius and let us omit from the series the terms, finite in number, for which $|\omega| < 2R$. Then clearly $|z - \omega| \geq |\omega|/2$ for all $z \in D$ and all $\omega$. In consequence, the series is, on $D$, dominated up to a factor $2^k$ by the series $\sum 1/|\omega|^k$ which, as we have seen, converges. The series $f_k$ is thus normally convergent on $D$ (after subtracting the terms for which $|\omega| < 2R$).

# §3. Bolzano-Weierstrass and Cauchy's criterion

### 9 – Nested intervals, Bolzano-Weierstrass, compact sets

The purpose of this n° is to prepare the proof of Cauchy's general criterion of convergence that we will establish in the next n°. But these preliminary results are, even so, also important.

It all rests on the following result:

**Theorem 10.** *Let* $(K_p)_{p \in I}$ *an arbitrary*[18] *family of compact nonempty intervals. Suppose that, for all* $p, q \in I$, *there exists an* $r \in I$ *such that* $K_r \subset K_p \cap K_q$. *Then the intersection of the* $K_p$ *is a compact nonempty interval.*

Let us write $K_p = [a_p, b_p]$ with $a_p$ and $b_p$ finite. The intersection of the $K_p$ is clearly the set of $x \in \mathbb{R}$ such that

$$a_p \leq x \leq b_p \text{ for every } p \in I.$$

If we put $a = \sup a_p$ and $b = \inf b_p$, the preceding relation is equivalent to $a \leq x \leq b$ by the definition of least upper bound and of greatest lower bound of a family of real numbers. It remains to prove that the compact interval $[a, b]$, the intersection of the $K_p$ as we have just seen, is nonempty, i.e. that $a \leq b$.

But the fact that, for all $p$ and $q$, the intersection $K_p \cap K_q$ is nonempty – it contains a $K_r$ – shows that $a_p \leq b_q$. Since $b_q$ majorises all the $a_p$, it majorises their least upper bound $a$ too. Since $a$ minorises all the $b_q$, it also minorises their greatest lower bound $b$, whence $a \leq b$, qed. Compare with the proof of Theorem 5 or, later, to the second proof of Theorem 13 (Cauchy's criterion).

**Corollary 1 (Principle of nested intervals).** *Let* $I_1 \supset I_2 \supset \ldots$ *be a decreasing sequence of nonempty compact intervals. Then the intersection of the* $I_n$ *is a nonempty compact interval.*

Obvious; the set of indices in Theorem 10 is here $\mathbb{N}$.

**Corollary 2.** *Let* $(J_i)_{i \in I}$ *be an infinite family of nonempty compact intervals. Then the intersection of* $J_i$ *is nonempty if and only if for all* $i_1, \ldots, i_n \in I$ *the intersection of the corresponding* $J_i$ *is nonempty.*

The condition of the statement is clearly necessary.
To show that it is sufficient let us put

---

[18] The fact that we name the elements of $I$ as $p$, $q$, $r$ does not mean that we are dealing with whole numbers: an arbitrary set is an arbitrary set.

$$K_F = \bigcap_{i \in F} J_i$$

for every *finite* subset $F$ of $I$. By hypothesis, the $K_F$ are, like the $J_i$, nonempty compact intervals. If, further, $F'$ and $F''$ are two finite subsets of $I$ and if $F = F' \cup F''$, then (associativity of intersections)

$$K_F = \bigcap_{i \in F' \cup F''} J_i = \bigcap_{i' \in F'} J_{i'} \cap \bigcap_{i'' \in F''} J_{i''} = K_{F'} \cap K_{F''},$$

so that the family of $K_F$, indexed by the set of all finite subsets of $I$, satisfies the hypothesis of Theorem 10. The intersection of all the $K_F$, i.e. of all the $J_i$, is thus a nonempty compact interval, qed.

If for example one has a countable family $(J_n)$, the hypothesis means that $J_1 \cap \ldots \cap J_n = I_n$ is nonempty for any $n$, since every finite subset of $\mathbb{N}$ is contained in a set $\{1, \ldots, n\}$. In this case, Corollary 2 follows directly from the previous one.

**Corollary 3 (Bolzano-Weierstrass Theorem).** *Every* bounded *sequence of complex numbers has a convergent subsequence.*

First we examine the case of a sequence $u(n)$, $n \in \mathbb{N}$, with real terms, so contained in a compact interval $I$ of radius $r$. We divide $I$ into two equal compact intervals (they have a point in common, but this does not matter) and, for each of them, we consider the set of $n \in \mathbb{N}$ such that $u(n)$ belongs to it. We obtain two subsets of $\mathbb{N}$ whose union is $\mathbb{N}$. One of these at least, say $\mathbb{N}_1$, is an infinite set; let us denote by $I_1$ that of the two halves of $I$ containing the $u(n)$ corresponding to $\mathbb{N}_1$. After this, let us divide again $I_1$ into two equal compact intervals; the preceding argument shows that, for one of these two halves, say $I_2$, the set $\mathbb{N}_2$ of $n \in \mathbb{N}_1$ such that $u(n) \in I_2$ is infinite. This done, let us again divide, $I_2$ into two equal halves, etc.

In this way we define a *decreasing* sequence of *compact* nonempty intervals $I_k \subset I$ and a *decreasing* sequence of *infinite* subsets $\mathbb{N}_k$ of $\mathbb{N}$ satisfying the following conditions:

(i)  $u(n) \in I_k$ for all $n \in \mathbb{N}_k$;
(ii) $I_k$ has radius $r/2^k$

where $r$ is the radius of $I$. By Corollary 1 above, the $I_k$ have a common point $u$, clearly unique by (ii).

Write $p_1$ for the least element of $\mathbb{N}_1$. Every element of $\mathbb{N}_2$ is $\geq p_1$ since $\mathbb{N}_2 \subset \mathbb{N}_1$; since $\mathbb{N}_2$ is infinite, it contains numbers strictly greater than $p_1$; write $p_2 > p_1$ for the least of these. Generally, write $p_k$ for the least of the numbers of $\mathbb{N}_k$ which is $> p_{k-1}$. Finally, write $v(k) = u(p_k)$, a subsequence of the given sequence.

Relation (i) above shows that $v(k) \in I_k$ for all $k$, and so satisfies $|v(k) - u| < r/2^k$. Thus the subsequence $v(k)$ converges to $u$, which completes the proof for real sequences.

The extension to complex sequences $u(n) = v(n) + i.w(n)$ is immediate: extract a convergent subsequence $v(p_n)$ from the sequence $v(n)$, then extract a convergent subsequence from the sequence $w(p_n)$ – not from the complete sequence $w(n)$. After two successive extractions of subsequences the real and imaginary parts of the new sequence converge, so the sequence does too, qed.

*Exercise:* prove the BW Theorem directly, using the decimal expansions of real numbers.

The BW Theorem leads to one of the most fundamental concepts in all analysis, that of *a compact subset* of $\mathbb{C}$; this is the term for any set $K \subset \mathbb{C}$ which is both *bounded and closed.* The compact sets possess, and are the only ones to possess, the BW property: *every sequence of points of $K$ has a subsequence converging to a point of $K$.* Proof in two stages:

(i) Every sequence of points of $K$ is, like $K$, bounded; one can therefore extract a convergent sequence from it; the limit of the latter belongs to $K$ just by the definition of *closed* sets given in n° 1: one cannot escape from $K$ by passing to a limit.

(ii) Suppose conversely that a set $K \subset \mathbb{C}$ possesses the property in question. First, $K$ is bounded, for if it were not one could, for all $n$, find an $u_n \in K$ such that $|u_n| > n$; it would be impossible to extract a convergent sequence from the sequence so obtained. On the other hand, $K$ is closed, since if a sequence $u(n) \in K$ converges to a limit $u \in \mathbb{C}$, then, by hypothesis, one can extract a sequence from it which converges to a limit *belonging to $K$;* and this must be $u$, whence $u \in K$, qed.

*The principle of nested intervals extends to every decreasing sequence of nonempty compact sets $K_n \subset \mathbb{C}$.* We need only choose a $u(n) \in K_n$ for all $n$ and extract a convergent subsequence $v(n) = u(p_n)$ from the sequence $u(n)$. Since $K_i \supset K_{i+1} \supset \ldots$ and $p_i \geq i$, we have $v(n) \in K_{p(n)} \subset K_n \subset K_i$ for all $n > i$, so that the limit of $v(n)$ belongs to each $K_i$, qed.

It is not even really necessary, in the statement above, to assume that the $K_n$ decrease; it is enough for the partial intersections $K_1 \cap K_2 \cap \ldots \cap K_n = H_n$ to be nonempty. These decrease and are again compact, since every intersection, finite or not, of closed sets is again closed, as we saw in n° 1. Then we apply the preceding result to the $H_n$.

We shall meet compact sets again later, mainly in the theory of integration. For the moment let us record the following fundamental result:

**Theorem 11.** *Let $K$ be a compact subset of $\mathbb{C}$ and let $f$ be a continuous map of $K$ into $\mathbb{C}$. Then the image $f(K)$ of $K$ under $f$ is compact.*

We need only show that $f(K)$ has the BW property. So let $y(n)$ be a sequence of points of $f(K)$; choose $x(n) \in K$ such that $y(n) = f(x(n))$; since $K$ is compact there exists a subsequence $x(p_n)$ which converges to an $x \in K$; clearly the sequence $y(p_n)$ then converges to $y = f(x) \in f(K)$, qed.

**Corollary.** *Let $f$ be a real function defined and continuous on a compact subset $K$ of $\mathbb{C}$. Then there are points $a, b \in K$ such that*

$$f(a) \leq f(x) \leq f(b) \text{ for all } x \in K.$$

The image $f(K)$ is certainly a compact subset of $\mathbb{R}$. It is bounded, so that $u = \inf(f(K))$ and $v = \sup(f(K))$ are finite. Since $f(K)$ is closed it must contain $u$ and $v$. So there are points $a, b \in K$ for which $f(a) = u$ and $f(b) = v$, qed.

The preceding corollary, which states that $f$ has an absolute minimum and maximum on $K$, can be generalised to the case of continuous functions with complex values: *for the equation $f(x) = w$ to possess an exact solution $x \in K$ for $w \in \mathbb{C}$ given, it is sufficient that it possesses a solution to within $10^{-n}$ for all $n \in \mathbb{N}$*, i.e. that there exists an $x_n \in K$ such that $|f(x_n) - w| < 10^{-n}$. For if so, then $w$ is adherent to $f(K)$, so belongs to this set, because $f(K)$ is compact and therefore closed.

**Theorem 12.** *Let $f$ be a complex-valued function, defined, continuous and injective on a compact set $K \subset \mathbb{C}$. Then $g : f(K) \to K$, the inverse map to $f$, is continuous.*

Let $b = f(a)$ be a point of $H = f(K)$, choose an $r > 0$ and consider the set $K'$ of $x \in K$ such that $d(a, x) \geq r$. This is closed and bounded like $K$, so compact. In consequence, $H' = f(K')$ is a compact subset of $\mathbb{C}$ (in fact of $H$) and in particular is closed. Since $f$ is injective, $H'$ does not contain $b$, so there exists an $r' > 0$ such that the ball $B(b, r')$ does not meet $H'$. It is clear that then

$$f(x) \in B(b, r') \Longrightarrow x \in B(a, r),$$

qed.

## 10 – Cauchy's general convergence criterion

**Theorem 13. (Cauchy's general convergence criterion for sequences)** *For a sequence $(u_n)$ of complex numbers to converge it is necessary and sufficient that for all $r > 0$*

(10.1)        $d(u_p, u_q) < r$    *for $p$ and $q$ sufficiently large.*

*First proof.* We use the BW theorem. By this there exists a subsequence $u(p_n)$ which converges to a limit $u$. For $r > 0$ given we thus have $|u(p_n) - u| < r$ for $n$ large. But since $p_n \geq n$ we also have, by hypothesis, $|u(p_n) - u(n)| < r$ once $n$ is sufficiently large. So $|u(n) - u| < 2r$ for $n$ large, qed.

*Second proof.* First we note that if the given sequence satisfies (1) then so do the sequences formed by the real and imaginary parts of $u_n$. So we can restrict ourselves to the case of a sequence of real numbers.

For all $n \in \mathbb{N}$, denote by $E_n$ the set of $u_p$ for which $p \geq n$, i.e. $E_n = \{u_n, u_{n+1}, \ldots\}$. These $E_n$ form a decreasing sequence of nonempty subsets of $\mathbb{R}$. Condition (1) shows further that the sequence $(u_n)$, and so the sets $E_n$, are bounded. Let us put

$$a_n = \inf E_n, \qquad b_n = \sup E_n, \qquad I_n = [a_n, b_n].$$

The $I_n$ are compact intervals.

Since $E_n \supset E_{n+1}$, any number which bounds $E_n$ above (resp. below) must also bound $E_{n+1}$ above (resp. below). So

$$a_n \leq a_{n+1} \leq b_{n+1} \leq b_n$$

and consequently $I_n \supset I_{n+1}$. The principle of nested intervals then shows the existence of a $u \in \mathbb{R}$ belonging to all the $I_n$. This is the required limit of the sequence $(u_n)$.

For: choose an $r > 0$ and suppose that (1) is satisfied for $p, q > N$. By the definition of $E_N$ we then have $|x - y| < r$ for all $x, y \in E_N$, whence $|a_N - b_N| \leq r$ since the greatest lower bound and least upper bound of a set are limits of elements of that set. Since $u \in I_N$ we thus have $|u - x| \leq r$ for all $x \in I_N$ and in particular for all $x \in E_N$. Hence $|u - u_n| \leq r$ for $n > N$, qed.

*Example 1.* Consider a closed set $F$ in the plane and, for all $x \in \mathbb{C}$, let

$$d(x, F) = \inf_{z \in F} d(x, z) = \inf |x - z|$$

be the distance from $x$ to $F$. First, this is a continuous function of $x$. For all $r > 0$ there exists a $z \in F$ such that $d(x, z) \leq d(x, F) + r$, whence, for all $y \in \mathbb{C}$,

$$d(y, z) \leq d(y, x) + d(x, F) + r$$

and so $d(y, F) \leq d(y, x) + d(x, F) + r$; since $r$ is arbitrary one may deduce that $d(y, F) \leq d(x, F) + d(y, x)$ and also the inequality obtained by interchanging $x$ and $y$; whence finally the relation

$$|d(x, F) - d(y, F)| \leq d(x, y)$$

which proves continuity, whether $F$ is closed or not.

When $F$ is *closed* then for any $x$ there exists at least one $a \in F$ for which $d(x, a) = d(x, F)$, i.e. a point at the minimum distance to $F$. This is obvious if $d(x, F) = 0$: there are then $a_n \in F$ such that $d(a_n, x)$ tends to 0, so that $x$ is adherent to $F$ and so belongs to $F$. If $d(x, F) = d > 0$, consider the intersection $K$ of $F$ and the closed ball $B(x, 2d)$; this is a closed bounded set, so compact. By the definition of a greatest lower bound, for any $n$ there exists an $a_n \in F$ such that $d(x, a_n) < d + 1/n$, so that $a_n \in K$ for $n$ large; by BW, one can extract from the sequence $(a_n)$ a subsequence converging to an $a \in K$, and since $\lim d(x, a_n) = d$ it is clear that $d(x, a) = d$, qed, since $F$ contains $K$.

There is no reason for the point $a$ to be unique in general. But there is an important particular case where Cauchy's criterion suffices to prove *the existence and the uniqueness* of $a$ simultaneously, namely the case where the closed set $F$ is *convex*, i.e. such that, for all $a, b \in F$, the line segment $[a, b]$ joining $a$ to $b$ is contained in $F$.

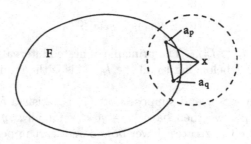

**fig. 11.**

To see this, we start from the identity

$$|u + v|^2 + |u - v|^2 = 2(|u|^2 + |v|^2),$$

easy to prove on writing $|z|^2 = z\bar{z}$ for all $z \in \mathbb{C}$. We deduce that $|u - v|^2/4 = (|u|^2 + |v|^2)/2 - |(u+v)/2|^2$. Now consider an $x \in \mathbb{C}$ and, for all $n$, choose an $a_n \in F$ such that $d(x, F)^2 \leq d(x, a_n)^2 \leq d(x, F)^2 + 1/n$. Since $F$ is convex, it contains the points $\frac{1}{2}(a_p + a_q)$ for all $p$ and $q$; on taking $u = x - a_p$ and $v = x - a_q$ in the above we obtain

$$|a_p - a_q|^2/4 \;=\; \left(|x - a_p|^2 + |x - a_q|^2\right)/2 - \left|x - \frac{1}{2}(a_p + a_q)\right|^2 \leq$$
$$\leq \; d(x, F)^2 + \frac{1}{2}(1/p + 1/q) - d(x, F)^2$$

since $\left|x - \frac{1}{2}(a_p + a_q)\right|^2 \geq d(x, F)^2$. Hence $|a_p - a_q|^2 \leq \frac{1}{2}(1/p + 1/q) < r$ for $p$ and $q$ large. Cauchy's criterion is thus satisfied and the sequence $(a_n)$ converges to a limit $a \in F$ since $F$ is closed, with clearly $d(x, a) = \lim d(x, a_n) = d(x, F)$. The preceding identities show the uniqueness of $a$ immediately: replace $u$ and $v$ by $x - a$ and $x - b$ where $a, b \in F$ are at the minimum distance from $x$.

This result and its proof extends to the Cartesian spaces $\mathbb{R}^p$ (where BW is again valid) and even to "Hilbert spaces" of infinite dimension to which, on the contrary, the BW Theorem does not extend. In the usual euclidean space of three dimensions one knows, for example, that if $F$ is a plane or a line, then the point of $F$ at the minimum distance to $x$ is precisely the orthogonal projection of $x$ onto $F$.

Cauchy's criterion extends naturally to functions of a "non-discrete" variable:

**Theorem 13'.** *Let $X$ be a subset of $\mathbb{C}$, let $a$ be an adherent point of $X$ and $f$ a scalar function defined on $X$. For $f$ to tend to a limit as $x \in X$ tends to $a$ it is necessary and sufficient that for all $r > 0$ there exists an $r' > 0$ such that, for $x', x'' \in X$,*

$$(10.2) \qquad x', x'' \in B(a, r') \Longrightarrow d[f(x'), f(x'')] < r.$$

*If $u$ is the limit value of $f$ at $a$ then $d[f(x), u] \leq r$ for $d(a, x) < r'$.*

There is an analogous result when, $X \subset \mathbb{R}$ not being bounded above, one lets $x$ tend to $+\infty$: for all $r > 0$ there must exist an $M$ such that

$$\{(x' > M) \ \& \ (x'' > M)\} \Longrightarrow |f(x') - f(x'')| < r.$$

The proof of Theorem 13' is practically identical to that of Cauchy's criterion for sequences; on the other hand one can deduce the former from Theorem 13' by choosing $X = \mathbb{N}$ and $f(x) = u_n$ for $x = n$. Condition (2) then means that $|u_p - u_q| < r$ for $p$ and $q$ sufficiently large. But it is simpler to deduce Theorem 13' from Theorem 13. To do this one chooses a sequence of points $x_n \in X$ tending to $a$; with the notation of (2), the ball (open or closed, it doesn't matter) $B(a, r')$ with centre $a$ and radius $r'$ contains $x_p$ and $x_q$ for $p, q$ large, whence

$$(10.3) \qquad d[f(x_p), f(x_q)] < r;$$

so the points $f(x_n)$ form a Cauchy sequence, so tend to a limit $u$. But for $x \in B(a, r')$ and $n$ large enough that $x_n \in B(a, r')$, $f(x)$ is equal to $f(x_n)$ to within $r$, so equal to $u$ to within $2r$, whence convergence. The inequality $d[f(x), u] \leq r$ for $d(x, a) < r'$ follows from observing that $d[f(x), f(y)] < r$ for all $y \in B(a, r')$ and passing to the limit[19] when $y$ tends to $a$.

Finally, there is a Cauchy criterion for uniform convergence. Here we use the concept of the distance between two functions as defined in n° 7:

$$(10.4) \qquad d_X(f, g) = \sup_{x \in X} |f(x) - g(x)|.$$

---

[19] The remarks at the beginning of n° 9 of Chap. II on limits of inequalities clearly apply also to functions more general than sequences.

**Theorem 13".** *Let $X$ be a set and $(f_n)$ a sequence of scalar functions defined on $X$. For it to converge uniformly on $X$, it is necessary and sufficient that for all $r > 0$ there exists an integer $N$ such that*

(10.5)                              $d_X(f_p, f_q) < r$   for all $p, q > N$.

The condition is obviously necessary: by definition, uniform convergence to $f$ means that $d_X(f_n, f)$ tends to 0, and since the triangle inequality applies as much to the distance $d_X$ as to the usual distance in $\mathbb{C}$, the inequality (5) is obtained exactly as in the classical case of a sequence of numbers.

Sufficiency: since for any $x$

$$|f_p(x) - f_q(x)| \leq d_X(f_p, f_q) < r \text{ for } p, q > N,$$

Theorem 13 shows that $\lim f_n(x) = f(x)$ exists for all $x \in X$, and, further, that $|f(x) - f_n(x)| \leq r$ for $n > N$. Since $N$ does not depend on $x$ one concludes that $d_X(f, f_n) \leq r$ for all $n > N$, whence uniform convergence (no hypothesis at all on $X$).

The same idea is at the base of Theorems 13, 13' and 13": for $f(x)$ to converge when $x$ tends to a limit $a$, finite or not, it is necessary and sufficient that, for all $r > 0$, the function $f(x)$ should be *constant to within $r$ on a neighbourhood of $a$.*

The second proof of Theorem 13 rests on axiom (IV) of Chap. II, but the theorem is in fact equivalent to it, i.e. allows us to give a proof of it, or, equivalently, as we have already seen, to prove Theorem 2 of Chap. II, n° 9 directly, for increasing sequences. For let $(u_n)$ be an increasing sequence bounded above. There is an index $p$ such that $u_p + 1$ majorises the sequence, otherwise there would exist a $u_q > u_p + 1$, then an $u_r > u_q + 1 > u_p + 2$, etc., and the sequence would not be bounded. For the same reason, there exists an index $q > p$ such that $u_q + 1/10$ majorises the sequence, then an index $r > q$ such that $u_r + 1/100$ majorises the sequence, etc. For $i, j > r$, $u_i$ and $u_j$ lie between $u_r$ and $u_r + 1/100$, whence $|u_i - u_j| < 1/100$, which is Cauchy's criterion for $1/100$. Etc. The given increasing sequence then converges, thanks to Theorem 13.

One could also use the BW Theorem directly, since it assures the existence of a convergent subsequence; it is clear that the full sequence, being increasing, must converge to the same limit.

Theorem 13 also allows one to show immediately that *all nonterminating decimal expansions actually correspond to a real number*: starting from rank $p$, all the terminating expansions obtained by truncating it are equal to within $10^{-p}$, and so the sequence formed by these terminating expansions satisfies Cauchy's criterion trivially.

Conversely, in the case of an arbitrary sequence, when, for all $p$, the $u_n$ of sufficiently large index are equal to each other to within $10^{-p}$; if decimal enumeration did not present the bizarre behaviour already noted in Chap. II one could deduce that, for $n$ sufficiently large, the decimal expansions of order $p$ of $u_n$ would be independent of $n$, which would render Cauchy's criterion obvious to those who believed it *a priori*, and wrongly, that a nonterminating decimal expansion "evidently" represents a real number. We have already made this remark at the end of n° 4 of Chap. II.

This may be the reason why Bolzano and Cauchy seem to have considered their criterion as too obvious to merit a proof, but it is too late to ask them. It might also be that they did not see its importance, not having used it much (for what is due to Cauchy, see the proof of Theorem 14 below); their successors were the ones who exploited this.

However it may be, these remarks show that we could have chosen Theorem 13 as axiom (IV) for $\mathbb{R}$. This would, further, be more justified, since, in contrast to the concept of least upper bound, it has a meaning in any metric space and, in contrast to the method of cuts, allows us to extend to this general case the construction of $\mathbb{R}$ from $\mathbb{Q}$.

One can indeed, as Georg Cantor did in 1872, and as a much less famous Frenchman, Charles Meray, did in 1869, use Cauchy sequences to *define* the real numbers starting from $\mathbb{Q}$. Meray remarked that the two propositions at the base of analysis, up to then considered as "axioms", are (i) the convergence theorem for increasing sequences, (ii) Cauchy's criterion. He then decided – like Cantor – to partition the Cauchy sequences formed by rational numbers into classes, considering two sequences $(u_n)$ and $(v_n)$ such that $u_n - v_n$ tends to 0 as equivalent, and to *define* an irrational number as being such a class: for example, $\sqrt{2}$ will be the class of all rational Cauchy sequences $> 0$ such that $\lim u_n^2 = 2$. Professor at Dijon, where he expounded these ideas to students fresh from school, Meray published a *Nouveau précis d'analyse infinitésimal* in 1872 – it was greatly enlarged after 1894 – which, as well as containing his construction of the real numbers, grounded all of analysis on power series (he introduced this term, and also the expression "radius of convergence") claiming that one meets no other useful or interesting functions in mathematics or in physics, already an idea of Lagrange's. An opinion which, in this late age, hardly risks finding unanimity since a large part of the guild is, on the contrary, in process of attacking the continuous functions without derivatives, Cantor's bizarre sets of real numbers, everywhere discontinuous functions that one can nevertheless integrate thanks to Emile Borel and Henri Lebesgue at the end of the century, etc., while the physicists are starting to meet non-analytic functions everywhere, for example in the theory of wave propagation, while expecting better, or worse, in the century to come. Further, Meray, who never read his contemporaries, had his own idiosyncratic language, nor did he have the inventive genius of a Weierstrass or of a Cantor, so his influence was negligible and his construction of the real numbers is attributed to the latter, even in France[20].

---

[20] An example of the "Matthew Effect in Science" studied by the founder of the American school of the sociology of sciences, Robert K. Merton, in an article

Some criticised his *Précis* for demanding of school-leavers a capacity for abstraction that they do not possess and forcing a brutal rupture from their old habits, to which Meray replied that "it was not [his] fault if they are bad to the point that makes a rupture necessary[21]" (in 1872 ...).

One sees that all these ideas – cuts *à la* Dedekind, increasing sequences, least upper bounds, decimal expansions, Cauchy sequences, nested intervals, BW, etc. – overlap each other, and consist of always repeating the same thing in various disguises, leaving one to choose, in practice, the most directly usable form. When once one has understood these ideas, one has made an enormous leap forwards in the comprehension of analysis: one has overtaken Newton and Euler, 1665–1750, Dedekind, Weierstrass, Cantor, etc. Do not worry if you do not assimilate them instantly and completely: it needed all of the XIX[th] century and the reflections of a good dozen of mathematicians of the highest order to elaborate it and even more in order to give it, in the XX[th] century, the almost perfect form and the enormous generality that it has acquired today.

## 11 – Cauchy's criterion for series: examples

Cauchy's criterion applies to series $\sum u_n$; one applies it to the sequence of partial sums $s_n$ of the series. Since

$$s_q - s_p = u_{p+1} + \ldots + u_q \quad \text{for } p \leq q,$$

we obtain the following result:

---

where, analysing the reward system for scientists, he comes to the same conclusion as Saint Matthew: "For whosoever hath, to him shall be given, and he shall have more abundance: but whosoever hath not, from him shall be taken away even that he hath". Merton bases himself on the distribution of Nobel Prize, on the order in which the authors of articles written in cooperation are cited, on interviews with scientists who stress that new ideas are much more easily accepted and recognised as such when they emanate from established scientists, etc. See *Science*, Vol. 159, 5/1/1968, or the collection of articles from Robert K. Merton, *The Sociology of Science* (U. of Chicago Press, 1973), a new discipline which has developed greatly recently and not always congenial to the scientists, who are deeply allergic to not uniformly admiring commentary on their activities, and even more, if possible, to see themselves subjected to the same methods of enquiry as the amazonian tribes or the "white collars" of the General Motors, their general theory being that scientists alone are able to understand their own activities ... See for example Bernard Barber, *Science and the Social Order* (Macmillan/Free Press, 1952), W. O. Hagstrom, *The Scientific Community* (Basic Books, 1965), D. Crane, *Invisible Colleges* (Chicago UP, 1972), J. R. Cole & S. Cole, *Social Stratification in Science* (Chicago UP, 1973), Harriet A. Zuckerman, *Scientific Elite: Studies of Nobel Laureates in the United States* (Chicago UP, 1974), Bruno Latour & Steve Woolgar, *Laboratory Life* (Sage Publications, 1979), etc. There is even at least one journal, *Social Studies of Science*, entirely devoted to the subject.

[21] See Dugac's thesis, pp. 81–88.

**Theorem 14 (Cauchy).** *The series $\sum u_n$ converges if and only if for every number $r > 0$ there exists an integer $N$ such that*

(11.1) $$|u_p + \ldots + u_q| < r \quad \text{for} \quad N < p \leq q.$$

Since the left hand side of (1) is smaller than the analogous expression for the series *with positive terms* $\sum |u_n|$, convergence of the latter will all the more imply (1), and so convergence of the initial series. So we again find, with Cauchy, the fact that *every absolutely convergent series converges*, but with quite another proof.

Cauchy's criterion also allows one to establish more subtle results than Leibniz' Theorem for alternating series.

**Theorem 15 (Dirichlet).** *Let $\sum u_n$ be a series whose partial sums are bounded and let $(v_n)$ be a sequence of positive numbers which decreases to 0. Then the series $\sum u_n v_n$ converges.*

Put
$$s_n = u_1 + \ldots + u_n, \qquad t_n = u_1 v_1 + \ldots + u_n v_n;$$
for $p < q$ (put $s_{p-1} = 0$ if $p = 1$): then

$$u_p v_p + \ldots + u_q v_q = (s_p - s_{p-1})v_p + \ldots + (s_q - s_{q-1})v_q =$$
$$= -s_{p-1} v_p + s_p(v_p - v_{p+1}) + \ldots + s_{q-1}(v_{q-1} - v_q) + s_q v_q.$$

Since $v_n \geq v_{n+1} \geq 0$, it follows that

$$|u_p v_p + \ldots + u_q v_q| \leq |s_{p-1}|v_p + |s_p|(v_p - v_{p+1}) + \ldots + |s_{q-1}|(v_{q-1} - v_q) + |s_q|v_q.$$

By hypothesis, $\sup |s_n| = M < +\infty$, whence

(11.2) $$|u_p v_p + \ldots + u_q v_q| \leq$$
$$\leq M \left[ v_p + (v_p - v_{p+1}) + \ldots + (v_{q-1} - v_q) + v_q \right] = 2M v_p;$$

since $v_p$ tends to 0 the left hand side is $< \varepsilon$ for $p$ large, whence Cauchy's criterion for $\sum u_n v_n$, qed.

In 1826 Abel established just the inequality (2) for $p = 1$ without assuming that the decreasing sequence $v_n$ tends to 0 nor drawing a conclusion as to the convergence of the series, yet this is clearly the essential point. In this more general case one again obtains the conclusion of Dirichlet's Theorem on strengthening the hypothesis imposed on the series $u_n$. Now $v_n$ certainly tends to a limit $v \geq 0$ so that, if the partial sums of the series $\sum u_n$ are bounded, the series $\sum u_n(v_n - v)$ will converge, by Theorem 15. To deduce the convergence of $\sum u_n v_n$ it then suffices to suppose the series $u_n$ convergent. Finally:

**Theorem 15' (Abel).** *Let $\sum u_n$ be a convergent series and $(v_n)$ a decreasing sequence of positive numbers. Then the series $\sum u_n v_n$ converges.*

We remark that if the partial sums of the series $u_1 + u_2 + \ldots$ are bounded then so are, for all $p$, those of the series $u_p + u_{p+1} + \ldots$ since they differ from the preceding by the number $u_1 + \ldots + u_{p-1}$, fixed for $p$ given; the preceding calculations thus apply to the "truncated" series $u_p v_p + \ldots$ even if the $v_n \geq 0$ decrease without necessarily tending to 0, a hypothesis which serves only to pass from (2) to Cauchy's criterion for $\sum u_n v_n$. Putting

$$(11.3) \qquad M_p = \sup_{n \geq p} |u_p + \ldots + u_n|$$

and applying the particular case $p = 1$ of (2) to the truncated series we thus find

$$(11.4) \qquad |u_p v_p + \ldots + u_q v_q| \leq 2 M_p v_p.$$

For given $p$ this estimate is valid for any $q > p$, so one can pass to the limit and, subject to Dirichlet's hypotheses, obtain the inequality

$$(11.5) \qquad \left| \sum_{n \geq p} u_n v_n \right| \leq 2 M_p v_p.$$

This can be used to prove the uniform convergence of certain series of functions:

**Corollary 1 (Abel[22]).** *Let $\sum a_n x^n$ be a power series with finite radius of convergence $R > 0$, and converging for $x = R$. Then the series converges uniformly on the interval $[0, R]$ (and, in fact, on all compacta contained in $] - R, R]$); its sum is continuous on $] - R, R]$.*

Let us put $x = Ry$, $u_n = a_n R^n$ and $v_n = y^n$, with $0 \leq y \leq 1$. The series $\sum u_n$ being convergent, its partial sums are bounded; the $v_n$ are positive and decrease, even for $y = 1$. Since $u_n v_n = a_n x^n$ we have, by (5),

$$\left| \sum_{n \geq p} a_n x^n \right| \leq 2 M_p y^p \quad \text{where} \quad M_p = \sup_{n \geq p} |a_p R^p + \ldots + a_n R^n|.$$

Since the series $u_n = a_n R^n$ converges we have $|u_p + \ldots + u_n| < \varepsilon$ for $p$ and $n$ large, and so $M_p \leq \varepsilon$ for $p > N$; since $y^p \leq 1$, we see that

$$p > N \Longrightarrow \left| \sum_{n \geq p} a_n x^n \right| \leq 2\varepsilon \quad \text{for all } x \in [0, R],$$

which means that the difference between the total sum and the partial sums of the power series converges to 0 uniformly on $[0, R]$, qed.

This result applies for example to the series $x - x^2/2 + x^3/3 - \ldots$ for $\log(1 + x)$, but the method used for this case in n° 8, example 4, requires definitely less ingenuity than Dirichlet's Theorem or its Corollary.

---

[22] There is an analogous result in the complex domain; see Remmert, *Funktionentheorie 1*, p. 94.

**Corollary 2.** *Let* $(v_n)$ *be a sequence of positive numbers decreasing to* $0$. *Then the series* $\sum v_n z^n$ *converges on the set*

(11.6)                    $$X : |z| \leq 1, \quad z \neq 1$$

*and, for any number* $r > 0$, *uniformly on the set*

(11.7)                    $$X(r) : |z| \leq 1, |1 - z| \geq r;$$

*its sum is continuous on* $X$.

Since $|z| \leq 1$ and $z \neq 1$ we have

$$|1 + z + \ldots + z^{n-1}| = |(1 - z^n)/(1 - z)| \leq 2/|1 - z|$$

for any $n$, whence convergence on $X$ by Dirichlet's Theorem (make $u_n = z^n$). Further, the inequality (2) shows, at the limit, that the remainder $r_p(z)$ of the series, is, in modulus, majorised by $4v_p/|1 - z|$; if one works on the set $X(r)$, equivalently, removes from the *closed* disc $|z| \leq 1$ a neighbourhood of the point 1 where the situation may become catastrophic if the series $v_n$ diverges, one then has

$$|r_n(z)| \leq 4v_p/r$$

on $X(r)$ for all $n > p$, whence

$$\|r_n\|_{X(r)} \leq 4v_p/r.$$

Now, for $r$ given, the right hand side is arbitrarily small for $p$ large, qed.

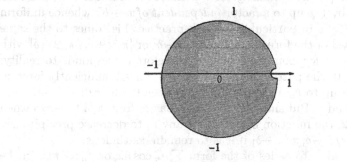

**fig. 12.**

*Exercise*: show that any compact $K \subset X$ is contained in an $X(r)$ and deduce from this that the series converges uniformly on $K$.

*Example 1.* Theorem 15 applies to certain Fourier series[23], for example to the square wave series $\cos x - \cos 3x/3 + \cos 5x/5$ etc. already considered in n° 9, example 5. On choosing

$$u_n = (-1)^{n-1}\cos(2n-1)x, \qquad v_n = 1/(2n-1),$$

it reduces to showing that the sums

(11.8)        $\cos x - \cos 3x + \cos 5x - \ldots + (-1)^{n-1}\cos(2n-1)x$

are bounded. Now the famous formula $2\cos a.\cos b = \ldots$ shows that

$$2\cos x.\cos x = 1 + \cos 2x, \quad 2\cos x.\cos 3x = \cos 2x + \cos 4x, \text{ etc..;}$$

on multiplying the sum (8) by $2\cos x$ one finds, thanks to the alternating signs, that

$$\cos x - \cos 3x + \cos 5x - \ldots + (-1)^{n-1}\cos(2n-1)x =$$
$$= \left[1 + (-1)^{n-1}\cos 2nx\right]/2\cos x,$$

whence

(11.9)      $\left|\cos x - \cos 3x + \ldots + (-1)^{n-1}\cos(2n-1)x\right| \le 1/|\cos x|$

for any $n$, of course on condition that $\cos x \ne 0$, i.e. that $x$ is not an odd multiple of $\pi/2$. Theorem 15 now shows that the series converges apart from at these values (and fails to converge at these excluded values ...) and that the remainder $r_p(x) = s(x) - s_p(x)$ satisfies, by (2) and $v_p = 1/(2p-1)$,

(11.10)                $|r_p(x)| \le 2/(2p-1)|\cos x|.$

On a set $K$ where $|\cos x|$ remains *greater* than a fixed number $> 0$, one then obtains a bound by $1/p$ up to a factor *independent of* $x \in K$: whence uniform convergence on $K$, so in particular – but in practice this comes to the same –, on every interval of the form $[-\pi/2+\delta, \pi/2-\delta]$ or $[\pi/2+\delta, 3\pi/2-\delta]$ with $\delta > 0$. The diagram for example 5 of n° 8 therefore corresponds to reality: for all $\delta$ and $r > 0$, the partial sum of order $n$ is, for all sufficiently large $n$, everywhere equal up to $r$ to the total sum between 0 and $\pi/2 - \delta$, similarly between $\pi/2+\delta$ and $\pi$. But since the total sum passes from $\pi/4$ to $-\pi/4$ when one traverses $\pi/2$, the function is forced, for any $n$, to decrease precipitously on the interval $[\pi/2 - \delta, \pi/2 + \delta]$ if it is to remain continuous.

One can generalise to series of the form $\sum a_n \cos nx$ or $a_n \sin nx$. In the first case, one observes, thanks again to high school trigonometry, proof that the latter is not entirely useless, that

---

[23] Hardly surprising. Dirichlet came to Paris in 1822 (he was then 17), and remained there for four years. Because of this he was up to date with Fourier's work and Cauchy's first research on the foundations of Analysis. He proved the first general result on the convergence of Fourier series in 1829.

(11.11) $$2\sin(x/2).(\cos x + \cos 2x + \ldots + \cos nx) =$$
$$= \sin(n + \tfrac{1}{2})x - \sin(x/2),$$

whence

$$|\cos x + \ldots + \cos nx)| \le 1/|\sin(x/2)|.$$

If the $a_n$ decrease to 0 then Theorem 15 and the inequality (2) show that the series converges uniformly on every set $K$ on which $|\sin(x/2)|$ remains greater than a fixed number $> 0$, so on every *compact* set not containing any multiple of $2\pi$: on such an interval $x$ remains distant from the points where $\sin(x/2)$ vanishes, as the graph of this function will demonstrate immediately (or because, if this were not the case, then one of points where $\sin x/2$ vanishes would be adherent to $K$, and so belong to $K$ since any compact set is closed).

In the case of the series $\sum a_n \sin nx$ one uses the identity

(11.12) $$2\cos(x/2).(\sin x + \sin 2x + \ldots + \sin nx) =$$
$$= \sin(n + \tfrac{1}{2})x - \sin(x/2)$$

and this time it is the relation $|\cos x/2| \ge m > 0$ which, satisfied on a compact $K$, ensures uniform convergence on $K$. To sum up:

**Corollary 3.** *Let $(a_n)$ be a decreasing sequence of positive numbers tending to 0. Then the series $\sum a_n \cos nx$ (resp. $\sum a_n \sin nx$) converges uniformly on every compact set not containing any even (resp. odd) multiple of $\pi$, and its sum is continuous away from these points.*

One can remember these conditions by observing that, in the first case, $\cos nx = 1$ if $x = 2k\pi$, and that then there is no reason for the series $\sum a_n$ to converge. For $x = (2k+1)\pi$ one has $\cos nx = (-1)^n$ and one obtains a convergent alternating series. In the case of the square wave series, the $a_n$ again tend to 0 but with alternating signs; the reader should have no trouble in formulating a variant of Corollary 3 in this case.

## 12 – Limits of limits

Consider, on a set $X \subset \mathbb{C}$, a sequence of functions $u_n(x)$ which converges uniformly on $X$ to a limit function $u(x)$. We saw in n° 5 that if all the $u_n$ are continuous at a point $a$ of $X$ then so is $u$. We can express this as

(12.1) $$\lim_{x \to a} \lim_{n \to +\infty} u_n(x) = \lim_{n \to +\infty} \lim_{x \to a} u_n(x).$$

This can be generalised:

**Theorem 16.** *Let $X$ be a subset of $\mathbb{C}$, let $a$ be an adherent point of $X$, and let $u_n(x)$ be a sequence of scalar functions defined on $X$. Suppose that*

*(i) $u_n(x)$ tends to a limit $c_n$ when $x$ tends to a,*
*(ii) $u_n(x)$ converges uniformly on $X$ to a limit $u(x)$.*

*Then $u(x)$ tends to a limit $c$ when $x \in X$ tends to a, and $c = \lim c_n$, i.e. (1) holds.*

If $a \in X$, the hypothesis (i) simply means that the $u_n$ are continuous at the point $a$, the case already examined. Suppose now that $a \notin X$ and consider the functions $v_n$ on the set $X' = X \cup \{a\}$ given by

$$v_n(x) = u_n(x) \text{ if } x \in X, \qquad v_n(x) = c_n \text{ if } x = a.$$

Hypothesis (i) means that the $v_n$ are continuous at the point $a$ of $X'$, as remarked in n° 2. The relation (1) will thus be a consequence of the theorem on the uniform limits of continuous functions if we can show that the $v_n$ converge to a limit function uniformly on $X'$, and not only on $X$.

It is enough to show that it satisfies Cauchy's criterion for uniform convergence. Now hypothesis (ii) shows that for any $r > 0$ there exists an integer $N$ such that

$$p, q > N \implies d_X(u_p, u_q) = d_X(v_p, v_q) \leq r.$$

When $x \in X$ tends to $a$, $|u_p(x) - u_q(x)|$ tends, for $p$ and $q$ given, to $|c_p - c_q| = |v_p(a) - v_q(a)|$, which is thus $\leq r$. So $d_{X'}(v_p, v_q) \leq r$, qed.

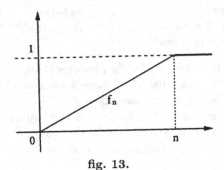

**fig. 13.**

When the hypothesis of uniform convergence is not satisfied, the conclusion of the preceding theorem may fail, a sign of a good theorem. Figure 13 indicates an example: the sequence $u_n(x)$ converges simply to 0 as $n \to +\infty$, while, for $n$ given, $u_n(x)$ tends to 1 as $x \to +\infty$; in this case, the left hand side of (1) is equal to 0 and the right hand side to 1.

Let us give an application of Theorem 16 which is essential to the theory of integration that we sketched in Chap. II, n° 11. We have already said that a function $\varphi$ defined on a bounded interval $I$ of $\mathbb{R}$ is a *step function* if one can dissect $I$ into a finite number of disjoint intervals (of any nature, some may reduce to a point) on each of which the function is constant, so that the graph of $\varphi$ reduces to a finite number of horizontal line segments and maybe some isolated points. We also said that a function $f$ defined on a bounded interval $I = (u, v)$ is *regulated* if it is the *uniform* limit of step functions on $I$ [see relation (11.2) of Chap. II], when we can give a meaning to the integral of $f$ on $I$ immediately.

**Corollary.** *Let $f$ be a scalar function defined on a bounded interval $I$ of $\mathbb{R}$. Suppose that for any $r > 0$ there exists a step function $\varphi$ on $I$ such that $|f(x) - \varphi(x)| < r$ for all $x \in I$ (i.e. that $f$ is regulated). Then $f$ has right and left limit values at all points of $I$, and the set of its points of discontinuity is countable.*

It suffices to remark that a step function possesses left and right limits at every point $a \in I$ and to apply the preceding theorem, replacing $I$ by the set $X$ of $x \in I$ such that $x < a$ (or $x > a$). Direct proof: for every step function $\varphi$ there exists an interval of the form $]a, a'[$ with $a' > a$ on which $\varphi$ is constant; the function $f$ of the corollary is thus, for any $r > 0$, constant to within $r$ on an interval of the same type, whence the existence of $f(a+)$ by Cauchy's criterion.

Let $(\varphi_n)$ be a sequence of step functions converging uniformly to $f$ and let $D_n$ be the (finite) set of discontinuities of $\varphi_n$; the union $D$ of $D_n$ is countable or finite (Chap. I). For $a \notin D$, all the $\varphi_n$ are continuous at $a$; similarly so is $f$, qed.

We shall see *à propos* integration (Chap. V, n° 7) that, for $I$ *compact*, the existence of left and right limits *characterises* the regulated functions. In particular, continuous and monotone functions are regulated. For this reason we shall now extend the definition of a regulated function to the case of an arbitrary interval $I$ by requiring that it should possess left and right limits at all points of $I$. A regulated function on $I$ will then be a uniform limit of step functions on all *compact* intervals $K \subset I$, but not necessarily on all of $I$. Compact convergence again ...

## 13 – Passing to the limit in a series of functions

The results of the preceding n° can be translated into the language of series of functions. Theorem 16 yields the following result:

**Theorem 17 (Passage to the limit under the $\sum$ sign).** *Let $T$ be a subset of $\mathbb{C}$, let $a$ be an adherent point of $T$ and let $\sum u(n, t)$ be a series of functions defined on $T$. Suppose that*

*(i)  for all $n$ the function $t \mapsto u(n, t)$ tends to a limit $u(n)$ as $t$ tends to $a$,*
*(ii) the series $\sum u(n, t)$ converges normally in $T$.*

*Then the sum $s(t)$ of the given series tends to a limit when $t$ tends to $a$, and*

(13.1)   $$\lim_{t \to a} \sum_{N} u(n,t) = \sum_{N} \lim_{t \to a} u(n,t) = \sum u(n),$$

*and further, the series $\sum u(n)$ is absolutely convergent.*

The existence of a convergent series with positive terms $\sum v(n)$ such that $|u(n,t)| \leq v(n)$ for all $n$ and $t$ shows that also, in the limit, $|u(n)| \leq v(n)$, whence the absolute convergence of the series of the limits $u(n)$.

As for the relation (1), one obtains it by observing that the sequence of partial sums $s_n(t) = u(1,t) + \ldots + u(n,t)$ converges to $s(t)$ *uniformly on $T$* (n° 8, Theorem 9), so that (Theorem 16)

$$\lim_{t \to a} \lim_{n \to \infty} s_n(t) = \lim_{n \to \infty} \lim_{t \to a} s_n(t);$$

the left hand side, the limit for $n \to \infty$ is simply, by definition, the total sum $s(t)$ of the series $\sum u(n,t)$; the right hand side, the limit as $t \to a$ is just the *partial* sum $s_n$ of the series of limits, so that the limit as $n \to \infty$ is the *total* sum $s$ of this, qed.

We can also prove this theorem directly. First, the absolute convergence of the series $\sum u(n)$ is obvious as we have seen above; let us write $s$ for its sum. For all $t$ and all $p \in \mathbb{N}$ we then have

$$|s(t) - s| \leq \sum_{n \leq p} |u(n,t) - u(n)| + \sum_{n > p} |u(n,t) - u(n)| \leq$$
$$\leq \sum_{n \leq p} |u(n,t) - u(n)| + 2 \sum_{n > p} v(n).$$

Choose an $r > 0$. For $p$ sufficiently large, $v(p+1) + v(p+2) + \ldots$ is $< r$, so that the second term is $< 2r$ for any $t$. Having chosen such a $p$, each of the $p$ terms of the first sum is $< r/p$ for $t$ sufficiently close to $a$, so that the said sum is itself $< r$ for $t$ sufficiently close to $a$. Adding, the difference $|s(t) - s|$ is thus $< 3r$ for all $t \in T$ sufficiently close to $a$, qed.

It is clear that if $X$ is a subset of $\mathbb{R}$, and not bounded above, one can also choose $a = +\infty$ in the preceding statements, respecting the usual conventions: replace "$t$ sufficiently close to $a$" by "$t$ sufficiently large".

Since the set $T$ of Theorem 17 is required only to be a subset of $\mathbb{C}$ – so as to give a meaning to the expression "when $t \in T$ tends to $a$" –, one can choose $T = \mathbb{N}$. Instead of a series $\sum u(n,t)$ whose general term depends on a "parameter" $t \in T$, one has a series $u(n,p)$ whose general term depends on the summation variable $n$ and on an integer $p \in \mathbb{N}$. When one is passing to a limit, and not summing over $p$, it is natural to consider that one is in the presence of a "sequence of series" and in consequence to denote the general term of series n° $p$ by $u_p(n)$. We shall state the result when we in need it in Chap. IV, n° 12, but the reader should have no trouble in formulating it now.

The notation of Theorem 17 was chosen to make manifest its analogy with
the theorems on "passing to the limit under the $\int$ sign" or on "dominated
convergence" in the theory of integration. These say, roughly, that if one
has a function $f(x,t)$ defined on $I \times T$, where $I$ is an arbitrary interval and
$T$ a subset of $\mathbb{R}$ or $\mathbb{C}$, if the function $x \mapsto f(x,t)$ is absolutely integrable
on $I$ for any $t \in T$, if it converges simply to a limit function $f(x)$ when
$t \in T$ tends to a limit $a$ *and if there exists* an integrable function $g(x)$ with
positive values satisfying

$$|f(x,t)| \leq g(x) \text{ for all } x \text{ and } t,$$

then

$$\lim_{t \to a} \int_I f(x,t)dx = \int_I f(x)dx.$$

This is the analogue of normal convergence when one replaces series, which
are sums of terms depending on a "discrete" variable, by integrals, which
are (or claim to be, *chez* Leibniz) sums of terms depending on a "continuous" variable: the variable of integration $x$ plays the rôle of the index of
summation $n$, the integral taken over $I$ replaces the sum taken over $\mathbb{N}$, the
concept of an "absolutely integrable" function replaces that of an absolutely convergent series, and finally the existence of an integrable function
$g(x)$ which "dominates" all the $f(x,t)$ plays the rôle of normal convergence.

# §4. Differentiable functions

## 14 – Derivatives of a function

We have not yet properly introduced the concept of the derivative of a function of a real variable, except in a rather schematic way in Chap. II, n° 4, though we used it in Chap. II *à propos* power series and the exponential and trigonometric functions. It is time to examine it more closely.

First let us recall what a derivative is. Consider a function $f$ with complex values defined on an interval[24] $I \subset \mathbb{R}$ not reducing to a single point. Given an $a \in I$, we examine the behaviour of $f(a + h)$ as $h$ tends to 0, implicitly assuming, in all that follows, that $a + h$ remains in $I$. If $f$ is continuous at the point $a$, then $f(a + h)$ tends to $f(a)$, in other words, the function $f$ is "almost constant" on a neighbourhood of $a$ as we explained profusely in n° 2.

But instead of approximating $f$ on a neighbourhood of $a$ by the *constant* function $x \mapsto f(a)$, one might try to approximate it by a function a little less simple, for example a *linear* function of the form $g(x) = cx + d$.

The least that one can ask is that it should be equal to $f$ at the point $a$, whence the condition $ca + d = f(a)$. Then

$$(14.1) \qquad g(x) = c(x - a) + f(a),$$

whence

$$(14.2) \qquad g(a + h) = ch + f(a).$$

It remains to choose the constant $c$ as well as possible.

Now the error committed in replacing $f$ by $g$ is given by

$$f(a + h) - g(a + h) = f(a + h) - f(a) - ch = h\left[\frac{f(a + h) - f(a)}{h} - c\right].$$

To minimise this we should choose $c$ so that the coefficient of $h$ is as small as possible and, even better, tends to 0 with $h$. This means that we must choose

$$(14.3) \qquad c = \lim_{\substack{h \to 0 \\ h \neq 0}} \frac{f(a + h) - f(a)}{h}.$$

When this limit exists – and, of course, it does not always do so –, we say that $c$ is the *derivative* of $f$ at $a$, denoted by $f'(a)$; so the best way of approximating $f$ on a neighbourhood of $a$ by a linear function is to choose the function

$$(14.4) \qquad y = f'(a)(x - a) + f(a);$$

---

[24] or more generally on an open subset $X$ of $\mathbb{R}$, since an open set is an interval on a neighbourhood of any of its points. For example, for the case of a rational fraction $f(x)/g(x)$, $X$ is the set of $x \in \mathbb{R}$ where $g(x) \neq 0$.

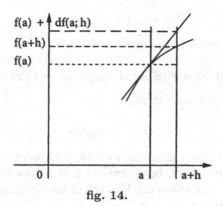

fig. 14.

this is the *linear tangent function* to $f$ at $a$, so called because (4) is the equation of the tangent line to the graph of $f$ at the point $(a, f(a))$ of the plane. Its homogeneous part in $h = x - a$ is called the *differential* of $f$ at $a$; it depends both on the point $a \in I$ and on an auxiliary variable $h \in \mathbb{R}$, whence the notation $df(a)$ to denote the *function* $h \mapsto f'(a)h$ and the notation

$$(14.5) \qquad df(a; h) = f'(a)h$$

to denote its value at $h$. This mode of presenting derivatives and differentials is already essentially to be found in Weierstrass.

If for example $f(x) = x$, then $f'(a) = 1$, so that the differential of $f$ is the function $h \mapsto h$; in other words[25]

$$(14.6) \qquad dx(a; h) = h$$

for all $a$ and $h$ real. Comparing with (5), we see that

$$(14.7) \qquad df(a; h) = f'(a)dx(a; h)$$

for every function $f$ that is differentiable at $a$; in a more condensed way:

$$(14.8) \qquad df(a) = f'(a)dx(a),$$

the product of the linear *function* $dx(a)$, i.e. $h \mapsto h$, by the *constant* $f'(a)$ (relative to $h$); and since in fact the differential $dx(a)$ does not depend on $a$, one may as well call it $dx$ for short, and obtain the formula

---

[25] Strictly speaking, one should, once and for all, give a name to the *identity function* $x \mapsto x$; the natural notation would be to put $i(x) = x$, but to use $i$ for a function having nothing to do with the complex numbers would certainly be risky. One might denote it by *id*, which would lead to interesting formulae such as $id \circ id = id$, $id' = 1$, $d(id)(a, h) = h$, etc. .... To avoid these follies, one writes $dx(a)$ for what one ought to write as $d(id)(a)$. This is an "abuse of notation" that we allow ourselves for other functions too; no one ever writes $d(\sin)(x)$ for the differential of the sine function at a point $x$; one writes simply $d\sin x$.

(14.9)                              $df(a) = f'(a)dx,$

or even, if $f$ is differentiable at any $x \in I$,

(14.10)          $df(x) = f'(x)dx$ or simply $df = f'(x)dx.$

Hence the traditional formulation

(14.11)                              $f'(x) = df/dx$

for derivatives; it has no meaning in this framework since $df$ and $dx$ are functions and not numbers, but everyone uses it, not only to follow in the tradition, but also, and above all, because of its convenience, particularly at the elementary level.

The inventor of the notation $dx$, $df$ and $df/dx$, namely Leibniz[26], a metaphysician, interpreted them in quite another way; for him it was neither a matter of a variable $h$ nor of a linear function. For Leibniz and for those who followed him, at least up to Cauchy, the symbol $dx$ represented an "infinitely small increase" in the variable $x$ and $df$ the "principal part", proportional to $dx$, of the increase

$$f(x + dx) - f(x)$$

of $f$ at the point $x$. These concepts, which rest on the "infinitely smalls" which no one has been able to define, have put too many people to useless worry, that one can only attribute them the rôle of an historical explanation the differential notation. Newton, who had a positive outlook in what concerned Mathematics, Astronomy, Physics, the minting of money and, in a lesser measure, the Bible and Alchemy, did not care for them since "they are not to be met in Nature".

But there is in fact an equivalent of the $dx$ of Leibniz in his *Treatise on the methods of series and fluxions*, composed in Latin in 1671 and never published, except in translation, in 1739 in England, and in 1740 in France by Buffon, in order to legitimise the posthumous glory of the great man, somewhat eclipsed by that of Leibniz and of Bernoulli, who did not wait forty

---

[26] See G. W. Leibnitz, *Naissance du calcul différentiel*, papers translated and annotated by Marc Parmentier (Paris, Vrin, 1989 or 1995). For the history of derivatives from Galileo to Cauchy, see the excellent resumé in Walter, *Analysis 1*, pp. 221–240. The classic by H. G. Zeuthen, *Geschichte der Mathematik im 16. und 17. Jahrhundert* (Teubner, 1903 or Johnson Reprint Co., 1966) is still very usable and weighs much less than the *blockbusters* (ten ton bombs developed at the end of the war, so, by extension: a book of 800 pages finding nevertheless 200 000 readers in the USA, subject permitting) of Moritz Cantor, useful for their abundance of detail.

or fifty years to publish[27]. To understand him, one needs to know that for Newton there was only one independent variable: time, an idea apparently suggested by Isaac Barrow and conceived in an age when the concepts of independent variable and of function[28] were still very badly understood, except in Mechanics and in Astronomy, where, precisely, everything depends on time. This never appeared explicitly in Newton's formulae, though ever present. All the other variables, the *fluents*, are functions of time, so that for Newton the equation $x^2 + y^2 = a^2$ represents a point moving as an *a priori* arbitrary function of time on a circumference of radius $a$. The crucial passage of his manuscript is the following[29]:

> The moments of the fluent quantities (that is, their indefinitely small parts, by addition of which they increase during each infinitely small period of time) are as their speeds of flow. Wherefore if the moment of any particular one, say $x$, be expressed by the product of its speed[30] $\dot{x}$ and an infinitely small quantity $o$ (that is, by $\dot{x}o$), then the moments of the others,

---

[27] Newton, at least initially, was not entirely responsble for these delays. It was clearly not his fault that his first papers were not published in the Transactions of the Royal Society; his efforts, several years later, to find a publisher for his *Method* fell through: the possible publishers had had their fingers burnt by their losses on the books of Wallis and Mercator, books which, unsurprisingly, were very far from being best-sellers. Moreover, Newton's publication of a report on his experiments on the decomposition of white light brought violent attacks from partisans of Descartes' theory (see Loup Verlet, *La malle de Newton*, Chaps. II and III). Profoundly averse to polemic and to "disputation", Newton developed a deep distrust of taking a public stance; his English students or disciples were those who expounded his mathematical works, with a delay of many years. Newton made only two important exceptions, in favour of his *Principia* (1687) and of his *Opticks* (1704); in the first he expounds the results of his research in Astronomy. This is the book which made him the hero of the French philosophers of the XVIII[th] century (and of scientists, French or not) despite the fact that his declared purpose was to demonstrate the Divine perfection of Creation. Having presented his "Mécanique Céleste" to Napoleon, who was surprised at seeing no mention of the name of God, Laplace replied that on the contrary his book proved that it was not necessary. The standard argument – if one needs a Creator to explain "Creation", how to explain the existence of a Creator who had not been created? – clearly remains valid.

[28] The word was introduced by Leibniz in 1694, but it was Jakob Bernoulli who gave it its present meaning, in 1698, in very particular situations. The idea that a function consists of arbitrarily associating a number $y$ to each number $x$ of an interval, without any other condition, waited for the XIX[th] century and Dirichlet, who invented the function equal to 1 for $x$ is rational and to 0 if not.

[29] D. T. Whiteside, *The Mathematical Papers of Isaac Newton* (CUP), III, p. 79–81. The passages between ( ) are in Newton.

[30] In fact, Newton wrote $p$, $q$, $r$ for what from 1691 he wrote $\dot{x}$, $\dot{y}$, $\dot{z}$. The modern notation $f'$ or $x'$ was invented by Lagrange. Newton's usage of the term "moment" to denote an infinitesimal increase of a fluent was inspired by the analogous term relative to time: an infinitely small period (his $o$). Newton naturally does not explain what is to be understood by $\dot{x}$, as though everyone could understand it intuitively.

$v, y, z, [\ldots]$, will be expressed by $\dot{v}o, \dot{y}o, \dot{z}o, [\ldots]$ seeing that $\dot{v}o, \dot{x}o, \dot{y}o$ and $\dot{z}o$ are to one another as $\dot{v}, \dot{x}, \dot{y}$ and $\dot{z}$.

Now, since the moments (say, $\dot{x}o$ and $\dot{y}o$) of fluent quantities ($x$ and $y$, say) are the infinitely small additions by which those quantities increase during each infinitely small interval of time, it follows that those quantities $x$ and $y$ after any infinitely small interval of time will become $x + \dot{x}o$ and $y + \dot{y}o$.

And Newton explains that, if one has a relation between $x$ and $y$, one can substitute $x + \dot{x}o$ and $y + \dot{y}o$ and calculate algebraically. In the relation

$$x^3 - ax^2 + axy - y^3 = 0,$$

for example, one obtains

$$\left(x^3 + 3\dot{x}ox^2 + 3\dot{x}^2o^2x + \dot{x}^3o^3\right) - \left(ax^2 + 2a\dot{x}ox + a\dot{x}^2o^2\right) +$$
$$+ \left(axy + a\dot{x}oy + a\dot{y}ox + a\dot{x}\dot{y}o^2\right) - \left(y^3 + 3\dot{y}oy^2 + 3\dot{y}^2o^2y + \dot{y}^3o^3\right) = 0.$$

But by hypothesis $x^3 - ax^2 + axy - y^3 = 0$, and when these terms are cancelled *and the rest is divided by o*, one obtains a relation that the reader can easily write, by calculating as crudely as can be, as did Newton himself. There will be terms not containing $o$ and terms which contain $o$:

> But further, since $o$ is supposed to be infinitely small so that it be able to express the moments of quantities, terms which have it as a factor will be equivalent to nothing in respect of the others. I therefore cast them out and there remains

$$3\dot{x}x^2 - 2a\dot{x}x + a\dot{x}y + a\dot{y}x - 3\dot{y}y^2 = 0.$$

Thus Newton, starting from a relation between his fluents $x$ and $y$, functions of time, obtained a relation between $x, y$ and their derivatives $\dot{x}$ and $\dot{y}$ with respect to time, that he called their *fluxions*, i.e., as he explained earlier, their "speeds of flow in time". What Newton denoted by $o$ and $\dot{x}o$ would be written as $dt$ and $x'(t)dt$ by Leibniz [up to the fact that Leibniz used the notation $dx$ and not the modern notation $x'(t)dt$] and gave rise to an identical calculus, up to notation. For a long time, maybe wrongly, these heuristic but very suggestive calculations have been replaced by rules that allow one to calculate the derivatives of any algebraic combination of functions, but the current method is not necessarily quicker from the moment (in the naive sense, and not in the sense that Newton meant in the preceding citation!) when one has understood these ideas of Newton's, or, equivalently, those of Leibniz. All this is theoretically incorrect since (i) infinitely small quantities "are never met in Nature", as Newton said much later to annoy Leibniz, (ii) the increment $x(t + h) - x(t)$ corresponding to a "small", but not infinitely small, increment $h$ of $t$ is not rigorously proportional to $h$. But nevertheless this all works perfectly so long as one takes a few precautions.

One can understand this more easily in the following way. In the formula (3), the difference between the right hand side and $c$ tends to 0 with $h$. Multiplying by $h$, one has a relation

(14.12)         $f(a + h) = f(a) + f'(a)h + o(h)$   when $h \to 0$,

where the symbol $o(h)$, with a *lower case o* like Newton, curiously – but here it is the $h$ multiplying $f'(a)$ which plays the rôle of his $o$ –, denotes (Chap. II, n° 4) a function about which one can say no more than that

(14.13)                            $\lim o(h)/h = 0$,

in other words such that, for any $r > 0$ there exists an $r' > 0$ for which

(14.14)                  $|h| < r' \Longrightarrow |o(h)| < r|h|$.

Conversely, the relation

$$f(a + h) = f(a) + ch + o(h) \quad \text{when } h \to 0,$$

where $c$ is a constant, proves that $f$ admits a derivative equal to $c$ at $a$; indeed, the preceding relation can be rewritten as

$$\frac{f(a + h) - f(a)}{h} - c = o(h)/h$$

with a right hand side which, by definition, tends to 0, as we have already explained in Chap. II, n° 4.

Like Newton's, this notation may serve to deduce a relation between the derivatives from a relation between two functions $x(t)$ and $y(t)$. Take for example the relation $x^2 + y^2 = 1$. Replacing $x(t)$ by $x(t + h)$, i.e. by $x(t) + x'(t)h + o(h)$, and $y(t)$ by $y(t + h) = \ldots$, one clearly finds [writing $x, x', \ldots$ instead of $x(t), x'(t), \ldots$] a relation

$$0 = 2xx'h + 2yy'h + \ldots$$

where the $\ldots$ denote terms containing $o(h)$ or even $o(h)^2$ as a factor. On dividing throughout by $h$ there follows a relation of the form

$$0 = 2xx' + 2yy' + \quad \text{terms which tend to 0 with } h.$$

So $2xx' + 2yy' = 0$, since these are the terms independent of $h$.

Newton's method is a little simpler since, following him, the two terms $x'(t)h + o(h)$ are condensed into just $\dot{x}o$; equivalently, one eliminates the residual term $o(h)$ and replaces $x$ by $x+x'h$, $y$ by $y+y'h$, etc. in the equations.

Following Leibniz, one replaces $x$ and $y$ by $x + x'dt$ and $y + y'dt$ (in fact by $x + dx$ and $y + dy$) and performs the same calculation, so obtaining, after division by $dt$, the relation $0 = 2x.dx/dt + 2y.dy/dt$ plus terms containing $dt$, i.e. which are "infinitely small" and so can be eliminated.

It remains to understand what, from Newton's point of view, plays the rôle of the derivative of a fluent $y$ in terms of another fluent $x$ on which it depends, $y = f(x)$. It is the ratio between an "infinitely small increment" of $x$ and the corresponding increment in $y$, i.e. what Newton denoted $\dot{x}o$ and $\dot{y}o$ and Leibniz $dx$ and $dy$; so

$$f'(x) = \dot{y}o/\dot{x}o = \dot{y}/\dot{x},$$

a relation whose resemblance to the notation $dy/dx$ of Leibniz shows well how close their ideas were, if not identical. Whence fifty or so years of quarrels about priority, Leibniz himself and his disciples being much more courteous than Newton's English partisans who accused the former of plagiarism – wrongly, as we have known since the historians have examined the masses of personal papers of Leibniz which slept in obscure corners of libraries for two centuries. Newton's papers were absorbed by the descendants of the one who had inherited them, namely an Earl of Portsmouth whose mother, having married a Lord Lymington, was the daughter of a niece of Newton, practically adopted by the great man when he set up in London and to whom he had entrusted the care of his establishment; she herself married John Conduitt MP, in 1717, and Newton lived with them until his death, so that they inherited his papers which they transmitted to their daughter who handed them on to Portsmouth. Conduitt was the deputy and became Newton's successor at the *Mint*, which Newton, leaving Cambridge, directed from 1695. He organised it with redoubtable efficiency and integrity; there he recast all the silver money, reduced to half of its value through the activities of sniders. In 1872, after a fire, the Portsmouth family gave Newton's scientific papers, in disorder and often damaged, to Cambridge. Most of the rest (theology, history, alchemy, etc.) were sold at auction and dispersed in the 1930s. Some purchasers, including libraries, refuse access. In contrast, the economist and former mathematician John Maynard Keynes bought some and brought them back to life by donating them to Cambridge.

Newton's calculations can be understood otherwise and in an ultra-modern way by introducing what one calls *dual numbers*. To define the complex numbers one introduced a mysterious symbol $i$ such that $i^2 = -1$ and calculated with expressions of the form $a + ib$ with $a, b \in \mathbb{R}$ using this. The dual numbers are similarly expressions $a + b\varrho$, where $a$ and $b$ are real (or even complex) numbers with which one calculates in the usual way, but requiring the symbol $\varrho$ to satisfy the relation $\varrho^2 = 0$, which is neither more nor less strange than $i^2 = -1$. Then

$$(a + b\varrho) + (c + d\varrho) = (a + c) + (b + d)\varrho,$$
$$(a + b\varrho)(c + d\varrho) = ac + (ad + bc)\varrho,$$

which allows us to define correctly, i.e. without recourse to the symbol $\underline{o}$, a dual number as a pair $(a, b)$ of complex numbers, the rules for manipulating these pairs being as in the formulae above. It all works out perfectly except that, in contrast to the complex numbers, the dual numbers do not form a field since $\underline{o}^2 = 0$.

If one has a polynomial $f(x) = \sum a_n x^n$ with complex coefficients, one can then define its "value" for any dual number $x + h\underline{o}$ by routinely applying the formula which defines $f$. Now

$$(x + h\underline{o})^n = x^n + nx^{n-1}h\underline{o} + \ldots = x^n + nx^{n-1}h\underline{o}$$

since the unwritten terms contain $\underline{o}^2$ as a factor. It follows that

$$f(x + h\underline{o}) = \sum a_n(x^n + nx^{n-1}h\underline{o}) = f(x) + f'(x)h\underline{o}$$

where $f'$ is the usual derivative of $f$.

This exactly is what Newton constantly did in calculating with his $x + \dot{x}o$ – do not confuse his $o$ with ours, and do not confuse ours with the expression $o(h)$ – according to the strict rules of algebra and then suppressing all terms containing a factor of at least $o^2$. The difference is that, for us, the symbol $\underline{o}$ represents not an "infinitely small increment" of time like the $o$ of Newton, but an element of a ring in which one calculates according to rules prescribed in advance.

This artifice, which allows one to algebraise the concept of derivative, was developed in the 1950s by Andre Weil in an incomparably more general and difficult framework – the "infinitesimal extensions of higher order of differential varieties" – in a work, unfortunately not published, where he writes $\varepsilon$ (why not?) for what we write $\underline{o}$. I didn't think to ask him if he had been inspired by reading Newton; but knowing his pronounced taste for the history of mathematics, the reply would hardly have been in doubt.

The reader may amuse himself by introducing, in the same spirit, another symbol $\underline{o}$ which, for example, satisfies only the relation $\underline{o}^4 = 0$, and "numbers" $x + h_1\underline{o} + h_2\underline{o}^2 + h_3\underline{o}^3$ with coefficients in $\mathbb{C}$ and, given a polynomial, calculate its "value" at such a number as a function of its derivatives at the point $x$ as above for the case of dual numbers. The reader who knows that this is the quotient of a ring by an ideal will notice that systems of this kind are just quotients of $\mathbb{C}[X]$, the ring of polynomials in one variable with complex coefficients, by the ideal of multiples of $X^2$ in the first case, of $X^4$ in the second, and of $X^p$ in the general case. The second person to give a perfectly correct definition of the complex numbers (the first was Hamilton), namely Cauchy in 1847, constructed them by considering the quotient of $\mathbb{R}[X]$ by the ideal of polynomials that are multiples of $X^2 + 1$, a method preferable by reason of the vast generalisations, unknown to Cauchy, to which it is susceptible, mainly to the algebraic extensions of commutative fields. See Serge Lang, *Algebra*, among many other possible references.

As we have already said (Chap. II, n° 11), there are close connections between the concepts of derivative and of integral: if one has a continuous function $f$ on an interval $I$ and if, for an $a \in I$, one denotes by $F(x)$ the integral of $f$ taken over the interval with end points $a$ and $x$, then $F'(x) = f(x)$; this is the "fundamental theorem" of the integral calculus (Chap. V). Consider, for example, for $x > 0$, the function $f(x) = 1/x$. We proved in Chap. II, n° 11, that, for $0 < a < b$,

$$\int_a^b f(x)dx = \log b - \log a.$$

Let us take $b = a + h$, with $h > 0$ for simplicity. Now $f(x)$ remains between $1/(a+h)$ and $1/a$ between $a$ and $b$; since the interval $(a, a+h)$ is of length $h$ the integral thus lies between $h/(a+h)$ and $h/a$, whence

$$1/(a + h) \le [\log(a + h) - \log a]/h \le 1/a.$$

As $h$ tends to 0, the outer terms converge to $1/a$. Similar calculation when $h$ tends to 0 through negative values. Thus we obtain the relation

(14.15)                           $$\log' x = 1/x$$

of Chap. II, n° 10, which we proved in another way in Chap. IV and which we accepted in Chap. II, n° 5, for $x = 1$, to obtain or suggest the formula

$$\log x = \lim n \left( x^{1/n} - 1 \right).$$

## 15 – Rules for calculating derivatives

If $f'(a)$ exists for every $a \in I$, we say that $f$ is *differentiable on $I$*; the map $f' : x \mapsto f'(x)$ is the *derived function* of $f$. It may happen that the latter is again differentiable, whence the concept of *second derivative* $f'' = (f')'$, and so on. If one can continue indefinitely one says that $f$ is *indefinitely differentiable* or *of class $C^\infty$* on $I$. More generally, $f$ is said to be *p times continuously differentiable* or *of class $C^p$* on $I$ if the derived functions $f', f'', \ldots, f^{(p)}$ exist and are continuous on $I$. There are more subtle concepts, but these suffice in the great majority of cases.

Derivatives obey simple rules that the reader surely knows already.

(D 1) *If $f$ and $g$ are differentiable at $a$ then so is their sum $s(x) = f(x) + g(x)$, and $s'(a) = f'(a) + g'(a)$.*

Indeed

(15.1)        $$\begin{aligned} f(a+h) &= f(a) + f'(a)h + o(h), \\ g(a+h) &= g(a) + g'(a)h + o(h), \end{aligned}$$

with two functions $o(h)$, not the same. On adding,

$$s(a + h) = s(a) + ch + o(h) + o(h)$$

where $c = f'(a) + g'(a)$. It remains to show that

(15.2)                          $$o(h) + o(h) = o(h),$$

a result which is obvious since on dividing the two sides by $h$ one has merely to show that the sum of two functions which tend to 0 also tends to 0. Hence (D 1).

From Newton's point of view, and with his notation: if one has two fluents $x$ and $y$ and their sum $z$, to the infinitely small increment $o$ of time there correspond increments $\dot{x}o$ and $\dot{y}o$ for $x$ and $y$, and so, for $z$, an increment $\dot{z}o = \dot{x}o + \dot{y}o = (\dot{x} + \dot{y})o$, whence $\dot{z} = \dot{x} + \dot{y}$ on dividing by $o$. The beauty of the argument is that it conjures away the proof – certainly a very easy one – of the *existence* of the derivative of $z$; one has only to *calculate*, as always with Newton and his successors for 150 years.

(D 2)  *If $f$ and $g$ are differentiable at $a$ then so is their product $p(x) = f(x)g(x)$, and $p'(a) = f'(a)g(a) + f(a)g'(a)$.*

Again one starts from the relations (1) and multiplies term-by-term:

$$p(a+h) = p(a) + ch + f(a)o(h) + g(a)o(h) + f'(a)g'(a)h^2 + $$
$$+ f'(a)ho(h) + g'(a)ho(h) + o(h)o(h)$$

where this time $c$ has the appropriate value for $p'(a)$. In view of (2), which extends to the case of any finite sum, it remains to verify that the product of an $o(h)$ function by a constant is again $o(h)$, that

$$h^2 = o(h),$$

that $ho(h) = o(h)$ [and even $= o(h^2)$] and that $o(h)o(h) = o(h)$ [and even $= o(h^2)$], which is clear.

In Newtonian style: at time $t + o$ the fluent $z = xy$ becomes

$$z + \dot{z}o = (x + \dot{x}o)(y + \dot{y}o) = z + (x\dot{y} + \dot{x}y)o + \dot{x}\dot{y}o^2,$$

whence

$$\dot{z}o = (x\dot{y} + \dot{x}y)o + \dot{x}\dot{y}o^2$$

and one obtains the desired result $\dot{z} = x\dot{y} + \dot{x}y$ on grouping the terms in $o$ and neglecting the "infinitely small" term $\dot{x}\dot{y}o^2$. No one had had this sort of idea before him, and his calculus is quicker than ours ...

The reader of the remark on the dual numbers of the preceding n° can interpret the rule (D 2) as follows. Given a function $f$ differentiable at the point $a$, let us define the "value" of $f$ at a dual number $a + h\underline{o}$ as the dual number

$$f(a) + f'(a)h\underline{o} = f(a + h\underline{o})$$

already found in the case of a polynomial [do not confuse this with the linear tangent function $h \mapsto f(a) + f'(a)h$: the values of the latter are ordinary numbers]. Then

$$p(a + h\underline{o}) = f(a + h\underline{o})g(a + h\underline{o}).$$

In other words, the value of a product (or of a sum, or of a quotient) of two functions is the product (or ... ) of their values.

(D 3) *If $f$ and $g$ are differentiable at $a$ and if $g(a) \neq 0$ then so is their quotient $q(x) = f(x)/g(x)$, and*

$$q'(a) = \frac{f'(a)g(a) - f(a)g'(a)}{g(a)^2}.$$

It is actually enough to prove this for $f(x) = 1$ and then to apply (D 2) to the product of $f$ and $1/g$. First we note that $g$, being differentiable and so continuous[31], is $\neq 0$ on a neighbourhood of $a$ in the interval of definition of $g$. Next, the second relation (1) shows that

(15.3)
$$\frac{1}{g(a+h)} - \frac{1}{g(a)} = \frac{1}{g(a) + ch + o(h)} - \frac{1}{g(a)} =$$
$$= \frac{-ch + o(h)}{g(a)^2 + cg(a)h + o(h)} =$$
$$= c'h\frac{1 + o(h)/h}{1 + c''h + o(h)}$$

where we have put $c = g'(a)$, $c' = -g'(a)/g(a)^2$ , $c'' = c/g(a)$ and have simplified the calculations by using the fact that a constant multiple of an $o(h)$ function is again $o(h)$. Now the formula

$$\frac{1}{1-x} = 1 + x + \frac{x^2}{1-x}$$

shows that

(15.4)
$$\frac{1}{1 + c''h + o(h)} = 1 - [c''h + o(h)] + \frac{h^2[c'' + o(h)/h]^2}{1 + c''h + o(h)}.$$

Consider the fraction on the right hand side. Its numerator is clearly $o(h)$ by reason of the factor $h^2$ and of the fact that the expression between [ ] tends to $c''$, so is $O(1)$. Its denominator tends to 1, so is *greater* than $1/2$ in absolute value for $|h|$ sufficiently small. The fraction itself is thus, for $|h|$ sufficiently small, *smaller* than twice its numerator, so is $o(h)$. The left hand side of (4) is therefore of the form $1 - c''h + o(h)$.

---

[31] If $g$ is differentiable in $a$, then clearly, by (5) and (7), we have an estimate of the form

$$|f(a + h) - f(a)| \leq (|f'(a)| + r) \cdot |h|$$

for $|h| < r'$, whence continuity and even more. This trivial result hardly rates being presented as a *bona fide* theorem.

On substituting this result in (3) we find

$$\frac{1}{g(a+h)} - \frac{1}{g(a)} = c'h\left[1 - c''h + o(h)\right] = c'h + o(h)$$

with $c' = -g'(a)/g(a)^2$, which concludes the proof.

In Newtonian style:

$$1/(x + \dot{x}o) = 1/x \times 1/(1 + \dot{x}o/x) = 1/x - \dot{x}o/x^2 + \ldots,$$

so that for $y = 1/x$ one finds $\dot{y} = -\dot{x}/x^2$. Again, much quicker!

The three preceding rules of calculus have an obvious consequence for functions which are everywhere differentiable on an interval $I$, or of class $C^p$ ($p \leq +\infty$): every reasonable algebraic operation (i.e. excluding division by 0, which restricts the set on which the result is defined) transforms functions of such a type into functions of the same type. For example, every rational function $f(x)/g(x)$, where $f$ and $g$ are polynomials, is $C^\infty$ on the set where $g(x) \neq 0$. It is also obvious that the derivation $f \mapsto f'$ transforms functions of class $C^p$ into functions of class $C^{p-1}$, and in particular the functions of class $C^\infty$ into functions of class $C^\infty$. If we write $C^p(I)$ for the set of functions of class $C^p$ on an interval $I$ of $\mathbb{R}$, we see that derivation can be considered as a map

$$D: C^p(I) \longrightarrow C^{p-1}(I).$$

A good example of a function or map defined on a set considerably vaster than $\mathbb{R}$, namely a "functional space".

In this circle of ideas, let us write

$$D^2: C^p(I) \longrightarrow C^{p-2}(I), \qquad D^3: C^p(I) \longrightarrow C^{p-3}(I), \text{ etc.}$$

for the maps obtained by differentiating an $f \in C^p(I)$ $2, 3, \ldots$ times, and consider an integer $n \leq p$. Leibniz' formula states that

$$
\begin{aligned}
(15.5) \quad D^n(fg) &= D^n(f).g + nD^{n-1}(f).D(g)/1! + \ldots + \\
&\quad + n(n-1)\ldots(n-k+1)D^{n-k}(f)D^k(g)/k! + \ldots + f.D^n(g)
\end{aligned}
$$

for $f$ and $g$ in $C^p(I)$. This bears an analogy to the binomial formula for $(x + y)^n$, having same coefficients[32]. It is proved in the same way too: one assumes the formula established for $n - 1$ and differentiates the two sides to obtain the formula for $n$; all one needs is to check the relation

$$\binom{n-1}{k-1} + \binom{n-1}{k} = \binom{n}{k}.$$

---

[32] On introducing the "divided derivatives" $D^{[n]} = D^n/n!$, Leibniz' formula can be written in the form $D^{[n]}(fg) = \sum D^{[k]}(f)D^{[n-k]}(g)$. Not yet the everyday standard.

between binomial coefficients. For example

$$(fg)''' = f'''g + 3f''g' + 3f'g'' + g'''.$$

After these algebraic rules, here are two very different formulae: for the composition of differentiable maps, and for the inverse map of a differentiable function.

(D 4) (Chain rule) *Let $I$ and $J$ be two intervals in $\mathbb{R}$, let $f$ be a map from $I$ into $J$, let $g$ be a map from $J$ into $\mathbb{C}$, and let $c(x) = g[f(x)]$ be the composite map of $I$ into $\mathbb{C}$. Assume that $f$ is differentiable at $a \in I$ and that $g$ is differentiable at $f(a) = b \in J$. Then $c$ is differentiable at $a$, and*

$$(15.5) \qquad c'(a) = g'(b)f'(a) = g'[f(a)].f'(a).$$

Recall the relation

$$f(a + h) = f(a) + f'(a)h + o(h)$$

and put $k = f'(a)h + o(h)$, whence $f(a + h) = f(a) + k = b + k$. It is clear that $k$ tends to 0 with $h$. Now

$$\begin{aligned} c(a + h) &= g[f(a+h)] = g(b+k) = g(b) + g'(b)k + o(k) = \\ &= c(a) + g'(b)[f'(a)h + o(h)] + o(k). \end{aligned}$$

The term $g'(b)o(h)$ is $o(h)$. Since, on the other hand,

$$k = h[f'(a) + o(h)/h]$$

and since, in this relation, the factor between [ ] tends, by definition of $o(h)$, to $f'(a)$ as $h$ tends to 0, one has $|k| \leq M|h|$ for $|h|$ small, for some constant $M > 0$. Every $o(k)$ function is thus also[33] $o(h)$, in particular the last term of the expansion of $c(a + h)$. Finally

$$c(a + h) = c(a) + g'(b)f'(a)h + o(h),$$

qed.

*Example 1.* Since we believe we know that the derivative of the function $\log x$ is $1/x$ by (14.15) or Chap. II, n° 10, Theorem 3, and since it is not very difficult to establish, by applying the general formulae of Chap. II, n° 19, that that of the function $\exp x = \sum x^{[n]}$ is $\sum x^{[n-1]} = \exp x$, we see that

---

[33] If $|k| < 10^{100}|h|$ for $|h|$ sufficiently small one can be sure that $|\varphi(k)| < 10^{-1000}|h|$ once $|\varphi(k)|$ is $< 10^{-1100}|k|$, which is the case for $|k| < r'$, so for $|h| < r = 10^{-100}r'$. More generally: if $\theta = O(\psi)$ and if $\psi = o(\varphi)$, then $\theta = o(\varphi)$. See the general rules of Chap. VI, n° 1, which follow directly from the definitions.

the derivative of the function $\log(\exp x)$ is $1/\exp x \times \exp x = 1$, which does not contradict the relation $\log(\exp x) = x$ established in n° 2, example 1. There we first showed that $\log(\exp x) = ax$ with a mysterious constant $a$; the preceding argument now shows that $a = 1$. Chap. IV will confirm this by other methods, all roads leading to this fundamental formula; we will even end up making a theorem out of it ...

An obvious consequence of (D 4), applying it repeatedly, is that if $f$ and $g$ are of class $C^p$ on $I$ and $J$, then the composite function $c = g \circ f$ is again of class $C^p$ on $I$.

The following rule completes Theorem 7 of n° 4 regarding the inverse of a continuous strictly monotone map:

(D 5) *Let $f$ be a continuous strictly monotone map of an interval $I \subset \mathbb{R}$ onto an interval $J \subset \mathbb{R}$ and let $g : J \longrightarrow I$ be the inverse map of $f$. Suppose that the function $f$ is differentiable at a point $a \in I$. If so, then $g$ possesses a derivative at the point $b = f(a)$ if and only if $f'(a) \neq 0$. And then $g'(b) = 1/f'(a)$.*

The condition is necessary, since the relation $g[f(x)] = x$ shows that

$$g'(y)f'(x) = 1 \qquad \text{if } y = f(x),$$

by the preceding rule; this forbids $f'$ from vanishing and even exhibits the only possible value of $g'$. It remains to show that, conversely, the condition $f'(a) \neq 0$ assures the existence of $g'(b)$.

We need to show that the ratio $[g(b+k) - g(b)]/k$ tends to $1/f'(a)$ as $k$ tends to 0. To do this, put

$$g(b+k) = a + h \quad \text{i.e.} \quad h = g(b+k) - g(b) = p(k),$$

whence $b + k = f(a + h)$. When $k$ tends to 0 so does $h = p(k)$, since $g$ is continuous. Now the relation

(15.6) $\qquad k = (b+k) - b = f(a+h) - f(a) = f'(a)h + o(h)$

shows that the ratio $k/h = k/p(k)$ tends to the *nonzero* limit $f'(a)$ when $k$ (and so $h$) tends to 0. In consequence, $p(k)/k$ tends to $1/f'(a)$. But $p(k)/k = [g(b+k) - g(b)]/k$, qed.

*Example 2.* Take $I = [-\pi/2, \pi/2]$ and $f(x) = \sin x$, a function which the reader surely knows to be continuous, and strictly increasing on $I$ (though not on $\mathbb{R}$), maps $I$ onto $J = [-1, 1]$, and everywhere has a derivative given by $f'(x) = \cos x$; we will justify all this in Chap. IV by much more rigorous methods than contemplating a sketch on a sheet of paper as Newton did, having had no other choice.

The inverse map $g : J \longrightarrow I$, traditionally written $g(y) = \arcsin y$, is differentiable at all points $y = \sin x$ where $\cos x \neq 0$, i.e. for $-1 < y < 1$ (strict inequalities). Its derivative at the point $y = \sin x$ is given by

$$g'(y) = 1/\cos x = 1/(1 - \sin^2 x)^{1/2} = 1/(1 - y^2)^{1/2}$$

since $\cos x > 0$ on $I$. On swapping the letters $x$ and $y$, we get

(15.7) $\qquad \arcsin'(x) = 1/(1 - x^2)^{1/2}$ for $-1 < x < 1$.

On the other hand, the function is not differentiable at $x = -1$ or $+1$ since, at the corresponding point of $I$, the derivative of $\sin x$ vanishes. The graph of $g$, symmetric in the diagonal with that of $f$, possesses a tangent at these two points, but they are vertical, which means that the ratio $[g(1 + k) - g(1)]/k$ tends to infinity. One could extend the concept of derivative to cover this kind of situation, but the slight benefits of such a generalisation would be very much less than the effort.

*Example 3.* Consider the function $f(x) = \exp x$ on $I = \mathbb{R}$, whence $f'(x) = f(x)$ as observed above; by n° 4, example 2, it maps $\mathbb{R}$ onto $\mathbb{R}_+^*$ and has an inverse $g$. Since $\exp' x = \exp x$ is nowhere zero the inverse function $g$ is differentiable and $g'(y) = 1/\exp' x$ if $y = \exp x$, which can also be written as $g'(y) = 1/y$. Surprising?

We conclude these generalities on derivatives with a few more remarks on notation.

First, the rules of calculus can be put in the form

(15.8) $\quad d(f + g) = df + dg, \quad d(fg) = gdf + fdg, \quad d(1/g) = -dg/g^2.$

The second relation, for example, means that at the point $x$, the differential $h \mapsto dp(x; h)$ of the product function $p = fg$ is given by

$$dp(x; h) = g(x)df(x; h) + f(x)dg(x; h);$$

for $x$ given this is an identity between linear functions of the auxiliary variable $h$ and not between functions of $x$.

The chain rule can similarly be written in a very simple form. Put $y = f(x)$ and $z = g(y) = h(x)$. Then

$$dz = g'(y)dy \text{ and } dy = f'(x)dx, \text{ whence } dz = g'(y)f'(x)dx$$

and so $h'(x) = g'[f(x)]f'(x)$ since $dz = h'(x)dx$. Following Leibniz and Co. one would write this quite simply in the seductive form $dz/dx = dz/dy.dy/dx$;

it must not be applied too mechanically since the points where the derivatives are to be calculated are not indicated. Formulae of this kind, not appearing in so convenient a form *chez* Newton, led to the success of Leibniz' system, relieving the users from having to think, though not, even nowadays, from sometimes writing stupidities.

There are also expressions analogous to $df$ or $dy/dx$ for the second differentials and derivatives ..., appreciably more subtle and, in fact, too subtle and useless to have had lasting success in the elementary theory of functions of one real variable, except as a convenient notation. Consider the linear function $df(x) : h \mapsto df(x; h)$ as being itself a function of $x$, namely $df : x \mapsto df(x)$; although the values of this function are not numbers[34], one can calculate its differential $d(df)(a)$ at a point $a$, and naturally one denotes it by $d^2 f(a)$; to find it one calculates

$$(15.9) \qquad\qquad df(a + k) - df(a),$$

an expression which denotes the function

$$(15.10) \qquad h \mapsto df(a + k; h) - df(a; h) = f'(a + k)h - f'(a)h;$$

as $k$ tends to zero

$$(15.11) \qquad\qquad f'(a + k) - f'(a) = f''(a)k + o(k),$$

so that (9) is the sum of the two following functions of $h$: (i) the function $h \mapsto f''(a)kh$, (ii) a function $o(k)h = o(kh)$. Thus the differential $d^2 f(a)$, which involves the variable $k$ for the first $d$ sign and the variable $h$ for the second, is given by

$$(15.12) \qquad d^2 f(a) : (h, k) \longmapsto f''(a)hk = f''(a)dx(a; h)dx(a; k)$$

by (14.6) or, in short, $d^2 f(a) = f''(a)dx(a)^2$ [up to the fact that the usual square of a function of $h$ is a function of $h$ and not of the pair $(h, k)$...]. Hence Leibniz' expression $d^2 y/dx^2$ to denote the second derivative of $y = f(x)$. Another explanation of the notation is that the second derivative is $df'/dx$, and since $f' = dy/dx$ one "clearly" finds $d^2 y/dx^2$. On this point, it would be considerably clearer once and for all to write $D$ for the derivation operator[35] $d/dx$, i.e. the map $f \mapsto f'$, and $D^2 = D \circ D$ for the map $f \mapsto f''$, as already said above.

---

[34] For readers who know the elements of linear algebra, these are linear forms on $\mathbb{R}$ considered as a real vector space of dimension 1. This set, endowed with the obvious algebraic operations (addition, product by a scalar $c \in \mathbb{R}$) is in turn a vector space of dimension 1. There is always a tendency to confuse it with $\mathbb{R}$ because a linear form on $\mathbb{R}$, i.e. a function $h \mapsto ch$, is characterised by a number $c \in \mathbb{R}$. But a real number is not a function defined on $\mathbb{R}$.

[35] The word "operator", which is also used in algebra ("linear operator") and elsewhere in analysis ("Laplace operator" for example) is synonymous with the words "function" and "map", although one frequently says "operators" without prescribing exactly the sets where they are defined or take their values.

All this may not be very clear at this level – it cannot be understood outside the framework of finite dimensional vector spaces –, but one can console oneself that the contemporaries and successors of Leibniz did not understand it either. Whence the abundance of polemics and philosophical dissertations on the subject until the appearance of "modern" mathematics in the XX$^{th}$ century, of functions with vector values defined on a subset of $\mathbb{R}^n$, of linear and multilinear functions, etc. (Chap. IX, §1). This point of view has been expounded with his habitual conciseness in Vol. I, Chap. VIII, § 12 of the *Treatise on Analysis* by Dieudonné, in Serge Lang, *Analysis I*, Chap. XVI (same remark), and in many other works.

## 16 – The mean value theorem

Although the concept of a derivative was transformed into a formidable instrument by Newton, Leibniz and their successors, it had appeared earlier, even though only implicitly, in Descartes, Fermat and Cavalieri. In Fermat it is linked to seeking the maxima and minima of a function. This can be understood immediately, since, if $f(x) \leq f(c)$ for all $x$ or even only on a neighbourhood of $c$, then the ratio $[f(c + h) - f(c)]/h$ which, in the limit, defines the derivative at $c$, is $\leq 0$ for $h > 0$ and $\geq 0$ for $h < 0$ sufficiently small: so it can tend only to 0, so long as $f$ is defined on a neighbourhood of $c$, and not only to the right or to the left of $c$.

As we saw above (n° 9, Corollary of Theorem 12), if we have a continuous function $f$ with real values on a *compact* interval $K = [a, b]$ then there actually exists a point in $K$ where the function attains a maximum (resp. a minimum). This is not enough to prove that, if $f$ is differentiable in $K$, its derivative vanishes at these points: if $f$ is monotone, for example linear, it attains its minimum and its maximum at the end points of $K$ and the preceding argument fails. To eliminate this objection it suffices to assume that $f(a) = f(b)$; if one denotes by $c'$ (resp. $c''$) a point of $K$ where $f$ is a maximum (resp. a minimum), and if these two points are the end points of $K$, then the function is everywhere between $f(a)$ and $f(b) = f(a)$, so is constant, in which case it is not difficult to prove that its derivative vanishes somewhere. If, however, $f$ is not constant, then at least one of the points $c'$, $c''$ is *interior* to $K$, and the standard argument applies: $f'$ vanishes somewhere between $a$ and $b$ and even at a point $c$ *interior* to $K$.

When $f(a) \neq f(b)$, one can reduce to the preceding case by replacing $f(x)$ by $g(x) = f(x) + ux$, where $u$ is a constant chosen so that $g(a) = g(b)$. This requires $f(a) + ua = f(b) + ub$, whence

$$g(x) = f(x) - [f(b) - f(a)]x/(b - a).$$

Since $g'(x) = f'(x) - [f(b) - f(a)]/(b - a)$, we finally obtain the following result:

**Theorem 18 (mean value theorem).** *Let $f$ be a real function defined and differentiable on an interval $I$. For any $a, b \in I$ there exists a number $c \in\ ]a, b[$ such that*

$$(16.1) \qquad\qquad f(b) - f(a) = (b-a)f'(c).$$

The geometric interpretation is obvious: there exists a point with abscissa $c$ where the tangent to the graph of $f$ is parallel to the "chord", of slope $[f(b) - f(a)]/(b-a)$, joining the points of the graph with abscissae $a$ and $b$. If one puts $b = a + h$ in (1) one finds

$$(16.1') \qquad\qquad f(a+h) = f(a) + f'(a + th)h$$

where $t$ (traditionally denoted $\theta$) is a number lying between 0 and 1 and of course depending on $a$ and $h$; compare with the relation $f(a + h) = f(a) + f'(a)h + o(h)$. A curious aspect of (1') is that, when $h$ tends to 0, the term $f'(a + th)$ tends to $f'(a)$; since $0 < t < 1$, this would be obvious if $f'$ were continuous at the point $a$, but we have not made this hypothesis; this shows that the points $a + th$ which appear in (1'), for example for $h = 1/n$, are not distributed at random ...

On this theme we remark that if $f'$ is not continuous, it is still not a *very* savage function – just enough to be well beyond the imagination of people who are not experts in set theory *à la* Cantor (Georg) and *à la* Baire: assuming for simplicity that $f$ is defined and differentiable on all $\mathbb{R}$, the function $f'$ is a *simple limit of continuous functions*, namely of

$$f_n(x) = n[f(x + 1/n) - f(x)].$$

Reread the end of n° 6.

We can generalise formula (1) by modifying the argument a little. Instead of looking for a function of the form $f(x) + ux$ which takes the same values at $a$ and $b$, let us choose any differentiable function $g(x)$ and consider the function $f(x) + ug(x) = h(x)$. We will have $h(a) = h(b)$ if $u = -[f(b) - f(a)]/[g(b) - g(a)]$. Since $h'(x) = f'(x) + ug'(x)$ and since $h'$ vanishes at some point $c \in\ ]a, b[$, we have $f'(c) + ug'(c) = 0$, and hence the formula

$$(16.1'') \qquad\qquad \frac{f(b) - f(a)}{g(b) - g(a)} = \frac{f'(c)}{g'(c)}$$

which, for $g(x) = x$, reduces to Theorem 18.

There is no result analogous to (1) for a complex-valued function: there is of course a $c'$ for its real part and a $c''$ for its imaginary part, but why should they be equal? In this case one replaces the formula (1) by an inequality – see below – at least as useful in practice. First, let us note the more important consequences of Theorem 18.

**Corollary 1.** *(i) Let $f$ be a complex-valued function defined and differentiable on an interval $I$; if $f'(x) = 0$ for all $x \in I$, then $f$ is constant on $I$. (ii) Let $f$ and $g$ be functions with complex values defined and differentiable on an interval $I$; if $f'(x) = g'(x)$ for all $x \in I$, then $f(x) = g(x) + C$ where $C$ is a constant.*

(i) is obvious from (1) in the case of a real-valued function; the complex case reduces to it by considering the real and imaginary parts of $f$. (ii) follows from applying (i) to $f - g$.

*Example 1.* From (14.15) or Theorem 3 of Chap. II, n° 10 we know that $\log' x = 1/x$, so that, on putting $f(x) = \log(1 + x)$,

$$(16.2) \qquad f'(x) = 1/(1 + x) = 1 - x + x^2 - x^3 + \dots$$

if $|x| < 1$. Consider its primitive series (Chap. II, n° 19)

$$(16.3) \qquad g(x) = x - x^2/2 + x^3/3 - x^4/4 + \dots,$$

which converges in the same disc. By Chap. II, n° 19, we know that $g$ is differentiable and that $g'(x)$ is the derived series of (3), i.e. precisely (2). Thus $g'(x) = \log'(1 + x)$, whence $\log(1 + x) = g(x)$ since the two sides are obviously zero for $x = 0$. So we obtain the formula

$$(16.4) \qquad \log(1 + x) = x - x^2/2 + x^3/3 - x^4/4 + \dots,$$

valid for $-1 < x \leq 1$ (see n° 8, example 4). This proof depends on a manifest subterfuge: confusing the derived *series* of a power series, in the purely formal sense introduced in Chap. II, n° 19, with the derived *function* of its sum, defined by passing to the limit. But we showed in Chap. II that, in the case of convergent power series, these two senses of the word "derived" were identical in the interior of the disc of convergence and, in particular, of the interval of convergence in $\mathbb{R}$.

*Example 2.* Let us seek the functions $f(x)$ defined and differentiable on an interval $I$ and proportional to their derivative:

$$f'(x) = cf(x)$$

where $c \in \mathbb{C}$ is a given constant. The series $\exp x = \sum x^{[n]}$ trivially satisfies the relation $\exp' x = \exp x$, so that the function $\exp(cx)$ satisfies the condition imposed on $f$. Since the exponential function never vanishes the function $g(x) = f(x)/\exp(cx) = f(x)\exp(-cx)$ is differentiable on $I$. We have

$$g'(x) = f'(x)\exp(-cx) - cf(x)\exp(-cx) = 0.$$

In consequence $g$ is constant, from which we see that the only solutions of the problem are the constant multiples of $\exp(cx)$.

**Corollary 2.** *Let f be a real function defined and differentiable on an interval I; for f to be increasing in I it is necessary and sufficient that $f'(x) \geq 0$ for all $x \in I$; for it to be* strictly *increasing, it suffices that $f'(x) > 0$ for all $x \in I$. If $f'(x) > 0$ in I and if f is of class $C^p$ on I, then the inverse map $g : f(I) \to I$ is of class $C^p$.*

The fact that $f'(x) \geq 0$ if $f$ is increasing is obvious. It is equally clear (Theorem 18) that $f$ is strictly increasing if $f'(x) > 0$, and then $f$ maps $I$ bijectively onto an interval $J = f(I)$ with an inverse map $g : J \to I$, differentiable according to the rule (D 5) of n° 15. Further, at corresponding points $x \in I$ and $y = f(x) \in J$, one has

$$g'(y) = 1/f'(x) = 1/f'[g(y)],$$

so that $g$ is of class $C^1$ if $f$ is; if $f$ is of class $C^2$, then $1/f'$ is of class $C^1$, and the preceding formula shows that $g'$ is too, so that $g$ is of class $C^2$, and so on.

Note that the derivative of a strictly increasing function can well vanish at some points, as is the case for the function $x^3$ on $\mathbb{R}$, for example.

Formula (1) has an important consequence when the derivative $f'$ is *bounded* on $I$. Since $|f'(c)| \leq \|f'\|_I$, the least upper bound over $I$ of the numbers $|f'(x)|$, we have

(16.5)            $$|f(b) - f(a)| \leq \|f'\|_I.|b - a|$$

for all $a, b \in I$. This is the *inequality of the mean* , , an expression whose significance stems from the theory of integration (Chap. V, n° 11). Note that it is not the uniform norm of $f'$ on $I$ which really appears; it is that of $f'$ over the compact interval $[a, b]$, and this is finite if, for example, $f'$ is continuous (n° 9, Theorem 11).

This argument supposes that $f$ has real values, but the result extends to functions with complex values, thanks to the following trick.

Suppose that one wants to prove that a given complex number $u$ is $\leq A$, where $A > 0$ is given. If such is the case, then $|zu| \leq A|z|$ for all $z \in \mathbb{C}$ and so

(16.6)            $$|\text{Re}(zu)| \leq A|z| \qquad \text{for all } z \in \mathbb{C}.$$

If, conversely, this condition is fulfilled, it applies to $z = \bar{u}$, whence $|u| \leq A$ since $\text{Re}(u\bar{u}) = \text{Re}(|u|^2) = |u|^2$.

This point established, let us return to a complex function $f$ that is everywhere differentiable on an interval $I$, and put $u = f(b) - f(a)$. For $z \in \mathbb{C}$ we then have

$$|\text{Re}(zu)| = |\text{Re}\{z[f(b) - f(a)]\}| = |\text{Re}[zf(b)] - \text{Re}[zf(a)]|.$$

The function $f_z(x) = \text{Re}[zf(x)]$ is differentiable, like $f$, and

$$f'_z(x) = \text{Re}[zf'(x)],$$

whence $|f'_z(x)| \le |zf'(x)| \le M|z|$ where $M = \|f'\|_I$. Since $f_z$ has real values Theorem 18 shows that

$$|\text{Re}(zu)| = |f_z(b) - f_z(a)| \le M(b - a)|z|$$

for any $z \in \mathbb{C}$. Thus $|u| \le M(b - a)$. To sum up:

**Corollary 3.** *Let $f$ be a complex-valued function defined and differentiable on an interval $I$. Suppose that $f'$ is bounded on all compact subsets $K \subset I$. Then*

(16.7) $$|f(b) - f(a)| \le \|f'\|_K (b - a)$$

*for all points $a, b \in I$, where $K = [a, b]$.*

Still with the hypotheses of Corollary 3, let us choose an arbitrary $u \in \mathbb{C}$ and apply (7) to the function $g(x) = f(x) - ux$. We find:

**Corollary 4.** *Under the hypotheses of Corollary 3 we have*

(16.8) $$|f(b) - f(a) - u(b - a)| \le (b - a) \sup_{a \le x \le b} |f'(x) - u|$$

*for all $a, b \in I$ and $u \in \mathbb{C}$.*

For example we can choose $u = f'(a)$, $b = a + h$, whence

(16.9) $$|f(a + h) - f(a) - f'(a)h| \le |h|. \sup_{0 \le t \le 1} |f'(a + th) - f'(a)|,$$

which, if $f'$ is *continuous* at $a$, expresses more precisely the $o(h)$ which would appear on the right hand side than if one assumed only the existence of $f'(a)$.

The inequality of the mean is valid for a much larger class of functions: it is enough to assume that $f$ is continuous and that the set of $x \in I$ where $f'(x)$ does not exist is *countable*. Since the problem is directly linked to the construction of a "primitive" $f$ of a regulated function $g$, i.e. of a function such that one has, according to taste, either $f'(x) = g(x)$ "apart from exceptions", or

$$f(x) - f(a) = \int_a^x g(t)dt,$$

it will be better to delay these useless subtleties to Chap. V.

## 17 – Sequences and series of differentiable functions

The classical analysts, notably Euler but also Fourier a good half century later, seem to have believed that, if a sequence or series of differentiable functions converges, one will, on differentiating term-by-term as for a finite sum, again obtain a convergent sequence or series. Fourier's own square wave series rebutted this conjecture; its sum is neither differentiable nor even continuous for $x = \pi/2$ and, moreover, differentiating it leads, as we have seen, to bewildering results. Fourier, in reality, was trying to represent the periodic function equal to 1 for $|x| < \pi/2$ and to $-1$ for $\pi/2 < |x| < \pi$ as a series of the form $a_1 \cos x - a_3 \cos 3x + a_5 \cos 5x - \ldots$, with coefficients to be determined. To do this he set $x = 0$ in the given series and in those obtained by differentiating term-by-term $ad$ $libitum$; he found the relations

$$a_1 - a_3 + a_5 - \ldots = 1$$
$$a_1 - 3^2 a_3 + 5^2 a_5 - \ldots = 0$$
$$a_1 - 3^4 a_3 + 5^4 a_5 - \ldots = 0$$

etc. So now he had to determine the coefficients from a system of an infinite number of linear equations in an infinite number of unknowns. To solve this, Fourier simplified it by replacing, for given $n$, the unknowns $a_{2n+3}, a_{2n+5}$, etc. by 0; working then with only the first $n$ equations he obtained a fully orthodox system which he solved explicitly by computations within the reach of a present day candidate to the Ecole polytechnique. He then made $n$ tend to infinity in his formulae, and thanks to the expansion of $\pi$ as an infinite product (Wallis' formula) finally found, up to the factor $4/\pi$, the square wave series $\cos x - \cos(3x)/3 + \cos(5x)/5 - \ldots$.

But had he wanted to $verify$ the result of his calculations he would have had to substitute these values in his infinite system of linear equations and would then have obtained the formulae

$$1 - 1/3 + 1/5 - 1/7 + \ldots = \pi/4$$
$$1 - 3^2/3 + 5^2/5 - 7^2/7 + \ldots = 0$$
$$1 - 3^4/3 + 5^4/5 - 7^4/7 + \ldots = 0;$$

the first one, Leibniz' formula, is correct, but the others are still more foolish than the marvels of the harmonic series. The mathematicians of the XIX[th] century have discovered even more bewildering counterexamples than this last: a sum or series of indefinitely differentiable functions, even if uniformly convergent, can have a sum not admitting a derivative $anywhere$. Bolzano constructed one, though this was not discovered until 1930. Riemann considered the series $\sum \sin(n^2 x)/n^2$ and tried, without success, to show that its sum has no derivative (in fact, in 1970 it was proved differentiable at certain points, for example $x = m\pi/n$ with $m$ and $n$ odd, but not if $x/\pi$ is irrational).

Weierstrass in his turn tried his luck and produced the series $\sum q^n \cos(a^n x)$, normally convergent for $0 < q < 1$, and proved that it has no derivative for an odd integer[36] $qa > 3\pi/2 + 1$; just imagine the effect produced on mathematicians who, up until then, thought that a continuous function was always differentiable except perhaps at a finite number of exceptional points, as was "proved" by naïve diagrams that anyone can draw on a sheet of paper. But no one has ever traced the graph of Weierstrass' function; you can of course trace those of the partial sums of the series; computers having been invented specifically to resolve this kind of practical problem, trajectories of a shell (Eckert and Mauchly, 1942–1945) or missiles for example, you can probably find the graphs on the Internet and download them onto your computer at six o'clock in the morning when the majority of Americans are taking their well earned rest after a hard day's work; but as to *seeing* what happens in the limit ... This would anyway not help: Weierstrass proved his theorem without waiting for the Web and, most likely, without wasting his time in sketching graphs on graph paper.

What was not understood was that the situation is governed by the convergence of the *derivatives* and not by that of the given functions, as the theory of integration will show yet more clearly:

**Theorem 19.** *Let* $(f_n)$ *be a sequence of functions defined and differentiable on an interval $I$. Suppose that*

*(i)  the sequence of derived functions $(f_n')$ converges uniformly to a limit $g$ on every compact $K \subset I$,*
*(ii) the sequence $(f_n(x))$ converges at some point of $I$.*

*Then the functions $f_n$ converge uniformly to a limit function $f$ on every compact $K \subset I$, and $f' = g$.*

Choose a $c \in I$ where $\lim f_n(c)$ exists. On replacing each function $f_n(x)$ by $f_n(x) - f_n(c)$, which does not change the derivatives and does not influence the possible uniform convergence of $f_n$, one can assume that $f_n(c) = 0$ for all $n$. Put $f_{pq}(x) = f_p(x) - f_q(x)$, whence $f_{pq}(c) = 0$. If $K \subset I$ is a compact interval containing $c$ and of length $m(K)$ then Corollary 3 above shows that

$$(17.1) \qquad |f_{pq}(x)| \le \|f_{pq}'\|_K |x - c| \le m(K) \|f_{pq}'\|_K$$

for all $x \in K$. This can be written as

$$(17.2) \qquad \|f_{pq}\|_K \le m(K) \|f_{pq}'\|_K$$

---

[36] See Walter, *Analysis 1*, p. 359, for a full treatment of a similar example, due to a Japanese at the beginning of the century (they "already" did mathematics, and the principal collaborator of the physician and biologist Paul Ehrlich, who discovered the first effective treatment for syphilis in about 1910, was also a Japanese).

or

$$(17.3) \qquad d_K(f_p, f_q) \le m(K) d_K(f_p', f_q').$$

Since the $f_n'$ converge uniformly on $K$ to $g$, they satisfy Cauchy's criterion for uniform convergence (n° 10, Theorem 12"); then (3) shows that the $f_n$ do so too, and therefore they converge uniformly on $K$.

It remains to prove that the limit $f$ of the $f_n$ is differentiable and that $f' = g$. This reduces to showing that, for all $a \in I$,

$$(17.4) \qquad \lim_{x \to a} \lim_{n \to \infty} \frac{f_n(x) - f_n(a)}{x - a} = \lim_{n \to \infty} \lim_{x \to a} \frac{f_n(x) - f_n(a)}{x - a}.$$

In fact, for the left hand side, the limit with respect to $n$ is $[f(x) - f(a)]/(x - a)$, so that the limit with respect to $x$, if it exists, is $f'(a)$. As to the right hand side, the limit with respect to $x$ is $f_n'(a)$, which exists, and the limit with respect to $n$, which also exists, is $g(a)$. The relation (4) therefore means that $f$ has a derivative equal to $g(a)$ at $a$.

It is now necessary to use Theorem 16 of n° 12:

Let $X$ be a subset of $\mathbb{C}$, let $a$ be an adherent point of $X$, and let $u_n(x)$ be a sequence of scalar functions defined on $X$. Suppose that (i) $u_n(x)$ tends to a limit $c_n$ when $x$ tends to $a$, (ii) $u_n(x)$ converges uniformly on $X$ to a limit function $u(x)$. Then $u(x)$ tends to a limit $c$ when $x \in X$ tends to $a$; and $c = \lim c_n$.

Choose $X = K - \{a\}$, where $K$ is the set of $x \in I$ such that $|x - a| \le r$, with $r > 0$ small enough for $K$ to be a compact interval[37], and put

$$\begin{aligned} c_n &= f_n'(a), \\ u_n(x) &= [f_n(x) - f_n(a)]/(x - a), \\ u(x) &= [f(x) - f(a)]/(x - a) \end{aligned}$$

for $x \in X = K - \{a\}$. Hypothesis (i) of Theorem 16 asserts the differentiability of $f_n$ at the point $a$. It remains to verify that $u_n(x)$ converges *uniformly on* $K$ to $u(x)$. Put $f_{pq} = f_p - f_q$. Now

$$u_p(x) - u_q(x) = [f_{pq}(x) - f_{pq}(a)]/(x - a)$$

and, by the mean value theorem,

$$|f_{pq}(x) - f_{pq}(a)| \le \|f_{pq}'\|_K \cdot |x - a|$$

for all $x \in X$, whence

$$|u_p(x) - u_q(x)| \le \|f_{pq}'\|_K = \|f_p' - f_q'\|_K$$

---

[37] There is no problem if $a$ is interior to $I$. If $a$ is, for example, the left end point of $I = [a, b)$, take $r < b - a$, so that then $K = [a, a + r] \subset [a, b[$.

for all $x \in X = K - \{a\}$. On taking the least upper bound of the left hand side for $x \in X$, we find

$$\|u_p - u_q\|_X \leq \|f_p' - f_q'\|_K$$

and since the $f_n'$ converge uniformly on $K$ by hypothesis, the $u_n$ satisfy Cauchy's criterion for uniform convergence on $X$. Whence hypothesis (ii) of Theorem 16, qed.

**Corollary.** *Let $\sum f_n(x)$ be a series of functions defined and differentiable on an interval $I$. Suppose that the series converges at a point of $I$ and that the derived series converges normally on every compact $K \subset I$. Then the given series converges normally on every $K \subset I$, its sum $s(x)$ is differentiable, and $s'(x) = \sum f_n'(x)$.*

We need only apply the theorem to the partial sums $s_n(x)$ of the given series, and observe that normal convergence of the derived series implies uniform convergence of the sequence $s_n'(x)$ on every compact set. Normal convergence of the given series follows from inequality (2).

*Example 1.* Let $f(x) = \sum a_n x^n$ be a power series converging for $|x| < R$, $R > 0$, and let us work on $I = ] - R, R[$. We know (Chap. II, n° 19) that the series $\sum n a_n x^{n-1}$ of derivatives also converges for $|x| < R$ and so converges normally on all compact intervals $K = [-r, r]$ with $r < R$. Theorem 19 then applies, confirming that one can differentiate a power series term-by-term, at least in the real domain.

*Example 2.* Consider a Fourier series

$$f(x) = \sum a_n \cos nx + b_n \sin nx$$

and suppose that the series $\sum n a_n$ and $\sum n b_n$ converge absolutely (examples: $a_n = 1/n^k$ with $k > 2$, or $a_n = q^n$ with $|q| < 1$, etc.) Since sine and cosine are $\leq 1$ in modulus, the series obtained by differentiating the given series term-by-term will converge normally, whence

$$f'(x) = \sum (n b_n \cos nx - n a_n \sin nx).$$

The square wave series clearly does not fit into this framework. If you consider the series $f(x) = \sum \sin(n^2 x)/n^2$ at which Riemann, Weierstrass, and surely many others, tilted without success up to 1930, the argument fails totally, and must do so, since $f(x)$ has a strong tendency not to be differentiable except at very exceptional points.

*Example 3.* Consider the series $\zeta(s) = \sum 1/n^s$ for $s > 1$. We shall see in Chap. IV, where we shall define real powers – they behave exactly like integral powers –, that the function $s \mapsto n^s = \exp(s.\log n)$ is differentiable and has derivative $n^s \log n$. The series of derivatives

$$\sum -(n^s \log n)/(n^s)^2 = -\sum (\log n)/n^s$$

is, like the Riemann series itself (n° 8, example 2), normally convergent on $[\sigma, +\infty[$ for any $\sigma > 1$. We shall show later (Chap. IV, n° 5) that for any real number $k > 0$ we have

$$\log x = o(x^k) \quad \text{when } x \to +\infty;$$

up to a constant factor one has $(\log n)/n^s \leq 1/n^{s-k} \leq 1/n^{\sigma-k}$ for $n$ large; by choosing $k > 0$ small enough that $\sigma - k > 1$, i.e. $0 < k < \sigma - 1$, one obtains normal convergence. So we can differentiate term-by-term, whence

$$\zeta'(s) = -\sum (\log n)/n^s$$

for $s > 1$, and we can repeat this operation *ad libitum*.

*Example 4.* In Chap. II, n° 21, we asked the question of whether, from the relation

$$(17.5) \qquad \pi \cot \pi x = 1/x + \sum x/n(x-n)$$

where one sums over all nonzero $n \in \mathbb{Z}$, one can, by differentiation, deduce the formula

$$(17.6) \qquad \pi^2/\sin^2 \pi x = \sum 1/(x-n)^2$$

where one sums over all the $n \in \mathbb{Z}$ without exception. For this, let us work on a compact interval of the form $K = [-p,p]$ with $p \in \mathbb{N}$ and split the series in (5) as the sum of terms for which $|n| \leq p$ and the sum of the other terms. The first has only a finite number of terms, so can be differentiated term-by-term, and one thus obtains, in (6), the corresponding partial sum. As for the sum of the terms for which the index satisfies $|n| > p$, it is, in (5), a series of functions defined and differentiable on all of $K$ and which converges everywhere on $K$. For Theorem 19 to be applicable it is thus enough to show that the series $\sum (x-n)^{-2}$, taken over those $n$ such that $|n| > p$, converges normally on $K$.

Now the equality $n = (n-x) + x$ shows that

$$|n| \leq |x-n| + |x| \leq |x-n| + p,$$

whence $|x-n| \geq |n| - p > 0$ and thus

$$1/(x-n)^2 \leq 1/(|n|-p)^2 = v(n).$$

The right hand side is independent of $x \in K$ and the series $\sum v(n)$, taken over the $n$ considered, converges, since its general term is equivalent to $1/n^2$. Whence normal convergence, qed. (This argument, modified a little, would in fact show that (6) converges normally on any compact subset of $\mathbb{C}$ not containing any $n \in \mathbb{Z}$).

Theorem 19 can be stated in terms of primitive functions. Consider a sequence of functions $f_n$ on the interval $I$ which converges uniformly to a limit $f(x)$ on every compact $K \subset I$. Suppose that each $f_n$ has a primitive, i.e. that there exists a differentiable function $F_n$ on $I$ such that $F_n'(x) = f_n(x)$ for all $x$. We can then apply Theorem 19 to the $F_n$ so long as the sequence $(F_n)$ converges for at least one point $a \in I$, as is the case, if, for example, $F_n(a) = 0$ for all $n$. We see then that $F_n$ converges uniformly on all compact $K \subset I$ to a limit $F$, that $F$ is differentiable, and that $F' = f$. In other words, $F$ is a primitive of $f$. We shall return to this result in more detail in Chap. V à propos the "fundamental theorem of the differential and integral calculus", i.e. of the relation

$$F(x) - F(a) = \int_a^x f(t)dt$$

between a function $f$ and its primitives.

Theorem 19 has another important corollary:

**Theorem 20.** *Let $(f_n)$ be a sequence of functions of class $C^p$ ($p \leq +\infty$) on an interval $I \subset \mathbb{R}$. Suppose that the functions $f_n$ and all their derivatives of order $\leq p$ converge uniformly on every compact $K \subset I$. Then the limit of the $f_n$ is of class $C^p$ in $I$ and*

$$(17.7) \qquad f^{(k)}(x) = \lim f_n^{(k)}(x)$$

*for all $k \leq p$, where $f(x) = \lim f_n(x)$.*

To see this we need only apply Theorem 19 repeatedly, first to the $f_n$, whence existence of $f'$ with (7) for $k = 1$, then to the $f_n'$, whence existence of $(f')' = f''$ with (7) for $k = 2$, etc.

Theorem 19 allows one to solve all sorts of more or less classical problems. Theorem 20 for $p = +\infty$ (indefinitely differentiable functions), and its generalisations to several variables, are the foundation of the theory of distributions of Laurent Schwartz (Chap. V, n° 34).

## 18 – Extensions to unconditional convergence

In so far as they concern only absolutely convergent series, all the results of this chapter about series of functions extend to the unconditional convergence of Chap. II; not very surprising, since this differs from classical absolute convergence only in appearance.

By reasoning on *a priori* arbitrary subsets of the set of indices of the family one can also give direct proofs which liberate one from having to choose bijections of $\mathbb{N}$ onto the set of indices in question.

First, normal convergence of a series $\sum u_i(t)$ of scalar functions defined on a set $X$ and indexed by a set $I$ means that there exists a family of numbers $v_i$, $i \in I$, satisfying

$$v_i \geq 0, \quad \sum v_i < +\infty \quad \text{and} \quad |u_i(t)| \leq v_i$$

for all $i$ and $t$. Equivalently, we may require

$$\sum \|u_i\|_X < +\infty.$$

The series $\sum u_i(t)$ is then unconditionally convergent for any $t$.

Suppose further that $X \subset \mathbb{C}$ and that, when $t \in X$ tends to a point $a$ adherent to $X$, each $u_i(t)$ tends to a limit $u_i$. Then clearly again $|u_i| \leq v_i$, whence the unconditional convergence of $\sum u_i$. If on the other hand one denotes by $s_F(t)$ (resp. $s_F$) the sum of $u_i(t)$ (resp. $u_i$) for $i \in F$ for any subset $F$ of $I$, and the corresponding total sums by $s(t) = s_I(t)$ and $s = s_I$, then

$$|s(t) - s| = \left| \sum_{i \in F} [u_i(t) - u_i] + \sum_{i \notin F} [u_i(t) - u_i] \right| \leq$$

$$\leq |s_F(t) - s_F| + \sum_{i \notin F} |u_i(t) - u_i| \leq |s_F(t) - s_F| + 2 \sum_{i \notin F} v_i;$$

now, for all $r > 0$ there exists a *finite* $F$ such that the last $\sum$ is $< r$ (Chap. II, n° 15); for this $F$ the difference $|s_F(t) - s_F|$ is $< r$ for $t$ sufficiently close to $a$ (limit of a *finite* sum of functions). Whence $|s(t) - s| < 3r$ and so $\lim s(t) = s$. In other words, Theorem 17 can now be written in the form

$$(18.1) \qquad \lim_{t \to a} \sum_{i \in I} u_i(t) = \sum_{i \in I} \lim_{t \to a} u_i(t).$$

It applies in particular to normally convergent series of continuous functions.

Similarly there is a theorem on term-by-term differentiation for unconditional convergence. This time one considers, as in Theorem 19, a series of functions $u_i(x)$ defined and differentiable on a compact interval[38] $K$, assumes that the given series converges somewhere, and, finally, that the series

---

[38] The theorems for compact convergence (i.e. uniformly on all compacta) for an arbitrary interval $I$ are no more general than the uniform convergence theorems for a compact interval: one applies the "particular" case to each arbitrary compact interval contained in $I$.

of derivatives $\sum u_i'(x)$ is dominated, as above, by a series $\sum v_i$. One needs to pass to the sums

$$s_F(x) = \sum_{i \in F} u_i(x), \qquad s(x) = \sum_{i \in I} u_i(x).$$

Choose an $a \in K$. The mean value theorem shows that

$$|u_i(x) - u_i(a)| \le m(K)\|u_i'\|_K$$

where $m(K)$ is the length of $K$. Since $\sum \|u_i'\|_K < +\infty$, the series $\sum[u_i(x) - u_i(a)]$ converges normally on $K$ for all $a$ and $x$. If the series $\sum u_i(a)$ converges unconditionally then the series $u_i(x)$ converges normally on $K$.

   To show that one can differentiate it like a finite sum, consider the functions

$$q_i(x) = [u_i(x) - u_i(a)]/(x - a),$$

as in the proof of Theorem 19. They are defined on $X = K - \{a\}$ and tend to a limit $u_i'(a)$ when $x$ tends to $a$. Corollary 3 of the mean value theorem shows on the other hand that

$$|q_i(x)| \le v(i),$$

so that the series $\sum q_i(x) = [s(x) - s(a)]/(x - a)$ converges normally on $X$. We can therefore pass to the limit when $x$ tends to $a$ as we saw above, which shows that the sum $s(x)$ is differentiable at $a$ and that $s'(a) = \sum u_i'(a)$, qed.

   Consider for example, instead of the Riemann series of the preceding n°, the double series $f(s) = \sum(m^2 + n^2)^{-s/2}$ with $s$ real $> 2$ to ensure convergence (Chap. II, n° 12). If we accept the formula $(a^x)' = a^x \log a$ of Chap. IV, the derived series is

$$g(s) = -\frac{1}{2}\sum \log(m^2 + n^2)/(m^2 + n^2)^{-s/2}.$$

We leave it to the reader to show that it converges normally for $s \ge \sigma > 2$ as do all the successive derived series. The problem is only to show unconditional convergence, because, the general term being a decreasing function of $s$, the series $g(\sigma)$ dominates the series $g(s)$ for all $s \ge \sigma$. You may be guided by the arguments of Chap. II, n° 12.

# §5. Differentiable functions of several variables

To conclude this chapter we shall generalise the results obtained in the preceding n° to functions of several real variables. We shall need them occasionally, mainly à propos holomorphic functions, and almost always for functions of two variables, i.e. defined on an open subset $U$ of $\mathbb{R}^2$ or of $\mathbb{C}$. The proofs will be presented so that they be extended immediately to the general case (see Vol. III, Chap. IX, §1). Apart from the particular case of holomorphic functions, we shall state few theorems, for the reason that *everything* contained in this § is fundamental and in general easy to remember, since they are direct generalisations of the results for a single variable.

In all that follows, when it is a question of the values taken by the variables or the functions considered, we shall make no distinction between the complex number $z = x + iy$ and the vector $(x, y)$ of $\mathbb{R}^2$; anyhow, this is how complex numbers are defined. Note that whenever we speak of *linear* functions or maps in $\mathbb{C}$, this will almost always be in the *real* sense, as in any vector space over $\mathbb{R}$; such a map is necessarily of the form

$$(*) \qquad\qquad (u, v) \longmapsto cu + dv$$

with real variables $u$ and $v$ and constant coefficients $c, d \in \mathbb{C} = \mathbb{R}^2$. This point of view generalises to any number of real variables subject to taking the coefficients $c$ and $d$ as vectors having real coordinates; to exhibit this fact we must separate the real and imaginary parts of $c$ and $d$ in the preceding formula, whence a formula which we write in terms of *real* matrices as

$$(**) \qquad\qquad \begin{pmatrix} u \\ v \end{pmatrix} \longmapsto \begin{pmatrix} \alpha & \beta \\ \gamma & \delta \end{pmatrix} \begin{pmatrix} u \\ v \end{pmatrix} = \begin{pmatrix} \alpha u + \beta v \\ \gamma u + \delta v \end{pmatrix}.$$

In $\mathbb{C}$, the linear functions in the *complex* sense are the functions $z \mapsto cz$, with $c \in \mathbb{C}$, since $\mathbb{C}$ is a vector space of dimension 1 over the field $\mathbb{C}$; in other words, these are the maps of the form $(*)$ satisfying the condition

$$(***) \qquad\qquad d = ic,$$

which allows us to exhibit $u + iv$ as a factor; in the form $(**)$ this means that $\delta = \alpha$, $\gamma = -\beta$. These $\mathbb{C}$-linear functions are also $\mathbb{R}$-linear, but too specialised to feature outside the theory of analytic or holomorphic functions.

## 19 – Partial derivatives and differentials

Let $f = f_1 + if_2$ be a complex-valued function defined and continuous on an open subset $U$ of $\mathbb{R}^2$; $f$ is also a map

$$(x, y) \longmapsto (f_1(x, y), f_2(x, y))$$

of $U$ into $\mathbb{R}^2$. For $(a, b) \in U$ given, the functions $x \mapsto f(x, b)$ and $y \mapsto f(a, y)$ are defined on a neighbourhood of $x = a$ and of $y = b$ respectively; in fact, the set of $x \in \mathbb{R}$ such that $(x, b) \in U$ is a open subset of $\mathbb{R}$: apply the definitions. If they have derivatives at $a$ and $b$ we shall denote them by $D_1 f(a, b)$ and $D_2 f(a, b)$ or, exceptionally, $f'_x(a, b)$ and $f'_y(a, b)$, and one meets everywhere the cumbersome Jacobi notation $\partial f / \partial x$. We shall say that $f$ is *of class* $C^1$ in $U$ if $D_1 f$ and $D_2 f$ exist for any $(a, b) \in U$ and are continuous on $U$. If $D_1 f$ and $D_2 f$ are in their turn of class $C^1$, in which case one says that $f$ is of class $C^2$, one can define the second partial derivatives

$$(19.1) \qquad \begin{aligned} D_1 D_1 f &= D_1^2 f = f''_{xx} = \partial^2 f / \partial x^2, \\ D_1 D_2 f &= f''_{yx} = \partial^2 f / \partial x \partial y, \end{aligned}$$

$D_2 D_1 f$, $D_2^2 f$, etc.

A point of view closer to that of n° 14 consists of extending the concept of the differential or of the linear tangent map to $f$ at $(a, b)$. Here one attempts to approximate $f(a + u, b + v) - f(a, b)$ by a $\mathbb{R}$-linear function of $(u, v)$, i.e. of the form $cu + dv$ with constants $c, d \in \mathbb{C}$, i.e. to write

$$f(a + u, b + v) = f(a, b) + cu + dv + \text{?}$$

where the error indicated by a ? must be "negligible" with respect to the dominating term $cu + dv$ when $u$ and $v$ tend to 0. The solution is that it must be negligible with respect to the *distance* of the point $(a, b)$ from the point $(a + u, b + v)$, i.e. with respect to $|u| + |v| \asymp (u^2 + v^2)^{\frac{1}{2}} = |u + iv|$. The preceding relation can then be written as

$$(19.2) \qquad f(a + u, b + v) = f(a, b) + cu + dv + o(|u| + |v|)$$

and implies

$$f(a + u, b) = f(a, b) + cu + o(|u|),$$

whence the existence of $D_1 f(a, b) = c$ and, similarly, of $D_2 f(a, b) = d$. When the condition (2) is satisfied one says that $f$ is *differentiable* at the point $(a, b)$; the linear function $(u, v) \mapsto cu + dv$ is called, depending on the author, the *differential* or the *linear tangent map* or the *derivative* of $f$ at $(a, b)$, and denoted by

$$(19.3) \qquad df(a, b) : (u, v) \mapsto D_1 f(a, b)u + D_2 f(a, b)v,$$

or $Df(a, b)$, or even $f'(a, b)$. We can now write (2) in a much more concise way:

$$(19.2') \qquad f(z + h) = f(z) + Df(z)h + o(h)$$

on putting $z = (a, b) \in U$, $h = (u, v) \in \mathbb{R}^2$, and agreeing to write $Df(z)h$ for the image of a vector $h \in \mathbb{R}^2$ under the *linear* map[39] $Df(z)$: $\mathbb{R}^2 \longrightarrow \mathbb{R}^2$; the expression $o(h)$ now clearly denotes a function of $h$ which is negligible with respect to the length $|h| \asymp |u| + |v|$ of the vector $h$.

The expression $df = f'(a)dx$ introduced in n° 14 extends immediately to functions of several variables. It is quite clear that at any point of the plane the differentials of the coordinate functions $(x, y) \mapsto x$ and $(x, y) \mapsto y$ are

$$dx(a, b) : (u, v) \longmapsto u, \quad dy(a, b) : (u, v) \longmapsto v.$$

As in n° 14, one can then write (3) in the form

$$(19.3') \qquad df(a, b) = D_1 f(a, b) dx(a, b) + D_2 f(a, b) dy(a, b),$$

a relation between linear functions of $(u, v)$. In short,

$$(19.3'') \qquad df = f'_x dx + f'_y dy.$$

This applies, for example, to the functions $z = x + iy$ and $\bar{z} = x - iy$, whence

$$dz = dx + idy, \qquad d\bar{z} = dx - idy,$$

which means that the differential of $(x, y) \mapsto \bar{z}$ at any point is the linear function $(u, v) \mapsto u - iv$. As we recalled above, a linear map of $\mathbb{R}^2$ into $\mathbb{R}^2$ can be represented by a $2 \times 2$ matrix. To obtain that of (3), i.e. of $Df(a, b)$, one has to calculate the real coordinates of the result, i.e., since $u$ and $v$ are real, replace the derivatives in (3) either by their real parts, or by their imaginary parts, which one obtains on performing the same operation on $f$. It is clear that on putting $(a, b) = z$ the matrix of $df(z)$ or $Df(z)$ is simply

$$(19.4) \qquad J_f(z) = \begin{pmatrix} D_1 f_1(z) & D_1 f_2(z) \\ D_2 f_1(z) & D_2 f_2(z) \end{pmatrix} = (D_i f_j(z));$$

it is called the *Jacobian matrix* of $f$ at $z$ and often we shall make no distinction between the linear map $Df(z)$ and the matrix (4). This causes no problem so long as one does not change the system of coordinates in $\mathbb{R}^2$, i.e. makes no distinction between a "geometric" vector $h$ with coordinates $u$, $v$ and the matrix

$$\begin{pmatrix} u \\ v \end{pmatrix}$$

associated with it.

---

[39] Recall that in linear algebra one frequently writes $Ah$, rather than $A(h)$, for the the value at the vector $h$ under a linear map $A$ (inheritance from of the age when one spoke of matrices instead of linear maps). Failing this convention, one would have to write $Df(z)(h)$ for what we write as $Df(z)h$. The notation $df(z; h)$ is also used often; see Vol. III, Chap. IX, § 1.

For functions of one variable, derivability and differentiability are equivalent properties; not so in the general case, for a good and simple reason. Consider what happens when one traverses a line passing through $(a, b)$, i.e. in the direction of a given vector $h = (u, v)$; by (2')

$$(19.5) \qquad f(z + th) = f(z) + t \cdot Df(z)h + o(t),$$

so that $f$ is differentiable in any direction originating at $(a, b)$, with

$$(19.5') \qquad Df(z)h = \frac{d}{dt} f(z + th) \quad \text{at } t = 0.$$

The existence at the point $(a, b)$ of derivatives in *all* the directions originating from $(a, b)$ is thus *necessary* to ensure differentiability. This condition, which may seem very strict, is still insufficient to ensure the differentiability or even only the continuity of $f$ at the point in question[40].

## 20 – Differentiability of functions of class $C^1$

These difficulties disappear when $f$ is of class $C^1$.

Suppose that $f$ is such a function and, following an idea of Euler's in another context, let us calculate

$$(20.1) \quad f(a + u, b + v) - f(a, b) =$$
$$= [f(a + u, b + v) - f(a, b + v)] + [f(a, b + v) - f(a, b)].$$

We need to compare this difference to $D_1 f(a, b)u + D_2 f(a, b)v = cu + dv$ for $u$ and $v$ small. For $b$ and $v$ given, the function $g(x) = f(x, b + v)$ has, by hypothesis, a derivative $g'(x) = D_1 f(x, b + v)$ for all $x$ near $a$, with $g'(a) = D_1 f(a, b + v)$. Then, by (16.8) or (16.9),

$$(20.2) \qquad |g(a + u) - g(a) - D_1 f(a, b + v)u| \le$$
$$\le |u| \cdot \sup_{0 \le t \le 1} |D_1 f(a + tu, b + v) - D_1 f(a, b + v)|.$$

Since $D_1 f$ is continuous at the point $(a, b)$, the difference $|D_1 f(a + tu, b + v) - D_1 f(a, b + v)|$ is $\le r$ for any $t \in [0, 1]$ if $|u|$ and $|v|$ are sufficiently small. The right hand side of (2) is thus $o(u)$ when $(u, v)$ tends to $(0, 0)$.

Similarly

$$(20.3) \qquad |f(a, b + v) - f(a, b) - D_2 f(a, b)v| \le$$
$$\le |v| \sup_{0 \le t \le 1} |D_2 f(a, b + tv) - D_2 f(a, b)|,$$

---

[40] Counterexample: $f(x, y) = x^2 y / (x^4 + y^2)$ if $(x, y) \ne (0, 0)$, $f(0, 0) = 0$. The function has derivatives in all directions at the origin [calculate the limit of $f(tu, tv)/t$ when $t \to 0$, paying particular attention to the case where $v = 0$]. It is not continuous at the origin since for all $a \in \mathbb{R}$, $f(x, ax^2)$ tends to (and is even equal to) $a/(1 + a^2)$ instead of tending to 0 as it would if $f$ were continuous. See the graph in $\mathbb{R}^3$ of the function in Hairer and Wanner, p. 303.

which is $o(v)$ when $(u, v)$ tends to $(0,0)$. In view of (19.4), one finds[41]

$$f(a+u, b+v) = f(a,b) + D_1 f(a, b+v)u + D_2 f(a,b)v + o(|u|+|v|);$$

but since $D_1 f$ is continuous, one has $D_1 f(a, b+v)u = D_1 f(a,b)u + o(u)$. In conclusion:

**Theorem 21.** *Let $f$ be a function of class $C^1$ on an open subset $U$ of $\mathbb{R}^2$. Then $f$ is differentiable at every point $(x,y) \in U$, and*

$$(20.4) \qquad f(x+u, y+v) =$$
$$= f(x,y) + D_1 f(x,y)u + D_2 f(x,y)v + o(|u|+|v|)$$

*when $u$ and $v$ tend to $0$.*

This result has an immediate consequence in the theory of analytic functions. We saw in Chap. II, n° 19 that if a function $f(z)$ is analytic on an open $U$ of $\mathbb{C}$, i.e. is expandable in a power series on a neighbourhood of any point of $U$, then it admits a derivative

$$(20.5) \qquad f'(z) = \lim_{h \to 0} \frac{f(z+h) - f(z)}{h}$$

in the *complex sense* where $h = u + iv$ tends to $0$ through complex values; and we deduced from this, by an immediate calculation, that $f$ then has continuous partial derivatives $D_1 f$ and $D_2 f$ satisfying Cauchy's identity $D_2 f = i D_1 f$, i.e. that $f$ is holomorphic. We stated that, conversely, every holomorphic function is analytic. We cannot establish this yet – this will be one of the aims of Chap. VII – but we can now make a step in the right direction:

**Corollary.** *Let $f$ be a function of class $C^1$ on an open subset $U$ of $\mathbb{C}$. If $f$ satisfies Cauchy's equation $D_2 f = i D_1 f$ (i.e. is holomorphic), then $f$ has a complex derivative (5) at all points $z \in U$, and*

$$(20.6) \qquad f' = D_1 f = -i D_2 f, \qquad df = f'(z)dz.$$

Indeed, the relation (4) can now be written

$$f(x+u, y+v) = f(x,y) + D_1 f(x,y)(u+iv) + o(|u|+|v|)$$

or, on putting $h = u + iv$ and $z = x + iy$,

$$f(z+h) = f(z) + D_1 f(z)h + o(h),$$

---

[41] Subject to showing that $o(u) + o(v) = o(|u|+|v|)$, which is clear, since, for all $r > 0$, the expressions on the left hand side are, for $|u|+|v|$ small, majorised by $r|u|$ and $r|v|$ respectively.

whence the existence of $f'$ and the first formula (6). The relation $df = f'dz$ is also immediate since

$$df = D_1 f.dx + D_2 f.dy = D_1 f(dx + idy) = f'dz,$$

qed.

This argument in fact shows that $f$ *is holomorphic if and only if its derivative maps in the real sense are* $\mathbb{C}$-*linear*, i.e. of the form

$$(u, v) \longmapsto c(u + iv)$$

with a constant $c \in \mathbb{C}$, and not only $\mathbb{R}$-linear: this is exactly what Cauchy's relations express.

We can express this fact in a more striking way by noting that the formulae

$$(20.7) \qquad\qquad dz = dx + idy, \qquad d\bar{z} = dx - idy$$

conversely give

$$(20.8) \qquad dx = (dz + d\bar{z})/2, \qquad dy = (dz - d\bar{z})/2i;$$

the differential $df = D_1 f dx + D_2 f dy$ of every differentiable function $f$, holomorphic or not, can thus always be put in the form

$$(20.9) \qquad\qquad df = \frac{\partial f}{\partial z} dz + \frac{\partial f}{\partial \bar{z}} d\bar{z}$$

where *by definition*

$$(20.10) \quad \partial f/\partial z = (D_1 f - i D_2 f)/2, \qquad \partial f/\partial \bar{z} = (D_1 f + i D_2 f)/2;$$

here these are pure notational conventions, not to be confused with the usual derivatives defined by passing to the limit in a quotient.

In the formula (9), $dz$ is a $\mathbb{C}$-linear function $(u, v) \mapsto c(u + iv)$, but $d\bar{z}$, of the form $(u, v) \mapsto c'(u - iv)$, is not. The holomorphic functions are thus characterised by the relation

$$\partial f/\partial \bar{z} = 0,$$

obviously equivalent to Cauchy's condition by (10). Then

$$\partial f/\partial z = f',$$

the derivative of $f$ in the complex sense.

Since the rules for calculating derivatives established in n° 15 apply to functions of several variables, with the same proofs for those who care to prove them, it is clear that *the sum, the product and the quotient of two holomorphic functions are again holomorphic*: it is enough to check that these algebraic operations do not violate Cauchy's condition. For example:

$$D_1(fg) = D_1 f.g + f.D_1 g = -i(D_2 f.g + f.D_2 g) = -iD_2(fg).$$

One may also observe that the formulae (15.8)

$$d(f+g) = df + dg, \quad d(fg) = gdf + fdg, \quad d(1/f) = -df/f^2$$

extend to functions of several variables; it is then clear that if the differentials of $f$ and $g$ are proportional to $dz$, then, similarly, so are those of $f+g$, $fg$ and $f/g$.

*Exercise.* Write a polynomial function $P$ in the form

$$P(x,y) = \sum a_{pq} z^p \bar{z}^q;$$

calculate $\partial P/\partial z$ and $\partial P/\partial \bar{z}$.

## 21 – Differentiation of composite functions

The formula for differentiating composite functions (the Chain Rule) generalises without a problem – apart from the notation – to $C^1$ functions. First consider the simplest, and very useful, case: we have a function

$$z \longmapsto g(z) = (g_1(z), g_2(z)) = g_1(z) + ig_2(z)$$

of class $C^1$ on an open $V \subset \mathbb{C}$ and a differentiable map $t \mapsto f(t) = (f_1(t), f_2(t))$ of an interval or, more generally, of an open subset[42] $U$ of $\mathbb{R}$ into $V$; we may then consider the composite function

$$p(t) = g \circ f(t) = g[f(t)] = g[f_1(t), f_2(t)],$$

with values in $\mathbb{C}$. For $t$ given and $h \in \mathbb{R}$ such that $t + h \in U$ let us put

$$f_1(t+h) = f_1(t) + u, \qquad f_2(t+h) = f_2(t) + v,$$

whence

$$u = f_1'(t)h + o(h) = ah + o(h), \qquad v = f_2'(t)h + o(h) = bh + o(h).$$

Since $g$ is differentiable at the point $f(t)$, with partial derivatives which we denote by $c$ and $d$, it follows that

$$\begin{aligned} p(t+h) &= p(t) + cu + dv + o(u) + o(v) = \\ &= p(t) + (ca + db)h + o(h) \end{aligned}$$

---

[42] Since an open set in $\mathbb{R}$ is, on a neighbourhood of each of its points, identical to an interval, everything that applies to functions defined on an interval applies to the general case.

since, $u$ and $v$ clearly being $O(h)$, with an upper case $O$, every function which is $o(u)$ or $o(v)$, with a lower case $o$, is $o(h)$. In consequence, $p$ is differentiable and

$$(21.1) \qquad p'(t) = \frac{d}{dt}g[f(t)] = D_1g[f(t)].f_1'(t) + D_2g[f(t)].f_2'(t).$$

Condensed proof: put $f(t+h) = f(t) + k$, whence

$$p(t+h) = g[f(t) + k] = p(t) + Dg[f(t)]k + o(k);$$

but $k = f'(t)h + o(h) = O(h)$, whence $o(k) = o(h)$ and

$$p(t+h) = p(t) + Dg[f(t)]f'(t)h + o(h),$$

which gives (1) again, in the form

$$(21.1') \qquad\qquad\qquad p'(t) = Dg[f(t)]f'(t),$$

the image of the vector $f'(t) \in \mathbb{R}^2$ under the linear map $Dg[f(t)] : \mathbb{R}^2 \to \mathbb{R}^2$ tangent to $g$ at the point $z = f(t)$.

In differential notation: the differential $dp = p'(t)dt$ of $p$ is obtained starting from the differential $dg = D_1g(z)dx + D_2g(z)dy$ of $g$, substituting $f(t)$ for $z = (x,y)$ in the derivatives, and replacing $dx$ and $dy$ by the differentials of the functions $f_1(t)$ and $f_2(t)$ that have been substituted for $x$ and $y$. The analogy with Leibniz' system is complete.

*Example 1.* If $g$ is *holomorphic* and if one considers $f$ as a function with values in $\mathbb{C}$ rather than in $\mathbb{R}^2$, so that $f = f_1 + if_2$, one finds

$$(21.2) \qquad\qquad\qquad \frac{d}{dt}g[f(t)] = g'[f(t)]f'(t)$$

since the relation $D_2g = iD_1g$ allows one to exhibit

$$(21.3) \qquad\qquad\qquad f_1'(t) + if_2'(t) = f'(t)$$

as a factor of the right hand side of (1); one could also use the fact that, for a holomorphic function, the map $Df(z)$ consists of multiplying each vector $h \in \mathbb{R}^2 = \mathbb{C}$ by the complex number $f'(z)$.

In terms of differentials: in $dg = g'(z)dz$, replace $z$ by $f(t)$ and $dz$ by $df = f'(t)dt$. One should pay attention to the fact that $f'$ is a derivative in the real sense of n° 14, while $g'$ is a derivative in the complex sense. The reader can also prove (2) directly, starting from the usual definition of $g'$ as the limit of a quotient.

*Example 2.* Suppose that $f_1(t) = x + tu$, $f_2(t) = y + tv$, i.e. $f(t) = z + th$, so that one is examining the behaviour of $g$ along a line passing through $(x,y) = z$. One finds

(21.4)    $\dfrac{d}{dt}g(x+tu,y+tv) =$

$$= D_1g(x+tu,y+tv)u + D_2g(x+tu,y+tv)v$$

or, in condensed notation,

(21.4')    $$\dfrac{d}{dt}g(z+th) = Dg(z+th)h.$$

Suppose that $z+th \in V$ for any $t \in [0,1]$; this means that the line segment joining the points $z$ and $z+h$ lies entirely in $V$, as is the case if $|h|$ is sufficiently small or, as well, if $V$ is *convex*. Let us apply Theorem 18, the mean value theorem, and its corollaries to the function $p(t) = g(x+tu,y+tv)$, in the form $p(1)-p(0) = \ldots$; Theorem 18 shows that if $g$ has *real* values, then there exists a $t \in [0,1]$ such that

(21.5)    $g(x+u,y+v) - g(x,y) =$

$$= D_1g(x+tu,y+tv)u + D_2g(x+tu,y+tv)v,$$

(21.5')    $$g(z+h) - g(z) = Dg(z+th)h;$$

if $g$ has complex values one finds only that

(21.6)    $|g(x+u,y+v) - g(x,y)| \le$

$$\le \sup_{0\le t\le 1} |D_1g(x+tu,y+tv)u + D_2g(x+tu,y+tv)v|,$$

or

(21.6')    $$|g(z+h) - g(z)| \le \sup|Dg(z+th)h|.$$

Now every linear map $A$ of $\mathbb{R}^2$ into $\mathbb{R}^2$ (or of $\mathbb{R}^p$ into $\mathbb{R}^q$) has a *norm* $\|A\|$, namely the least number $M \ge 0$ such that $\|Ah\| \le M\|h\|$ for all vectors $h$ in the domain space; if one uses the usual Pythagorean norm, and if

(21.7)    $$A\begin{pmatrix} u \\ v \end{pmatrix} = \begin{pmatrix} au + bv \\ cu + dv \end{pmatrix},$$

so that $a$, $b$, $c$, $d$ are the coefficients of the matrix of $A$, one finds easily, from the Cauchy-Schwarz inequality in $\mathbb{R}^2$, that

(21.8)    $$\|A\| = (a^2 + b^2 + c^2 + d^2)^{\frac{1}{2}} \asymp |a| + |b| + |c| + |d|.$$

If $A = Dg(z)$ as above, the (real) coefficients of the matrix of $A$ are the real and imaginary parts of $D_1g(z)$ and $D_2g(z)$, whence

(21.9)    $$\|Dg(z)\|^2 = |D_1g(z)|^2 + |D_2g(z)|^2.$$

In any case, the formulae used to measure the "length" of a vector and thus the norm of a linear map or matrix are of little importance, so long as they are "equivalent" to the Pythagorean norm (i.e. that the ratio between the two measures of the length of $h$ should, for any $h$, lie between fixed constants $> 0$), since one uses formulae of the type (6') only to estimate *orders of magnitude* and not for totally exact and explicit calculations.

One can write (6') in the practically as useful form

$$(21.10) \qquad |g(z+h) - g(z)| \leq |h|. \sup \|Dg(z+th)\|.$$

As in Corollary 3 of Theorem 18, (6) allows one to estimate the left hand side in terms of bounds of the derivatives. If for example one confines oneself to examining what happens on a set $K \subset V$ which is both compact and convex, and if one puts $z = (x,y) = x + iy$, $z' = (x+u, y+v)$, one obtains the inequality

$$(21.11) \qquad |g(z') - g(z)| \leq \|Dg\|_K . |z' - z|$$

for all $z, z' \in K$, where

$$\|Dg\|_K = \sup_{z \in K} \|Dg(z)\| = \sup(|D_1 g(z)|^2 + |D_2 g(z)|^2)^{\frac{1}{2}},$$

adopting the standard formula for measuring lengths in $\mathbb{R}^2$.

From this one deduces that *if* $Dg(z) = 0$ *for all* $z \in V$, the open set where $g$ is defined, then $g(z') = g(z)$ so long as $z$ and $z'$ are close enough for $V$ to contain the line segment $[z, z']$ and that therefore $g$ *is constant if $V$ is connected*, i.e. if any two points of $V$ can be joined one to the other by a broken line entirely contained in $V$ (Chap. II, n° 20).

Again for $g$ with complex values, the relation (16.9) applied to the function $t \mapsto g(x+tu, y+tv)$ between 0 and 1 yields the inequality

$$(21.12) \qquad |g(x+u, y+v) - g(x,y) - [D_1 g(x,y)u + D_2 g(x,y)v]| \leq$$
$$\leq \sup_{0 \leq t \leq 1} \big| [D_1 g(x+tu, y+tv) - D_1 g(x,y)] u +$$
$$+ [D_2 g(x+tu, y+tv) - D_2 g(x,y)] v \big|$$

or, in condensed notation,

$$(21.12') \quad |g(z+h) - g(z) - Dg(z)h| \leq \sup |Dg(z+th)h - Dg(z)h|$$
$$\leq |h|. \sup \|Dg(z+th) - Dg(z)\|.$$

These relations are as fundamental as the mean value theorem and its corollaries for functions of one variable.

Now consider the more complicated case, where, instead of replacing the variables $x$ and $y$ by functions of a single real variable, one replaces them by functions of two variables. This means that one starts from an open set $U \subset \mathbb{C}$ and a map

$$f : (x,y) \longmapsto (f_1(x,y), f_2(x,y))$$

of class $C^1$ from $U$ into $V$, whence a new composite function

$$p(x,y) = g\left[f_1(x,y), f_2(x,y)\right]$$

defined on $U$ or, in condensed notation, $p(z) = g[f(z)]$. It is easy to see that $p$ is $C^1$, and immediate to calculate its derivatives. These are obtained by letting $x$ vary while $y$ remains constant, and *vice versa*; one finds oneself again in the simplest situation just explained. On applying (1) either to $x \mapsto f(x,y)$, or to $y \mapsto f(x,y)$, one then obtains

$$(21.13) \qquad \begin{aligned} D_1 p(z) &= D_1 g[f(z)] D_1 f_1(z) + D_2 g[f(z)] D_1 f_2(z), \\ D_2 p(z) &= D_1 g[f(z)] D_2 f_1(z) + D_2 g[f(z)] D_2 f_2(z); \end{aligned}$$

these formulae show that the derivatives of $p$ are continuous, qed.

One can also start again from scratch using condensed notation as in (12'). For $z \in U$ given and $h \in \mathbb{R}^2$ small, Theorem 21 shows that

$$\begin{aligned} f(z+h) &= f(z) + Df(z)h + o(h) = f(z) + k, \\ p(z+h) &= g[f(z) + k] = p(z) + Dg[f(z)]k + o(k). \end{aligned}$$

Since $k = Df(z)h + o(h) = O(h) + o(h) = O(h)$, everything that is $o(k)$ is also $o(h)$, and, on the other hand,

$$Dg[f(z)]k = Dg[f(z)]Df(z)h + Dg[f(z)]o(h) = Dg[f(z)]Df(z)h + o(h).$$

Consequently $p$ is differentiable, with

$$Dp(z)h = Dg[f(z)]Df(z)h,$$

the value of the linear map $Dg[f(z)]$ at the vector $Df(z)h$, itself the value of the linear map $Df(z)$ at the vector $h$. In other words,

$$(21.13') \qquad Dp(z) = Dg[f(z)] \circ Df(z),$$

the *composition* of the linear maps $Dg[f(z)]$ and $Df(z)$; this is absolutely the same proof – and the same formula – as for rule (D 4) of n° 15. One often omits the sign $\circ$ when dealing with linear maps, so writes (13) in the form

$$Dp(z) = Dg[f(z)]Df(z).$$

We can of course make all this explicit in terms of $f_1, \ldots, g_2$. We need only know how to calculate the matrix of a product of linear maps as a function of those of the factors. Whence, here,

$$
(21.14) \qquad \begin{pmatrix} D_1p_1 & D_2p_1 \\ D_1p_2 & D_2p_2 \end{pmatrix} =
$$

$$
= \begin{pmatrix} D_1g_1 & D_2g_1 \\ D_1g_2 & D_2g_2 \end{pmatrix} \begin{pmatrix} D_1f_1 & D_2f_1 \\ D_1f_2 & D_2f_2 \end{pmatrix} =
$$

$$
= \begin{pmatrix} D_1g_1D_1f_1 + D_2g_1D_1f_2 & D_1g_1D_2f_1 + D_2g_1D_2f_2 \\ D_1g_2D_1f_1 + D_2g_2D_1f_2 & D_1g_2D_2f_1 + D_2g_2D_2f_2 \end{pmatrix}
$$

where the partial derivatives are calculated at the values of the variables indicated in (13). In the form (13') the argument and the result extend to any number of variables and even to functions defined on, or with values in, Banach spaces of infinite dimension, spaces where the purely algebraic concept of coordinates with respect to a "base" is unknown or, more exactly, pathological (the coordinates of a vector may even then not be continuous functions of the vector).

*Example 3.* Suppose that $g$ is holomorphic and regard $f$ as a complex-valued function. Using Cauchy's relation (20.6) one finds

$$
D_1p(z) = g'[f(z)]D_1f(z), \qquad D_2p(z) = g'[f(z)]D_2f(z).
$$

If, further, $f : U \to V$ is itself holomorphic, i.e. satisfies $D_2f = iD_1f$, one clearly finds the same relation between the derivatives of $p$. If $f$ and $g$ are holomorphic, their tangent maps are $\mathbb{C}$-linear; their composition is then forced to be so too. In consequence:

**Theorem 22.** *If $f : U \to \mathbb{C}$ and $g : V \to U$ are holomorphic, then the composite function $p = g \circ f : U \to \mathbb{C}$ is holomorphic, and*

$$
(21.15) \qquad p'(z) = g'[f(z)]f'(z),
$$

in other words, just the same formula as in $\mathbb{R}$, already obtained in Chap. II, n° 22, Theorem 17, for analytic functions. Here $f'(z)$, $g'(f(z))$ and $p'(z)$ are derivatives of holomorphic functions, i.e. *complex numbers*.

## 22 - Limits of differentiable functions

Consider a sequence of functions $f_n$ of class $C^1$ on an open subset $U$ of $\mathbb{C}$. Suppose that the two derived sequences $(D_1f_n)$ and $(D_2f_n)$ converge uniformly on every compact $K \subset U$ to limits $g_1$ and $g_2$, necessarily continuous. If one wants to generalise the result from the theory of one variable (Theorem 19), one will have at least to assume the existence of $\lim f_n(a, b)$ for some $(a, b) \in U$, and even to assume $U$ *connected*[43]; to simplify the proof, we

---

[43] because if $U$ is the union of two open *disjoint* nonempty $U'$ and $U''$, then what happens at a point of $U'$ can have no influence on what happens in $U''$; the same problem as with analytic continuation in Chap. II, n° 19.

shall assume that the sequence $(f_n)$ converges *simply* on $U$, which makes the connectedness hypothesis on $U$ redundant, and, in practice, is always verified from the start. With these hypotheses:

**Theorem 23.** *Let $(f_n)$ be a sequence of functions of class $C^1$ on an open subset $U$ of $\mathbb{R}^2$. Suppose that (i) the $f_n$ converge simply on $U$ to a limit function $f$, (ii) the derivatives $D_1 f_n$ and $D_2 f_n$ converge to limit functions $g_1$ and $g_2$ uniformly on every compact $K \subset U$. Then $f$ is of class $C^1$ and $Df = (g_1, g_2)$, i.e. $D_1 f = g_1$ and $D_2 f = g_2$, or*

$$D_i(\lim f_n) = \lim D_i f_n \qquad (i = 1, 2).$$

We mimic the proof of Theorem 19, introducing the functions $f_{pq} = f_p - f_q$. Let us work on a compact disc $K \subset U$. Since $K$ is convex we have

$$(22.1) \qquad |f_{pq}(z') - f_{pq}(z)| \leq \|Df_{pq}\|_K |z' - z|$$

on $K$, by (21.11). But $Df_{pq} = Df_p - Df_q$, and since, by hypothesis, the derivatives converge uniformly[44] on $K$, the uniform norms appearing on the right hand side of (1) are $\leq r$ for $p$ and $q$ large, hence so is the left hand side since $|z' - z|$ is majorised by the diameter of $K$. Since $\lim f_n(z')$ exists for all $z'$, for example at the centre $a$ of $K$, one also has $|f_{pq}(a)| \leq r$ for $p$ and $q$ large, whence, by the usual process of dividing $\varepsilon$ in four, an inequality

$$|f_{pq}(z)| \leq r \qquad \text{for all } z \in K$$

once $p, q > N$. This means that $\|f_p - f_q\|_K \leq r$, whence *uniform* convergence of $f_n$ on $K$, by Cauchy's criterion. We extend to the case of an arbitrary compact set by an argument which will be expounded in Chap. V, n° 6 (Corollary 2 of the Borel-Lebesgue Theorem[45]) and which presupposes no more than the definition of open and compact sets.

It remains to show that we obtain the derivatives of $f = \lim f_n$ by taking the limits of those of $f_n$. Choose an $(a, b) \in U$ and consider the functions $x \mapsto f_n(x, b)$. They are defined on an open subset of $\mathbb{R}$, are $C^1$, and converge to $x \mapsto f(x, b)$; their derivatives $D_1 f_n(x, b)$ converge uniformly on every compact set to the function $x \mapsto g_1(x, b)$ since the point $(x, b)$ describes a compact set in $\mathbb{C}$ when $x$ runs through a compact set in $\mathbb{R}$. Theorem 19 now assures

---

[44] The values of the $Df_p$ are linear maps from $\mathbb{R}^2$ into $\mathbb{R}^2$, but one can clearly still speak of uniform convergence since the "norm" of a linear map allows one to measure the "distance" between two such maps: $d(A, B) = \|A - B\|$. Equivalently, one can reason with the coefficients of Jacobian matrices.

[45] namely: for a sequence of functions defined on $U$ to converge uniformly on all compacta $K \subset U$, it is (necessary and) sufficient that, for all $a \in U$, there exists a disc $D \subset U$ with centre $a$ on which the sequence converges uniformly: *local* character of compact convergence.

us that $D_1 f(x, b) = g_1(x, b)$. Same argument switching the rôles of $x$ and $y$. The function $f$ is thus $C^1$ since $g_1$ and $g_2$ are $C^0$, etc.

One can also start from the relation (21.12') applied to the $f_n$. If one remains in a compact disc $K \subset U$ this shows that

$$(22.2) \qquad |f_n(z + h) - f_n(z) - Df_n(z)h| \le$$
$$\le |h| \cdot \sup_{0 \le t \le 1} \|Df_n(z + th) - Df_n(z)\| ;$$

now the linear maps $Df_n$ converge to the linear map $g$. If one could deduce from (2), by passing to the limit, that

$$(22.3) \qquad |f(z + h) - f(z) - g(z)h| \le |h| \cdot \sup \|g(z + th) - g(z)\|,$$

then differentiability of $f$ and the values of its derivatives would follow, since, $g(z)$ being continuous, the sup is $o(1)$ and the right hand side is $o(h)$.

So there remains the passage to the limit in (2). This poses no problem for the left hand side. As to the right hand side, for given $x$, $y$ and $h$ we have a function $\varphi_n(t) = \|Df_n(z + th) - Df_n(z)\| \ge 0$ defined on $[0, 1]$ and converging to a limit $\varphi(t) = \|g(z+th)-g(z)\|$; the $\varphi_n$ even converge *uniformly* on $[0, 1]$ because of the hypotheses on the derivatives of $f_n$. It is therefore enough to establish the following result:

**Lemma.** *Let $(\varphi_n)$ be a sequence of real functions defined and bounded above on a set $X$; suppose that the $\varphi_n$ converge* uniformly *on $X$ to a limit $\varphi$. Then $\varphi$ is bounded above on $X$ and*

$$(22.4) \qquad \sup_{x \in X} \varphi(x) = \lim_{n \to \infty} \sup_{x \in X} \varphi_n(x).$$

For $r > 0$ given and $n > N(r) = N$, one has $\varphi(x) \le r + \varphi_n(x)$ *for any* $x \in X$. In consequence,

$$n > N \implies \sup_{x \in X} \varphi(x) \le r + \sup_{x \in X} \varphi_n(x)$$

since the right hand side majorises *all* the values of the function $r + \varphi_n$, so also those of $\varphi$. This already shows that $\varphi$ is bounded above, like the $\varphi_n$. Uniform convergence also shows that $\varphi_n(x) \le r + \varphi(x)$ for any $x$ for $n$ large. If then $N$ is sufficiently large, one finds also that

$$n > N \implies \sup_{x \in X} \varphi_n(x) \le r + \sup_{x \in X} \varphi(x).$$

Combining these two results gives

$$\left| \sup_{x \in X} \varphi(x) - \sup_{x \in X} \varphi_n(x) \right| \le r$$

for $n$ large, qed.

Hence the passage to the limit which transforms (2) into (3).

We leave it to the reader to extend this limit theorem for derivatives to the case of arbitrary functions of class $C^p$.

Finally we examine the case of holomorphic functions. If the $f_n$ are holomorphic in $U$, it suffices to assume compact convergence of the complex derivatives $f'_n$ to a limit $g$. It is clear that Cauchy's formulae then also apply to the limit $f$ of $f_n$, which is thus holomorphic and satisfies $f' = g$. Despite appearances, this result is of no interest, because, when dealing with *holomorphic* or analytic functions, *the compact convergence of $f_n$* (and even much less) *implies that of the derivatives*, as we shall see in Chap. VII; one uses this "miraculous" result constantly. It generalises, not without a little more work – say a century to obtain the general case starting from the particular case of holomorphic or harmonic functions – to the solutions of the much larger class of "elliptic" linear partial differential equations of arbitrary order.

In all cases, it is clear that if the $f_n$ satisfy a partial differential equation of some reasonable kind, for example

$$\left( f''_{xx} - f''_{yy} \right)^2 + f'_x = \sin x + \log y,$$

and if the derivatives of order $\leq 2$ converge, then the limit function again satisfies the same equation. But in a case of this kind, the convergence of solutions does not imply that of their derivatives nor that the limit is again a solution, unless one adopts the point of view of Schwartz' theory of distributions, an artifice which, despite its usefulness, neither permits one to prove false theorems nor exempts one from proving true theorems.

## 23 – Interchanging the order of differentiation

Consider a holomorphic function $f$, and assume that it is of class $C^2$ – hardly a restrictive hypothesis since we shall learn that $f$ is analytic and so $C^\infty$ –, so that its derivative $f'$ is of class $C^1$. If $f$ is analytic, so also is $f'$, and it seems to follow that $f'$ is holomorphic. Might one prove this directly from Cauchy's relation?

Since $f' = D_1 f = -iD_2 f$ and so

$$D_1 f' = D_1^2 f = -iD_1 D_2 f, \qquad D_2 f' = D_2 D_1 f,$$

Cauchy's relation $D_1 f' = -iD_2 f'$ for $f'$ reduces to $D_1 D_2 f = D_2 D_1 f$. It is also satisfied trivially by the nonholomorphic function $x^p y^q$, for example.

Hence arises the general problem of interchanging the order of differentiation

(23.1) $$D_1 D_2 f = D_2 D_1 f$$

where, of course, one no longer supposes $f$ holomorphic. To establish this fundamental relation when it is true, assume for a start that $f$ is of class $C^1$

– which clearly will not be enough – on the open set $U$ where it is defined. For $(a, b)$ given, consider again, this time in the context where Euler used it, the expression

$$(23.2) \qquad [f(a + u, b + v) - f(a + u, b)] - [f(a, b + v) - f(a, b)]$$

where $|u|$ and $|v|$ remain so small that, in all that follows, one needs to consider values of $f$ only on a compact disc $K \subset U$ with centre $(a, b)$. For $a$, $b$, $u$ and $v$ given, (2) can be written $g(1) - g(0)$, where

$$(23.3) \qquad\qquad g(t) = f(a + tu, b + v) - f(a + tu, b).$$

By the results of n° 16

$$(23.4) \qquad\qquad |g(1) - g(0) - g'(0)| \leq \sup |g'(t) - g'(0)|$$

where the sup is taken over $t \in [0, 1]$. But (chain rule)

$$\begin{aligned} g'(t) \;&=\; D_1 f(a + tu, b + v)u - D_1 f(a + tu, b)u = \\ &=\; [D_1 f(a + tu, b + v) - D_1 f(a, b)]u - [D_1 f(a + tu, b) - D_1 f(a, b)]u. \end{aligned}$$

If one assumes that $D_1 f$ is *differentiable* at the point $(a, b)$, then the first difference between [ ] is, by definition, equal to

$$D_1 D_1 f(a, b)tu + D_2 D_1 f(a, b)v + o(|tu| + |v|);$$

the second difference between [ ] is similarly equal to

$$D_1 D_1 f(a, b)tu + o(|tu|).$$

So we have

$$g'(t) = D_2 D_1 f(a, b)uv + o(|tu| + |v|)u,$$

and in particular

$$g'(0) = D_2 D_1 f(a, b)uv + o(|v|)u,$$

whence $|g'(t) - g'(0)| = o(|tu| + |v|)u$. Since

$$(|tu| + |v|)|u| \leq (|u| + |v|)^2,$$

it follows that

$$\sup |g'(t) - g'(0)| = o[(|u| + |v|)^2].$$

The relation (4) can then be written

$$\begin{aligned} g(1) - g(0) \;&=\; [D_1 f(a, b + v) - D_1 f(a, b)]u + o[(|u| + |v|)^2] = \\ &=\; [D_2 D_1 f(a, b)v + o(v)]u + o[(|u| + |v|)^2]; \end{aligned}$$

since $|uv| \leq (|u|^2 + |v|^2)/2$ we get

(23.5)  $\qquad (2) = D_2 D_1 f(a,b) uv + o[(|u| + |v|)^2].$

But instead of calculating with the function (3), one may also, if $D_2 f$ is differentiable at $(a,b)$, use the function

$$h(t) = f(a+u, b+tv) - f(a, b+tv).$$

Then one finds

(23.6)  $\qquad (2) = D_1 D_2 f(a,b) uv + o[(|u| + |v|)^2].$

Comparing (5) and (6), we see that

$$[D_2 D_1 f(a,b) - D_1 D_2 f(a,b)] uv = o[(|u| + |v|)^2];$$

one can apply this result to $u = tu_0$, $v = tv_0$ where $u_0$ and $v_0$ are nonzero constants and where $t$ tends to 0; then one finds that

$$[D_2 D_1 f(a,b) - D_1 D_2 f(a,b)] u_0 v_0 t^2 = (|u_0|^2 + |v_0|^2) o(t^2),$$

and deduces the interchangeability of the order of differentiation on dividing the two sides by $t^2$ and making $t$ to 0, which yields a right hand side tending to 0, qed.

The hypotheses used in the proof[46], namely that *the functions $D_1 f$ and $D_2 f$ are differentiable at $(a,b)$*, are in particular satisfied if $f$ is of class $C^2$, by far the most important case in applications. In general, the subtleties that one meets in the case of a single real variable rarely generalise, or, when they do, often involve far too great a *cost/efficiency* ratio to motivate the mathematicians. The important problems posed in the theory of differentiable functions of several variables (differential topology for example) are of another nature.

The most obvious consequence – and the most important – of (1) is that, *if $f$ is of class $C^n$, $n \geq 2$, then every partial derivative of $f$ of order $\leq n$ can be written in the form $D_1^p D_2^q f$ with $p + q \leq n$*; it is unnecessary to write expressions such as $D_1^3 D_2^2 D_1 D_2^4 D_1^2 f$.

Another consequence: *if a function $f$ of class $C^\infty$ is holomorphic, then so similarly are its successive derivatives $f', f'' = (f')'$, etc.,* which are simply $D_1 f$, $D_1^2 f$, etc. So in this case

$$D_1^p D_2^q f = i^q f^{(p+q)}.$$

Since we shall eventually learn that a holomorphic function is analytic, these results are not particularly sensational.

---

[46] which I have borrowed from Dieudonné's *Treatise on Analysis*, Vol. 1, Chap. VIII, n° 12; he never cites his sources, in this case W.H. Young, beginning of the XX$^{\text{th}}$ century. It goes without saying that, in Dieudonné, one is working in Banach spaces, which, strictly speaking, does not change the proof at all. You will find a very different proof in Chap. V, n° 12, Theorem 14; but this uses the integral calculus and assumes the existence *and continuity* of $D_2 D_1 f$ and $D_1 D_2 f$.

## 24 – Implicit functions

The last classical results, and palpably more difficult to establish than the preceding ones, are the "local inversion" or "implicit function" theorems. In the second case, one aims to show that, given a *real* function $F$ of class $C^1$ on an open set $U \subset \mathbb{R}^2$, and a point $(a, b) \in U$ where $F(a, b) = 0$, then – subject to an apparently inoffensive condition – one can find a real function $y = f(x)$ of class $C^1$ on a neighbourhood of $a$, with $f(a) = b$ and such that

$$(24.1) \qquad\qquad F[x, f(x)] = 0$$

on a neighbourhood of $a$. In the first case, one is given a function $f : G \longrightarrow \mathbb{R}^2$ of class $C^1$ on an open set $G \subset \mathbb{R}^2$ and tries to provide an "inverse" map, at least on a neighbourhood of a given point $(a, b)$.

It is generally a good idea to think before attempting a proof. In the simplest case, $F(x, y) = x - g(y)$ where $g$ is real and of class $C^1$, with $a = g(b)$; the solution $f$ of (1) must then satisfy $x - g[f(x)] = 0$, so that the problem consists of constructing an inverse map of $g$, at least *locally*, the one -variable version of the local inversion problem. Clearly there cannot be a solution unless $g$ is injective on a neighbourhood of $b$. By n° 4, Theorem 7 bis, this forces $g$ to be strictly monotone on a neighbourhood of $b$, and so its derivative $g'(y)$ must have constant sign. If $f$ is differentiable then $g'[f(x)].f'(x) = 1$, which again forbids $g'$ from vanishing on a neighbourhood of $b$, so forces it to be either everywhere $> 0$, or everywhere $< 0$ on a neighbourhood of $b$; since $g'$ is continuous, it is enough that $g'(b) \neq 0$. In this case, $g$ is strictly monotone on an open interval $I$ with centre $b$, so admits an inverse $f$ on $J = f(I)$, an open interval containing $a = g(b)$, and rule (D 5) of n° 14 shows that it is of class $C^1$, since $f'(x) = 1/g'[f(x)]$, the reciprocal of a continuous function everywhere $\neq 0$. We emphasis the fact that, if one knows only that $g'(b) \neq 0$, these arguments are valid only *on a neighbourhood of b* – to be precise, on the largest open interval containing $b$ where $g'$ does not vanish. If for example one considers the function $g(x) = x^3$ on $\mathbb{R}$, which maps $\mathbb{R}$ bijectively onto $\mathbb{R}$, it admits a $C^1$ inverse on the interval $x > 0$ (namely $x^{1/3}$) or on $x < 0$ (namely $-|x|^{1/3}$), but not on any neighbourhood of $0$ where the graph of the inverse map has a vertical tangent. One can clearly not hope for more when one progresses to functions of several variables.

In the general case of the equation (1), the existence of a continuous derivative for the desired solution $f$ would entail

$$(24.2) \qquad\qquad D_1 F[x, f(x)] + D_2 F[x, f(x)]f'(x) = 0$$

by (21.1) applied to $F[x, f(x)]$. If $D_2 F$ vanishes at $(a, b)$ but $D_1 F$ does not, there is again no hope of obtaining a derivative $f'(a)$ even under the hypothesis that $f$ exists. This is what happens in the simplest cases, for example that of the function $F(x, y) = x^2 + y^2 - 1$ at the point $(a, b) = (1, 0)$; the two obvious solutions, namely $f(x) = \pm(1 - x^2)^{\frac{1}{2}}$, are not defined on any *neigh-bourhood* of the point $a = 1$ - only for $-1 \leq x \leq 1$ - and, even worse, are not differentiable at $x = 1$: their graphs, i.e. the upper and lower semicircles, here again, have vertical tangents at this point. In this example, the situation is on the contrary excellent at all points $(a, b)$ where $D_2F(a, b) = 2b \neq 0$, i.e. for $-1 < a < 1$. If $b > 0$, the formula

$$y = (1 - x^2)^{\frac{1}{2}}$$

defines a $C^1$, and even $C^\infty$, function on $]-1, 1[$, which satisfies (1); if $b < 0$, one changes the sign of $y$. It is thus prudent to work at a point $(a, b)$ where

(24.3) $$F(a, b) = 0, \qquad D_2F(a, b) \neq 0$$

simultaneously.

In the first case (existence of a local inverse of a map $f$ from an open subset $G$ of $\mathbb{C}$ into $\mathbb{C}$), if one puts $f = (f_1, f_2)$ with $f_1$ and $f_2$ real, one has to show – modulo some hypotheses ... – that the map

(24.4) $$(x, y) \longmapsto (f_1(x, y), f_2(x, y))$$

of $G$ into $\mathbb{C} = \mathbb{R}^2$ admits an inverse map

$$(u, v) \longmapsto g(u, v) = (g_1(u, v), g_2(u, v))$$

on an open neighbourhood $U$ of a given point $c = (a, b)$, and, precisely, that
(i) $f$ maps $U$ *bijectively* onto an *open* neighbourhood $V$ of the point $f(c)$,
(ii) the inverse map $g : V \to U$ is $C^1$, exactly as in the case of functions of one real variable. If one assumes the problem solved, then the relation $g[f(z)] = z$ shows, by the chain rule (21.13'), that

(24.5) $$Dg[f(z)] \circ Df(z) = 1,$$

on $U$, where $1 = id$ is the identity map $h \mapsto h$, the derivative of the identity map $z \mapsto z$. It is therefore necessary for $Df(z)$ to be *invertible* at all $z \in U$, and in particular at the point $c = (a, b)$ in question.

This condition can be made more explicit by replacing the linear maps figuring in (5) with their Jacobian matrices:

(24.6) $$\begin{pmatrix} D_1g_1 & D_2g_1 \\ D_1g_2 & D_2g_2 \end{pmatrix} \begin{pmatrix} D_1f_1 & D_2f_1 \\ D_1f_2 & D_2f_2 \end{pmatrix} = \begin{pmatrix} 1 & 0 \\ 0 & 1 \end{pmatrix}$$

or

(24.6') $$D_1g_1.D_1f_1 + D_2g_1.D_1f_2 = 1, \quad D_1g_2.D_1f_1 + D_2g_2.D_1f_2 = 0,$$
(24.6") $$D_1g_1.D_2f_1 + D_2g_1.D_2f_2 = 0, \quad D_1g_2.D_2f_1 + D_2g_2.D_2f_2 = 1,$$

the derivatives of $f_i$ being calculated at the point $c$ and those of $g_i$ at the point $f(c)$.

Now for a matrix

$$\begin{pmatrix} a & b \\ c & d \end{pmatrix}$$

to be invertible, it is necessary and sufficient that its determinant $ad - bc$ be $\neq 0$. In the present case, i.e. of the Jacobian matrix (19.4) of $f$, this is the expression $D_1 f_1(z) D_2 f_2(z) - D_2 f_1(z) D_1 f_2(z)$; one calls this the *Jacobian* of the map $f$ at the point $z = (x, y)$, from the name of its inventor (formula for change of variables in a multiple integral), notation $J_f(z)$, or the *functional determinant* of the pair of functions $f_1, f_2$, denoted classically by

$$(24.7) \qquad \frac{D(f_1, f_2)}{D(x, y)} = D_1 f_1(x, y) D_2 f_2(x, y) - D_2 f_1(x, y) D_1 f_2(x, y) = J_f(z).$$

Suppose for example that $f_1 + i f_2 = f$ is holomorphic. Cauchy's relations

$$f' = D_1 f_1 + i D_1 f_2 = -i(D_2 f_1 + i D_2 f_2) = D_2 f_2 - i D_2 f_1$$

signify precisely that the second matrix (6) is of the form

$$(24.8) \qquad \begin{pmatrix} a & b \\ -b & a \end{pmatrix},$$

so that (7) is

$$(24.9) \qquad [D_1 f_1(z)]^2 + [D_2 f_1(z)]^2 = |f'(z)|^2.$$

Since the inverse of (8) is

$$\begin{pmatrix} a' & b' \\ -b' & a' \end{pmatrix} \quad \text{where } a' = a/(a^2 + b^2), \quad b' = -b/(a^2 + b^2),$$

the first matrix figuring in the left hand side of (6) is then also of the form (8), which means that, if it exists, the function $g_1 + i g_2$ satisfies Cauchy's conditions, i.e. is holomorphic like $f$. *Inverting a holomorphic function leads to a holomorphic function* if what we hope is indeed correct.

After these explanations, we shall first examine the problem of local inversion – it will provide the solution of the other almost *gratis*, as we shall see – and establish the following result:

**Theorem 24 ("local inversion").** *Let $G$ be an open subset of $\mathbb{R}^2$ and let $f$ be a map of class $C^1$ of $G$ into $\mathbb{R}^2$. Suppose that the derivative map $Df(c)$ is invertible at a point $c \in G$. Then there is an open $U \subset G$, containing $c$, which $f$ maps bijectively onto an open $V$ and such that the inverse map $g : V \to U$ is of class $C^1$ on $V$.*

One can clearly assume that the point $c \in \mathbb{C}$ considered is the origin $(0,0)$ and that $f$ maps it to $(0,0)$. Since the linear map $Df(0) = A$ is by hypothesis invertible, one can replace $f$ by $A^{-1} \circ f$, the composition of $f$ and of the linear map $A^{-1}$. The linear tangent map to the new function at a point $z \in G$ is clearly $A^{-1}Df(z)$, the product or composition of two linear maps, so is the identity map when $z = 0$. If one can invert the function $z \mapsto A^{-1}f(z)$ locally there will exist a function $g$ such that $A^{-1}f[g(z)] = z$ on a neighbourhood of $0$, whence $f[g(z)] = Az$. Since the invertible linear map $A$ transforms every neighbourhood of $0$ into a neighbourhood of $0$, one has $f\left[g\left(A^{-1}z\right)\right] = z$ on a neighbourhood of $0$, so that the required local inverse of $f$ is the function $z \mapsto g\left(A^{-1}z\right)$.

We may thus reduce to inverting $f$ under the hypothesis that $f(0) = 0$ and $Df(0) = 1$. The proof divides into several parts, the essential argument being that which allowed us to solve an equation of the form $f(x) = x$ in $\mathbb{R}$ (Chap. II, n° 16).

(a) Putting $f(z) = z + p(z)$, one has $p(0) = 0$, $Dp(0) = 0$; since $p$ is $C^1$ there is thus an $r > 0$ such that

$$(24.10) \qquad |z| \leq r \Longrightarrow \|Dp(z)\| \leq \frac{1}{2},$$

so that (inequality of the mean)

$$(24.11) \quad \{|z'| \leq r \ \& \ |z''| \leq r\} \implies |p(z') - p(z'')| \leq \frac{1}{2}|z' - z''|$$
$$\implies |f(z') - f(z'')| \geq \frac{1}{2}|z' - z''|.$$

We conclude that the map $f$ is *injective* on the closed ball $B(r) : |z| \leq r$, and that the inverse map

$$g : f(B(r)) \longrightarrow B(r)$$

is *continuous*: it satisfies $|g(\zeta') - g(\zeta'')| \leq 2|\zeta' - \zeta''|$.

(b) Let us show that $f(B(r)) \supset B(r/2)$, i.e. that the equation $\zeta = f(z) = z + p(z)$ has a, necessarily unique, solution $z \in B(r)$, for all $\zeta$ such that $|\zeta| \leq r/2$. To do this we write $\zeta - p(z) = z$ and apply the *method of successive approximations*:

$$z_0 = 0, \quad z_1 = \zeta - p(z_0) = \zeta, \quad z_2 = \zeta - p(z_1) = \zeta - p(\zeta), \ldots .$$

First we have to verify that the construction can be continued indefinitely without leaving the ball $B(r)$. But, by (10) or (11), it is clear that

$$\{|\zeta| \leq r/2 \ \& \ |z| \leq r\} \Longrightarrow |\zeta - p(z)| \leq |\zeta| + |p(z)| \leq r/2 + r/2 = r.$$

So once we have $z_0, z_1, \ldots, z_n \in B(r)$ we find $z_{n+1} = \zeta - p(z_n) \in B(r)$: there are no obstructions.

This established, the inequalities

$$|z_{m+1} - z_m| = |p(z_m) - p(z_{m-1})| \le \frac{1}{2} |z_m - z_{m-1}| \le \dots,$$

show that

$$|z_{m+1} - z_m| \le |z_1 - z_0| / 2^m = |\zeta| / 2^m,$$

whence, for $p < q$,

$$|z_q - z_p| = |z_q - z_{q-1} + \dots + z_{p+1} - z_p| \le |\zeta| \left(2^{-q+1} + \dots + 2^{-p}\right) \le 2^{-p+1} |\zeta|.$$

The sequence $(z_n)$ satisfies Cauchy's criterion, so converges to a limit $z \in B(r)$. Since $p$ is continuous, $z_{n+1} = \zeta - p(z_n)$ converges to $\zeta - p(z)$, but also to $z$. Hence $z = \zeta - p(z)$, i.e. $f(z) = \zeta$ , which proves that $f(B(r)) \supset B(r/2)$.

   (c) From this we deduce that for all $a$ where $Df(a)$ is invertible, i.e. where $J_f(a) \neq 0$, the image under $f$ of a ball with centre $a$ contains a ball with centre $b = f(a)$. Since $f$ is $C^1$ its Jacobian is a continuous function of $z$ so that the relation $J_f(z) \neq 0$ defines an open $\Omega \subset G$. By part (b) of this proof $f(\Omega)$ contains a ball of centre $f(z)$ for each $z \in \Omega$; the image under $f$ of any open $U \subset \Omega$ is thus open.

   (d) Returning to point (b) and choosing $r$ sufficiently small for $Df(z)$ to be invertible for all $z \in B(r)$, we see that $f$ maps the open ball $U : |z| < r$ onto an *open* $V$ containing 0. Point (a) shows that the map $f : U \to V$ is bijective and that the inverse map $g : V \to U$ is continuous.

   (e) Finally we show that $g$ is $C^1$ on $V$. Consider a point $\zeta \in V$, a vector $k$ small enough that $\zeta + k \in V$, and put

$$g(\zeta) = z, \qquad g(\zeta + k) = z + h,$$

whence $z, z + h \in U$ and $\zeta = f(z), \zeta + k = f(z + h)$. (11) then shows that $|k| = |f(z + h) - f(z)| \ge \frac{1}{2} |h|$, so that every $o(h)$ function is also $o(k)$. Since

$$k = f(z + h) - f(z) = Df(z)h + o(h),$$

we have $Df(z)h = k + o(h) = k + o(k)$ and so

$$g(\zeta + k) - g(\zeta) = h = Df(z)^{-1}(k + o(k)) = Df(z)^{-1}k + o(k)$$

because $Df(z)^{-1}$ does not depend on $k$. This proves that $g$ is differentiable and that

(24.12) $$Dg(\zeta) = Df(z)^{-1} = Df [g(\zeta)]^{-1},$$

the inverse of the linear tangent map to $f$ at the point $g(\zeta) = z$. Since $g$ and $Df$ are continuous, so similarly is $\zeta \mapsto Df[g(\zeta)]$. Since the determinant of this map [i.e. of the Jacobian matrix of $f$ at the point $g(\zeta)$] is everywhere

$\neq 0$ in $V$, the inverse is a continuous function of $\zeta$ by virtue of the classical formulae for solving a system of two (or 314159) linear equations in two (resp. ...) unknowns. The function $g$ is therefore of class $C^1$ on $V$, which completes the proof of Theorem 24.

**Corollary 1.** *Let $G$ be an open subset of $\mathbb{R}^2$ and $f : G \to \mathbb{R}^2$ a map of class $C^p$ $(p \geq 1)$ such that $J_f(z) \neq 0$ for all $z \in G$. Then the image under $f$ of every open subset $U \subset G$ is open. If $f$ is injective then the inverse map $f^{-1} : f(G) \to G$ is of class $C^p$.*

The first statement is point (c) of the proof of Theorem 24. Since the relation $J_f(z) \neq 0$ shows that $f$ has an inverse of class $C^1$ on a neighbourhood of each $z \in G$, it is clear that if $g$ has a global inverse this must also be of class $C^1$. The fact that it must be of class $C^p$ like $f$ is immediate, since, applying the Chain Rule repeatedly to the relation $g[f(z)] = z$, one sees that the derivatives of $f^{-1}$ can be calculated by dividing polynomials in the derivatives of $f$ by a power of $J_f(z)$, and then substituting $g(\zeta)$ for $z$ in the result.

**Corollary 2.** *Let $f$ be a holomorphic function on an open subset $U$ of $\mathbb{C}$ and let $a$ be a point of $G$ where $f'(z) \neq 0$. Then there is an open set $U \subset G$ containing $a$ such that: (i) $f$ maps $U$ bijectively onto an open subset $V$ of $\mathbb{C}$, (ii) the inverse map $g : V \to U$ is holomorphic on $V$. If $f'(z) \neq 0$ for every $z \in G$ and if $f$ is injective, then $f(G)$ is open and $f^{-1}$ is holomorphic.*

As we showed even before Theorem 24 that $J_f(z) = |f'(z)|^2$ and that the inverse of $f$, if it exists, must be holomorphic, the first statement is its translation to this particular case. The second is then obvious, because the inverse of a $\mathbb{C}$-linear map is again $\mathbb{C}$-linear.

The hypothesis that $f$ is globally injective, which is indispensable (set theory!) if there is to be a global inverse map, *does not follow* from the condition $f'(z) \neq 0$ as is shown by the example of the map $z \mapsto z^n$ of $\mathbb{C}^*$ into $\mathbb{C}$; in this case we certainly have $f'(z) \neq 0$ everywhere, but, as we shall see in Chap. IV, the equation $z^n = \zeta$ has $n$ distinct roots for every $\zeta \in \mathbb{C}^*$. The function $f(z) = \exp$ on $G = \mathbb{C}$ again has $f'(z) = \exp z \neq 0$ everywhere, but the equation $\zeta = \exp z$ has an infinite number of roots for every $\zeta \neq 0$, which makes it impossible to define a true *function* $\mathrm{Log}\,\zeta$ on $\mathbb{C}^*$; we will return to these examples in Chapter IV, while waiting to revisit them with more technique in Vol. III, Chap. VIII. We shall then prove a new "miraculous" property of *holomorphic* maps, namely that the hypothesis that $f'(z) \neq 0$ is superfluous in proving the second statement of Corollary 2: if $f$ is globally or even locally injective, then $f'(z) \neq 0$.

Let us now pass on to the problem of implicit functions: we are given a *real* function $F$ of class $C^1$ on an open subset $G$ of $\mathbb{C}$ and we seek a function $f$

of class $C^1$ satisfying $f(a) = b$, $F[x, f(x)] = c$, on a neighbourhood of a point $(a, b)$ where $F(a, b) = c$ and $D_2F(a, b) \neq 0$.

So consider the map

$$\Phi : (x, y) \longmapsto (x, F(x, y))$$

of $G$ into $\mathbb{C}$. It is $C^1$ and its Jacobian at $z = (x, y)$ is

$$\begin{vmatrix} 1 & 0 \\ D_1F(z) & D_2F(z) \end{vmatrix} = D_2F(z).$$

We can therefore apply Theorem 24 to it at every $z \in G$ where $D_2F(z) \neq 0$. There must therefore be an open neighbourhood $U \subset G$ of $(a, b)$ such that (i) $\Phi$ maps $U$ bijectively onto an open neighbourhood $V$ of the point $\Phi(a, b) = (a, c)$, (ii) the inverse map $\Psi : V \to U$ is $C^1$.

Since $\Phi$ transforms $(x, y)$ to $(x, F(x, y))$, $\Psi$ transforms $(x, F(x, y))$ to $(x, y)$. For $\zeta = (\xi, \eta)$ we thus have $\Psi(\zeta) = (\xi, \psi(\xi, \eta))$ with a real $C^1$ function $\psi$. Since

$$\Psi \circ \Phi : (x, y) \longmapsto (x, F(x, y)) \longmapsto (x, \psi[x, F(x, y)]),$$

the relation $\Psi \circ \Phi = id$ can be written

$$\psi[x, F(x, y)] = y.$$

In the same way, since

$$\Phi \circ \Psi : (\xi, \eta) \longmapsto (\xi, \psi(\xi, \eta)) \longmapsto (\xi, F[\xi, \psi(\xi, \eta)]),$$

the relation $\Phi \circ \Psi = id$ can be written

$$F[\xi, \psi(\xi, \eta)] = \eta;$$

These relations are valid for $(x, y) \in U$ and $(\xi, \eta) \in V$ respectively. The first provides immediately the (one and only) solution of $F(x, y) = c$ on a neighbourhood of $(a, b)$, namely

$$y = \psi(x, c) = f(x).$$

This result is valid for all $x$ such that $(x, c) \in V$, the open set on which $\Psi$ is defined, i.e. for $x$ in the open set $V(c)$, the "section" of $V$ at the height $c$; and $(x, f(x))$ is the only solution of $f(x, y) = c$ such that $(x, y) \in U$. In conclusion:

**Theorem 25.** *Let $G$ be an open subset of $\mathbb{R}^2$, let $F$ be a real function defined and of class $C^p$ on $G$, and let $(a, b)$ be a point of $G$ where $D_2F(a, b) \neq 0$. If $c = F(a, b)$ then there exists an open neighbourhood $U \subset G$ of $(a, b)$ and an open neighbourhood $V$ of $(a, c)$ possessing the following properties: (i) the map $(x, y) \mapsto (x, F(x, y))$ is a bijection of $U$ onto $V$; (ii) for all $(x, z) \in V$ the equation $F(x, y) = z$ possesses one and only one solution $y = \psi(x, z)$ such that $(x, y) \in U$; (iii) the function $\psi$ is of class $C^p$ on $V$.*

The interest of this result is not only that it shows the existence of a solution $y$ of $F(x, y) = z$ for $z = c$, which was the original problem; but also it shows that this solution is a $C^p$ function of $x$ *and* of the right hand side $z$ of the equation to be solved, granted of course that $z$ is sufficiently close to $c = F(a, b)$ and that one restricts oneself to looking for solutions for which $(x, y)$ is sufficiently close to $(a, b)$.

Since Theorem 25 is a direct consequence of the local inversion theorem it must be possible to adapt its proof to obtain a direct proof of Theorem 25. Though we will not obtain all the conclusions of Theorem 25, let us show how one can solve $F(x, y) = c$.

To simplify the notation, we first reduce to the case where $a = b = c = 0$ and where $D_2 F(0, 0) = 1$, by replacing $F$ by $[F(a + x, b + y) - c] / D_2 F(a, b)$, a function which we again denote by $F$, which vanishes for $x = y = 0$ and satisfies $D_2 F(0, 0) = 1$. If we put $c = D_1 F(0, 0)$ then

$$F(x, y) = cx + y + o(|x| + |y|)$$

since $D_2 F(0, 0) = 1$. Now consider the function

$$G(x, y) = y - F(x, y - cx),$$

defined and of class $C^1$ on a neighbourhood of $(0, 0)$; we have

$$G(x, y) = y - (cx + y - cx) + o(|x| + |y|) = o(|x| + |y|),$$

on a neighbourhood of the origin, so

(24.13)                          $D_1 G(0, 0) = D_2 G(0, 0) = 0.$

This done, suppose we have found a function $g(x)$ of class $C^1$ on a neighbourhood of $0$ such that

(24.14)                          $G[x, g(x)] = g(x);$

we then have $g(x) - F[x, g(x) - cx] = g(x)$, so that $f(x) = g(x) - cx$ will be a $C^1$ solution of $F[x, f(x)] = 0$.

The construction of $g$ is then, for $x$ given, to define a sequence of points $y_n = y_n(x)$ by

(24.15)        $y_0 = 0, \quad y_1 = G(x, y_0) = G(x, 0), \quad y_2 = G(x, y_1), \ldots$

and to verify that the $y_n$ converge to the desired solution $g(x)$. This proof too can be divided into several stages.

(a) Since $DG(0, 0) = 0$, there exists a compact interval $I = [-r, r]$ such that $G$ is defined on $K = I \times I$ and satisfies $\|DG(z)\| \leq \frac{1}{2}$ on $K$, i.e. $\|DG\|_K \leq \frac{1}{2}$. Then

(24.16)     $$|G(z') - G(z'')| \le \frac{1}{2}|z' - z''|$$

on $K$, whence it follows (put $z' = (x, y)$ and $z'' = 0$) that

$$\{|x| \le r \ \& \ |y| \le r\} \Longrightarrow |G(x, y)| \le r.$$

Examining the construction (15) of the $y_n$, we see that the definition of $y_n$ will meet no obstruction if $|x| \le r$.

(b) By (16) we have

(24.17)     $$|G(x, y') - G(x, y'')| \le \frac{1}{2}|y' - y''|$$

on $K$, whence

$$|y_{n+1} - y_n| \le |y_n - y_{n-1}|/2 \le |y_{n-1} - y_{n-2}|/2^2 \le \dots;$$

the $y_n = y_n(x)$ satisfy Cauchy's criterion, so converge on $I$ to a solution $y = g(x)$ of $G(x, y) = y$; it is unique on $I$, since from $G(x, y') = y'$ and $G(x, y'') = y''$ it follows that $|y' - y''| \le \frac{1}{2}|y' - y''|$. And since furthermore, by (17),

$$|y_n - y| = |G(x, y_{n-1}) - G(x, y)| \le |y_{n-1} - y|/2 \le \dots$$

for any $x \in I$, we see that

$$|y_n(x) - g(x)| \le |y_0(x) - g(x)|/2^n = |g(x)|/2^n \le r/2^n$$

for all $x \in I$, which proves that the approximations $y_n(x)$ converge uniformly on $I$ and therefore that $g$ is *continuous*.

(c) It remains to verify that $g(x)$ is $C^1$. We can do this either by using the hypothesis that $G$ has real values, in other words the fact that we have restricted ourselves to a map of the type $\mathbb{R} \times \mathbb{R} \to \mathbb{R}$, or, and this is more subtle, by an argument valid for every map of the type $E \times F \to F$, where $E$ and $F$ are vector spaces of finite dimension, or even Banach spaces.

**First method.** For $x, x + h \in I$, let us put $g(x) = y$ and $g(x + h) = y + k$. Since $g(x) = G[x, g(x)]$ for any $x \in I$, we have

$$\begin{aligned} k &= G(x + h, y + k) - G(x, y) = \\ &= D_1 G(x + th, y + tk)h + D_2 G(x + th, y + tk)k \end{aligned}$$

for a certain $t \in [0, 1]$ since $G$ is real (mean value theorem). Since $|D_2 G(x + th, y + tk)| \le \frac{1}{2} < 1$, we can solve with respect to $k$, whence, replacing $y$ by $g(x)$,

$$\frac{k}{h} = \frac{D_1 G[x + th, g(x) + tk]}{1 - D_2 G[x + th, g(x) + tk]};$$

and since $g$ is continuous by the point (b) of the proof, $k$ tends to 0 with $h$; the functions $D_1 G$ and $D_2 G$ being continuous on $K$ and $D_2 G[x, g(x)]$ being $\neq 1$, we can pass to the limit in the preceding relation, whence the existence and the value of

$$(24.18) \qquad \lim \frac{k}{h} = g'(x) = \frac{D_1 G[x, g(x)]}{1 - D_2 G[x, g(x)]}.$$

**Second method.** Since we already know that $g$ is continuous, $k$ tends to 0 with $h$. Since $G$ is $C^1$, (21.12) or (12') shows that

$$k = D_1 G(x, y)h + D_2 G(x, y)k + o(|h| + |k|).$$

Since $1 - D_2 G(x, y)$ is invertible[47] and independent of $h$ and $k$, it follows that

$$(24.19) \quad k = \gamma h + o(|h| + |k|) \quad \text{where } \gamma = [1 - D_2 G(x, y)]^{-1} D_1 G(x, y).$$

For $|h|$ and so $|k|$ sufficiently small, we then have

$$|k - \gamma h| \leq \varepsilon(|h| + |k|),$$

whence $|k| \leq (|\gamma| + \varepsilon)|h| + \varepsilon|k|$ and consequently $(1 - \varepsilon)|k| \leq (|\gamma| + \varepsilon)|h|$. Choosing $\varepsilon = \frac{1}{2}$, for example, we deduce that $k = O(h)$. The relation (19) then shows that $k = \gamma h + o(h)$, whence the existence of $\lim k/h = \gamma$ where, very luckily, $\gamma$ has the value already found by the first method ...
    (18) shows that $g$ is $C^1$, so the solution of $F[x, f(x)] = c$ is so too. That $f$ is of class $C^p$ if $F$ is, follows from the formula

$$(24.20) \qquad f'(x) = -D_1 F[x, f(x)]/D_2 F[x, f(x)]$$

and from the fact that $D_2 F$ does not vanish on the square $I \times I$ considered at the beginning of the proof: if $F$ is $C^p$ and if one has already shown that $f$ is $C^k$ with $k < p$, then (20) shows that $f'$ is similarly $C^k$ and so $f$ is $C^{k+1}$, qed.

    Theorem 25 allows one to show that, under certain conditions, the set $C \subset G$ of solutions of $F(x, y) = c$ is an excellent "curve" possessing, at each of its points, a tangent varying in a continuous way as a function of the point considered. The hypothesis to make is that the derivatives $D_1 F$ and $D_2 F$ are never *simultaneously* zero at the points of $C$; example: $F(x, y) = x^2 + y^2$, with $c > 0$ arbitrary; examples to the contrary: $xy = 0$, $x^2 + y^3 + y^2 = 0$, etc., cases where the derivatives are zero at $(0, 0)$.

---

[47] Abstract proof: if, on a Banach space, we have a linear map $A$ such that $\|A\| < 1$, then $1 - A$ is invertible and we even have

$$(1 - A)^{-1} = 1 + A + A^2 + \dots$$

Banal proof in the case in question: $1 - D_2 G(x, y) \neq 0$.

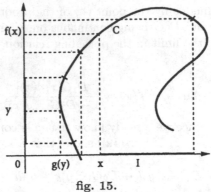

**fig. 15.**

If indeed one has $D_2F \neq 0$ at a point $(a,b)$ of $C$, then the part of $C$ contained in a sufficiently small neighbourhood of $(a,b)$ is just the graph of a function $y = f(x)$, of class $C^p$ like $F$. If it is $D_1F$ which does not vanish, it is the graph of a function $x = g(y)$ of class $C^p$ (permute the rôles of $x$ and $y$). The trivial case of the equation $x^2 - y^2 = 0$ shows that at the point $(0,0)$ of the "curve", there exist two arcs ("branches") which intersect again at the origin with distinct tangents, so that, on a neighbourhood of the origin, $C$ is not the graph of a single function $y = f(x)$ or $x = g(y)$. Vast generalisations to the spaces $\mathbb{R}^n$, up to the study of "singular" points, including, and to start with, when $F(x,y)$ is a polynomial ("algebraic varieties"), present difficulties incommensurable with what happens in the plane and which Newton already knew, more or less.

Theorem 24 on the other hand allows one to define *curvilinear coordinates*, as they used to be called. Consider a map $f : U \to \mathbb{C}$ of class $C^\infty$ (for simplicity), which has an invertible derivative $Df(z)$ at each point $z$ of $U$, and which, moreover, is *globally* injective. The image $V$ of $U$ under $f$ is then an open set, and the inverse map $g$ of $V$ onto $U$ is of class $C^\infty$, as one sees by applying the theorem on a neighbourhood of any point of $U$. If one puts

(24.21) $$\xi = f_1(x,y), \qquad \eta = f_2(x,y),$$

whence inversely

$$x = g_1(\xi,\eta), \qquad y = g_2(\xi,\eta),$$

one sees that knowledge of the point $(\xi,\eta)$ of $V$ determines the point $(x,y)$ entirely, and that a function of $(x,y)$ is of class $C^r$ if and only if its expression in terms of $\xi$ and $\eta$ is of class $C^r$. The functions (21) may therefore be considered as "coordinates" of the point $(x,y)$. But clearly the relations $\xi = $ const. or $\eta = $ const. no longer define lines as in the case of standard Cartesian coordinates; they define curves to which the remarks following Theorem 25 apply. Since the matrix

$$\begin{pmatrix} D_1 f_1 & D_1 f_2 \\ D_2 f_1 & D_2 f_2 \end{pmatrix}$$

of $Df(x, y)$ is invertible, the derivatives of $f_1$ (resp. $f_2$) are never *simultaneously* zero. In consequence, the relations $\xi = const.$ and $\eta = const.$ define $C^\infty$ curves having a continuously varying tangent at each point, etc.

Consider for example the *polar coordinates* of $\mathbb{C}$, which consist of assigning to each $z \in \mathbb{C}$ its modulus $r = |z|$ and its argument, given by $\tan \theta = y/x$, whence $z = r(\cos \theta + i. \sin \theta)$. If one considers the map

$$g : (r, \theta) \longmapsto (r \cos \theta, r \sin \theta)$$

of $\mathbb{R}^2$ into $\mathbb{R}^2$ (one can allow negative values of $r$), it is clear that it is $C^\infty$ and surjective. Its Jacobian is

$$\begin{vmatrix} \cos \theta & \sin \theta \\ -r \sin \theta & r \cos \theta \end{vmatrix} = r$$

and so $\neq 0$ on the open set $V = \mathbb{C}^*$, whose image under $g$ is the open set $U = \mathbb{C}^*$. The map $g$ is not *globally* injective, but one can apply Theorem 24 *locally*. One sees that if, for a point $z_0 \neq 0$ of $\mathbb{C}$, one chooses values $r_0, \theta_0$ of the polar coordinates of $z_0$, then there exists a neighbourhood $U$ of $z_0$ on which one can choose a determination of the polar coordinates for all $z \in U$ so that it reduces to $(r_0, \theta_0)$ at $a$ and that $r$ and $\theta$ are $C^\infty$ functions of $z$ (i.e. of the Cartesian coordinates $x$ and $y$ of $z$) in $U$. To define $C^\infty$ polar coordinate functions of $z$ *globally* one has to work on an open $V \subset \mathbb{C}^*$ such that the map $(r, \theta) \mapsto (r \cos \theta, r \sin \theta)$ of $V$ into $\mathbb{C}^*$ is *injective*, for example the open set defined by the *strict* inequalities

$$r > 0, \quad \alpha < \theta < \alpha + 2\pi,$$

where $\alpha$ is given; the image of $V$ is then the open set $U$ obtained by deleting from $\mathbb{C}^*$ the half-line originating at the origin making the angle $\alpha$ with $0x$.

The only real problem is to choose $\theta$ as a function of $z$, since by choosing $r > 0$ conventionally, i.e. $r = |z| = (x^2 + y^2)^{1/2}$, one obtains an excellent $C^\infty$ function on $\mathbb{C}^*$. The deep study of the pseudo-function $\theta = \arg(z)$, on the other hand, is not as obvious as one tends to believe *a priori*; it is one of the elementary situations where real topological problems arise, on which apprentice mathematicians have a good chance of committing errors not only of calculation, but of comprehension: how to attribute an inverse to a map which has none? We shall return to this in detail at the end of Chap. IV.

However it may be, it is useful to know the relations between the derivatives of a function $f$ with respect to Cartesian coordinates $x$, $y$ and its derivatives with respect to polar coordinates $r$, $\theta$. It suffices to use (21.13) or its translation into the language of differentials.

First one writes

$$df = f'_x.dx + f'_y.dy$$

and differentiates the relations $x = r \cos \theta$, $y = r \sin \theta$:

$$dx = \cos \theta.dr - r \sin \theta.d\theta, \qquad dy = \sin \theta.dr + r \cos \theta.d\theta.$$

Hence
$$df = f'_r.dr + f'_\theta.d\theta$$

with

$$
\begin{aligned}
f'_r &= \cos \theta f'_x + \sin \theta f'_y, \\
f'_\theta &= -r \sin \theta f'_x + r \cos \theta f'_y.
\end{aligned}
$$

# Appendix to Chapter III

## Generalisations[48]

### 1 – Cartesian spaces and general metric spaces

In algebra one defines vector spaces over any field, for example $\mathbb{Q}$, $\mathbb{R}$ or $\mathbb{C}$. The simplest examples, and the most important in analysis, are $\mathbb{R}^p$ and $\mathbb{C}^p$ whose elements, which one calls "points" or "vectors" according to the circumstances, are ordered sequences

$$x = (x_1, \ldots, x_p)$$

of $p$ real or complex numbers, the coordinates of $x$ (Chap. I, n° 4); one speaks of a real or complex *Cartesian space* when it is unnecessary to specify the *dimension p*. What we call in analysis a "function of several real (or complex) variables" is just a function, in the general sense of the term, defined on a subset of a $\mathbb{R}^p$ (or $\mathbb{C}^p$). We propose, in this appendix, to show very succinctly how the definitions and results of this chapter extend to these and to other much more general spaces.

First, inspired by Pythagoras' theorem[49], we define the *Euclidean distance* $d(x, y) \geq 0$ between two points of $\mathbb{R}^p$ or $\mathbb{C}^p$ by the formula

(1.1) $$d(x, y)^2 = |x_1 - y_1|^2 + \ldots + |x_p - y_p|^2 ;$$

on introducing the *norm* or length

$$\|x\| = d(0, x) = \left( |x_1|^2 + \ldots + |x_n|^2 \right)^{1/2}$$

of a vector, we then have

$$d(x, y) = \|x - y\|.$$

Again we have the same triangle inequality

---

[48] This appendix is a reference text for use as needed in the following chapters.
[49] For the needs of analysis one might well choose the simpler formula

$$d'(x, y) = |x_1 - y_1| + \ldots + |x_p - y_p|$$

given that $d'(x, y)/p < d(x, y) < d'(x, y)$ for all $x, y$.

(1.2) $$\|x + y\| \leq \|x\| + \|y\|$$

as in $\mathbb{R}^2$ or $\mathbb{C}$. To see this, we define the *scalar product* of two vectors $x$ and $y$ of $\mathbb{C}^p$ by the formula

(1.3)    $(x \mid y) = x_1 \overline{y_1} + \ldots + x_p \overline{y_p}$,    whence    $\|x\| = (x \mid x)^{1/2}$,

and verify that the number

$$(zx + y \mid zx + y) = (x \mid x)z\bar{z} + (x \mid y)z + \overline{(x \mid y)}\bar{z} + (y \mid y)$$

is $\geq 0$ for any $z \in \mathbb{C}$; if $(x \mid x) \neq 0$, we can choose

$$z = -\overline{(x \mid y)}/(x \mid x),$$

and obtain the *Cauchy-Schwarz inequality*

(1.4)    $$|(x \mid y)|^2 \leq (x \mid x)(y \mid y) = \|x\|^2 \cdot \|y\|^2,$$

immediately: it remains valid if $(x \mid x) = 0$ since then $x = 0$. From this it follows that

$$
\begin{aligned}
\|x + y\|^2 &= (x + y \mid x + y) = (x \mid x) + (x \mid y) + \overline{(x \mid y)} + (y \mid y) = \\
&= \|x\|^2 + 2\mathrm{Re}(x \mid y) + \|y\|^2 \leq \|x\|^2 + 2\|x\| \cdot \|y\| + \|y\|^2,
\end{aligned}
$$

whence (2).

*Metric spaces* provide a much more general framework to which one can extend almost all of what we have said about $\mathbb{R}$ or $\mathbb{C}$. By definition, such a space is a pair $(X, d)$ – one writes simply $X$ when there is no ambiguity as to $d$ – formed by a set $X$ and a function

$$d : X \times X \longrightarrow \mathbb{R}_+$$

which allows us, by convention, to define the *distance* of two "points" $x, y \in X$. We ask only for it to satisfy the obvious conditions:

(1.5)    $d(x, y) \geq 0$,    $d(x, y) = 0 \Longleftrightarrow x = y$,    $d(x, y) = d(y, x)$,

$$d(x, y) \leq d(x, z) + d(z, y).$$

The simplest example consists of taking an arbitrary subset of a Cartesian space for $X$ – no matter which – and defining the distance by the formula[50] (1); more generally, every subset of a metric space can be considered as a metric space in itself. But there are also much less obvious spaces

---

[50] One can also be less brutal. On the surface of the Earth one does not define the distance between Paris and Heidelberg by measuring the straight line joining these two points through the Earth. If, in a Cartesian space, one has a sufficiently "smooth" "surface" $S$ (of dimension possibly $> 2$), one measures the distance between $a, b \in S$ by considering all the curves joining $a$ and $b$ on $S$, measuring their lengths, and taking their greatest lower bound. In good cases there will exist a curve of minimum length ("geodesic"), the arc of a great circle in the case of a sphere.

whose elements are, for example, functions $f$ defined on a more or less arbitrary set $M$ (*functional spaces*) and satisfying certain conditions. On the set $X = \mathcal{B}(M)$ of all *bounded* maps $f$ of $M$ into $\mathbb{C}$, the expression

$$(1.6) \qquad d_M(f,g) = \sup |f(x) - g(x)| = \|f - g\|_M$$

satisfies conditions (5) as we saw in Chap. III, n° 7. On the space $C^p(K)$ of functions of class $C^p$, $p$ finite, on a compact interval $K \subset \mathbb{R}$, one can use the expression

$$(1.7) \qquad d(f,g) = \|f - g\|_K + \|f' - g'\|_K + \cdots + \|f^{(p)} - g^{(p)}\|_K;$$

distances of this kind feature in the theory of distributions (Chap. V, n° 34). Integration provides numerous examples; on $C^p(K)$ one can for example use the distance

$$\sum_{0 \leq i \leq p} \int_K \left| f^{(i)}(x) - g^{(i)}(x) \right| dx,$$

the simplest case being $p = 0$ (continuous functions).

The construction of $\mathbb{R}^p$ from $\mathbb{R}$ extends to metric spaces. Consider two metric spaces $X'$ and $X''$ and the product $X' \times X''$, formed by pairs $(x', x'')$ with $x' \in X'$ and $x'' \in X''$ (Chap. I, n° 4). For two such pairs, let us put[51]

$$(1.8) \qquad d[(x', x''), (y', y'')] = d'(x', y') + d''(x'', y''),$$

where the distance functions $d'$ and $d''$ given on $X'$ and $X''$ appear on the right hand side. It is immediate to check that in this way one again obtains a distance function on $X = X' \times X''$. Hence the concept of *Cartesian product* of two metric spaces, which extends in an obvious way to an arbitrary finite number of such spaces.

---

[51] The reader who likes to complicate his existence may prefer the function given by

$$d\left[(x', y'), (x'', y'')\right]^2 = d\left(x', x''\right)^2 + d\left(y', y''\right)^2.$$

This gives the metric a perfectly unnecessary Pythagorean touch, since the two versions of the "product" metric on $X' \times X''$ define the same open sets, the same convergent sequences, the same continuous functions, etc. The reason is that, if $a$ and $b$ are real positive numbers, one always has

$$a^2 + b^2 \leq (a + b)^2 \leq 2(a^2 + b^2).$$

This is a particular case of the concept of equivalent metrics on a set $X$; this is what one calls two distance functions whose ratios lie between two fixed numbers $> 0$. In $\mathbb{C}$ for example, one does not change the definitions in respect of convergence if one agrees to take as an "open ball with centre $a$" the interior of any square with centre $a$; it is not even necessary to assume its sides parallel to the coordinate axes. The essential is that such a "ball" contains all the points of $\mathbb{C}$ sufficiently close to $a$ and *vice versa*.

## 2 – Open and closed sets

In a metric space, as in $\mathbb{R}$ or $\mathbb{C}$, one can define an *open ball* $B(a,r)$ with centre $a$ by the strict inequality $d(a,x) < r$, where $r$ is a number $> 0$. The weak inequality $d(a,x) \leq r$, with $r \geq 0$, defines a *closed ball*, including for $r = 0$ if one maintains that every intersection of closed balls with centre $a$ is again a closed ball. Although, in $\mathbb{R}$, the intersection of two open balls with arbitrary centres $a'$ and $a''$ is again an open ball, this is not so in the general case, nor even in $\mathbb{R}^2$; the triangle inequality, however, shows that if $a \in B(a',r') \cap B(a'',r'')$ there exists a ball $B(a,r)$ contained in this intersection. The analogous statement for closed balls is false, even in $\mathbb{R}$. The definitions of Chap. II, n° 3 in respect of what happens on a neighbourhood of a point $a$ or at infinity extends immediately to metric spaces. An assertion $P(x)$ involving a variable $x \in X$ is *true on a neighbourhood* of an $a \in X$ if there exists an open ball $B$ with centre $a$ such that $x \in B \implies P(x)$. It is *true at infinity* if, having chosen a $c \in X$, there exists a number $R$ such that $d(c,x) > R \implies P(x)$; the choice of $c$ hardly matters. Since the intersection of a finite number of open balls with centre $a$ is again an open ball with centre $a$, it is clear that if fifteen assertions are separately valid on a neighbourhood of $a$, they are simultaneously valid on a neighbourhood of $a$.

A yet more convenient way of expressing this is to introduce, with N. Bourbaki, the technical, not naïve, concept, of a *neighbourhood of a point* $a$ in a metric space $X$. By definition, it is any set containing an open ball with centre $a$; it may reduce to this ball or spill over *ad libitum*. It is clear that

(i)   any set containing a neighbourhood of $a$ is again a neighbourhood of $a$,
(ii)  the intersection of a finite number of neighbourhoods of $a$ is a neighbourhood of $a$,
(iii) any neighbourhood of an $a \in X$ is a neighbourhood of all $b \in X$ sufficiently close to $a$,

because an open ball $B(a)$ with centre $a$ contains an open ball with centre $b$ for any $b \in B(a)$, so is a neighbourhood of $b$.

Now, the most radical way of expressing that an assertion $P(x)$, where $x$ varies in $X$, is "true for all $x$ sufficiently close to $a$" is to say: the set of $x \in X$ such that $P(x)$ is true is a neighbourhood of $a$. In this form, the $r$, the $\varepsilon$, the $\delta$, etc. have disappeared.

As in $\mathbb{R}$ or $\mathbb{C}$, the points $a$ of $X$ may, relative to a set $E \subset X$, be divided into three categories: the *interior points* of $E$ ($E$ contains an open ball with centre $a$), the *exterior points* of $E$ (they are interior to $X - E$) and the *boundary points* (every open ball with centre $a$ meets $E$ and $X - E$). The points which are not exterior to $E$ are called *adherent* to $E$; one writes $\bar{E}$ for the set of adherent points the *closure* of $E$.

These definitions lead immediately, as in the case of $\mathbb{R}$ or $\mathbb{C}$, to the concepts of open set and of closed set in a metric space $X$:

$E$ *is open*  $\iff$  every $x \in E$ is interior to $E$,

$F$ *is closed*  $\iff$  every $x$ adherent to $F$ belongs to $F$.

In consequence, $E$ is open if and only if $X - E$ is closed. Note that with these definitions the set $E = X$ is both open and closed, so similarly is the empty set: in the latter case, all is satisfied since there is nothing to satisfy. In a metric space, the open balls in the sense just defined are determined by an inequality of the form $d(a, x) < r$, while the closed balls are defined by $d(a, x) \leq r$, as one may verify using the triangle inequality.

The open or closed sets have the same properties as in $\mathbb{R}$ or $\mathbb{C}$ in respect of unions and intersections, proved in the same way: the union of an arbitrary family of open sets is an open set; the intersection of a finite family of open sets is an open set, etc.

One should pay attention to the fact that the concept of open or closed set is relative to a given "ambient" metric space $X$. If $X$ is itself a subspace of a larger metric space $Y$, a set $E \subset X$, to start with $X$ itself, can very well be open in $X$ without being so in $Y$; there is no problem if $X$ is open in $Y$. In the general case, the open sets in $X$ are (exercise!) the sets $X \cap U$, where $U$ is open in $Y$.

Even though these definitions have been inspired by familiar situations, it is prudent not to let oneself be mystified by "obvious" geometrical images. To give an example of what can happen, take for $X$ the set $\mathbb{Q}$ of rational numbers, with the everyday distance $d(x, y) = |x - y|$. For every $a \in \mathbb{Q}$ there then exist balls in $\mathbb{Q}$ with centre $a$ which are simultaneously open and closed, namely all the open balls $B(a, r)$ where $r$ is an *irrational* number. For $B(a, r)$ is the set of $x \in \mathbb{Q}$ (and not of $x \in \mathbb{R}$) such that $|x - a| < r$; every $x \in \mathbb{Q}$ adherent to $B(a, r)$ satisfies $|x - a| \leq r$, but since $r$ is irrational one cannot have $|x - a| = r$; so then $|x - a| < r$, whence $x \in B(a, r)$, qed.

In normal *classical* analysis one most often has to do with *connected* spaces, i.e. in which the only simultaneously open and closed sets are the whole set $X$ itself and the empty set; more general are the *locally connected spaces*, those in which every point has an open connected neighbourhood. It is clear that every open subset of $\mathbb{C}$ is locally connected.

All the same there are very strange examples. If one considers the case of a gyroscope whose axis of rotation is fixed at its lower extremity $I$, then its upper end $s$, oscillating periodically, describes a curve $C$ traced on a sphere of centre $I$ and contained between two horizontal sections of the sphere. With great luck the trajectory of $S$ will be a good closed curve (so compact) with maybe some multiple points or cusps; for this it is necessary that after completing an integral number of rotations about the vertical the gyroscope should come back to the same position with the same velocity. But in the general case the trajectory is not closed and it can well happen that the intersection of the curve $C$ with every neighbourhood of every point of the sphere lying between these horizontal limits is the union of a countable number of pairwise disjoint arcs. If one defines the distance between two points of $C$ in the usual way one then sees that $C$ has no connected neighbourhoods.

## 3 – Limits and Cauchy's criterion in a metric space; complete spaces

Everything we said in n° 5 of Chap. II about scalar sequences extends without modification to sequences of points of a Cartesian or metric space $X$: it suffices to replace expressions of the form $|x - y|$ everywhere by the distance function $d(x, y)$, which we have already often done to accustom the reader. The concept of limit of a sequence is thus defined by the relation

$$(3.1) \qquad \lim x_n = a \Longleftrightarrow \lim d(x_n, a) = 0$$

which brings one back to limits in $\mathbb{R}$. One sees immediately that

$$(3.2) \qquad \lim x_n = a \ \& \ \lim y_n = b \Longrightarrow \lim d(x_n, y_n) = d(a, b)$$

on using the triangle inequality (or, in this particular case, the quadrilateral ...).

If $E$ is a subset of $X$ the adherent points of $E$ are clearly the limits of convergent sequences of points of $E$. It follows that $E$ is closed if and only if every limit of points of $E$ belongs to $E$.

If for example $X$ is the space $\mathcal{B}(M)$ of functions $M \longrightarrow \mathbb{C}$ defined and bounded on a set $M$, with the distance $d_M(f, g) = \|f - g\|_M$, then convergence of a sequence $(f_n)$ to a limit $f$ is precisely *uniform convergence* on $M$, since, for all $r > 0$, the relation $d_M(f, g) \leq r$ means that $|f(x) - g(x)| \leq r$ *for all* $x \in M$. When $M \subset \mathbb{C}$ [or more generally if $M$ is itself a metric space, see below], one can consider the subset $C^0(M)$ of $\mathcal{B}(M)$ formed by the continuous scalar functions on $M$. Theorem 8 of Chap. III, n° 5, says that if a sequence of functions $f_n \in C^0(M)$ converges uniformly on $M$, i.e. converges in the metric space $\mathcal{B}(M)$, then the limit function is again continuous, i.e. still belongs to $C^0(M)$. Conclusion: $C^0(M)$ is a closed subset of the metric space $\mathcal{B}(M)$. If $M$ is a compact interval in $\mathbb{R}$, the set $E = C^p(M)$ of functions of class $C^p$ in $M$ is not closed in $X = C^0(M)$, since a uniform limit of differentiable functions is not necessarily differentiable. But Theorem 27 of Chap. V, n° 27, will show that every element of $X$ is a limit of elements of $E$ (one can even restrict oneself to considering polynomial functions, by Weierstrass' theorem). When every element of a metric space $X$ is the limit of elements of a given set $E \subset X$, i.e. when $\bar{E} = X$, one says that $E$ is *everywhere dense* in $X$, a concept introduced by Cantor for $X = \mathbb{R}$; for example, $\mathbb{Q}$ is everywhere dense in $\mathbb{R}$.

The definition of a limit extends no less immediately to scalar functions (i.e. with values in $\mathbb{C}$) defined on a subset $E$ of a metric space $X$: $f(x)$ tends to $b \in \mathbb{C}$ when $x \in E$ tends to $a \in X$ if, for all $r > 0$, there exists an $r' > 0$ such that

(3.3)          $x \in E$   and   $d(a, x) < r' \implies d[b, f(x)] < r.$

In this form, the definition extends even to maps into another metric space $Y$ since it uses only distances. The principle is always the same: the distance measures both the error in the function and the error in the variable, the error in the function must be $< r$ so long as the error in the variable is sufficiently small, i.e. $< r'$ in our usual language.

In the case of a Cartesian space, it is clear that for a sequence $(u_n)$ of points of $\mathbb{R}^p$ to converge to a limit $u$ it is necessary and sufficient that, for any $i$, the $i^{\text{th}}$ coordinate of $u_n$ should converge to the $i^{\text{th}}$ coordinate of $u$. This extends trivially (i.e. directly from the definitions) to any Cartesian product $X_1 \times \ldots \times X_n$ of metric spaces.

In the particular case of $\mathbb{R}^p$, the general Cauchy criterion remains valid: for a sequence $(u_n)$ to converge, it is necessary (in any metric space) and is sufficient (in $\mathbb{R}^p$) that, for all $r > 0$,

(3.4)          $d(u_p, u_q) < r$ for $p$ and $q$ sufficiently large.

To see this, one remarks first that if $x$ and $y$ are two vectors in $\mathbb{R}^p$, their coordinates satisfy $|x_i - y_i| \leq d(x, y)$ for any $i$. It is then clear that, for all $i$, the coordinates of index $i$ of $u_n$ form a Cauchy sequence in $\mathbb{R}$, so converge to limits which are clearly the coordinates of the desired vector $\lim u_n$.

Though Cauchy's criterion is valid in $\mathbb{R}$ and so in $\mathbb{R}^p$, it is not always valid in an arbitrary metric space $X$ as already shown by the case of $X = \mathbb{Q}$ endowed with the usual distance. The metric spaces in which Cauchy's criterion is valid are called *complete*. $\mathbb{R}$ is complete, $\mathbb{Q}$ is not, which explains the terminology.

The space $C^0(K)$ of continuous functions on a compact interval $K \subset \mathbb{R}$, endowed with the distance $d_K(f, g)$ of uniform convergence, is complete: this follows from Theorem 13" of Chap. III, n° 10 and the fact that a uniform limit of continuous functions is continuous (Chap. III, n° 5, Theorem 5). Likewise, and for more trivial reasons, the space $\mathcal{B}(M)$ of bounded functions on a set $M$ is complete with respect to the uniform metric on $M$. If on the other hand one endows the space $C^p(K)$ considered above with the distance $\|f - g\|_K$ of uniform convergence, one does not obtain a complete space since a uniform limit of differentiable functions may very well not be differentiable. To make $C^p(K)$ a complete space one has to use the distance (1.7); if a sequence $(f_n)$ satisfies Cauchy's criterion relative to the latter, it is clear that the derived sequences $f_n^{(i)}$ converge uniformly (Cauchy's criterion for uniform convergence) and Theorem 20 of Chap. III, n° 17, then shows that the sequence $(f_n)$ converges in the metric of $C^p(K)$.

There is a procedure for "embedding" any metric space $X$ in a complete metric space $\hat{X}$; it generalises the construction of $\mathbb{R}$ from $\mathbb{Q}$ invented by

Cantor and Meray (Chap. III, end of n° 10). To do this, one considers the set $S$ of *all* Cauchy sequences in $X$ – it is contained in the set of maps of $\mathbb{N}$ into $X$ – and one defines an equivalence relation $R$ on $S$ by declaring that two Cauchy sequences $(x_n)$ and $(y_n)$ are *equivalent* if $\lim d\,(x_n, y_n) = 0$; see the end of Chap. I, n° 4, for the general concept of equivalence, specifically invented for the kind of situation examined here; the conditions of the general definition are satisfied here by reason of the triangle inequality

$$d\,(x_n, z_n) \leq d\,(x_n, y_n) + d\,(y_n, z_n)\,.$$

Now to every Cauchy sequence we associate its equivalence class, the subset of $S$ formed by all the sequences equivalent to the given sequence. The set of these classes, i.e. the quotient set $S/R$, is by definition $\hat{X}$. The fact that an element of $\hat{X}$ can be a very complicated set is nothing abnormal and, quite to the contrary, confirms the basic postulate of Chap. I: every mathematical object is a set, even when this is not immediately obvious, so *a fortiori* in the present case.

If two Cauchy sequences $(x_n)$ and $(y_n)$ in $X$ define two elements $x$ and $y$ of $\hat{X}$ one puts

$$(3.5) \qquad\qquad\qquad d(x, y) = \lim d\,(x_n, y_n)\,.$$

Since $d\,(x_p, x_q)$ and $d\,(y_p, y_q)$ are $< r$ for $p$ and $q$ large, the relation

$$|d\,(x_p, y_p) - d\,(x_q, y_q)| \leq d\,(x_p, x_q) + d\,(y_p, y_q)$$

shows that the sequence of $d\,(x_n, y_n)$ satisfies Cauchy's criterion, whence the existence of the limit (5); similar inequalities show that it is independent of the Cauchy sequences chosen in $X$ to "represent" $x$ and $y$. The fact that $d(x, y) = 0 \iff x = y$ does not have to be proved: it is the very definition of equality between *classes* of Cauchy sequences; the other properties of distances are obvious. The "embedding" of $X$ into $\hat{X}$ is obtained by associating to each $x \in X$ the class of the Cauchy sequence $(x, x, x, \ldots)$, i.e. the set of sequences $(x_n)$ which converge to $x$ in $X$. In practice, one makes no distinction between an element of $X$ and its class in $\hat{X}$ (which confirms the fact that, once a mathematical object is defined as too complicated a set, one stops thinking of this psychologically inhibiting "detail" ... ).

It is also clear that, for $x, y \in X$, the distance between $x$ and $y$ as calculated in $\hat{X}$, for example starting from the sequences $(x, x, x, \ldots)$ and $(y, y, y, \ldots)$, is the same as in $X$.

To calculate the distance between an $x \in \hat{X}$ and a $y \in X$ one chooses a Cauchy sequence $(x_n)$ in the class $x$; since $y$ corresponds to the sequence with general term $y$, the definition (5) shows that $d(x, y) = \lim d\,(x_n, y)$; in particular, one has $d\,(x, x_p) = \lim_n d\,(x_n, x_p)$, a result $< r$ for $p$ large since $(x_n)$ is a Cauchy sequence; the distance from $x$ to the points $x_p \in X$ thus tends to 0, which shows that, in $\hat{X}$, one has $x = \lim x_p$: every $x \in \hat{X}$ is thus a limit of points of $X$, as every real number is the limit of rational numbers.

It remains to prove that $\hat{X}$ is a *complete* metric space; an exercise consisting of applying the definitions and the triangle inequality, as in every case where one knows no more.

## 4 – Continuous functions

The concept of continuity extends immediately to functions defined on a metric space $X$ (so also to functions defined on any subset $E$ of $X$) and with values in another metric space $Y$. Such a function $f$ is said to be continuous at a point $a$ of $E$ if it satisfies these clearly equivalent conditions:

(C1) for every $r > 0$ there exists an $r' > 0$ such that, for $x \in E$,

$$d(a, x) < r' \implies d[f(a), f(x)] < r;$$

(C2) for every ball $B \subset Y$ with centre $f(a)$ there exists a ball $B' \subset X$ with centre $a$ such that $f(B' \cap E) \subset B$;

(C3) $f(x)$ tends to $f(a)$ when $x$ tends to $a$:

$$\lim_{x \to a} f(x) = f(a).$$

A function or map defined on $X$ is said to be continuous on $X$, or continuous for short, if it is so at every point of $X$.

There is a simple relation between the concepts of continuous function and of an open set: *a map $f$ of $X$ into $Y$ is continuous if and only if, for every open $V$ in $Y$, the inverse image $U = f^{-1}(V)$ of $V$ under $f$ is open in $X$.*

Necessity of the condition: let $a \in U$ and $b = f(a) \in V$. If $V$ is open it contains a ball $B = B(b, r)$. If $f$ is continuous at $a$ by (2) there exists a ball $B'$ with centre $a$ such that $f(B') \subset B$. One has $B' \subset U$ by definition of $U = f^{-1}(V)$ (Chap. I), so that $U$ is open.

Sufficiency of the condition: consider an $a \in X$, put $b = f(a)$, and consider an open ball $B = B(b, r)$ in $Y$. Let $U$ be its inverse image under $f$. By hypothesis, $U$ is open. Since $U$ contains $a$, it also contains a ball $B'$ with centre $a$. One has $f(B') \subset B$, whence continuity by (C2).

The preceding result immediately yields a result as trivial as it is fundamental: *Let $X$, $Y$ and $Z$ be metric spaces, $f$ a map of $X$ into $Y$ and $g$ a map of $Y$ into $Z$. Suppose that $f$ and $g$ are continuous. Then the composite map $h = g \circ f$ of $X$ into $Z$ is continuous.*

For let $W$ be open in $Z$. The set $V = g^{-1}(W)$ is open in $Y$ since $g$ is continuous. Since $f$ is also continuous, $U = f^{-1}(V)$ is open in $X$. Now $U = h^{-1}(W)$, qed.

More precisely: *Let $X$, $Y$ and $Z$ be metric spaces, $f$ a map of $X$ into $Y$ and $g$ a map of $Y$ into $Z$. Suppose that $f$ is continuous at $a \in X$ and $g$ is continuous at $f(a) = b \in Y$. Then the map $h = g \circ f$ is continuous at $a$.*

For let $c = h(a) = g(b)$ and choose an $r > 0$. Since $g$ is continuous at $b$ there exists an $r' > 0$ such that

$$d(b, y) < r' \implies d[c, g(y)] < r.$$

Since $f$ is continuous at $a$ there is an $r'' > 0$ such that

$$d(a, x) < r'' \text{ implies } d[b, f(x)] < r'.$$

It is clear that then the relation $d(a, x) < r''$ implies $d[c, h(x)] < r$, qed.

One verifies immediately that if $f'$ and $f''$ are functions defined on metric spaces $X'$ and $X''$, with values in metric spaces $Y'$ and $Y''$, and continuous at $a' \in X'$ and $a'' \in X''$, then the map

$$(x', x'') \longmapsto (f'(x'), f''(x''))$$

of $X' \times X''$ into $Y' \times Y''$ is continuous at $(a', a'')$. If one has a further continuous map $p : Y' \times Y'' \longrightarrow Y$ into another metric space $Y$, then the map $(x', x'') \longmapsto p[f'(x'), f''(x'')]$ is continuous at $(a', a'')$, a vast and trivial generalisation of theorems on the sums, products, upper and lower envelopes, etc. of continuous functions.

Similarly one can extend the concept of *uniform convergence* to the maps of a metric space $X$ into another metric space $Y$ and show that a uniform limit of continuous maps is again continuous. To do this, one defines the distance between two maps $f$, $g$ of $X$ into $Y$ by the formula

$$(4.1) \qquad\qquad d_X(f, g) = \sup d[f(x), g(x)].$$

Apart from a detail of no importance in this context – the distance can be infinite –, all the properties of a distance are clearly satisfied. Then uniform convergence can be expressed, as in Chap. III, by the relation $\lim d_X(f_n, f) = 0$ or by the fact that, for any $r > 0$, there exists an integer $N$ such that

$$n > N \implies d[f_n(x), f(x)] \leq r \text{ for any } x \in X.$$

The fact that a uniform limit of continuous functions is again continuous is proved exactly as we proved it for $X = Y = \mathbb{C}$.

Note moreover that one can replace $X$ by any subset $E$ of $X$ in definition (1), and define the distance $d_E(f, g)$ on $E$, so also uniform convergence on $E$.

If these abstract trivialities have forced the reader to reflect, of which I am not convinced[52], they will at least have served for that.

---

[52] Maybe there will be readers who will ask why we did not develop the concepts expounded in Chap. II and III in this general framework from the outset. Reply: though they may appear trivial to you after you have read these chapters, they would have seemed incomprehensible if you had met them at the outset.

## 5 – Absolutely convergent series in a Banach space

The concept of series has no meaning in a metric space unless one has a way of defining the necessary algebraic operations, as for example in the case of Cartesian spaces. Everything we have said about series in n° 6, 14, 15 and 18 of Chap. II extends without the least modification to those whose terms are vectors or points of a Cartesian space. This is so in particular in the case, by far the most important, of absolute or unconditional convergence: convergence of the scalar series $\sum \|u_n\|$ implies, for all $i$, that of the series formed by the $i^{\text{th}}$ coordinates of $u_n$ since $|x_i| \leq \|x\|$ for all $x \in \mathbb{R}^p$, so also that of the vector series $\sum u_n$.

More generally, let $E$ be a vector space over $\mathbb{R}$ or $\mathbb{C}$ and suppose we are given a *norm* on $E$, i.e. a map

$$x \longmapsto \|x\| \geq 0$$

satisfying the following conditions:

(5.1) $$\|x\| = 0 \Longleftrightarrow x = 0, \qquad \|\lambda x\| = |\lambda|.\|x\|,$$

$$\|x + y\| \leq \|x\| + \|y\|$$

where $\lambda$ is any scalar, real or complex according to the case. This is, for example, the case of the length of a vector in $\mathbb{R}^n$, of the norm of uniform convergence on the space of bounded scalar functions on any set, of the norm on a separated pre-Hilbert space (see below), etc. A vector space endowed with such a function norm is called a *normed vector space*.

Putting $d(x, y) = \|x - y\|$ defines a distance on $E$ and allows us to apply all the above. Since we also have a concept of addition in $E$ we can consider not only sequences, but also series. A series $\sum u_n$ of elements of $E$ will be said to be *convergent* with sum $s \in E$ if $\lim (u_1 + \ldots + u_n) = s$, i.e. if

$$\lim \|s - (u_1 + \ldots + u_n)\| = 0.$$

Writing $s_n$ for the sum of its $n$ first terms, a necessary condition for convergence is, as in the case of $\mathbb{R}$ or $\mathbb{C}$, that for all $r > 0$ one has $d(s_p, s_q) < r$ for $p$ and $q$ sufficiently large, in other words

(5.2) $$\|u_p + \ldots + u_q\| < r \quad \text{for } p \text{ and } q \text{ sufficiently large.}$$

Now the left hand side is, by the triangle inequality, bounded by the corresponding expression for the scalar series $\sum \|u_n\|$. If the latter converges then the given vector series will satisfy Cauchy's criterion. To deduce convergence from this, one clearly has to assume that $E$ is complete.

A complete normed vector space is called a *Banach space*, after one of the most gifted of the many Polish mathematicians who, during the first half of the XX$^{\text{th}}$ century[53], considerably advanced Logic, Set Theory, Topology, Functional Analysis, etc. The foremost theorem – this is a lot to say ... – is that, in a Banach space, every absolutely convergent series converges.

The calculations of n° 1 which led us to the Cauchy-Schwarz inequality are purely formal; the fact that they concern "vectors" in $\mathbb{R}^p$ or $\mathbb{C}^p$ (rather than integrable functions, for example, as we shall see in Chap. V, n° 5 or in Chap. VII, n° 7) is immaterial; only the algebraic rules of calculating with scalar products are important here. So we are led to a vast generalisation, and to define a *pre-Hilbert space* as follows: this is a vector space $\mathcal{H}$ over $\mathbb{C}$, in general of infinite dimension, in which one is given a "scalar product" satisfying the obvious conditions:

(H 1): $(x \mid y)$ is a linear function of $x$ for $y$ given;
(H 2): $(y \mid x) = \overline{(x \mid y)}$;
(H 3): $(x \mid x) \geq 0$ for all $x$.

The space is said to be *separated* if it satisfies the condition

(H 4): $(x \mid x) = 0$ implies $x = 0$.

Again we have the Cauchy-Schwarz inequality, a norm $\|x\| = (x \mid x)^{1/2}$ which can vanish for $x \neq 0$ if $\mathcal{H}$ is not separated, and a triangle inequality that one writes in the form (1), or as $d(x,y) \leq d(x,z) + d(z,y)$, where one defines $d(x,y) = \|x - y\| = (x - y \mid x - y)^{\frac{1}{2}}$. It might happen that the relation $d(x,y) = 0$ no longer implies $x = y$: axiom (H 4) serves to ensure this. A separated pre-Hilbert space is thus a normed vector space; if it is complete, one says that it is a *Hilbert space*.

In contrast, if $K \subset \mathbb{R}$ is a compact interval, the space $C^0(K)$ endowed with the distance

$$d_2(f,g) = \left( \int |f(x) - g(x)|^2 \, dx \right)^{1/2}$$

induced by the scalar product

$$(f \mid g) = \int_K f(x)\overline{g(x)}dx$$

or, even more simply, by the distance $d_1(f,g) = \int |f(x) - g(x)|dx$, *is not* complete; similarly if one considers the set of all Riemann integrable functions on $K$. On the other hand, the Lebesgue theory leads to complete spaces.

---

[53] The situation changed with the war: the Jews emigrated or were exterminated and the "Aryans" who survived after 1945 suffered materially very difficult conditions for a long time, and had hardly any contacts except with the Soviet mathematicians (and they too ...), not nothing, but not replacing the rest of the world. Similar situation in Hungary. In their period of greatness the mathematicians of these two "small" countries were, together, much more "modern" than the majority of their French, English, American, or even German, counterparts.

*Example 1.* Consider a set $X$ (in practice, finite or countable) and the set $L^2(X)$ of functions $f(x)$ defined on $X$, with complex values, and such that

(5.3) $$\sum |f(x)|^2 < +\infty$$

where of course we are dealing with unconditional convergence; compare with the space $L^1(\mathbb{Z})$ of Chap. II, n° 18, example 3. If $f, g \in L^2(X)$, then, for every finite subset $F$ of $X$,

(5.4) $$\left(\sum_F |f(x) + g(x)|^2\right)^{1/2} \le \left(\sum_F |f(x)|^2\right)^{1/2} + \left(\sum_F |g(x)|^2\right)^{1/2}$$

by the Cauchy-Schwarz inequality for a finite sum. Passing to the limit over $F$, one deduces that the series $\sum |f(x) + g(x)|^2$, extended over all $X$, converges, so that $f + g \in L^2(X)$. Since the product of an $f \in L^2(X)$ by a constant is clearly in $L^2(X)$, one concludes that $L^2(X)$, endowed with usual operations on scalar functions, is a complex vector space. Further, the series

(5.5) $$(f \mid g) = \sum f(x)\overline{g(x)}$$

converges absolutely or unconditionally for all $f, g \in L^2(X)$, as seen by applying Cauchy-Schwarz to its finite partial sums. This scalar product clearly satisfies the preceding conditions, including (H 4). The corresponding norm is denoted by

(5.6) $$\|f\|_2 = \left(\sum |f(x)|^2\right)^{1/2}.$$

The formulae are the same as in $\mathbb{C}^p$: the values $f(x)$ play the rôle of "coordinates" of the "vector" $f$. But the vector space $L^2(X)$ is of infinite dimension if $X$ is infinite.

The space $L^2(X)$ is *complete*. Consider a Cauchy sequence $(f_n)$ in this space. Since obviously $|f_p(x) - f_q(x)| \le \|f_p - f_q\|_2 = d(f_p, f_q)$ for any $x \in X$, the given functions $f_n(x)$ converge simply to a limit function $f(x)$ by Cauchy's criterion in $\mathbb{C}$, and clearly $|f(x)|^2 = \lim |f_n(x)|^2$ for all $x \in X$. For every finite subset $F$ of $X$ we then have

$$\sum_{x \in F} |f(x)|^2 = \lim_{n \to \infty} \sum_{x \in F} |f_n(x)|^2 \le \lim_{n \to \infty} \sum_{x \in X} |f_n(x)|^2 = \lim_{n \to \infty} \|f_n\|^2 ;$$

the last limit exists because $\big|\, \|f_p\| - \|f_q\| \,\big| \le \|f_p - f_q\|$ in any normed vector space.

The series $\sum |f(x)|^2$ is therefore convergent, whence $f \in L^2(X)$ and $\|f\|_2 \le \lim \|f_n\|_2$. In this inequality we replace the sequence $(f_n)$ by the sequence $(f_n - f_p)$, for a given $p$: it also satisfies Cauchy's criterion, and converges simply to $f - f_p$, so we obtain $\|f - f_p\|_2 \le \lim_n \|f_n - f_p\|_2$;

but this limit is $< r$ for $p$ large by definition of a Cauchy sequence. So $\lim \|f - f_p\|_2 = 0$, qed.

One can show that this procedure yields *all* Hilbert spaces[54].

*Example 2.* Let $X$ be an interval in $\mathbb{R}$ and consider the complex vector space $L(X)$ of continuous functions of compact support on $X$ (Chap. V, n° 31; if $X$ is compact, this is the set of all continuous functions on $X$). For two such functions, put

$$(5.7) \qquad (f \mid g) = \int f(x)\overline{g(x)}dx$$

where one integrates over $X$. The axioms (H 1), ..., (H 4) are again satisfied, (H 4) because one restricts to continuous functions. The corresponding norm is

$$\|f\|_2 = \left( \int |f(x)|^2 \, dx \right)^{1/2}.$$

You may clearly replace $dx$ by any positive Radon measure $d\mu(x)$ on $X$, and $X$ by any locally compact subset of $\mathbb{C}$ (Chap. III, n° 31). One thus obtains a *non complete* pre-Hilbert space. Lebesgue integration allows us to construct true Hilbert spaces by this type of procedure.

These two types of space play a large rôle in the theory of Fourier series and, in fact, are the origin of the concept of pre-Hilbert space; it needed fully a quarter of a century to pass from these "concrete" pre-Hilbert spaces, whose elements are functions, to the "abstract" Hilbert spaces whose theory (J. von Neumann), much richer than that of Banach spaces, dates from the end of the 1920s and later became the subject of much research.

Analysis provides an inexhaustible stock of examples of Banach spaces. The most obvious is the set of bounded scalar functions on a set $X$, endowed with the norm $\|f\|_X$ of uniform convergence. A less trivial example is the vector space $C^p(K)$ of functions of class $C^p$, $p$ finite, on a compact interval $K \subset \mathbb{R}$, endowed with the norm

$$(5.8) \qquad \|f\| = \|f\|_K + \|f'\|_K + \cdots + \|f^{(p)}\|_K,$$

a finite expression, since, on a compact set, a continuous function is necessarily bounded (Chap. III, n° 9, Theorem 11). As we showed in n° 3, with the help of Chap. III, n° 17, Theorem 20, this space is complete.

## 6 – Continuous linear maps

In practice, almost all the theorems about infinite dimensional Banach spaces concern *continuous linear* maps[55] $f : E \longrightarrow F$ from one Banach space into

---

[54] More precisely: if $\mathcal{H}$ is an Hilbert space there exists a set $X$ and a linear bijective map of $\mathcal{H}$ onto $L^2(X)$ which preserves the scalar products (an "isomorphism" of Hilbert spaces).

[55] There are exceptions, for example the fixed point theorem for the equation $f(x) = x$ when $f$ satisfies a condition $\|f(x') - f(x'')\| \leq q\|x' - x''\|$ with $q < 1$.

another. The continuity of such a map translates into a simple inequality. Since $f(0) = 0$, there is a number $c > 0$ such that $\|x\| \leq c$ implies $\|f(x)\| \leq 1$. Now, for all $x \in E$, one has $\|\lambda x\| \leq c$ if one chooses $\lambda = c/\|x\|$; since

$$f(\lambda x) = |\lambda| f(x) = c.f(x)/\|x\|,$$

one has $c\|f(x)\|/\|x\| \leq 1$ for all $x$, in other words

(6.1) $$\|f(x)\| \leq M.\|x\|$$

where $M = 1/c$ is a constant $> 0$ independent of $x$. Conversely, (1) implies

$$d[f(x), f(y)] = \|f(x) - f(y)\| = \|f(x - y)\| \leq M.\|x - y\|$$

for all $x, y \in E$, whence continuity. The latter, for *linear* maps, is thus equivalent to the existence of an inequality of the form (1). The least possible constant $M$ is called the *norm* of $f$, written $\|f\|$.

If $f$ is bijective it possesses an inverse map $g$, clearly linear like $f$. If $g$ is continuous there is a relation of the form $\|g(y)\| \leq M'\|y\|$, with a constant $M' > 0$. Since every $y \in Y$ is of the form $f(x)$ with $x = g(y)$, the relation in question can also be written as $\|x\| \leq M'.\|f(x)\|$. So we conclude that if a bijective continuous linear map $f : E \longrightarrow F$ has a continuous inverse $g$ then

(6.2) $$m.\|x\| \leq \|f(x)\| \leq M.\|x\|$$

with two *strictly* positive constants $m = 1/M'$ and $M$.

*Exercise* – A linear map satisfying (2) is injective, maps $E$ onto a *closed* vector subspace $H = f(E)$ of $F$ and the inverse map $H \longrightarrow E$ is continuous. Note in passing that a *closed* vector subspace of a Banach space is itself a Banach space (and, more generally, that every closed set in a complete metric space is itself a complete metric space).

Let us give a simple example of a continuous linear map using Chap. V. Consider, on a compact interval $K \subset \mathbb{R}$, the spaces $E = C^1(K)$ and $F = C^0(K)$, $E$ being endowed with the norm

(6.3) $$\|x\| = \sup_{t \in K} |x(t)| + \sup_{t \in K} |x'(t)| = \|x\|_K + \|x'\|_K$$

and $F$ with the norm

(6.4) $$\|y\| = \sup_{t \in K} |y(t)| = \|y\|_K.$$

$E$ and $F$ become Banach spaces in this way, since they are complete, as we have seen above. Now consider the map $f : E \longrightarrow Y$, which, to each function $x \in E$, associates its derivative $f(x) = x'$ [it is not without reason that, contrary to our habits, we have denoted by $t$ the real variable on which $x$

and $x'$ depend; here is an example of a case where the "functions" become "variables" ]. It is clear that $f$ is linear, and continuous since $\|f(x)\| = \|x'\|_K \leq \|x\|$, by (3).

Is the map $f$ bijective? It would first have to be injective, i.e. $f(x) = 0$ must imply $x = 0$; but $f(x) = 0$ means that the function $x(t)$ has an identically zero derivative; so it is constant, but not necessarily zero. One must therefore eliminate the constants from $E$, which can be done by imposing the supplementary condition $x(a) = 0$ on the functions of $E$, for an $a \in K$ chosen once and for all. Having so modified $E$, $f$ becomes injective. It is then surjective since a continuous function $y(t)$ on $K$ always admits (Chap. V, n° 12) a primitive $x(t)$ such that $x'(t) = y(t)$, $x(a) = 0$.

We can therefore consider the inverse map $g : F \longrightarrow E$ of $f$. It associates to each $y \in F$ the primitive $x = g(y) \in E$ which vanishes at $t = a$, namely

$$(6.5) \qquad x(t) = \int_a^t y(u)du.$$

The map $g$ is continuous because both

$$\|x\|_K \leq m(K)\|y\|_K \text{ and } \|x'\|_K = \|y\|_K,$$

whence $\|x\| \leq \left(1 + m(K)\right)\|y\|$.


There are important theorems in the theory of Banach spaces. The first is the Hahn-Banach theorem, which, at the lowest level, assures the existence, in any Banach space $E$, of nontrivial *continuous linear forms* (clf), i.e. of linear maps $f : E \longrightarrow \mathbb{R}$ or $\mathbb{C}$ which are continuous and not identically zero; example: $E = C^0(K)$, where $K$ is a compact subset of $\mathbb{R}$ or $\mathbb{C}$; the clf are exactly, by definition, the Radon measures on $K$ (Chap. V, n° 30). There are several variants of the theorem.

(i)   For every nonzero $a \in E$ there exists a clf $f$ such that $f(a) \neq 0$. More generally: if $F$ is a closed vector subspace of $E$ and if $g$ is a clf on $F$, then there exists a clf on $E$ such that $f = g$ in $F$.

(ii)  Every *closed* vector subspace $F$ of $E$ can be defined by a family (in general infinite) of equations $f_i(x) = 0$ where the $f_i$ are clf[56] on $E$.

---

[56] For example take $E = C^0(K)$, where $K$ is a compact interval in $\mathbb{R}$, and for $F$ the set of functions $x \in E$ which are uniform limits on $K$ of polynomials; it is a closed vector subspace (closure of the set of polynomials) of $E$. Since the clf on $E$ are the (complex) Radon measures on $K$, the Weierstrass approximation theorem – which affirms that $F = E$ – is thus equivalent to the following assertion: the only measure $\mu$ on $K$ such that $\mu(p) = 0$ for all polynomials, i.e. $\int t^n d\mu(t) = 0$ for all $n \in \mathbb{N}$, is $\mu = 0$.

(iii) (for spaces over $\mathbb{R}$) Every *closed convex* subset $C$ of $E$ can be defined by a family of inequalities $f_i(x) \geq a_i$ where the $f_i$ are clf and the $a_i$ are real constants (in other words, $C$ is an intersection of "closed half spaces").

One cannot prove these theorems *in full generality* without using the methods of "transfinite induction" in one place or another; these are founded on the ordinals and the axiom of choice of Chap. I, n° 9; they were not presented because totally unnecessary at the level of this treatise. Nevertheless these theorems lead to very important concrete results in analysis when applied to functional spaces. In this case, the existence of a not identically zero clf is always obvious in practice, but properties (i), (ii) and (iii) are not.

Another important result. Consider two Banach spaces $E$ and $F$ and endow their Cartesian product $E \times F$ with the obvious vector structure and with the norm $\|(x,y)\| = \|x\| + \|y\|$. One obtains a new Banach space. Consider now a *linear* map $h : E \longrightarrow F$; the graph of $h$, i.e. the set $H \subset E \times F$ of pairs $(x, h(x))$, is then clearly a vector subspace of $E \times F$. The *closed graph theorem* (Banach-Steinhaus, two interwar Poles) says that, for $h$ to be *continuous*, it is necessary and sufficient that its graph be a *closed* vector subspace of $E \times F$. The necessity of the condition is immediate, but the converse requires ingenious arguments which we do not reproduce, and which use Baire's Theorem mentioned in Chap. III, end of n° 6, note 12.

Let us explain the interest of the theorem by considering a linear map $g : E \longrightarrow F$ which is both continuous and bijective. It must have an inverse map $h : F \longrightarrow E$, such that

$$x = h(y) \iff y = g(x).$$

It is clear that $h$ is linear like $g$ and that the graph $H$ of $h$ is the image of the graph $G$ of $g$ under the "symmetry" $(x, y) \longmapsto (y, x)$ (Chap. I, n° 5). This clearly transforms the closed subsets of $E \times F$ into closed subsets of $F \times E$ since it preserves distances.

Since $g$ is continuous, it is clear that $G$ is closed; so similarly is $H$. Conclusion: a bijective continuous linear map of one Banach space onto another is "bicontinuous", i.e. possesses a continuous inverse map, i.e. is an homeomorphism. A reassuring result, but it depends on the closed graph theorem.

The example of the map $x \longmapsto x'$ of $C^1(K)$ into $C^0(K)$ does not bring out the power of the closed graph theorem, since, in this case, one can verify it directly by more elementary methods. But it is otherwise when one considers functions of several variables and maps of the "linear differential operator" kind, for which one does not *a priori* have a formula as simple as (5); it is on the contrary, thanks to the closed graph theorem (and, generally, to the theory of distributions) that one can establish the existence of a formula of this type and, sometimes, write it explicitly.

The reader may never have seen a discontinuous *linear* map; there is no risk of meeting them in the Cartesian spaces of finite dimension, even less in $\mathbb{R}$ with the schoolboy functions $x \mapsto ax$. Be reassured: no one has ever *seen* them in the case of true Banach spaces, *complete*[57] and of infinite dimension: it is impossible to construct one using "natural" procedures. To obtain one requires the use of methods – transfinite induction again – which, while affirming the *existence* of such functions in the sense of logicians (and even of mathematicians ...), never permit an *effective construction* because they require uncountably many arbitrary choices. The problem is the same as that of proving the existence of a "base" for any vector space of infinite dimension, i.e. of a family $(e_i)_{i \in I}$ of vectors such that every $x \in E$ can be expressed in a unique way as a *finite* linear combination of vectors $e_i$; this result applies to Banach spaces and allows one to construct linear maps or forms by purely algebraic procedures; but, in infinite dimensions, the coordinates of an $x \in E$ with respect to such a base, though linear functions of $x$, are never continuous, so that algebraic bases are of no use in analysis. One may conclude from this that all the linear maps of one Banach space into another *that can be constructed explicitly by the standard methods of analysis*, using "formulae", are continuous. This metamathematical statement does not, however, exempt one from proofs.

## 7 – Compact spaces

In the case of a subset $X$ of $\mathbb{C}$ the concept of compactness has several possible equivalent formulations (Chap. V, n° 6):

(DEF)  *X is closed and bounded in* $\mathbb{C}$;

(BW)  *from every sequence of points of $X$ one can extract a subsequence which converges to a point of* $X$;

(BL)  *from every covering of $X$ by open sets one can extract a finite covering of* $X$.

In an arbitrary metric space $X$, these properties may all fail, and, also, they are not always equivalent.

---

[57] If one forgets this "detail", the situation changes: take $C^1(K)$ with the norm $\|f\|_K$ of uniform convergence and the map $x \mapsto x'(a)$, where $a \in K$ is given. To say that it is continuous means that uniform convergence of a sequence of differentiable functions implies the convergence of their derivatives; wrong. But $C^1(K)$ endowed with this "bad" norm is not complete.

The first is meaningless unless $X$ is embedded in an "ambient" space: $X$ is always closed in $X$. The condition that $X$ be "bounded" means that

(7.1)
$$\sup_{x \in X} d(a, x) < +\infty$$

for some $a \in X$ (and so for all $a \in X$), but one needs more to ensure (BW) or (BL). On the other hand it is clear that (BW) implies (1), as in the classical case.

Moreover,

$$(\text{BW}) \implies X \text{ is complete},$$

since if a Cauchy sequence $(x_n)$ in $X$ contains a convergent subsequence, it is obvious that it converges to the same limit.

Since, for all $r > 0$, the open balls $B(x, r)$ centred at the points of $X$ cover $X$, (BL) implies the following property:

(PC)  *For every $r > 0$, $X$ is the union of a finite number of open balls of radius $r$.*

Neither does this property, *precompactness*, suffice to ensure compactness: in $\mathbb{C}$ any bounded set satisfies it trivially (see, in Chap. V, n° 6, beginning of the proof of BL). But if $X$ is *complete*, then (PC) implies both (BW) and (BL).

Proof of (BW): let $(x(n))$ be a sequence of points of $X$; by applying (PC) for $r = 1$ one obtains an open ball $B_1$ of radius 1 in $X$ which contains infinitely many terms of the given sequence, i.e. a subsequence of the given sequence; then, using (PC) for $r = 1/2$, one finds a ball $B_2$ of radius $1/2$ which contains infinitely many terms of the first subsequence, so a subsequence extracted from it. Pursuing this "construction", one obtains balls $B_n$ of radius $1/n$ and integers $p_1 < p_2 < \ldots$ such that $x(p_k) \in B_n$ for all $k > n$. Now $d[x(p_k), x(p_h)] < 2/n$ for $k, h > n$; whence a Cauchy sequence, which converges if $X$ is complete.

Proof of (BL): this reduces to showing that if $(U_i)$ is a covering of $X$ by open sets then there exists an $r > 0$ such that, for all $x$, the ball $B(x, r)$ is contained in one of the $U_i$; but since (PC) implies BW as we have just seen, one can argue as in Chap. V, n° 6: there is nothing to change in the proof.

The right definition then consists of declaring that a metric space is *compact* if it is precompact and complete. One can show without difficulty that conversely (BW) or (BL) implies compactness, in other words that these three properties are *equivalent* (see Dieudonné, *Treatise on Analysis*, Vol. 1, III.16).

It is almost obvious that all we have said in Chap. III and V about compact sets in $\mathbb{C}$ extends to the general case. First, a subset $K$ of a metric space $X$ will be called compact if it is so as a metric space "in itself"; clearly this forces $K$ to be closed and bounded in $X$, but this condition is not sufficient if one imposes no other hypothesis on $X$; for example, the unit ball $\|x\| \le 1$ of a Banach space $X$, though closed and bounded in $X$, is not compact except in finite dimensions, a famous theorem of F. Riesz (Dieudonné, Vol. 1, V.9).

The image of a compact space under a continuous map is a compact set and the inverse map, if it exists, is continuous. On a compact space, every continuous function (with values in a metric space) is uniformly continuous and bounded; if it is real, it attains its maximum and its minimum. In a compact space, uniform convergence is a property of a local nature. The product $X \times Y$ of two compact spaces $X$ and $Y$ is compact [use (BW)]. In a compact space, any closed set is compact and conversely; the union of a finite number of such sets is compact [use (BW)]. Etc.

One can also, more generally, define *locally compact* spaces by the condition that every point possesses a compact neighbourhood.

## 8 – Topological spaces

We can generalise further by introducing *topological spaces*, the final and definitive culmination of all these theories. This is an object formed by a set $X$ and a set of subsets of $X$, conventionally called "open", among which must be the set $X$ and the empty set, and required only to satisfy the two standard properties of open sets in $\mathbb{R}$ or $\mathbb{C}$: *every union of open sets is open; every finite intersection of open sets is open.*

The concept of continuity is introduced as follows. Given two topological spaces $X$ and $Y$ and a map $f : X \longrightarrow Y$, one says that $f$ is continuous at $a \in X$ if, for any open $V$ in $Y$ containing $b = f(a)$, there exists an open $U$ in $X$ containing $a$ such that

$$x \in U \Longrightarrow f(x) \in V.$$

Further, a sequence of points $x_n$ of $X$ converges to a limit $a \in X$ if, for every open set $U$ containing $a$, one has $x_n \in U$ for all sufficiently large $n$. A point $a$ will be said to be interior to a set $E \subset X$ if $E$ contains an open set containing $a$. Etc.

A quite general method for defining a topology on a set $X$ consists of specifying a family $(d_i)_{i \in I}$ of maps of $X \times X$ into $\mathbb{R}_+$, each having the properties of a distance with the exception of the relation $d_i(x, y) = 0 \Longrightarrow x = y$. The rôle of open balls with centre $a \in X$ is then taken by the sets defined by a *finite number* of inequalities $d_i(a, x) < r_i$, the open sets then being defined as in metric spaces.

For example, take $X = C^0(\mathbb{R})$ and, for all $k \in \mathbb{N}$, put

$$(8.1) \qquad d_k(f, g) = \sup_{|x| \leq k} |f(x) - g(x)| = \nu_k(f - g)$$

where

(8.2)                    $$\nu_k(f) = \sup_{|x| \le k} |f(x)|$$

is the norm of uniform convergence on the compact interval $[-k, k]$; for a given $k$, the relation $d_k(f, g) = 0$ shows that $f(x) = g(x)$ for $|x| \le k$, but not that $f = g$; to be sure that $f = g$ one must have $d_k(f, g) = 0$ for *every* $k$. Given an $f \in X$, a real number $r > 0$ and an integer $k > 0$, let us write $B_k(f, r)$ for the subset of $X$ defined by the inequality

$$d_k(f, g) < r.$$

We remark in passing that $d_{k+1}(f, g) \ge d_k(f, g)$ always, and so $B_{k+1}(f, r) \subset B_k(f, r)$ for all $k$, $f$ and $r$; a *finite* intersection of "balls" with centre $f$ and of not necessarily equal "radii" thus again contains a ball. In consequence, a subset $U$ of $X$ is open if and only if, for every $f \in U$, there exist a $k$ and an $r$ such that $B_k(f, r) \subset U$. For a sequence of functions $f_n \in X$, convergence to $f$ in this topology means simply that, on every interval of the form $[-k, k]$, and so more generally on every compact set $K \subset \mathbb{R}$, the sequence $f_n(x)$ converges to $f(x)$ uniformly on $K$: this expresses the fact that every ball $d_k(f, g) \le r$ with centre $f$ contains $f_n$ for all $n$ sufficiently large. This mode of "compact convergence" is weaker that uniform convergence on $\mathbb{R}$ as we have already remarked on several occasions.

One could in fact define the topology of $C^0(\mathbb{R})$ – and more generally of all topological spaces $X$ endowed, as above, with a finite or countable family of pseudodistances $d_k(x, y)$ – by means of a single distance. Such a function $d(x, y)$ on a set $X$ has values in $[0, +\infty[$ ; if one replaces it by the function

(8.3)                    $$d'(x, y) = \varphi[d(x, y)]$$

where $\varphi : \mathbb{R}_+ \longrightarrow \mathbb{R}_+$ is continuous, increasing, and satisfies $\varphi(0) = 0$, $\varphi(t) > 0$ for t> 0, $\varphi(s+t) \le \varphi(s) + \varphi(t)$ for all $s$ and $t$, one again obtains a distance on $X$. It is almost obvious that, for all $a \in X$, every $d$-ball with centre $a$ contains a $d'$-ball with centre $a$ and *vice versa*. One can thus submit the given $d_k$ to this type of modification without changing the topology (i.e. the open sets) of $X$. The function $\varphi$ may be bounded – choose $\varphi(t) = t/(1+t)$ –, so one may assume that, for example, $d_k(x, y) \le 1$ for all $x, y \in X$. The formula

(8.4)                    $$d(x, y) = \sum d_k(x, y)/2^k$$

then obviously defines a true distance[58] on $X$ and the topology of $X$ is identical to the one obtained using this distance. The inequality $d_k(x,y) \leq 2^k d(x,y)$ shows that the set $d(a,x) < r/2^k$ contains the set $d_k(a,x) < r$. As on the other hand one has $d_k(a,x) \leq 1$, the relations $d_1(a,x) < r_1, \ldots$, $d_k(a,x) < r_k$ imply

$$d(a,x) < r_1/2 + \ldots + r_k/2^k + 1/2^k$$

(remainder of a geometric series). Choosing $k$ sufficiently large and the $r_i$ sufficiently small, one then sees that the ball $d(a,x) < r$ is contained in a finite intersection of balls $d_i(a,x) < r_i$, whence the identity of the topologies.

This result applies for example to $C^0(\mathbb{R})$, but do not believe that it makes this space into a *normed* vector space: the modification (3) that one has to make to the functions (1) to make the series (4) converge, though preserving the fact that $d_k(f,g)$ depends only on $f - g$, destroys the homogeneity property $\nu_k(\lambda f) = |\lambda| \nu_k(f)$; the function

$$\nu(f) = \sum \varphi \left[ \nu_k(f) \right] / 2^k$$

from which the distance (4) is derived by subtraction, is not an actual norm; it is easy to show that in fact there is no norm on $C^0(\mathbb{R})$ which, alone, can define the same topology as the family of functions (1).

The above construction of a topology on $C^0(\mathbb{R})$ can be generalised. To define a topology on a real or complex vector space $E$ one chooses a family $(\nu_i)_{i \in I}$ of *semi-norms*, i.e. of functions with positive values possessing all the properties of a norm except perhaps the fact that, for $i$ given, the relation $\nu_i(x) = 0$ implies $x = 0$. The functions $d_i(x,y) = \nu_i(x - y)$ then define a topology on $E$ for which the translations $x \longmapsto x + a$ and the homotheties $x \longmapsto \lambda x$ are continuous. A neighbourhood of 0 is then a set which contains all the solutions of a *finite* number of inequalities of the form $\nu_i(x) < r_i$, with $r_i > 0$. These topological vector spaces are called *locally convex* since every neighbourhood of 0 (or, by translation, of any other point) contains a convex neighbourhood of 0, this because

$$\nu_i[(1-t)x + ty] \leq (1-t)\nu_i(x) + t\nu_i(y)$$

for $0 \leq t \leq 1$. These are the only topological vector spaces which play a rôle in analysis, mainly *via* the theory of distributions. In particular, the Hahn-Banach Theorem of n° 6 extends to these spaces.

---

[58] At least if one assumes that $d_k(x,y) = 0$ for all $k$ implies $x = y$, which will ensure that $d(x,y) = 0 \implies x = y$. Geometrically, this means that if $x \neq y$, there exists a neighbourhood $V$ of $x$ and a neighbourhood $W$ of $y$ such that $V \cap W = \emptyset$ [choose an index $k$ such that $d_k(x,y) = r > 0$ and define $V$ and $W$ by the inequalities $d_k(x,z) < r/3$ and $d_k(y,z) < r/3$]. One says then that the topological space considered is *separated*, a necessary and sufficient condition for every sequence to have at most *one* limit.

# IV – Powers, Exponentials, Logarithms, Trigonometric Functions

The general theorems of Chapters II and III have allowed us, in passing, to establish most of the principal properties of the elementary functions which crop up everywhere in analysis. In this chapter we shall go over it all systematically. One can do this in various ways, each as instructive as the other. A first method (n° 1 to 8) consists of erecting the theory from a minimum of knowledge, in particular using neither the theory of power series nor the function exp, particularly its addition formula. A second, contrasting, method, starts from these and goes much further. In particular it allows us to construct a rigorous analytic theory of the trigonometric functions. A third method would be to define the function $\log x$ as an integral and from this deduce its properties as well as those of the exponential functions.

## §1. Direct construction

### 1 – Rational exponents

The first problem to resolve is to define the expression $x^s$ for every number $x > 0$ (*strict* inequality) and every real (and even, later, complex) exponent $s$. The construction is effected in several stages: first one treats the case where $s \in \mathbb{Z}$, then the case where $s \in \mathbb{Q}$ and finally the general case. Naturally we want eventually to obtain *strictly positive* expressions $x^s$ satisfying the "obvious" rules

$$(1.1) \qquad x^s x^t = x^{s+t}, \quad (x^s)^t = x^{st}, \quad x^s y^s = (xy)^s,$$

valid when the exponents are integers of any sign, a case which we assume familiar to the reader.

If $s = p/q$ is rational, the rules (1) entail

$$(1.2) \qquad (x^{p/q})^q = x^p.$$

Since $x^{p/q}$ must be real and $> 0$, we conclude that $x^{p/q}$ must be the[1] $> 0$ $q^{\text{th}}$ root of the number $x^p > 0$ whose existence and uniqueness were established in Chap. II, n° 10, example 3 and again in Chap. III, n° 4, as a consequence of the intermediate value theorem. There is no need to recapitulate it here.

To have the right to write $x^{p/q}$ for the positive $q^{\text{th}}$ root of $x^p$ we need to establish that it depends only on the ratio $p/q$, i.e. that if $a$, $b$, $c$, $d$ are integers, with $b$ and $d$ nonzero, then

$$(1.3) \qquad a/b = c/d \Longrightarrow (x^a)^{1/b} = (x^c)^{1/d},$$

having agreed to write $x^{1/n}$ for the positive $n^{\text{th}}$ root of a number $x > 0$. In fact, we shall even show that

$$(1.3') \qquad (x^a)^{1/b} = (x^{1/b})^a = (x^c)^{1/d} = (x^{1/d})^c.$$

To do this, we raise each of these four numbers to the power $bd$. We find in turn, from the rules (1) for integer exponents and from the definition of roots:

$$\left(\left(\left(\left(x^a\right)^{1/b}\right)^b\right)^d\right) = \left(\left(x^a\right)^d\right) = x^{ad},$$

$$\left(\left(\left(x^{1/b}\right)^a\right)^{bd}\right) = \left(\left(x^{1/b}\right)^{abd}\right) = \left(\left(\left(x^{1/b}\right)^b\right)^{ad}\right) = x^{ad},$$

$$\left(\left(\left(\left(x^c\right)^{1/d}\right)^d\right)^b\right) = \left(\left(x^c\right)^b\right) = x^{bc},$$

$$\left(\left(\left(x^{1/d}\right)^c\right)^{bd}\right) = \left(\left(x^{1/d}\right)^{bcd}\right) = \left(\left(\left(x^{1/d}\right)^d\right)^{bc}\right) = x^{bc}.$$

Since the relation $a/b = c/d$ is equivalent to $ad = bc$, the four results obtained are equal, and when *positive* numbers become equal after being raised to a nonzero integer power, it is because they already were, by virtue of the uniqueness of $n^{\text{th}}$ roots.

The expression $x^s$ having now been defined for $x \in \mathbb{R}_+^*$ and $s \in \mathbb{Q}$, we have to prove that it satisfies the rules (1). To establish the first, one puts $s = a/q$ and $t = b/q$ with $a$, $b$, $q$ integers (reduction to the same denominator), whence $s + t = (a + b)/q$, and raises the two sides to the power $q$; the left hand side becomes $(x^s)^q (x^t)^q = x^a x^b = x^{a+b}$ as does the right hand side. To prove the second rule (1), one raises it all to the power $q^2$; $(x^s)^t$ becomes $(x^s)^{bq} = ((x^s)^q)^b = (x^a)^b = x^{ab}$, and $x^{st}$ also becomes $x^{ab}$ since $st = ab/q^2$. Finally, to establish the relation $x^s y^s = (xy)^s$, one puts $s = p/q$ and raises it all to the power $q$; the calculation is obvious in this case.

---

[1] In English and *a fortiori* in mathematics, the usage of the definite article "the" implies the existence *and the uniqueness* of the object thus referred to. The article "the *(plural)*" means "all the" ("the Negros are lazy", "the French are racists"); we need to be prudent when employing them. The indefinite article (expressed by the empty string in English) means "some" ("there are lazy Negros" and "French racists"). Theoretically these articles correspond to the difference between the logical quantifiers "for all" and "there exists".

Since we have always $x^0 = 1$, we deduce from the rules (1) that

(1.4)
$$x^{-s} = 1/x^s;$$

and, since $x^1 = x$,

(1.5)
$$(x^{1/s})^s = (x^s)^{1/s} = x \quad \text{for } x > 0 \text{ and } s \in \mathbb{Q},$$

similarly, a result which, for $s = n$ (an integer), justifies the expression $x^{1/n}$ for $n^{\text{th}}$ roots. Formula (5) also shows that, for $s$ rational and not just an integer, the maps $f : x \mapsto x^s$ and $g : x \mapsto x^{1/s}$ are mutually inverse.

Finally we note that $1^n = 1$ for all $n \in \mathbb{Z}$, so also

(1.6)
$$1^s = 1 \quad \text{for all } s \in \mathbb{Q}.$$

The power functions $x \mapsto x^s$, defined for the moment only for $s \in \mathbb{Q}$, are *continuous* on the set $x > 0$ where they are defined and, for $s$ nonzero, *strictly monotone*.

Continuity is established as follows: if $s = a/b$ one composes the map $x \mapsto x^a$, clearly continuous since $a$ is integer, with the map $x \mapsto x^{1/b}$, inverse to $x \mapsto x^b$ and so also continuous, by the general result of Chap. III, n° 4, Theorem 7; and the composition of continuous maps is a continuous function.

The fact that the functions $x \mapsto x^n$ are strictly monotone for $n \in \mathbb{Z}$ nonzero, and therefore their inverses $x \mapsto x^{1/n}$ are too, shows similarly that the function $x^s$ is strictly increasing for $s > 0$ and decreasing for $s < 0$.

## 2 – Definition of real powers

Instead of considering $x^s$ as a function of $x$ for a given $s \in \mathbb{Q}$ one can also fix an $a > 0$ and consider the function $s \mapsto a^s$ on $\mathbb{Q}$. Except for $a = 1$ one again obtains a strictly monotone function; to be precise:

(2.1)
$$s < t \Longrightarrow \begin{cases} a^s < a^t & \text{if } 1 < a \\ a^s > a^t & \text{if } 0 < a < 1. \end{cases}$$

On putting $t = s + u$, so that $a^t = a^s a^u$, it is enough to see that

(2.2)
$$u > 0 \Longrightarrow a^u > 1 \text{ if } a > 1 \text{ or } a^u < 1 \text{ if } a < 1.$$

But since $1 = 1^u$, (2) follows from the fact that the function $x \mapsto x^u$ is strictly increasing for $u > 0$, as we saw above.

After these preliminaries, let us move on to the definition of $x^s$ for $s$ real. It would be regrettable if the relations (1) were not to extend to arbitrary real exponents. We need to prove the following result:

**Theorem 1.** *Let a be a strictly positive real number. Then there is one and only one monotone map $f : \mathbb{R} \longrightarrow \mathbb{R}_+^*$ such that*

$$(2.3) \qquad\qquad f(s) = a^s \ \text{ for all } s \in \mathbb{Q}.$$

*The function $f$ is continuous and it is the only continuous function satisfying (3).*

Let us assume that $a > 1$, the case where $a < 1$ reducing to it by the formula $a^s = (1/a)^{-s}$ and the case $a = 1$ being trivial. The desired function $f$ must clearly be increasing as it is already so on $\mathbb{Q}$.

To establish existence and uniqueness, we first remark that, if $f$ exists, we must have

$$(2.4) \qquad\qquad a^u \le f(s) \le a^v \ \text{ for } u, v \in \mathbb{Q}, \ u < s < v.$$

If we consider the increasing function $s \mapsto a^s$ on $\mathbb{Q}$ (and not on $\mathbb{R} \dots$), then Chap. III, n° 3 shows that it has left and right limits at all $s \in \mathbb{R}$, which we denote by[2]

$$(2.5) \qquad\qquad f_-(s) = \lim_{\substack{u \to s-0 \\ u \in \mathbb{Q}}} a^u, \quad f_+(s) = \lim_{\substack{v \to s+0 \\ v \in \mathbb{Q}}} a^v,$$

where the expression $u \to s - 0$ means that $u$ tends to $s$ while remaining $< s$. These limits are also least upper bounds and greatest lower bounds. With this notation, the relation (4) is equivalent to

$$(2.6) \qquad\qquad f_-(s) \le f(s) \le f_+(s)$$

since $f(s)$ must majorise all the $a^u$ for $u < s$ and minorise all the $a^v$ for $v > s$.

In conclusion, we do not yet know $f(s)$ for $s$ real nonrational, but we do know bounds between which $f(s)$ must lie, namely the expressions (5). If we can prove them equal, then (6) will determine $f(s)$ without any ambiguity: for then

$$f(s) = f_-(s) = f_+(s).$$

Now if $u$ and $v$ are rational numbers such that $u < s < v$, then definition (5) shows on the one hand that

$$(2.7) \qquad\qquad a^u \le f_-(s) \le f_+(s) \le a^v$$

---

[2] It is clear that $f_-(s)$ and $f_+(s)$ are precisely the left and right limits of the desired function $f$, in the sense of Chap. III, n° 3. But this assumes that the function $f$ has been constructed. To avoid logical ambiguities we adopt this notation to denote numbers which, *at the end of the proof*, will indeed be the left and right limits $f(s+)$ and $f(s-)$, in fact equal to $f(s)$, of the function $f$ that we are now beginning to construct.

and, on the other hand, that one can choose $u$ and $v$ so that the differences $f_-(s) - a^u$ and $a^v - f_+(s)$ are arbitrarily small. So it all reduces to proving the following lemma, which will also, a little later, assure us of the continuity of the map $s \mapsto a^s$ on $\mathbb{R}$ (see also Chap. III, n° 11, (IV bis)):

**Lemma.** *Let $a$ and $s$ be two real numbers; assume that $a > 1$. Then, for any number $r > 0$, there exist rational numbers $u$ and $v$ such that*

$$(2.8) \qquad u < s < v, \qquad a^v - a^u \leq r.$$

For any $n \in \mathbb{N}$ we can indeed find a number $u_n$ in $\mathbb{Q}$ such that $u_n < s < u_n + 1/n = v_n$. Then

$$(2.9) \qquad a^{v_n} - a^{u_n} = a^{u_n}\left(a^{1/n} - 1\right) \leq f_-(s) . \left(a^{1/n} - 1\right)$$

by (7). Now $\lim a^{1/n} = 1$ as we saw in Chap. II, n° 5, example 7, where we inoffensively anticipated the existence of $n^{\text{th}}$ roots. The left hand side of (9) is thus $\leq r$ for $n$ large, qed.

We may now *define* the number

$$(2.10) \qquad f(s) = f_-(s) = f_+(s)$$

unambiguously, and it remains to establish that it satisfies all the conditions we require.

First, $f(s) = a^s$ for $s \in \mathbb{Q}$. We see this[3] from the lemma above, valid for $s \in \mathbb{Q}$ as for $s \in \mathbb{R}$: in $\mathbb{Q}$, we already know that

$$u < s < v \Longrightarrow a^u < a^s < a^v,$$

and, since $f_-(s)$ and $f_+(s)$ also lie between $a^u$ and $a^v$, the lemma shows that $a^s = f_-(s) = f_+(s)$.

On the other hand $f(s) < f(t)$ for all real $s$ and $t$ such that $s < t$. To see this, consider *rational* numbers $u$ and $v$ such that $s \leq u < v \leq t$; we have $f_+(s) \leq a^u$ and $a^v \leq f_-(t)$ by definition of these expressions, whence $f(s) = f_+(s) \leq a^u < a^v \leq f_-(t) = f(t)$, hence the result. (The reader will be interested scrupulously to distinguish the strict from the weak inequalities in these proofs ...)

The uniqueness of the monotone function $f$ was shown above.

Since we now know that $f(u) = a^u$ for $u \in \mathbb{Q}$, we can replace $a^u$ and $a^v$ by $f(u)$ and $f(v)$ in the lemma above; on the interval $[u, v]$, i.e. on a neighbourhood of $s$, the function $f$, being monotone, lies between $a^u$ and $a^v$. It is therefore equal to $f(s)$ to within $r$ on a neighbourhood of $s$, whence the continuity of $f$.

---

[3] This is not obvious: the definition of $f(s)$ for real $s$ involves only rational exponents different from $s$, *even if $s \in \mathbb{Q}$.*

Finally, if a continuous function of $s \in \mathbb{R}$, say $g$, coincides with $a^s$ on $\mathbb{Q}$, i.e. if the relation $f(s) = g(s)$ is satisfied for all $s \in \mathbb{Q}$, it must also do so for all $s \in \mathbb{R}$, since every $s \in \mathbb{R}$ is the limit of points of $\mathbb{Q}$. Whence Theorem 1.

Let us venture a remark *à propos* the lemma above. Consider the function $a^x$ on the set $\mathbb{Q}$ and make $x$ tend to an $s \in \mathbb{R}$; if one wanted to show *a priori* that it converged to a limit – this would be another method of defining $a^s$ –, one would have to check that Cauchy's criterion of Chap. III, n° 10, Theorem 13', held, in other words that, for all $r > 0$,

$$(2.11) \qquad |a^x - a^y| < r \text{ for } x, y \in \mathbb{Q} \text{ sufficiently near to } s.$$

This is precisely what the lemma assures: if $u$ and $v$ are chosen according to the lemma, it is clear, since the function is increasing in $\mathbb{Q}$, that (11) holds so long as $x, y \in [u, v]$. This argument could be used to prove directly the existence in $\mathbb{R}$ of a unique *continuous* function equal to $x \mapsto a^x$ for $x$ rational: its value at $s \in \mathbb{R}$ has to be the limit of $a^x$ when $x \in \mathbb{Q}$ tends to $s$. It remains to check that this limit is truly a continuous function of $s$, as the reader may show easily.

## 3 – The calculus of real exponents

Given a number $a > 0$, the function $f$ whose existence and the uniqueness are assured by Theorem 1 is called the *exponential function of base a*. From now on we write it in the usual form $f(x) = a^x$, or in the form

$$f(x) = \exp_a(x)$$

for a reason which will soon appear. The meaning of the symbol $a^x$ may appear obvious from the notation we have adopted, but in reality this is justified only by the constructions and arguments of the preceding n°; there is certainly nothing obvious in an expression such as $\pi^\pi$, the limit of the sequence whose successive terms are

$$\pi^3, \qquad \pi^{3,1} = \left(\pi^{31}\right)^{1/10}, \qquad \pi^{3,14} = \left(\pi^{314}\right)^{1/100}, \quad \text{etc.}$$

It is no more obvious that the expression $a^x$ satisfies the index laws irresistibly suggested to those who take their desires for reality or confuse causes with effects: the rules of calculus explain the notation – introduced in the case of rational exponents by the mathematicians of the XVII$^{th}$ century, mainly John Wallis and Newton –, and not conversely. One has to justify them in this general framework.

Let us first prove the formula

$$(I) \qquad\qquad x^s x^t = x^{s+t}.$$

To do this one chooses sequences $(s_n)$ and $(t_n)$ of *rational* numbers which converge to $s$ and $t$. Then

$$x^s = \lim x^{s_n}, \quad x^t = \lim x^{t_n}. \quad x^{s+t} = \lim x^{s_n + t_n} = \lim x^{s_n} x^{t_n}$$

by rule (I) for rational exponents, qed.

The rule

(II) $$(x^s)^t = x^{st}$$

can be established in two stages.

Suppose first that $t \in \mathbb{Q}$ and choose a sequence of numbers $s_n \in \mathbb{Q}$ which converges to $s$. Then $s_n t$ converges to $st$ and in consequence we have

(3.1) $$x^{st} = \lim x^{st_n}$$

since the exponential functions are continuous. Moreover, and for the same reason,

$$x^s = \lim x^{s_n};$$

now we already know that the "power function" $y \mapsto y^t$ is continuous on $\mathbb{R}_+^*$ for $t \in \mathbb{Q}$; it follows that

$$\begin{aligned}(x^s)^t &= (\lim x^{s_n})^t = \lim \left[(x^{s_n})^t\right] \text{ (continuity of } y \mapsto y^t \text{ for } t \in \mathbb{Q}) = \\ &= \lim(x^{s_n t}) \text{ (because } s_n, \, t \in \mathbb{Q}) = x^{st}\end{aligned}$$

by (1), whence (II) for $s \in \mathbb{R}$ and $t \in \mathbb{Q}$.

The general case remains. It is enough to approximate $t$ by rational $t_n$ and to pass to the limit in the relation

$$(x^s)^{t_n} = x^{st_n};$$

since, for $s$ given, the function $t \mapsto (x^s)^t$ is continuous, being an exponential function of $t$, the left hand side tends to $(x^s)^t$; since $st_n$ converges to $st$, the right hand side tends, for the same reason, to $x^{st}$. Whence (II) in the general case.

Rule (II) shows in particular that, for any nonzero $s \in \mathbb{R}$,

(III) $$(x^s)^{1/s} = x$$

so that the maps $x \mapsto x^s$ and $x \mapsto x^{1/s}$ of $\mathbb{R}_+^*$ onto $\mathbb{R}_+^*$ are mutually inverse bijections.

We have still to establish the formula

(IV) $$(xy)^s = x^s y^s.$$

Here again it is enough to approximate $s$ by $s_n \in \mathbb{Q}$ and to apply (IV) to the $s_n$; the continuity of the exponential functions immediately validates the passage to the limit.

## 4 – Logarithms to base $a$. Power functions

For $a = 1 + b > 1$, it is clear that $a^n > 1 + nb$ increases indefinitely as $n \in \mathbb{Z}$ tends to $+\infty$; since $a^{-n} = 1/a^n$, it is no less obvious that $a^n$ tends to $0$ as $n$ tends to $-\infty$; the results will be the opposite for $0 < a < 1$. The functions $s \mapsto a^s = \exp_a(s)$ are continuous, so map $\mathbb{R}$ onto *intervals* necessarily contained in $\mathbb{R}_+^*$, and so must map $\mathbb{R}$ *onto* $\mathbb{R}_+^*$. Being strictly monotone they admit inverse maps that are no less continuous and strictly monotone (Chap. III, n° 4, Theorems 6 and 7). The inverse map of $x \mapsto a^x$ is called the *logarithm to base* $a$, and is denoted by[4] $x \mapsto \log_a x$; thus, by definition,

$$(4.1) \qquad y = \log_a x \iff x = a^y \text{ for all } x > 0.$$

The formula $a^u a^v = a^{u+v}$ shows that

$$(4.2) \qquad \log_a xy = \log_a x + \log_a y;$$

on putting $\log_a x = u$ and $\log_a y = v$ one has $x = a^u$, $y = a^v$, whence $xy = a^{u+v}$ and in consequence $\log_a xy = u + v$. Also

$$(4.3) \qquad \log_a 1 = 0, \qquad \log_a a = 1$$

since $a^0 = 1$ and $a^1 = a$. Likewise, formula (II) of the preceding n° shows that

$$(4.4) \qquad \log_a(x^s) = s . \log_a x \text{ for } x > 0, \ s \in \mathbb{R}$$

since on putting $u = \log_a x$ we have $x = a^u$, whence $x^s = a^{su}$, so that the left hand side is equal to $su = s . \log_a x$. In other words,

$$(4.5) \qquad x^s = a^{s . \log_a x} = \exp_a(s . \log_a x).$$

Finally, the logarithmic functions are proportional to each another. For if we put $y = \log_a x$ and $z = \log_b x$ with $a, b \neq 1$, then

$$x = a^y = b^z;$$

and since there exists a $c \in \mathbb{R}$ such that $b = a^c$ we also have

$$x = (a^c)^z = a^{cz},$$

whence $y = cz$, in other words $\log_a x = c . \log_b x$ for all $x$. In fact $c = \log_a b$ (let $x = b$) and $c = 1/\log_b a$ (let $x = a$), which yields the more specific formula

---

[4] In principle one should write $\log_a(x)$ for what one writes as $\log_a x$. The older traditions rarely being the best – what concerns us here dates from an age when the notation $f(x)$ had not even been invented –, it falls to us to reestablish the correct functional notation when needed.

(4.6)                 $\log_a(x) = \log_a(b) \cdot \log_b(x) = \log_b(x)/\log_b(a)$.

Note finally, that since any real number can be written in the form $\log_a(b)$ for $a$ given, every function proportional to the function $\log_a(x)$ is itself a logarithmic function.

These fascinating formulae serve no purpose in (mathematical) life, the only fundamental functions being $\exp(x)$ and the Napierian log to be discussed later. That $\log_{10}$ is useful in numerical calculations does not rebut this remark.

The *power function*

$$f(x) = x^s$$

where $s \in \mathbb{R}$ is a constant and where the variable $x$ takes all real strictly positive values (except for $s \in \mathbb{N}$, a case it profits not to dilate on ...), has properties that it is fastidious but obligatory to rehearse. Unless it is expressly mentioned to the contrary, we assume that $s \neq 0$ in what follows.

It is strictly increasing for $s > 0$ and strictly decreasing for $s < 0$. Suppose that $x < y$, whence $y = xa$ with $a > 1$ and that in consequence $f(y) = f(x)f(a)$; since $f(x) > 0$, it is enough to show that

(4.7)                 $a^s > 1$ for $a > 1$ and $s > 0$.

But this again can be written as $a^s > a^0$, and follows from the fact that the function $\exp_a$ is strictly increasing for $a > 1$.

The power functions $x^s$ are, moreover, continuous on $\mathbb{R}^*_+$. It is clear that the image of $\mathbb{R}^*_+$ under such a function is the *interval* $\mathbb{R}^*_+$ since, for all $y > 0$, the equation $y = x^s$ possesses one (and only one) solution, namely $y = x^{1/s}$. Since the power functions are strictly monotone, their continuity then follows from Chap. III, n° 4, Theorem 7: for a strictly monotone function $f$ defined on an *interval* $I$, continuity is equivalent to the fact that $f(I)$ is again an *interval*.

## 5 – Asymptotic behaviour

It is indispensable – though not to the rest of this chapter – to have an exact idea of the behaviour of the preceding functions at the end points of their intervals of definition, and even more to know how to compare their orders of magnitude.

Consider first the function $a^x$ as $|x|$ increases indefinitely. The results are the following (omitting the trivial case $a = 1$):

(5.1)                 $$\lim_{x \to +\infty} a^x = \begin{cases} +\infty & \text{if } a > 1 \\ 0 & \text{if } a < 1, \end{cases}$$

(5.2)                 $$\lim_{x \to -\infty} a^x = \begin{cases} 0 & \text{if } a > 1 \\ +\infty & \text{if } a < 1, \end{cases}$$

We have already shown that, for $a > 1$ for example, $a^n$ tends to $+\infty$ (resp. 0) as $n \in \mathbb{Z}$ tends to $+\infty$ (resp. $-\infty$); the function being increasing, (1) and (2) follow trivially.

At the end points of their interval of definition $x > 0$ the power functions tend to easily calculated limits. First,

$$(5.3) \qquad \lim_{x \to 0, x > 0} x^s = \begin{cases} 0 & \text{if } s > 0 \\ +\infty & \text{if } s < 0. \end{cases}$$

Assume for example that $s > 0$. We have to show that, for all $r > 0$, we have $x^s < r$ for $x > 0$ sufficiently small; but on writing $r$ in the form $a^s$, this inequality is equivalent to $x < a$ since the function $x^s$ is strictly increasing. The case $s < 0$ reduces to the previous one, since $x^s = 1/x^{-s}$.

We have the same results, inverted, when $x$ increases indefinitely:

$$(5.4) \qquad \lim_{x \to +\infty} x^s = \begin{cases} +\infty & \text{if } s > 0 \\ 0 & \text{if } s < 0. \end{cases}$$

If $s > 0$ we have to verify that $x^s > M$ for $x$ large, which is clear since this relation is equivalent to $x > M^{1/s}$.

The log functions also have a simple behaviour:

$$(5.5) \qquad \lim_{x \to +\infty} \log_a x = +\infty, \qquad \lim_{x \to 0} \log_a x = -\infty \text{ if } a > 1.$$

It is enough to establish the first, the second reducing to this because $\log x = -\log(1/x)$. Now we know that the function $\log_a x$ is increasing, $> 0$ for $x > 1$, and that $\log_a(x^n) = n \log_a x$. This relation shows that $\log_a x$ takes arbitrarily large values for $x > 1$, and since it increases, it is forced to grow indefinitely. For $0 < a < 1$ one inverts the results, because $\log_{1/a} x = -\log_a x$.

Now let us try to compare the growth of these various functions as $x$ increases indefinitely. The results are very simple. First, the following formulae for those that grow indefinitely with $x$:

$$(5.6) \qquad a^x \; = \; o(b^x) \text{ when } x \to +\infty \text{ if } 1 < a < b;$$
$$(5.7) \qquad x^s \; = \; o(a^x) \qquad\qquad\qquad \text{if } s > 0, 1 < a;$$
$$(5.8) \qquad x^s \; = \; o(x^t) \qquad\qquad\qquad \text{if } 0 < s < t;$$
$$(5.9) \qquad \log_a x \; = \; o(x^s) \qquad\qquad\quad \text{if } s > 0, a > 0.$$

The first means that the ratio $a^x/b^x = c^x$, where $c = a/b$, tends to 0 at infinity; obvious since $c < 1$. The third means that $x^s/x^t = x^{s-t}$ tends to 0; obvious since $s - t < 0$.

To establish (7), one may assume that $s$ is an integer $p$ (if not, choose a $p > s$, whence $x^s < x^p$). Putting $a = b^p$, where again $b > 1$, we have only to show that $x^p/b^{px} = (x/b^x)^p$ tends to 0. So it is enough to examine the ratio

$x/b^x$. On putting $b = 1 + c$ with $c > 0$ and writing $n$ for the integer part of $x$, so that $n \leq x < n + 1$, we have

$$b^x \geq b^n = (1 + c)^n > n(n - 1)c^2/2$$

by the binomial formula, whence $x/b^x \leq 2(n + 1)/n(n - 1)c^2$, an expression which tends to 0 as $n$, i.e. $x$, increases indefinitely.

Finally, (9) reduces to (7) since, on putting $\log_a x = y$, we have $x = a^y$ and $x^s = a^{sy} = b^y$ with $b = a^s$, whence $(\log_a x)/x^s = y/b^y$, which tends to 0 if $b > 1$ i.e. if $a > 1$; for $a < 1$ use $\log_a x = -\log_{1/a} x$.

The results are similar for the functions which tend to 0 when $x \to +\infty$:

(5.10)    $a^x = o(b^x)$   when $x \to +\infty$ if $0 < a < b < 1$;

(5.11)    $a^x = o(x^s)$        if $0 < a < 1, s < 0$;

(5.12)    $x^s = o(x^t)$        if $s < t < 0$.

Relation (10) is proved like (6). (11) is obtained by putting $s = -t$ and $a = 1/b$, so that $a^x/x^s = x^{-t}/b^x$ which tends to 0 by (7). Finally, (12) is proved like (8).

We have not written the trivial relations between a function which tends to 0 and a function which tends to $+\infty$.

What happens when $x \to -\infty$ can be deduced immediately from the preceding, and concerns only the exponential functions since the others presuppose $x > 0$. We put $x = -y$ and apply the results for the case where $y \to +\infty$. For example,

(5.13)        $b^x = o(a^x)$ when $x \to -\infty$ if $0 < a < b$.

We have still to consider their behaviour on a neighbourhood of $x = 0$, and this reduces to the behaviour at infinity on putting $x = 1/y$. We find

(5.14)        $x^s = o(x^t)$    when $x \to +0$ if $s > t$;

(5.15)      $\log_a x = o(1/x^s)$        if $s > 0$.

Some of these formulae may be written in terms of limits:

(5.16)              $\lim_{x \to +\infty} x^s a^{-x} = 0$    if $a > 1$, $s \in \mathbb{R}$,

(5.17)            $\lim_{x \to +\infty} x^{-s} \log_a x = 0$    if $s > 0$, $a > 0$,

(5.18)              $\lim_{x \to 0} x^s \log_a x = 0$    if $s > 0$, $a > 0$.

For example, (17) shows that $(\log_a n)/n^s$ tends to 0 to infinity for any $s > 0$, and (18) shows that if you multiply $\log x$, which tends to $-\infty$ as $x$ tends to 0, by a function which tends to 0 as *slowly* as $x^{1/100\,000\,000}$, the result tends to 0.

These formulae allow us to understand why Cauchy's function[5]

(5.19)     $$f(x) = \exp(-1/x^2), \quad x \neq 0, \quad f(0) = 0$$

is indefinitely differentiable on $\mathbb{R}$, *including at* $x = 0$, that all its derivatives vanish at the origin, and therefore that it cannot be represented by Maclaurin's formula. The Chain Rule and the relation $\exp' = \exp$, obvious from the power series as we have already said several times, show first, that for $x \neq 0$, the function $f$ has successive derivatives given by

$$f'(x) = 2f(x)/x^3, \quad f''(x) = 4f(x)/x^6 - 6f(x)/x^4, \ldots$$

so of the form $f^{(k)}(x) = p_k(1/x)\exp(-1/x^2)$ where $p_k$ is a polynomial. Putting $1/x = y$ we get

$$f^{(k)}(x) = p_k(y)/\exp(y^2);$$

when $x$ tends to 0, $y^2$ tends to $+\infty$ and since every power, integer or not, of $y$ is negligible with respect to $\exp(y^2)$ by (16), we see that each derivative $f^{(k)}(x)$ tends to 0 with $x$. To show that $f$ is $C^\infty$ even on a neighbourhood of 0, it thus is enough to show directly that the successive derivatives of $f$ all vanish at 0.

First, $f'(0) = \lim f(x)/x = \lim \exp(-1/x^2)/x = 0$ as we have just seen. Thus the first derivative $f'$ exists for all $x$ and is continuous, including at 0. The second derivative, if it exists, is the limit of the ratio $f'(x)/x = 2f(x)/x^4$, so we find a zero second derivative, and so on indefinitely.

The situation will be the same for the function

(5.20)     $$g(x) = \exp(-1/x^2) \text{ for } x > 0, \quad = 0 \text{ for } x \leq 0.$$

This time the derivatives are all zero for $x \leq 0$. If all $C^\infty$ functions were analytic, aeroplanes could never take off; they would roll *eternally* on their take-off strips, assumed rigorously flat, and, in flight, would be incapable of changing direction or altitude, since if a function is real-analytic on an interval $I$ and constant on a neighbourhood of a point $t$ of $I$, it is constant on $I$. It is the existence of functions such as (20) which allows one to join $C^\infty$ functions (for example constants) defined on closed disjoint intervals perfectly *smoothly*. This remark will play an important rôle in Chap. V, n° 29, when we shall prove the existence of $C^\infty$ functions having arbitrarily prescribed successive derivatives at a given point.

## 6 – Characterisations of the exponential, power and logarithmic functions

For any $a > 0$, the exponential function $f(x) = a^x$ satisfies the identity

---

[5] Here we use the fact, proved later, that the function $\exp(x) = 1 + x + \ldots$ is a particular function $\exp_a(x)$, with $a = e = \exp(1) > 1$. The following argument can be applied equally to every other exponential function of base $a > 1$.

(6.1) $$f(x+y) = f(x)f(y);$$

this "addition formula", which we have already met à propos the series $\exp x$ (Chap. II, n° 22), analogous to those for the trigonometric functions but simpler, is an example of what is called a *functional equation* ; for the function $f(x) = x^s$ we have an analogous relation

(6.2) $$f(xy) = f(x)f(y),$$

for a linear function $f(x) = ax$ we have

(6.3) $$f(x+y) = f(x) + f(y),$$

for the logarithmic functions we have

(6.4) $$f(xy) = f(x) + f(y),$$

etc. As we shall see, these relations *characterise* the functions in question so long as one also insists that they be reasonable, i.e. continuous or monotone, and, if necessary, real-valued. This is already to be found in Cauchy's *Cours d'analyse*.

**Theorem 2.** *Every real-valued function defined on $\mathbb{R}$ which is not identically zero, satisfies* (1), *and is monotone or continuous, is an exponential function.*

Since $f(x) = f(x/2 + x/2) = f(x/2)^2$ we have $f(x) \geq 0$ for all $x$. We even have strict inequality, since if we had $f(c) = 0$ it would follow that $f(x) = f(c)f(x-c) = 0$ for any $x \in \mathbb{R}$. Since $f(0+x) = f(0)f(x)$, we deduce that $f(0) = 1$.

Let us put $f(1) = a$, an excellent idea if we want to prove that $f(x) = a^x$ for some $a$. Now $f(2) = a^2$, then $f(3) = f(2+1) = a^3$, and more generally $f(n) = a^n$ for all $n \in \mathbb{N}$. For $n$ negative we write

$$1 = f(0) = f(n + (-n)) = f(n)f(-n) = f(n)a^{-n},$$

whence again $f(n) = a^n$.

Let $x = p/q$ be a rational number, with $q > 0$. The relation (1) shows that

$$f(qx) = f(x + \ldots + x) = f(x) \ldots f(x) = f(x)^q,$$

whence $f(x)^q = f(p) = a^p$. Since $f(x) > 0$, it follows that

$$f(x) = (a^p)^{1/q} = a^{p/q} = a^x$$

for all $x \in \mathbb{Q}$. If $f$ is continuous or monotone, we can then apply Theorem 1 and conclude that $f(x) = a^x$ for all $x \in \mathbb{R}$, qed.

The reader will easily prove that in fact the relation

(6.5)             $f(nx) = f(x)^n$ for all $x \in \mathbb{R}$ and all $n \in \mathbb{Z}$

is enough to characterise the exponential functions among the real continuous or monotone functions.

**Theorem 3.** *Every real-valued function defined on $\mathbb{R}$ which satisfies $f(x+y) = f(x) + f(y)$, and is monotone or continuous, is a linear function.*

Put $f(1) = a$, and use (3) to show that $f(x) = ax$ first for $x \in \mathbb{N}$, then for $x \in \mathbb{Z}$, then for $x \in \mathbb{Q}$. The function being monotone or continuous, the formula extends immediately to arbitrary real values of $x$.

Here, as above, the relation $f(nx) = n.f(x)$ leads to the same result[6]. Note further that this remains valid if $f$ has complex values; but one must then assume $f$ continuous since the other possible hypothesis is then meaningless.

**Theorem 4.** *Every real-valued function, defined for $x > 0$, which is not identically zero, satisfies $f(xy) = f(x)f(y)$, and is monotone or continuous, is a power function.*

Such a function in fact has positive values since $f(x) = f(x^{1/2}x^{1/2}) = f(x^{1/2})^2$. It can never vanish, since $f(c) = 0$ would imply that $f(x) = f(c)f(x/c) = 0$ for all $x > 0$.

Now let us choose an $a > 0$, not 1, and put $g(x) = f(a^x)$. The relation (2) transforms immediately into $g(x+y) = g(x)g(y)$. Moreover, the function $g$, the composition of $f$ and of an exponential function, is continuous. By Theorem 1 there exists a $b > 0$ such that $g(x) = b^x$. On the other hand, there exists an $s \in \mathbb{R}$ such that $b = a^s$, namely $s = \log_a b$. Hence

$$f(a^x) = g(x) = b^x = (a^s)^x = a^{sx} = (a^x)^s,$$

and, since any number $y > 0$ can be put in the form $a^x$, we conclude that $f(y) = y^s$ for all $y > 0$, qed.

*Exercise.* Find a direct proof of Theorem 4.

**Theorem 5.** *Every real-valued function defined for $x > 0$ which is not identically zero, satisfies $f(xy) = f(x) + f(y)$, and is monotone or continuous, is a logarithmic function.*

---

[6] There are solutions of the equation $f(x+y) = f(x) + f(y)$ which are neither continuous nor monotone. The relation in question certainly implies $f(cx) = cf(x)$ for $c \in \mathbb{Q}$, so that, if one considers $\mathbb{R}$ as a vector space (of infinite dimension!) over $\mathbb{Q}$, every "linear form" on $\mathbb{R}$ satisfies this condition. To construct one effectively, one would have to choose a "base" of $\mathbb{R}$ over $\mathbb{Q}$, requiring set theoretic constructions calling on the theory of transfinite numbers or the axiom of choice. These solutions are interesting only as curiosities.

Choose an $a > 0$ different from 1 and put $g(x) = \exp_a[f(x)]$, in other words $g = \exp_a \circ f$. Then

$$g(xy) = \exp_a[f(x) + f(y)] = \exp_a[f(x)] \cdot \exp_a[f(y)] = g(x)g(y).$$

Like the functions $f$ and $\exp_a$, the function $g$ is continuous or monotone. It is therefore a power function by the preceding theorem, in other words, there exists an $s \in \mathbb{R}$ such that $a^{f(x)} = x^s$ for all $x > 0$, whence

$$f(x) = \log_a(x^s) = s \cdot \log_a x.$$

But every function proportional to a logarithmic function is itself of the same type as we saw at the end of n° 3, whence Theorem 5.

Theorem 5 applies clearly, and above all, to the function

$$\log x = \lim n \left( x^{1/n} - 1 \right)$$

of Chap. II, n° 10, Theorem 3, and Theorem 2 applies to the function $\exp x = \sum x^n/n!$ for which we established the addition formula in Chap. II, n° 22; these functions are indeed continuous and monotone. Compare with Chap. III, n° 2, example 1, where we proved in the same way that $\log[\exp(x)]$ is proportional to $x$. We shall return to this in the second part of this chapter, but may note now that if $\exp(x) = a^x$, then necessarily $a = \exp(1) = e = \sum 1/n!$. And if one knows that $\log[\exp(x)] = x$, as we have shown (same reference), then $\log = \log_e$.

## 7 – Derivatives of the exponential functions: direct method

Let $f(x) = \exp_a x = a^x$ be an exponential function. Suppose that we have established the existence of $f'(0)$. Then, for all $x \in \mathbb{R}$,

$$\begin{aligned} f(x+h) \quad &= \quad f(x)f(h) = f(x)[f(0) + f'(0)h + o(h)] \\ &= \quad f(x) + f(x)f'(0)h + o(h) \end{aligned}$$

as $h \to 0$, whence the existence of $f'(x) = c.f(x)$, with $c = f'(0)$; from this it is clear that $f$ has successive derivatives of all orders, and that $f^{(p)}(x) = c^p.f(x)$ for any $p$, so this function is indefinitely differentiable in the sense of Chap. III, n° 15.

We therefore have to show that, for all $a > 0$, the ratio

(7.1)                                    $(a^h - 1)/h$

tends to a limit when $h$ tends to 0. We restrict to the case where $a > 1$, as the other case can be treated similarly (or reduced to it).

Let us start by making $h$ tend to 0 through positive values. We shall see that the ratio (1) is an *increasing*[7] function of $h$ for $h > 0$. Since it is bounded below by 0 (we know that $a^h > 1$ for $a > 1$ and $h > 0$), the existence of the limit as $h$ decreases to 0 will be assured by the general theorems (Chap. III, n° 3) on monotone functions.

We now need to show that

(7.2) $$0 < u \le v \Longrightarrow \frac{a^u - 1}{u} \le \frac{a^v - 1}{v},$$

in other words, that, for $u$ and $v$ real,

(7.3) $$0 < u \le v \text{ implies } va^u - ua^v \le v - u.$$

It is enough to do this for $u, v$ *rational*. Approximating $u$ and $v$ by sequences $(u_n)$ and $(v_n)$ of rational numbers, we have indeed

$$\lim (v_n a^{u_n} - u_n a^{v_n}) = va^u - ua^v$$

by the continuity of the functions in question; if the relation (3) can be established for $u_n$ and $v_n$, it then holds in the limit for $u$ and $v$.

Assuming $u$ and $v$ rational, put $u = p/n$, $v = q/n$ with $p, q, n$ integers, $n > 0$, and $p < q$ since $u < v$. On dividing by $n$ the relation (2) becomes

$$\frac{a^{p/n} - 1}{p} \le \frac{a^{q/n} - 1}{q}.$$

Putting $a^{1/n} = b > 1$, this can again be written

(7.4) $$\frac{b^p - 1}{p} \le \frac{b^q - 1}{q}.$$

This has reduced (2) to the case where the exponents $u$ and $v$ in the formula (2) are *integers*.

Now put $b = 1 + c$ with $c > 0$. The algebraic binomial formula immediately yields the relation

$$\frac{b^p - 1}{p} = c + (p-1)c^{[2]} + (p-1)(p-2)c^{[3]} + \ldots + (p-1)\ldots 1\, c^{[p]}$$

and a similar relation for $q$. If $p < q$, it is clear that the coefficient of $c^{[k]} = c^k/k!$ in the formula relative to $p$ is smaller than the corresponding coefficient in the formula relative to $q$. What is more, the latter contains more terms than the first. Since all these terms are positive, the relation (4) is proved, so also (2). [A similar argument to that in Chap. II, n° 10, example 2.]

---

[7] For $h$ of the form $1/n$, this is the result which allowed us to define the function log in Chap. II, n° 10, Theorem 3.

So we see that the ratio (1) truly has a limit, say $c$, as $h$ tends to 0 through values $> 0$. Now consider what happens when $h$ tends to 0 through negative values. We can put $h = -k$ with $k > 0$, whence

$$\frac{a^h - 1}{h} = \frac{1/a^k - 1}{-k} = a^{-k}\frac{a^k - 1}{k};$$

when $h$ tends to 0 through negative values, the factor $a^{-k}$ tends to 1, and the last fraction to $c$ by the first case. So we find a left limit equal to the right limit, which completes the proof of the existence of the derivative at the origin, and so everywhere, of the function $a^x$.

These calculations have demonstrated that there is a formula

(7.5) $$(a^x)' = L(a).a^x$$

with a factor $L(a)$ independent of $x$, but of course depending on $a$. This is the derivative at the origin, so that

(7.6) $$L(a) = \lim_{h \to 0, h \neq 0} (a^h - 1)/h.$$

Since $(ab)^x = a^x b^x$, the product rule for differentiation shows that

$$L(ab)(ab)^x = L(a)a^x b^x + a^x L(b)b^x,$$

whence $L(ab) = L(a) + L(b)$, curiously. Moreover, one can take $h = 1/n$ and let the integer $n$ tend to infinity; one finds

(7.7) $$L(a) = \lim n \left( a^{1/n} - 1 \right) \quad \text{for all } a > 0.$$

In other words, $L(x) = \log x$ and in consequence

(7.8) $$(a^x)' = \log(a)a^x.$$

Even though we carefully banished the Napierian log in the preceding n°s, we cannot prevent it from returning at the gallop; this shows the privileged status of the so-called *natural* log relative to the log to any other base. Further, formula (6) is much more general than the definition (7) of the log since one can now substitute for $1/n$ anything tending to 0 in a "discrete" or "continuous" way. In the author's youth (he no longer enters competitions nor judges them, nor does he know what happens nowadays)[8] this remark gave rise to abundantly exploited traps. Consider for example the sequence

$$u_n = \left( n^2 + 3n - 100 \right) \left[ 2^{(n^2-1)/(3n^4+5)} - 1 \right] = c_n \left( 2^{h_n} - 1 \right);$$

---

[8] The university examinations suffice. Peano once said that "relations between students and professors would be excellent if it were not for the examinations, which force students to consider their masters as potential judges".

the exponent $h_n$ tends to 0, so that on replacing the factor $c_n$ by $1/h_n$ one will find a sequence $v_n$ which tends to $\log 2$. Now $u_n = c_n h_n v_n$, and since $c_n h_n$ is a rational fraction in $n$ whose terms of highest degree have ratio $1/3$, one concludes that $\lim u_n = \log(2)/3$. Confronted by this kind of challenge, a thoughtless or a somewhat slow young man (there were no young girls) might be deprived of the ineffable privilege of wearing a cocked hat and a sword on the Champs Elysées on national holidays[9], not to mention the incalculable consequences to the course of his career.

## 8 – Derivatives of exponential functions, powers and logarithms

Formula (7.8) also allows us to calculate the derivatives of the other functions defined above:

**Theorem 6.** *The exponential, logarithmic and power functions are indefinitely differentiable, and:*

(8.1) $\qquad (a^x)' = \log(a).a^x, \qquad (a^x)'' = \log(a)^2.a^x$, etc.

(8.2) $\qquad \log_a'(x) = 1/\log(a)x, \qquad \log_a''(x) = -1/\log(a)x^2$, etc.

(8.3) $\qquad (x^s)' = sx^{s-1}, \qquad (x^s)'' = s(s-1)x^{s-2}$, etc.

The exponential functions have been dealt with in the preceding n°. From this one can settle the case of any logarithmic function $g = \log_a$.

First of all, it is easy to guess *a priori*, up to a few details, the value of its derivative, assuming we have proved its existence. Let us start from the relation

$$g(ax) = g(x) + g(a)$$

and, for $a$ fixed and $x$ varying, let us calculate the derivatives of the two sides. That of the left hand side is $ag'(ax)$ and that of the right hand side is $g'(x)$. So $ag'(ax) = g'(x)$ for all $a$ and $x$, whence $ag'(a) = g'(1)$; since $a$ is arbitrary we can write this result in the form

(8.4) $$g'(x) = g'(1)/x.$$

One might also, as in the preceding case, remark that

(8.5) $\qquad g(x+h) - g(x) = g(1 + h/x) = g'(1)h/x + o(h).$

But we still have to justify the existence of the derivative. To do this we use rule (D 5) of Chap. III, n° 15, concerning the derivatives of inverse functions.

---

[9] Speaking of what he calls his "incorporation" into the Ecole polytechnique in 1951, a disabused nucleocrat wrote: "This was the day that I fully understood with horror that [Polytechnique] was a boarding school under military discipline". Yves Girard, *Un neutron entre les dents* (Paris, Ed. Rive droite, 1997), p. 18.

The logarithm function $g = \log_a$ is of course, by definition, the inverse of the exponential function $f = \exp_a$, which is differentiable as often as one wishes. The existence of $g'$ at a point $b > 0$ thus amounts to the fact that the derivative $f'$ is not zero at the point $a = g(b)$. And this is so, because $f'$ is proportional to $f$, which never vanishes[10].

In conclusion, we see that the function $\log_a x$, the inverse of $x \mapsto a^x = \exp_a(x)$, is differentiable and that its derivative at a point $x$ is the reciprocal of the derivative of the function $y \mapsto \exp_a y$ at the point $y = \log_a x$, i.e. is equal to $1/\log(a)a^y = 1/\log(a)x$ as stated.

To deduce that the logarithmic functions are indefinitely differentiable, and to calculate their successive derivatives, it is better, even if the reader has known how differentiate $1/x$ for a long time, to proceed first to the case of the power functions.

The function $f(x) = x^s$ is given by the formula

$$f(x) = \exp_a(s. \log_a x)$$

for any $a > 0$, as we have seen in (4.5); in other words, it is obtained by composing the two functions $x \mapsto s. \log_a x$ and $x \mapsto \exp_a(x)$ which we now know to be differentiable. Rule (D 4) of Chap. III, n° 15, now shows that $f$ is differentiable and that[11]

$$
\begin{aligned}
f'(x) &= \exp_a'(s. \log_a x).(s. \log_a x)' = \\
&= \log(a). \exp_a(s. \log_a x).s/x \log(a) = \log(a)x^s.s/x \log(a),
\end{aligned}
$$

whence, as stated,

(8.8)    $$(x^s)' = sx^{s-1} \qquad \text{for } x > 0, \ s \in \mathbb{R},$$

a formula well known for integer $s$, but valid for all real exponents (and even complex exponents, as we shall see).

On iterating this result, one sees that the function $x^s$ possesses derivatives of all orders, namely[12] $sx^{s-1}$, $s(s-1)x^{s-2}$, $s(s-1)(s-2)x^{s-3}$, etc.

---

[10] One still ought to check that $\log a \neq 0$. But if this were not the case, then the derivative $f'$ would be identically zero and the function $f(x) = a^x$ would be constant (mean value theorem).

[11] An expression such as $\exp_a'(s. \log_a x)$ denotes the value at the point $s. \log_a x$ of the function $\exp_a'$, the derivative of $\exp_a$. For consistency, we should write $(s. \log_a)'(x)$ instead of $(s. \log_a x)'$ ...

[12] Newton's binomial formula

$$(1 + x)^s = 1 + sx + s(s-1)x^{[2]} + s(s-1)(s-2)x^{[3]} + \ldots$$

stated in Chap. II, n° 6, thus reduces to the MacLaurin formula for the function $(1 + x)^s$ whose derivatives at $x = 0$ are 1, $s$, $s(s-1)$, etc. Less surprising if one knew in advance that the said function is analytic, but this is by no means obvious from its definition. We shall establish this later in another way.

In particular, this is the case for $s = -1$, and since $\log'_a(x) = 1/x \cdot \log a$, we conclude that the logarithmic functions are indefinitely differentiable, like the exponential functions and powers.

These results are absolutely fundamental, even though we still lack the most important, those concerning the functions log (without indices ...). The second part of this chapter will allow us to fill this gap in, and to go much further in conjuring up a *Deus ex machina:* the addition formula for the exponential *series*.

# §2. Series expansions

## 9 – The number $e$. Napierian logarithms

In Chap. II, n° 22, we showed that the exponential series

$$(9.1) \qquad \exp(z) = 1 + z/1! + z^2/2! + \ldots = \sum z^n/n! = \sum z^{[n]}$$

satisfies the relation

$$(9.2) \qquad \exp(x + y) = \exp(x).\exp(y)$$

for all $x, y \in \mathbb{C}$ and in particular for $x$ and $y$ real. It is clearly continuous on $\mathbb{R}$ (and even on $\mathbb{C}$) since it is analytic. In view of Theorem 2 above, there must exist a number $e > 0$ such that

$$(9.3) \qquad \exp(x) = \exp_e(x) = e^x$$

for all $x \in \mathbb{R}$, with necessarily

$$(9.4) \qquad e = \exp(1) = \sum 1/n! = 1 + 1 + 1/2! + 1/3! + \ldots.$$

Thus

$$(9.5) \qquad e^x = 1 + x/1! + x^2/2! + \ldots$$

where $e$ is the number (4), an *a priori* miraculous result if one defines $e^x$ by (1). The "miracle" stems from the addition formula for the series for exp.
Formula (8.1) can be then be written

$$(9.6) \qquad \exp'(x) = \log(e).\exp x.$$

But the formula in Chap. II, n° 19, for differentiating power series, namely that

$$f(x) = \sum c_n x^{[n]} \implies f'(x) = \sum c_{n+1} x^{[n]},$$

shows that

$$(9.7) \qquad \exp'(x) = \exp x$$

as we have already said N times. It follows that

$$(9.8) \qquad \log e = 1.$$

And since the function $\log x = \lim n(x^{1/n} - 1)$ satisfies the functional equation $\log(xy) = \log x + \log y$ (Chap. II, n° 10), is increasing and not identically zero, there is a number $a > 0$ such that $\log x = \log_a x$ (Theorem 5), namely *the* number such that $\log a = 1$. In other words, we see that Napier's function $\log$ is simply $\log_e$, i.e. the inverse map of $\exp_e = \exp$:

(9.9)          $$y = \log x \iff x = e^y \iff x = \exp y = \sum y^n/n!,$$

clearly a fundamental result, as is the relation

(9.10)          $$\log' x = 1/x$$

that one obtains on putting $a = e$ in (8.2), and already proved otherwise in Chap. II, n° 10.

We have also proved (9) in Chap. III, n° 2 and (10) at the end of n°  14, using the integral

$$\log b - \log a = \int_a^b dt/t$$

of Chap. II, n° 11. Starting from here and from the formula $\exp' = \exp$, one sees that the derivative of the function $\exp(\log x)$, or of $f(x) = \log(\exp x)$, is equal to 1. We must then have $f(x) = x$ up to an additive constant, by Chap. III, n° 16, Corollary 1 of the mean value theorem, whence $f(x) = x$ since $f(0) = 0$.

You will find another method in n° 10.

The functions log and exp thus satisfy all the identities obtained in n° 4; in particular,

(9.11)          $$x^s = e^{s.\log x} = \sum (s.\log x)^n/n!$$

for all $x > 0$ and all $s \in \mathbb{R}$. Since, on the other hand, the functions $\log_a$ are all mutually proportional, they are proportional to the Napierian log:

(9.12)          $$\log_a(x) = \log(x)/\log(a)$$

since the left hand side is 1 for $x = a$.

Further, all the exponential functions can be expressed in terms of the single function exp: if we replace $x$ by $a$ and $s$ by $x$ in (11), we have

(9.13)          $$\exp_a(x) = \exp(x.\log a).$$

In other words, *there are no exponential functions apart from the functions* $x \mapsto e^{cx}$ where c is an arbitrary real constant, and no logarithmic functions other than the functions $x \mapsto c.\log x$. From now on you may forget $\log_a$ and the other $\exp_a$ without any worry: they will never appear, except, with $a = 10$, in numerical calculations.

## 10 – Exponential and logarithmic series: direct method

To obtain the main result of the preceding n° – the fact that the functions log and exp are inverses of one another –, we drew on the direct construction of the expressions $a^x$ and on the theorems characterising the logarithmic and exponential functions by their functional equations.

One might do without all this by starting from the functional equation for the series exp and reconstituting it all by this route. In fact, we could have done this from the end of Chap. II, but it was already quite long ...

Let us now pretend not to have read n°s 1 to 9 of this Chap. IV and let us start from the relation

(10.1) $$\exp(x).\exp(y) = \exp(x+y),$$

valid for all $x, y \in \mathbb{C}$. It has immediate consequences.

First, (i) the function $\exp(z)$ never vanishes, even for $z$ complex, (ii) $\exp(x)$ is strictly positive for $x$ real, (iii) the function exp is strictly increasing, since the series $\exp x = 1 + x/1! + x^2/2! + \ldots$ shows that $x > 0 \Longrightarrow \exp x > 1$ (iv) we have

$$\exp(px/q)^q = \exp(x)^p$$

for $x \in C$ and $p, q$ integers. In particular, $\exp(px/q) = \exp(x)^{p/q}$ for $x$ real. It is unnecessary to detail these obvious properties again.

Secondly, the exponential function is *analytic* on $\mathbb{C}$, not only by reason of the general result of Chap. II, n° 19, but, more simply, because for any $a \in \mathbb{C}$ it has a power series expansion

$$\exp z = \exp(a)\exp(z-a) = \exp(a)\sum(z-a)^n/n!.$$

In particular, the function exp is continuous on $\mathbb{C}$, so on $\mathbb{R}$; since it takes arbitrarily large values on $\mathbb{R}$, $[\exp n = \exp(1)^n$ with $\exp 1 > 1]$ and arbitrarily close to 0 [since $\exp(-x) = 1/\exp x$], and since it is strictly increasing, the function exp maps $\mathbb{R}$ bijectively onto $\mathbb{R}_+^*$.

Thirdly, the exponential function is identical to its derivative, and so to all its successive derivatives as shown by Chap. II, n° 19, on the derived series defined by the general algorithm or as the traditional limit of the quotient. This result applies even for complex $z$.

An obvious calculation shows more generally that, for $t \in \mathbb{C}$, the function

$$f(z) = \exp(tz) \quad \text{satisfies} \quad f'(z) = tf(z)$$

i.e. is proportional to its derivative, whence

$$f^{(n)}(z) = t^n.f(z)$$

for any $n \in \mathbb{N}$. This allows us to solve "linear differential equations with constant coefficients". Suppose for example that we want to find the (or, at this level of the exposition, some) functions $y = f(x)$ satisfying

$$y'''' - 3y''' + 3y'' - 3y' + 2y = 0.$$

We are tempted to look for solutions of the form $\exp(tx)$; the parameter $t$ must clearly verify the algebraic equation

$$t^4 - 3t^3 + 3t^2 - 3t + 2 = 0;$$

this can be rewritten as $(t-1)(t-2)(t^2+1) = 0$, so that we immediately obtain four solutions of the given differential equation: $\exp x$, $\exp 2x$, $\exp ix$ and $\exp(-ix)$ and, more generally, all the functions

$$y = c_1 \exp(x) + c_2 \exp(2x) + c_3 \exp(ix) + c_4 \exp(-ix)$$

with arbitrary complex constants, since clearly every sum of solutions is again a solution. It is not obvious, but it is true, that all the solutions of the given differential equation can be found in this way.

We can now establish the fundamental result linking the functions exp and log:

**Theorem 7.**

(10.2)          $\log(\exp x) = x$ *for all $x$ real,*

(10.3)          $\exp(\log y) = y$ *for all $y$ real $> 0$.*

Recall the proof already given (Chap. III, n° 2, example 1). By definition,

(10.4)          $\log y = \lim n(y^{1/n} - 1)$

for all $y > 0$. Since $\exp(x)^{1/n} = \exp(x/n)$ for $x \in \mathbb{R}$ and $n$ an integer, it follows that

$$
\begin{aligned}
\log(\exp(x)) &= \lim n[\exp(x)^{1/n} - 1) = \lim n[\exp(x/n) - 1] = \\
&= \lim \frac{\exp(x/n) - 1}{1/n} = x.\lim \frac{\exp(x/n) - \exp(0)}{x/n}.
\end{aligned}
$$

When $n$ increases indefinitely, $h = x/n$ tends to 0, so that the quotient $[\exp(h) - \exp(0)]/h$ tends to the derivative of the function exp at $x = 0$, i.e. to 1 since we have seen above that $\exp' = \exp$ or, more simply, because $\exp(h) - 1 = h + h^2/2! + \ldots \sim h$ when $h$ tends to 0. Whence, in the limit, $\log(\exp(x)) = x$ for all $x \in \mathbb{R}$. Since the function exp is a bijection of $\mathbb{R}$ onto $\mathbb{R}^*_+$ it follows that the function $\log : \mathbb{R}^*_+ \longrightarrow \mathbb{R}$ is its inverse map, whence the theorem.

In this way we again find that the function $\log x$ is continuous (Chap. II, n° 10), differentiable, and that

(10.5)          $\log' x = 1/x$.

This follows from rule (D 5) of Chap. III, n° 15 and the relation $\exp' = \exp$. It follows from (5) that the function $\log(1 + x)$ has derivative

$$1/(1 + x) = 1 - x + x^2 - \ldots$$

for $x \in \ ] - 1, 1[$, and, since this power series is itself the derivative of $x - x^2/2 + x^3/3 - \ldots$, one concludes, as in Chap. III, n° 16, example 1, that

(10.6)     $\log(1 + x) = x - x^2/2 + x^3/3 - \ldots \ \text{for } x \in \ ] - 1, 1]$

(the case $x = 1$ is obtained by passing to the limit).

In its turn formula (5) implies a result which we shall extend in the next n° to the case where $x$ is complex:

(10.7)                  $\exp(x) = \lim(1 + x/n)^n$

for all $x \in \mathbb{R}$ and even

(10.7')          $\exp(x) = \lim(1 + hx)^{1/h} \ \text{when } h \to 0.$

Indeed

$$\log\left[(1 + hx)^{1/h}\right] = \log(1 + hx)/h = x\frac{\log(1 + hx) - \log 1}{hx},$$

and since $hx$ tends to 0, the fraction in the third term tends to the derivative of the function log at $x = 1$, i.e. to 1, by (5). In consequence, $\log\left[(1 + hx)^{1/h}\right]$ tends to $x$, whence, by continuity,

$$\exp(x) = \lim \exp\left\{\log\left[(1 + hx)^{1/h}\right]\right\} = \lim(1 + hx)^{1/h},$$

which proves (7').

To conclude, let us remark that the functions log and exp can be used to *define* the general exponentials $a^x$ and to establish their properties. The addition formula for the function log shows immediately that $\log(a^n) = n.\log a$ for $n \in \mathbb{Z}$, i.e. that

$$a^n = \exp(n.\log a).$$

One deduces that

$$\exp\left[\log(a)p/q\right]^q = \exp(p.\log a) = a^p,$$

whence

$$(a^p)^{1/q} = \exp[\log(a)p/q].$$

From this it is natural to *define*

(10.8)                  $a^x = \exp(x.\log a)$

for all $x \in \mathbb{R}$, or by

(10.9)                  $\log(a^x) = x.\log a.$

Continuity of the function $a^x$ is obvious, and the index laws are easily obtained:

$$
\begin{aligned}
a^x a^y &= \exp(x.\log a)\exp(y.\log a) = \exp[(x+y).\log a] = a^{x+y},\\
a^x b^x &= \exp(x.\log a)\exp(x.\log b) = \exp[x(\log a + \log b)] = \\
&= \exp[x.\log(ab)] = (ab)^x,\\
(a^x)^y &= \exp[y.\log(a^x)] = \exp(yx.\log a) = a^{xy}.
\end{aligned}
$$

The interest of definition (8) is that *it retains a meaning for x complex* and so allows one to define the *complex powers*, i.e. the expression $a^z$, assuming $a$ real $> 0$; definition (9), on the other hand, is unusable since the log of a complex number has no meaning (or, as we shall show later, is defined only up to addition of a multiple of $2\pi i$). The first and third rules (1.1) for real exponents extend to complex powers with the same proof, but the second raises the problem of defining a complex power of a complex number, so one cannot use it without precautions except when $a \in \mathbb{R}_+^*$, $x \in \mathbb{R}$, $y \in \mathbb{C}$. This is one of the traps in the subject ...

Definition (8) also allows us to recover – and to extend to complex powers – the general formula for differentiation of power functions. For $s \in \mathbb{C}$ given, and $x > 0$, whence $x^s = \exp(s.\log x)$, one finds, on applying the chain rule [Chap. III, n° 15, rule (D 4)], that $(x^s)' = \exp'(s.\log x)(s.\log x)' = \exp(s\log x)s/x = x^s(s/x)$, i.e.

$$
(10.10) \qquad\qquad (x^s)' = sx^{s-1} \qquad (x > 0,\ s \in \mathbb{C})
$$

as in the case where $s$ is real. This seductive proof leaves something to be desired, since the general formula $g'(f(a))f'(a)$ of Chap. III, n° 15 assumes that $f$ has *real* values, while here $f(x) = s.\log x$ has *complex* values if $s$ is not real. Very luckily, with great foresight, we showed in Chap. III, n° 21, example 1, equ. (21.2), that if $g$ is a *holomorphic* function on an open $V$ in $\mathbb{C}$ and $f$ is a differentiable map on an interval $U$ of $\mathbb{R}$ inside $V$, then the derivative of the composite function $g[f(x)]$ exists and is given by the standard formula on condition that we interpret $g'$ as the complex derivative of $g$. The function $g(z) = \exp(z)$ being analytic on $V = \mathbb{C}$ and identical to its derivative in the complex sense, and the map $f : x \longmapsto s.\log x$ of $U = \mathbb{R}_+^*$ into $V$ also being differentiable as many times as one wants, even if the coefficient $s$ is complex, the proof of (10) is still valid for $s \in \mathbb{C}$.

Further, for $a > 0$ given, the function

$$
a^z = \exp(z.\log a) = \sum z^n \log^n(a)/n!
$$

is a power series, so an analytic function of $z$ whose complex derivative is given by

$$
(10.11) \qquad\qquad (a^z)' = \log(a)a^z
$$

as one sees on differentiating the power series term-by-term.

Note finally that with this definition of complex powers, we have

$$\exp(z) = e^z \quad \text{for all } z \in \mathbb{C},$$

which often allows us to simplify the notation.

## 11 – Newton's binomial series

**Theorem 8.** *We have*

$$(11.1) \qquad (1+z)^s \;\; = \;\; N_s(z) = 1 + sz + s(s-1)z^2/2! + \ldots =$$
$$= \;\; \sum s(s-1)\ldots(s-n+1)z^{[n]}$$

*for $-1 < z < 1$ and $s \in \mathbb{C}$.*

Since

$$\log(1+z) = z - z^2/2 + z^3/3 - \ldots$$

for $z \in \, ]-1,1[$ and $a^s = \exp(s.\log a)$ (a theorem for real $s$, the definition of $a^s$ for $s$ complex and $a > 0$), Theorem 8 becomes a particular case of the following result:

**Theorem 8 bis.** *We have*

$$(11.2) \quad \exp\left[s(z - z^2/2 + z^3/3 - \ldots)\right] = \sum_{n \geq 0} s(s-1)\ldots(s-n+1)z^{[n]}$$

*for $s, z \in \mathbb{C}$ and $|z| < 1$.*

There are several methods of proving these theorems, all of them instructive. Let us start with the best, which presents the drawback or the advantage – everything depends on the point of view . . . – of using the general theorems of Chap. II, n° 19 on complex-analytic functions, and settles the question in a way that was probably beyond the scope of anyone before Weierstrass' systematic study of analytic functions.

**First proof** (general case). Consider the two sides of (2) as functions of $z$ with $s \in \mathbb{C}$ given. These are *analytic* functions on the disc $D : |z| < 1$.

Consider first the left hand side $f(z) = \exp[sL(z)]$. On putting[13]

$$(11.3) \qquad L(z) = z - z^2/2 + \ldots,$$

$L$ is analytic on $D$ and exp is analytic on $\mathbb{C}$, so the composition of these two functions is analytic (Chap. II, n° 22, Theorem 17).

---

[13] We hold back from writing $\log(1 + z)$, an expression which we have not yet defined for $z$ complex and which, when we do so later, may take infinitely many values.

**First variation.** We look for the Maclaurin series of $f$ at the origin. We have (same reference)

$$f'(z) = \exp'[sL(z)]sL'(z) = sf(z)L'(z).$$

Since $L'(z) = 1 - z + z^2 - \ldots = (1 + z)^{-1}$ on $D$ it follows that

(11.4) $$(1 + z)f'(z) = sf(z)$$

on $D$. Differentiating this relation $n - 1$ times, using Leibniz' formula, we find

$$(1 + z)f^{(n)}(z) + (n - 1)f^{(n-1)}(z) = sf^{(n-1)}(z),$$

from which the numbers $a_n = f^{(n)}(0)$ satisfy $a_n = (s - n + 1)a_{n-1}$. Since $a_0 = 1$, one finds $a_n = s(s-1)\ldots(s-n+1)$; the Maclaurin series $\sum f^{(n)}(0)z^{[n]}$ of $f$ at the origin is thus Newton's series. Since $f$ is analytic, it is represented by its Maclaurin series on a neighbourhood of 0 (Chap. II, n° 19); the two sides of (2) coinciding on a neighbourhood of 0, they are equal on all the disc $D$ where they are analytic (Chap. II, n° 20: analytic continuation), qed.

**Second variation.** Consider Newton's power series $g(z)$. This also is an analytic function on $D$. Its derivative is obtained by the usual procedure [Chap. II, equ. (19.4)], whence

$$g'(z) = s + s(s - 1)z + s(s - 1)(s - 2)z^{[2]} + \ldots$$

On multiplying by $1 + z$, and taking account of the relation

$$\binom{s-1}{n} + \binom{s-1}{n-1} = \binom{s}{n}$$

between the binomial coefficients, we again obtain the differential equation

(11.4') $$(1 + z)g'(z) = sg(z).$$

Since $f(z)$ never vanishes, being an exponential, the function $g(z)/f(z) = h(z)$ is analytic on $D$ (Chap. II, n° 22) and its derivative can be calculated by the standard formula. The relations (4) and (4') now show that $h'(z) = 0$ and thus that all the successive derivatives of $h$ are zero. Now we showed in Chap. II, n° 20 (principle of analytic continuation) that, on an open connected set, for example $D$, such a function is *constant*[14] – one even needs much less (vanishing of successive derivatives at a single point). Since $h(0) = 1$, we have $f(z) = g(z)$ on $D$, qed.

---

[14] Instead of invoking the principle of analytic continuation, one might observe that $h$, being analytic, is holomorphic and that the relation $h' = 0$ means that its first partial derivatives with respect to the real coordinates $x$ and $y$ of $z$ are zero; it then remains to cite Chap. III, n° 21, equ. (21.11), which generalises the mean value theorem.

Note that, for $s = 1$, we obtain the formula

$$\exp(z - z^2/2 + \ldots) = 1 + z, \quad |z| < 1.$$

This is proved even more easily: the two sides are analytic on $D$; they coincide for real $z$ since $\log(1 + z) = z - z^2/2 + \ldots$ for $z \in \, ] - 1, 1[$; by analytic continuation, they are identical on $D$.

**Second proof** $(z \in \mathbb{R}, \; s \in \mathbb{R})$. This more elementary proof uses the functional equation for the exponential functions, but imposes restrictions on $s$ and $z$. Here one considers the series (1) as a function of $s$ for given $z \in \, ]-1, 1[$, and no longer as a function of $z$ for $s$ given.

Choose numbers $z$, $s$ and $t$, complex for the moment, with $|z| < 1$ to ensure (Chap. II, n° 16) the absolute convergence of the series

$$(11.5) \qquad N_s(z) \;=\; 1 + sz + s(s-1)z^2/2! + \ldots = \sum c_n(s)z^n,$$

$$(11.6) \qquad N_t(z) \;=\; 1 + tz + t(t-1)z^2/2! + \ldots = \sum c_n(t)z^n.$$

For $s, t \in \mathbb{N}$, their sums are equal to $(1 + z)^s$ and $(1 + z)^t$ by the *algebraic binomial formula*, so that, in this case,

$$(11.7) \qquad N_s(z).N_t(z) = N_{s+t}(z).$$

We shall see that this formula persists for any complex $s$ and $t$ and $z \in \mathbb{C}$, $|z| < 1$. The argument which follows is due to Euler (1774).

Let us apply the multiplication rule for power series (Chap. II, n° 22). We find an absolutely convergent power series $\sum c_n(s,t)z^n$ with

$$(11.8) \qquad c_n(s,t) = \sum_{p+q=n} c_p(s)c_p(t).$$

It thus reduces to proving the identity

$$(11.9) \qquad c_n(s+t) = \sum c_p(s)c_q(t).$$

First of all it is clear that the coefficients figuring in (5) and (6) are polynomials in $s$ and $t$ respectively, so that the left hand side of (9) is a polynomial in $s$ and $t$. The relation (9) to be proved is thus an identity between two polynomials in $s$ and $t$, an identity which we know, by the algebraic binomial formula, to be valid when one gives the variables $s$ and $t$ *positive integer* values. By subtraction, it thus all reduces to proving the following general result, where here one calls the variables $x$ and $y$ instead of $s$ and $t$:

**Principle of continuation of algebraic identities.** *Let*

$$P(x,y) = \sum a_{ij} x^i y^j$$

be a polynomial in two[15] variables with complex coefficients. Suppose that $P(x, y) = 0$ for all $x, y \in \mathbb{N}$. Then $a_{ij} = 0$ for all $i$ and $j$, and therefore $P(x, y) = 0$ for all $x, y \in \mathbb{C}$.

Grouping together the terms of $P$ containing the same power of $y$, we have $P(x, y) = \sum P_j(x)y^j$ with polynomials $P_j(x)$ in $x$. Give $x$ a value $n \in \mathbb{N}$ and consider $P(n, y) = \sum P_j(n)y^j$. By hypothesis, this polynomial in $y$ vanishes for $y \in \mathbb{N}$; so it has *infinitely many roots*. Now one of the more elementary results in the theory of algebraic equations in one unknown is that an equation of degree $\leq 152$ possesses no more than $152$ roots, except of course if all its coefficients are zero. So $P_j(n) = 0$ for all $n \in \mathbb{N}$, and this, for the same reason, proves that all the coefficients of all the $P_j$, i.e. those of $P$, are zero, qed.

The formula (7), which we now can write

$$(11.10) \qquad N_s(z)N_t(z) = N_{s+t}(z) \quad \text{for } s, t, z \in \mathbb{C}, \ |z| < 1,$$

is what will establish the theorem for $z$ and $s$ real. First, for $z$ given, the function $s \mapsto N_s(z)$ satisfies the addition formula for the exponential functions and, for $s$ and $z$ real, has real values. Since $N_1(z) = 1 + z$, to establish the relation

$$(11.11) \qquad N_s(z) = (1 + z)^s \quad \text{for } -1 < z < 1, \ s \in \mathbb{R},$$

it is enough (n° 6, Theorem 2) to show that, for $z \in \, ]-1, 1[$ given, the left hand side is a *continuous* function of $s$ on $\mathbb{R}$. We shall show that this is the case on $\mathbb{R}$, and even for complex $z$, $|z| < 1$.

The terms of Newton's series are indeed continuous functions of $s$. It thus is enough to show that it converges *normally* on any disc $|s| \leq R$ (Chap. III, n° 8, Theorem 9). But on such a disc

$$(11.12) \qquad |s(s - 1) \ldots (s - n + 1)z^n/n!| \leq$$
$$\leq R(R + 1) \ldots (R + n + 1)|z|^n/n! = v_n,$$

the general term of a series with positive terms independent of $s$; the latter converges for $|z| < 1$ since

$$v_{n+1}/v_n = (R + n + 2)|z|/(n + 1)$$

tends to $|z|$ as $n$ increases; whence normal convergence and (11), qed.

**Third proof** [general case; uses (10)]. To establish the formula

$$(11.13) \qquad \exp[sL(z)] = \sum c_n(s)z^n$$

for $s$ and $z$ complex, $|z| < 1$, we shall first show that, for $z \in D$ given, the two sides are *holomorphic* functions of $s$ on $\mathbb{C}$, i.e. that, considered as functions

---

[15] The case of any number of variables is treated in the same way.

of the real and imaginary parts of $s$, they are $C^1$ and satisfy the Cauchy condition $D_1 f = -iD_2 f$ (Chap. II, n° 19; see also n°s 19 to 22 of Chap. III). This is obvious for the left hand side

$$f(s) = \exp[sL(z)] = \sum L(z)^n s^{[n]}$$

since it is analytic on $\mathbb{C}$. The case of the right hand side is a little less easy[16].

We showed in Chap. III, n° 22, as a consequence of Theorem 23 on term-by-term differentiation of a series of $C^1$ functions (in the real sense) on an open subset $U$ of $\mathbb{C}$, that if these functions are holomorphic and if the series of derivatives in the complex sense converges uniformly (for example, normally) on any compact $K \subset U$, then the sum of the series is again holomorphic, and its derivative is obtained by differentiating the series term-by-term. Although we then declared this result to be of "no interest", it will serve us here. The reader is asked to consider the following arguments as a simple exercise since we have already obtained the result above in a much easier way.

It is clear that the complex derivative of the function $c_n(s) = s(s-1)\ldots(s-n+1)/n!$ is

$$c'_n(s) = \sum s(s-1)\ldots(s-p)'\ldots(s-n+1)/n!.$$

For $|s| \leq R$, one thus has

$$|c'_n(s)| \leq \sum R(R+1)\ldots 1 \ldots (R+n+1)/n!;$$

since $1 \leq R + p$ for all $p \geq 0$ so long as $R \geq 1$, and since the sum considered has $n$ terms, we have

$$|c'_n(s)| \leq nR(R+1)\ldots(R+n+1)/n!.$$

When we differentiate the series

$$g(s) = \sum c_n(s)z^n$$

term-by-term with respect to $s$ we find a series whose general term is majorised in modulus on the disc $|s| \leq R$ by

---

[16] We have seen above that, for $z$ given, Newton's series converges normally on every disc $|s| \leq R$. Since its terms are analytic functions of $s$ – polynomials –, the general theorems of Chap. VII show, if we deploy them, that the sum of the series is analytic on $\mathbb{C}$. As for proving "without knowing anything" that Newton's series is an everywhere convergent power series in $s$, this would demand intricate explicit calculations. Note also that, for the rest of the proof, one need only know that the power series of $f(s)$ is, like every power series, holomorphic on its disc of convergence; Theorem 23 of Chap. III, n° 22 on sequences of $C^1$ functions would allow one instead to give a direct proof, since on differentiating a power series in $s$ with respect to the real or imaginary coordinates of $s$ one finds the derived power series, up to a factor equal to 1 or to $i$; it is then enough to know that the latter has the same radius of convergence as the given series.

$$nR(R+1)\ldots(R+n+1)|z|^n/n! = w_n.$$

Since $w_{n+1}/w_n = (R+n+2)|z|/n$ tends to $|z| < 1$, the series converges, hence, on differentiating Newton's series $g$ term-by-term with respect to $s$, we find a series normally convergent on all compact subsets of $\mathbb{C}$.

It follows that $g$ is a holomorphic function of $s$ for all $z \in D$ and that one can calculate $g'(s)$ by differentiating the series term-by-term.

Let us calculate this derivative. We know by (10) that $g(s+t) = g(s)g(t)$ for $s,t \in \mathbb{C}$; differentiating this formula with respect to $t \in \mathbb{C}$ for $s$ given, we find $g'(s+t) = g(s)g'(t)$ and in particular

$$g'(s) = g(s)g'(0) = g(s) \sum c'_n(0)z^n.$$

Since $c_n(s) \doteq s(s-1)\ldots(s-n+1)/n!$ is a polynomial, $c'_n(0)$ is the coefficient of $s$ in its expansion, i.e. the term independent of $s$ in the polynomial $c_n(s)/s = (s-1)\ldots(s-n+1)/n!$, i.e. the value of this polynomial for $s = 0$, whence

$$c'_n(0) = (-1)^{n-1}(n-1)!/n! = (-1)^{n-1}/n.$$

So we find, for $|z| < 1$,

$$g'(0) = z - z^2/2 + z^3/3 - \ldots = L(z),$$

an anticipated miracle from which it follows that $g'(s) = L(z)g(s)$. But the function $f(s) = \exp[sL(z)]$ also satisfies $f'(s) = L(z)f(s)$. Since $f$ never vanishes, we deduce that the *holomorphic* function $g(s)/f(s)$ (Chap. III, end of n° 20) has an identically zero derivative on $\mathbb{C}$. It is therefore constant (Chap. III, n° 21, a result valid, on an open connected subset of $\mathbb{R}^2$, for every function $C^1$ whose two first partial derivatives are zero). Since $f(0) = g(0) = 1$ we conclude that $f = g$, which accomplishes, in every sense of the term, a proof that might be considered too baroque, but which nevertheless yields the general result with the minimum of means: the theorems of Chap. III, § 5 on functions of two real variables.

Let us now give some examples of applications of Newton's formula.

If $s \in \mathbb{Z}$ the formula is also valid for any $z \in \mathbb{C}$ if $s > 0$ (algebraic binomial formula). For $s$ a negative integer and $z$ complex, $|z| < 1$, the relation $N_s(z)N_{-s}(z) = 1$ shows that $N_s(z) = 1/(1+z)^{-s} = (1+z)^s$, whence the formula again. For $z$ not real and $s = p/q$ rational but not an integer, it is more prudent to write only that $N_{p/q}(z)^q = (1+z)^p$, given that a complex number has, as we shall see in n° 14, several $q^{\text{th}}$ roots of which, *a priori*, none is more "natural" that the others, in contrast to what happens for a real positive number.

*Example 1.* Replace $s$ by $-s$ with $s \in \mathbb{N}$. After a short calculation we obtain the formulae

$$(11.14) \quad \frac{1}{(1+z)^s} = 1 - sz + s(s+1)z^2/2! -$$

$$- s(s+1)(s+2)z^3/3! + \ldots \quad (s \in \mathbb{N}, |z| < 1).$$

$$(11.15) \quad \frac{1}{(1-z)^s} = 1 + sz + s(s+1)z^2/2! + \ldots =$$

$$= \sum s(s+1)\ldots(s+n-1)z^{[n]}.$$

We recover the geometric series when $s = 1$, and, when $s = 2, 3, \ldots$ the series

$$(11.16) \quad (1-z)^{-2} = 1 + 2z + 3z^2 + 4z^3 + 5z^4 + \ldots =$$

$$= \sum (n+1)z^n \quad (|z| < 1)$$

$$(11.17) \quad (1-z)^{-3} = 1 + 3z + 6z^2 + 10z^3 + 15z^4 + \ldots =$$

$$= \sum \frac{(n+1)(n+2)}{2} z^n$$

etc., already met in Chap. II, n° 19, formulae (12) to (15).

*Example 2.* This was Newton's first triumph; it would be carved on his tomb in Westminster Abbey (maybe even the general case), but a historian declared at the beginning of the last century that the inscription was no longer visible, if it had ever existed. Reinstating it would contribute to the cultural development of tourists and of the local churchgoers. On taking $s = 1/2$ we obtain $N_{1/2}(z)^2 = 1 + z$, so Newton's series is one of the two square roots of $1 + z$. For $z$ real, one finds the *positive* square root of $1 + z$ and on identifying the coefficients we obtain

$$(11.18) \quad (1+z)^{1/2} =$$

$$= 1 + z/2 + \frac{1}{2}\left(\frac{1}{2}-1\right)z^2/2! + \frac{1}{2}\left(\frac{1}{2}-1\right)\left(\frac{1}{2}-2\right)z^3/3! + \ldots$$

$$= 1 + z/2 - z^2/8 + z^3/16 - 5z^4/128 + \ldots \quad (|z| < 1)$$

or again

$$(11.19) \quad (1-z)^{1/2} = 1 - z/2 - \sum_{p \geq 2} \frac{1.3\ldots(2p-3)}{2.4\ldots 2p} z^p.$$

The formula remains valid for $z \in \mathbb{C}$ so long as one prefixes the sign $\pm$ to the right hand side, or else decides that the latter represents, by definition, the symbol $(1-z)^{1/2}$ for $|z| < 1$, which amounts to attributing a privileged rôle to one of the two possible square roots[17]; this is what we have already done in the real case when selecting the positive root.

---

[17] In contrast, it is impossible to define a *continuous* (still less, analytic) function $f(z)$ on all $\mathbb{C}$, or even only on $\mathbb{C}^*$, such that $f(z)^2 = z$ for all $z$. This is possible only on certain open ("simply connected") subsets of $\mathbb{C}^*$, for example, as here, on the open disc with centre 1 and radius 1.

Newton did not know explicitly, or did not seek to discover, the *general* formula for multiplying power series; it would not have cost him much trouble[18] and he would surely simply have shrugged his shoulders if one had made such a remark to him. He was clearly capable, as he wrote to Leibniz in 1676, of verifying that

$$(1 + x/2 - x^2/8 + x^3/16 - 5x^4/128 + \ldots) \cdot$$
$$\cdot (1 + x/2 - x^2/8 + x^3/16 - 5x^4/128 + \ldots) =$$
$$= (1 + x/2 - x^2/8 + x^3/16 - 5x^4/128 + \ldots) +$$
$$+ (x/2 + x^2/4 - x^3/16 + x^4/32 - \ldots) +$$
$$+ (-x^2/8 - x^3/16 + x^4/64 - \ldots) +$$
$$+ (x^3/16 + x^4/32 \ldots) + (-5x^4/128 - \ldots) + \ldots$$

which reduces to $1 + x$ modulo terms of degree $> 4$. The reader who cares to verify this modulo terms of degree $> 8$ may continue the calculation in the same way; Newton once remarked that his brain had never worked so well as when he was twenty ...

He was also capable, so as to confirm his formulae, of calculating $(1+x)^{1/2}$ by extracting the square root of $1 + x$ as in commercial arithmetic; again the analogy with decimal numbers, his inspiration for power series. The practice of this sport[19] having now been lost, we shall not insist on this point. More simply, if one postulates the existence of a formula

$$(1 + x)^{1/2} = 1 + a_1 x + a_2 x^2 + a_3 x^3 + \ldots,$$

one must have

$$
\begin{aligned}
1 + x &= \left(1 + a_1 x + a_2 x^2 + a_3 x^3 + \ldots\right)\left(1 + a_1 x + a_2 x^2 + a_3 x^3 + \ldots\right) = \\
&= 1 + 2a_1 x + \left(a_1^2 + 2a_2\right) x^2 + \left(2a_1 a_2 + 2a_3\right) x^3 + \ldots,
\end{aligned}
$$

whence $a_1 = 1/2$, $1/4 + 2a_2 = 0$ i.e. $a_2 = -1/8$, etc. This method allows one to check that the first terms of (10) are correct, but it seems difficult to

---

[18] The only obstacle would have been the absence of convenient notation. Indices, the $\sum$, the notation $f(x)$, "bound" and "dummy" variables etc. were introduced only later, hence in Newton's time all calculations were performed in a totally explicit way. In fact, it was Newton who was first to write the general formula for the binomial coefficients, even for positive integer exponents; the "Pascal triangle", which allows one to calculate them one-by-one, was not known before him. And even Newton limited himself, later, to writing $s(s-1)(s-2)\ldots/1.2.3\ldots$ without specifying the last factors. The first modern notations, for example $\int f(x)dx$, are due to Leibniz, a philosopher and logician, and to the Bernoullis who were in contact with him. One cannot overemphasise the rôle that well-chosen notation has had, and continues to have, on the progress of mathematics.

[19] John Wallis, in the books from which Newton learned analysis, is reputed to have consecrated a night of insomnia to calculating in his head the square root of a number of fifty digits to about fifty digits and to have dictated the result to his secretary in the morning.

divine the general term from it; Newton, in fact, did so, starting from the law of formation of the first coefficients and extrapolating what he knew for the case of an exponent $s \in \mathbb{N}$.

*Example 3.* Taking $s = -1/2$, we find the square root of $1/(1+z)$, denoted $(1+z)^{-1/2}$ even for $z \in \mathbb{C}$. On identifying the coefficients we get

$$(11.20) \qquad (1+z)^{-1/2} = 1 - z/2 + 3z^2/8 - 5z^3/16 + \\ + 35z^4/128 - \ldots \quad (|z| < 1),$$

or, in different notation,

$$(11.21) \qquad \frac{1}{(1-z)^{1/2}} = 1 + \frac{1}{2}z + \frac{1.3}{2.4}z^2 + \frac{1.3.5}{2.4.6}z^3 + \ldots \quad (|z| < 1).$$

## 12 – The power series for the logarithm

We showed in Chap. III, n° 16, example 1, and again in n° 10, that the formula

$$\log' x = 1/x$$

leads to

$$\log(1+x) = x - x^2/2 + x^3/3 - \ldots$$

for $-1 < x < 1$. We also know that the formula remains valid for $x = 1$ because of the continuity of the left hand side and of the fact that the series on the right hand side converges uniformly on $[0, 1]$ (Chap. III, n° 8, example 4).

One can give a proof of this series expansion which is more complicated and, besides its historical interest, is a beautiful exercise in passing to the limit in a sequence of series.

Formula (11.7) shows that $N_{1/p}(x) = (1+x)^{1/p}$ for $-1 < x < 1$ and $p$ a nonzero integer. This relation can be written, by the definition of $N_s$, as

$$(12.1)\ (1+x)^{1/p} = $$
$$= 1 + x/p + (1/p)(1/p - 1)x^2/2! + (1/p)(1/p - 1)(1/p - 2)x^3/3! + \ldots$$
$$= 1 + x/p - (1 - 1/p)x^2/p.2 + (1 - 1/p)(2 - 1/p)x^3/p.2.3 - $$
$$- (1 - 1/p)(2 - 1/p)(3 - 1/p)x^4/p.2.3.4 + \ldots,$$

assuming $-1 < x < 1$. [For example

$$(1.5)^{1/8} = 1 + 1/16 - 7/512 + 105/24576 - 2415/1572864 + \ldots,$$

an example of an alternating series with decreasing terms which even converges absolutely, since $|0.5| < 1$.] (1) shows that

$$(12.2) \qquad p\left[(1+x)^{1/p} - 1\right] = $$
$$= x - (1 - 1/p)x^2/2 + (1 - 1/p)(1 - 1/2p)x^3/3 - $$
$$- (1 - 1/p)(1 - 1/2p)(1 - 1/3p)x^4/4 + \ldots.$$

The left hand side tends by definition to $\log(1+x)$ as $p$ increases indefinitely; now the factors $1 - 1/p$, $1 - 1/2p$, etc. figuring in (2) tend to 1. It follows "obviously", as Halley, the comet man, and Euler remarked, that

$$\log(1 + x) = x - x^2/2 + x^3/3 - \ldots$$

But here we have a passage to the limit in a sum of *infinitely* many terms.
    To justify this, we put

(12.3)  $u_p(n) = (-1)^n (1 - 1/p)(1 - 1/2p)\ldots(1 - 1/np)x^{n+1}/(n+1)$,

so that

(12.4)  $$p\left[(1+x)^{1/p} - 1\right] = \sum_n u_p(n).$$

We have now to deal with a series whose terms are functions of an integer $p$, in other words, what we called a "sequence of series" in Chap. III, end of n° 13. We know that, for all $n$,

(12.5)  $$\lim_{p \to \infty} u_p(n) = x^{n+1}/(n+1) = u(n)$$

exists, that $|u_p(n)| \leq |u(n)|$ for all $p$ and $n$, and we would like to be able to deduce that

(12.6)  $$\lim_{p \to \infty} \sum_n u_p(n) = \sum_n u(n) = \sum_n \lim_{p \to \infty} u_p(n).$$

The crucial point in the proof will be the fact that, for $|x| < 1$, the series $\sum u_p(n)$ is dominated by an absolutely convergent series *independent of $p$*, namely $v(n) = |u(n)|$.

**Theorem 9 (dominated convergence for series).** *Let $\sum_n u_p(n)$, $p \in \mathbb{N}$, be a sequence of series. Suppose that (i) $u_p(n)$ tends to a limit $u(n)$ when $p \to +\infty$, (ii) (normal convergence) there exists a convergent series with positive terms $v(n)$ such that*

(12.7)  $$|u_p(n)| \leq v(n) \qquad \text{for all } p \text{ and } n.$$

*Then the series $\sum u(n)$ converges absolutely, and (6) holds.*

This statement is in fact a particular case of Theorem 17 of Chap. III, n° 13, on passing to the limit in a sequence of normally convergent series in a variable $t \in T \subset \mathbb{C}$: here $T = \mathbb{N}$, $u(n,t) = u_t(n)$ and $t$ tends to $a = +\infty$. But one can also proceed directly.
    First, the absolute convergence of the series $\sum u(n)$ is obvious since in the limit we have $|u(n)| \leq v(n)$ for all $n$. Now let $s = \sum u(n)$ and $s_p = \sum u_p(n)$. For all $p$ and all integers $N > 0$ we have

$$|s_p - s| \leq \sum_n |u_p(n) - u(n)| = \sum_{n \leq N} |u_p(n) - u(n)| + \sum_{n > N} |u_p(n) - u(n)| \leq$$

$$\leq \sum_{n \leq N} |u_p(n) - u(n)| + 2 \sum_{n > N} v(n).$$

Take an $r > 0$. The second sum, which does not depend on $p$, is $< r$ for all sufficiently large $N$ since the series $\sum v(n)$ converges. Choosing such an $N$, the first sum tends to 0 as $p \to \infty$ since so do the $N$ differences which compose it; it is therefore $< r$ once $p$ exceeds a suitably chosen integer $p_N$. Then $|s_p - s| < 3r$ for $p$ sufficiently large, qed.

We can now state the result formally:

**Theorem 10.** *We have*

(12.8)                    $\log(1 + x) = x - x^2/2 + x^3/3 - \dots$

*for* $-1 < x \leq 1$.

Newton and Mercator, who were the first to obtain this result, in about 1665–1668, proceeded a little differently: it was already known that $\log(1+x)$ is the integral from 0 to $x$ of the function $1/(1 + x) = 1 - x + x^2 - \dots$; since the integral of a sum (though it be infinite ...) of functions is the sum of the integrals of the latter, and since it was already known how to integrate $x^m$ for $m$ an integer, thanks to Fermat, Cavalieri and Wallis, the series appeared of its own accord.

The crucial point here, naturally, is the connection between logarithms and the area of the hyperbola. It would have been obvious to Napier himself if he had known the "fundamental theorem of the differential and integral calculus", which enables one to calculate the area bounded by a curve $y = f(x)$ if one knows an $F$ for which $f(x) = F'(x)$. Let us explain the history of the subject.

Following the inventor of logarithms, one considers a point $y$ which moves from[20] $10^7$ to 0 at a speed inversely proportional to the time $x$, in other words satisfying the relation $y'(x) = -10^7/x$; the connection with the area of the hyperbola is obvious to us now, but not, of course, to his contemporaries and immediate successors. Furthermore, his tables, and those of Briggs to base 10, were not considered as mathematical theories: they were aids to numerical calculation, intended, doubtless, mainly for the astronomers in Napier's case (he wrote in Latin); according to Edward Wright, who translated them into English with a dedication to the *East India Company* and to Briggs, who completed them and transformed them into log to base 10, they are above all for the use of navigators, who worried little about the Latin but adopted the log with enthusiasm.

---

[20] Napier constructed a table of the function $\log \cos x$, and since the use of decimal fractions was only in its infancy, he tabulated $-10^7 \log \cos x$ in order to obtain positive integral values for log.

Their diffusion to London occurred with the greatest of ease[21]. On his death in 1579, a top-flight London merchant and financier, Sir Thomas Gresham, bequeathed the profits of his shops to the City and to the *Mercers' Company* to provide an endowment for a college to be administered by the merchants, and where the teaching, open to all, would be conducted on principles diametrically opposed to those of Oxford and of Cambridge. He certainly did not forget a chair of Divinity, clearly the most prestigious, intended to combat papist errors or heresies; Napier himself, author of a very popular book on the Apocalypse, had proclaimed the Pope none other than the Antichrist and, when the Invincible Armada menaced England, suggested purely theoretical tanks and submarines to the Queen; the North Sea was more efficacious. But apart from several courses in Latin intended for foreigners, most of the rest were in English and oriented towards practical or useful themes: navigation, geography, cartography, practical arithmetic and geometry, modern theories of medicine, juridical problems. The first professor of geometry at Gresham, from 1594 to 1620, was the very Henry Briggs (1561–1631) who popularised decimal numbers. Like Edward Wright, he had a close relationship with William Gilbert, the initiator of the first scientific, i.e. experimental, studies on magnetism; Briggs calculated the first tables of magnetic declination while artisans of the same circle invented the instruments. For his part, Wright explained the theory and use of the maps which the Flamand Gerard Mercator published from 1538, and wrote a book entitled *Certaine Errours in Navigation detected*; Briggs wrote a *Treatise on the North-West Passage* and navigators paid homage to him (maybe ironically) by baptising *Brigges his Mathematickes* an isle of the "passage" which, they were sure, joined the Atlantic to the Pacific by the North (correct, but a vast programme of exploration for the time). Briggs was the first to recognise the interest of Napier's logarithms and immediately taught them at Gresham College. All this, as one sees, is quite far from pure mathematics, but contributed no less strongly to launch the scientific movement in England which led to the creation of the *Royal Society*. Add that the calculation of log tables poses problems of interpolation very close to elementary differential calculus, and even, if one believes some authors, to the binomial formula for exponent 1/2.

As to the mathematicians, they slowly discovered the interest of log[22] for analysis. The first almost to have seen – although he did not say it explicitly – the relation between the area of the hyperbola and the log was the Belgian Jesuit Grégoire de Saint-Vincent in an *Opus geometricum quadraturae circuli and sectionum coni* of 1647 which claims above all to resolve the quadrature of the circle. He shows essentially that if, in the hyperbola of figure 1, the

---

[21] For what follows, see the superb book of Christopher Hill, *Intellectual Origins of the English Revolution* (OUP, 1965 or Panther Books, 1972).

[22] For what follows, see D. T. Whiteside, *Patterns of Mathematical Thought in the later Seventeenth Century* (*Archive for the History of Exact Sciences*, Vol. 1, 1961, pp. 179–388); but read with care, for at this date the author still has some problems with the mathematics ...

segments $DE$ and $PQ$ are in the ratio $q^m$, and the segments $HI$ and $KC$ in the ratio $q^n$, then the ratio between the areas $DEQP$ and $HIKC$ is equal to $m/n$, which comes down to saying that when the segment $EQ$ takes the values of a *geometric* progression, the area $DEPQ$ varies according to an *arithmetic* progression, an idea that one will find again in Newton, who does not however appear to have read Saint-Vincent, in contrast to Leibniz. Napier would have immediately deduced that this area is proportional to the log of $EQ$ since this is precisely his method of constructing log.

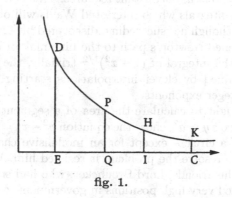

fig. 1.

Saint-Vincent's method is the better since, instead of decomposing the segment $EQ$ using an arithmetic progression like Archimedes or Cavalieri, he decomposes into segments $(q^i, q^{i+1})$; the ordinate of the point of the hyperbola $y = 1/x$ with abscissa $q^i$ is equal to $1/q^i$, the area of the slice of the hyperbola contained between the verticals $q^i$ and $q^{i+1}$ is contained between

$$\left(q^{i+1} - q^i\right)/q^i \quad \text{and} \quad \left(q^{i+1} - q^i\right)/q^{i+1},$$

i.e. between $q-1$ and $(q-1)/q$; since there are $m$ intervals $(q^i, q^{i+1})$ between $E$ and $Q$, the area of hyperbola $EQPD$ must lie between $m(q-1)$ and $m(q-1)/q$; further, that of $ICKH$ lies between $n(q-1)$ and $n(q-1)/q$, which yields the proportionality sought. As we saw at the end of n° 11 of Chap. II, this calculation can yield much more: it is not difficult, three hundred years later, to be cleverer than a Jesuit of 1647.

Then one finds the English. In the innocent hope of obtaining a beautiful explicit formula for calculating $\pi$, John Wallis calculated total areas, i.e. the integrals over the interval $(0,1)$ of the curves $y = (1 - x^{1/m})^n$, where $m$ and $n$ are natural integers: it is enough to expand and take account of the fact, which he extrapolates starting from several already known particular cases, that for $r$ rational the integral of $x^r$ on the interval $(0,1)$ is equal to $1/(r+1)$. This procedure unfortunately does not apply to calculating the area of the circle, since in this case $n = 1/2$ is not integer. Wallis' idea, which Newton exploited successfully, was to draw up a table of integrals for integer $n$, and observe the simple relations between the neighbouring terms, then to interpolate the cases $n = 1/2$, $3/2$, etc. starting from these relations. A natural idea to him: during the English Civil War Wallis rendered eminent service to Cromwell's party by decrypting a quantity of ciphered messages exchanged by the other camp; now, when one speaks of "ciphered" messages, one is almost always dealing, then as now, with replacing letters or groups of letters by numbers according to generally very simple rules, and one needs to discover them; "interpolation" and "extrapolation" are operations to which the decrypters have constant recourse. This does not prove that the calculation of the integrals which interested Wallis will obey the same rules, but he tried, and, though not succeeding, discovered his famous formula for $\pi$; in this way he opened Newton's path to the binomial formula which enabled him to calculate the integral of $(1 - x^2)^{1/2}$ trivially, the same formula that Newton also obtained by clever interpolations starting from the binomial coefficients for integer exponents.

Wallis also sought to calculate the area of a segment of hyperbola, but, instead of the curve $xy = a^2$, chose the equation $y^2 - x^2 = a^2$ which so closely resembles that of a circle – except for an inoffensive change of sign – that one might hope to resolve the problem; it resisted him. Wallis spoke of this in 1655 to one of his friends, Lord Brouncker, who had studied mathematics at Oxford, occupied very high positions in government, was interested in the sciences, and became the first president of the *Royal Society* in 1662. The latter calculated the area of the hyperbola $xy = 1$ contained between $x = 1$ and $x = 2$ by the usual method, but in a markedly more ingenious way: he decomposed $[1, 2]$ into equal intervals of length $1/2^n$ and grouped the partial areas obtained to conclude that $\log 2 = 1/1.2 + 1/3.4 + 1/5.6 + \dots$.

For his part, the Italian Pietro Mengoli apparently published the series in Bologna, in a book of 1659 where he gives an almost rigorous definition of the log of a rational number, inspired by the area of the hyperbola, though he did not say this explicitly. To define $\log(a/b)$ where $a$, $b$ are integers such that $b < a$, he introduced, up to notation, the sums

$$L_n^-(a/b) = \sum_{bn < p \leq an} 1/p, \qquad L_n^+(a/b) = \sum_{bn \leq p < an} 1/p$$

whose origin is clear: one considers on the axis $Ox$ the points of the form $p/n$ with $p$ integer, restricts to those which lie between $a$ and $b$, i.e. such that $bn \leq p \leq an$, and then approximates the area of the hyperbola $y = 1/x$ lying between the verticals at $a$ and $b$ by the vertical rectangles having for bases the intervals of length $1/n$ in question, etc.

This done, Mengoli proved that these sums possess the following properties: (i) as $n$ increases, the sums $L^-$ increase and the sums $L^+$ decrease, (ii) the first are smaller than the second (obvious), (iii) the difference between the second and the first tends to 0 as $n$ increases (obvious). He then *defined* $\log(a/b)$ by imposing the condition that it satisfy $L_n^-(a/b) \leq \log(a/b) \leq$

$L_n^+(a/b)$ for any $n$, an impeccable idea in view of properties (i), (ii) and (iii) above. Having done this, he proved that if $a/b$, $c/d > 1$ are two fractions, then

$$L_{nd}^-(a/b) + L_{na}^-(c/d) = L_n^-(ac/bd),$$

which (for us, with our modern notation ...) is easy: in the first expression the index of summation $p$ varies between $bnd$ and $and$, in the second between $dna = and$ and $cna$, and in third between $bdn$ and $acn = cna$. As $n$ increases, the sequence $L_{nd}^-(a/b)$, extracted from the sequence $L_n^-(a/b)$, tends to the same limit $\log(a/b)$ as the latter, with a similar remark for the sequence relative to $c/d$. The preceding relation then shows Mengoli that his definition satisfies the condition $\log(ac/bd) = \log(a/b) + \log(c/d)$. Finally, the limits of the "Riemann" sums provides him less attractive series expansions, though, for $a = 2$, $b = 1$, they lead to the alternating harmonic series. In this case one must seek the limit of the sum

$$\frac{1}{n} + \frac{1}{n+1} + \ldots + \frac{1}{2n-1} = \left(1 + \frac{1}{2} + \ldots + \frac{1}{2n-1}\right) - \left(1 + \frac{1}{2} + \ldots + \frac{1}{n-1}\right)$$

clearly equal to $1 - 1/2 + 1/3 - 1/4 + \ldots - 1/(2n-2) + 1/(2n-1)$, which, in the limit, yields Brouncker's series.

It is hard not to deduce from these arguments that Mengoli, and those of his contemporaries who performed the same kinds of calculation, already had a clear enough idea of what a real number and an integral are.

After Mengoli, we arrive at Newton and Mercator. From 1664, reading the *Géométrie* of Descartes in the Latin translation published by a Dutch algebraist, and, even more, reading Wallis' *Arithmetica Infinitorum*, led Newton to reflect, and a year later to the first case of his binomial formula. At the same time, and by the same method – expand in a power series and integrate term-by-term – to attack the hyperbola $y = 1/(1+x)$, he wrote $y = 1 - x + x^2 - x^3 + \ldots$ and deduced, from the rule $x^m \longrightarrow x^{m+1}/(m+1)$ which he found in Wallis but, in contrast to him, *applied to an arbitrary abscissa* $x$, that the area contained between the verticals 0 and $x$ is equal to $x - x^2/2 + x^3/3 - x^4/4 + \ldots$ This result for $x = 1$ yields Brouncker's formula, though Newton does not mention this.

The mystery is that none of the authors whom we have mentioned above, except perhaps Saint-Vincent, made explicit the connection between calculating this area and the log of the tables. Newton contented himself, in a manuscript of 1667, with affirming in a phrase that "*the areas are to their abscissae as the logarithms (i.e. when the abscissae increase in geometric progression, the areas increase in arithmetic progression)*". The idea was doubtless already a part of the English mathematical folklore of the time.

And Newton began to embark on numerical calculations of which, later, he would be little ashamed in front of Leibniz. Successively he took $x = 1/10, 2/10, -1/10, -2/10, 1/100$, etc. in his series and calculated the results to 52 decimal places, committing several small errors. This is not as difficult as

one might believe; 27 terms of the series suffice[23], $x$ is 1 or 2 divided by 10, 100
or 1000, so that the numbers to be calculated diminish as the degree increases
– the calculation, impeccably presented in a triangular form, occupying half
a page, is reproduced in the *Mathematical Papers* edited by Whiteside – and
the decimal fractions to be calculated are periodic: for $x = 1/10$ for example,
we have

$$x^{17}/17 = 0.0\ldots080(5882352941176470)(5882352941176470)(58823$$

as the happy possessors of pocket calculators can verify if the latter are precise
enough, which is doubtful.

These calculations, though apparently futile, in fact lead to more interesting results, namely the log of 2, 3, 5, 7, 11, 13, 17 and 37. It is clear that his
series cannot provide them, but there are more intelligent methods, namely
the relations

$$\begin{aligned}
2 &= 1.2 \times 1.2/0.8 \times 0.9, \quad 3 = 1.2 \times 1.2 \times 1.2/0.8 \times 0.8 \times 0.9, \\
5 &= 2 \times 2/0.8, \quad 10 = 2 \times 5, \quad 11 = 10 \times 1.1, \quad \text{etc.}
\end{aligned}$$

found again in Euler. They furnish the desired log in stages, starting from
those already calculated, or from similar log: we have $37 = 1000 \times 0.999/27$,
whence

$$\log 37 = 3.\log 2 + 3.\log 5 + \log(1 - 1/1000) - 3.\log 3,$$

the log of $1 - 1/1000$ being calculated easily by the series. One might imagine
that, to check his calculations, Newton compared them to the available tables
of logarithms; but apparently not, which can be understood in view of his
prodigious advances on their authors, whom, in any case, he seems not to
have read. See Houzel, *Analyse mathématique*, pp. 79-82.

Newton did not imagine that at 23 years old he could have made discoveries capable of interesting the mathematicians; all his work remained in
his drawer. But in 1668, when he was again in Cambridge, there appeared
in London the *Logarithmotechnia* of the German Mercator (whose real name
was Kaufmann, "merchant") established in London. The series $\log(1 + x)$
appears there, and even the series $x^2/2 - x^3/2.3 + x^4/3.4 - \ldots$ representing
the area of the function $y = -\log x$ contained between its asymptote and the
vertical at $x$ ($x < 1$).

Newton found himself, for the first but not the last time, in the classical
situation of the scientist, a debutant moreover, who sees a colleague publish
results which he had found earlier but kept to himself; he had not shown them
even to Barrow, his patron, the *Lucasian Professor of Mathematicks*. Newton decided to speak to him, and was recommended to present his results in

---

[23] Newton calculated separately the sum of the terms of even degree and of the
terms of odd degree; by addition and subtraction of the results he obtained
simultaneously $\log 1.1$ and $\log 0.9$.

Latin, which Newton accomplished in a few days; the manuscript, *De Analysi per aequationes numero terminorum infinitas*, was in Barrow's possession at the beginning of July 1669. Newton, who still lacked confidence, asked him to send it, without naming the author, to John Collins, an official in the English government who was greatly interested in science (Collins, not the government) and maintained an extensive correspondence with many scholars of the age. Collins made a copy – the original was returned to Newton –, and managed to extract the name of the author from Barrow in August "*who, with an unparalleled genius, has made very great progress in this branch of mathematics*" and showed his copy to Lord Brouncker, president of the *Royal Society*, and to other English and continentals. The obvious thing to do to safeguard Newton's priority was to publish his manuscript in the *Philosophical Transactions (PT) of the Royal Society* or, at least, to list it in the Reports of its meetings, and moreover Brouncker published his series for log 2 in 1668, in the same PT; nothing of kind happened; according to Moritz Cantor (III, p. 68) neither Collins nor Brouncker had judged that the contents of the paper was worth so protecting ... Others mention Newton's tendency not to publish work as yet uncompleted; it remained so, since after his first success Newton distanced himself from mathematics, having judged it too arid[24].

Nevertheless, the impact of the manuscript on Barrow was such – "*I am only a child beside him*" – that, already wanting to return to theology, much more prestigious, he resigned his post and recommended Cambridge to offer it to Newton who, in the autumn of 1669, aged 27, obtained a chair and lodgings for life!

The contents of the *De Analysi* – fifteen or so pages – is prodigious. Newton started by stating, with examples, the rules of calculus of areas (i.e. of integrals) for functions expandable in series of rational powers of $x$: one replaces $x^m$ by $x^{m+1}/(m+1)$ and adds the results. In the case where $y$ is given by a formula involving divisions and square roots one reduces to the preceding case by standard operations: the quotient of two series of powers of $x$, for example $y = (2x^{1/2} - x^{3/2})/(1 + x^{1/2} - 3x)$ is calculated with the help of the algorithm for decimal division, and, in the example that we have just mentioned, yields,

$$y = 2x^{1/2} - 2x + 7x^{3/2} - 13x^2 + 34x^{5/2} + \ldots;$$

the square root is similarly obtained by applying the decimal algorithm, so that if, for example, $y = (a^2 - x^2)^{1/2}$, one finds

$$y = a - x^2/2a - x^4/8a^3 - x^6/16a^5 - \ldots,$$

which explained how Newton guessed – or confirmed – the first, crucial, line of the table of coefficients of the functions $(1 - x^2)^{n/p}$.

---

[24] On Newton, see the DSB, the large biography by Richard Westfall, *Never at Rest: a Biography of Isaac Newton* (CUP, 1980), A. Rupert Hall, *Isaac Newton. Adventurer in Thought* (Cambridge, 1992) and Loup Verlet, *La malle de Newton* (Gallimard, 1993) already mentioned.

He then explains, by the example $y^3 - 2y - 5 = 0$, his method (inspired by Viète and Oughtred) for finding the roots of an equation. Here there is a visible change of sign, and so a root of the function, between 2 and 2, 1. Newton put $y = 2 + p$, whence $p^3 + 6p^2 + 10p - 1 = 0$; since the square and the cube are small relative to $p$, we have, approximately, $p = 1/10$. Now one puts $p = 1/10 + q$, whence a new equation $q^3 + 6.3q^2 + 11.23q + 0.061 = 0$, or almost $11.23q + 0.061 = 0$, whence $q = -0.0054 + r$, with a new equation for $r$, etc. In the case of the relation $y^3 + a^2y - 2a^3 + axy - x^3 = 0$ he explains the method which we mentioned in Chap. II, n° 22, and, from the expansion found, deduces a formula for calculating the area bounded by the curve and the verticals 0 and $x$, namely $ax - x^2/8 + x^3/192a + \ldots$, "*an expansion which approximates the truth more rapidly as $x$ becomes smaller*". He shows how by inverting the series for $\log(1 + x)$ one can calculate the abscissa corresponding to a given value $y$ of the area, and so finds in passing the first terms of the exponential series, though without realising, and for a good reason, that he had just discovered what would be the most important function in analysis up to the present day. He obtained the length of the arc of the semi-circle $x^2 + y^2 = x$, $y > 0$, contained between the point $(1, 0)$ and the point of ordinate $y$ (it is clearly $\frac{1}{2} \arcsin y$ since the radius of the circle is $\frac{1}{2}$) by first calculating the derivative of the arc with respect to $y$ (by the geometric infinitesimal argument that had become standard) and expanding that by his binomial formula; by "integrating" he obtained the power series for $\arcsin y$ which, later, he inverted to find that of $\sin x$ and, by the same type of argument, of $\cos x$.

To follow the history would take us too far; everything that he expounded in the manuscript of 1669 is taken up again and developed in the *De Methodis Serierum et Fluxionum* of 1670–1671 where, this time, he explains in detail the rôle of "fluents" and "fluxions" in his calculus, as we have indicated in Chap. III, n° 14. And he had other occasions to explain himself during his correspondence with Leibniz in 1676–1677.

## 13 – The exponential function as a limit

We saw in n° 10 that

$$\exp x = \lim(1 + x/n)^n = \lim(1 + hx)^{1/h}$$

for $x$ *real*. The dominated convergence theorem enables one to go further by imitating the calculation which led us to the series for $\log(1 + x)$:

**Theorem 11.** *Let* $(z_p)$ *be a sequence of complex numbers tending to a limit $z$. Then*

$$(13.1) \qquad \lim (1 + z_p/p)^p = \sum_{n=0}^{\infty} z^n/n! = \exp(z),$$

*and in particular*

(13.2) $$\exp(z) = \lim(1 + z/p)^p$$

*for any* $z \in \mathbb{C}$.

The algebraic binomial formula shows that generally

$$
\begin{aligned}
(1 + u/p)^p &= 1 + p(u/p) + p(p-1)u^2/p^2.2! + \dots \\
&\quad + p(p-1)\dots(p-p+1)u^p/p^p.p! \\
&= \sum_{0 \le n \le p} (1 - 1/p)(1 - 2/p)\dots[1 - (n-1)/p]\, u^n/n!,
\end{aligned}
$$

(13.3)

whence

(13.4) $$(1 + z_p/p)^p = \sum_{n \le p}(1 - 1/p)(1 - 2/p)\dots z_p^{[n]} = \sum u_p(n).$$

As $p$ increases indefinitely, the term $z_p^{[n]}$ in this expression tends to $z^{[n]}$ and its coefficient, a product of $n - 1$ factors all tending to 1, tends to 1, so that the general term of (4) tends to that of the series $\exp(z)$. If one considers the right hand side of (4) as a series whose terms of index $n > p$ are all zero and if one observes that the numbers $1 - 1/p$, $1 - 2/p$, etc. all lie between 0 and 1, it is clear that

$$|u_p(n)| \le |z_p|^{[n]}$$

for all $n$ and $p$. But since the sequence $(z_p)$ converges there exists a number $M > 0$ such that $|z_p| \le M$ for all $p$, whence $|u_p(n)| \le M^{[n]} = v(n)$, the general term of a convergent series *independent of* $p$, namely $\exp M$. The hypotheses of Theorem 9 are therefore satisfied, whence the relation (1), qed.

Theorem 11 – which depends only on the *algebraic* binomial formula and on the dominated convergence theorem – might form the starting point of another mode of exposition of the properties of the functions exp and log. First, it allows us to *define* $\exp z$ by the formula (2) and to *prove* that $\exp z = \sum z^n/n!$, which reverses the logical order adopted up to the present. We now have to show that

$$\exp(x + y) = \exp(x)\exp(y)$$

for all $x, y \in \mathbb{C}$. To see this, consider the product

$$(1 + x/p)^p(1 + y/p)^p = \left(1 + \frac{x + y + xy/p}{p}\right)^p = (1 + z_p/p)^p$$

where $z_p = x + y + xy/p$. As $p$ increases indefinitely, the left hand side tends to $\exp(x).\exp(y)$, and as $z_p$ tends to $x + y$, the right hand side tends to $\exp(x + y)$ by Theorem 11, qed.

Theorem 11 also provides an express proof of the relation $\exp(\log x) = x$ for $x$ real. Put

$$n(x^{1/n} - 1) = u_n,$$

whence

$$x = (1 + u_n/n)^n.$$

The sequence $(u_n)$ converges by definition to $\log x$; Theorem 11 now assures us that $(1 + u_n/n)^n$ tends to $\exp(\log x)$, qed.

Theorem 11 enables us, and this is very useful, to determine all reasonable functions with *complex* values that satisfy the addition formula

(13.5)     $$e(x + y) = e(x)e(y).$$

We have to distinguish two cases.

**Corollary 1.** *Let* $e : \mathbb{R} \longrightarrow \mathbb{C}$ *be a solution of (5), not identically zero, possessing a derivative at* $x = 0$. *Then*

$$e(x) = \exp(cx) \qquad with\ c = e'(0).$$

To see this, we first note, once again, that (5) implies $e(0) = 1$, then that $e(nx) = e(x)^n$ for $n \in \mathbb{N}$ and so also

(13.6)     $$e(x/n)^n = e(x).$$

By definition of the derivative, the quotient

(13.7)     $$\frac{e(x/n) - 1}{x/n} = x_n$$

tends to a limit $c$. But (6) and (7) show that

$$e(x) = (1 + xx_n/n)^n.$$

Since $xx_n$ tends to $cx$, Theorem 11 shows that the right hand side tends to $\exp(cx)$; so it is equal to $e(x)$ for all $n$, qed.

Another proof: since $[e(x + h) - e(x)]/h = e(x)[e(h) - 1]/h$, we see that $e(x)$ is everywhere differentiable and that $e'(x) = ce(x)$ where $c = e'(0)$. Now the function $\exp(cx)$ possesses the same property. We deduce immediately that the function $e(x)/\exp(cx)$ has a zero derivative, so is constant and in fact equal to 1 (put $x = 0$).

**Corollary 2.** *Let* $e : \mathbb{C} \longrightarrow \mathbb{C}$ *be a solution of (5), not identically zero, possessing a derivative in the complex sense at* $z = 0$. *Then*

(13.8)     $$e(z) = \exp(cz) \qquad with\ c = e'(0)$$

*for all* $z \in \mathbb{C}$.

The proof is strictly the same; the only difference is that now $x = z$ can be complex, so that in (7) the number $h = z/n$ tends to 0 through complex and not real values, which explains the necessity of giving the concept of derivative the meaning which enabled us to show in Chap. II, n° 19 that a power series is always differentiable.

We again emphasise the fact that, in contrast to the characterisation of exponential functions given in n° 6, Theorem 2, Corollary 1 applies to functions with *complex* values; the usefulness of this generalisation will appear clearly in the following n° *à propos* trigonometric functions.

Corollary 2, for its part, yields a fourth proof of Newton's binomial formula in the general case. Since we know that $N_{s+t}(z) = N_s(z)N_t(z)$ for $s$, $t$ and $z$ *complex*, we have to apply the corollary to the function

$$(13.9) \qquad e(s) = 1 + sz + s(s-1)z^2/2! + \ldots = N_s(z)$$

where $z$ is given, of course with $|z| < 1$. It reduces to proving that the function (9) has a derivative *in the complex sense* at $s = 0$. Now

$$(13.10) \quad \frac{e(s) - e(0)}{s} = z + (s-1)z^2/2! + (s-1)(s-2)z^3/3! + \ldots;$$

for $z$ given, $|z| < 1$, the right hand side, considered as a function of $s$, is a normally convergent series in all the disc $|s| < R$ as we have seen in n° 11 (second proof) for the binomial series itself. Its terms are continuous functions of $s$. So, similarly, is its sum, so that, when $s \in \mathbb{C}$ tends to 0, the left hand side tends to the value of the right hand side at $s = 0$, namely

$$(13.11) \qquad e'(0) = z - z^2/2 + z^3/3 - \ldots.$$

We can therefore apply Corollary 2, so obtaining Theorem 8 bis again:

$$(13.12) \qquad \exp\left[s(z - z^2/2 + z^3/3 - \ldots)\right] =$$
$$= N_s(z) = \sum s(s-1)\ldots(s-n+1)z^n/n!.$$

It is not surprising that, two hundred years after Newton, one of the more famous of Conan Doyle's characters should be considered as an eminent mathematician by virtue of his profound works on "Newton's binomial".

Let us return to Corollary 2 and scrutinize its hypothesis. Consider, in a general manner, a function $e$ on $\mathbb{C}$ satisfying $e(s+t) = e(s)e(t)$ for all $s, t$. For $z = x + iy$ with $x$ and $y$ real, one then has $e(z) = e(x)e(iy)$. It is clear that each of functions $x \mapsto e(x)$ and $y \mapsto e(iy)$ satisfies the eternal functional equation on $\mathbb{R}$. If we then assume $e$ continuous, the least we can do, then Corollary 1 shows that $e(x + iy) = \exp(ax)\exp(by) = \exp(ax + by)$ with complex constants $a$ and $b$, and it is clear that every function of this type satisfies the functional equation on $\mathbb{C}$.

For the function $\exp(ax + by)$ to be of the form $\exp(cz)$, it is necessary and sufficient that $b = ia$. But if one differentiates the function $\exp(ax + by)$ with respect to $x$ (resp. $y$), one finds it again, multiplied by $a$ (resp. $b$); the relation $b = ia$ is thus, in this case, just Cauchy's relation $f'_y = if'_x$ of Chap. II, n° 19. As in the case of the functions $\exp(cz)$, we have to impose the same condition on $e(z)$. But, if $e(z)$ is differentiable in the complex sense at $z = 0$, the formula $e(z+h) = e(z)e(h)$ shows that it is so everywhere, with $e'(z) = e'(0)e(z)$, which proves furthermore that $e'$ is continuous. In other words, Corollary 2 characterises the *holomorphic* solutions of our functional equation.

## 14 – Imaginary exponentials and trigonometric functions

Let us return to the function $\exp(z)$ for $z \in \mathbb{C}$. We said in Chap. II, n° 14, example 2, that the resemblance between the exponential series and those which represent the functions $\sin x$ and $\cos x$ led Euler to note that

$$(14.1') \qquad \qquad \exp(ix) \;=\; \cos x + i \sin x$$
$$(14.1'') \qquad \qquad \exp(-ix) \;=\; \cos x - i \sin x$$

for $x \in \mathbb{R}$. The proof of these formulae takes a few lines but presupposes the power series for the trigonometric functions and even the definition of these latter; but no considerations of elementary geometry will provide us with such recondite formulae. Try for example to understand why

$$\pi^2/2! - \pi^4/4! + \pi^6/6! - \ldots = 2.$$

This is obvious if you know that the left hand side represents $1 - \cos \pi$, but what if you only have sketches on a sheet of paper?

Corollary 1 of Theorem 11 provides a more economical "proof" of (1) because it relies on much more elementary properties of trigonometric functions than their mysterious series expansions, which will follow from them.

**Theorem 12.** *Let $c(x)$ and $s(x)$ be two real valued functions defined on $\mathbb{R}$, possessing the following properties: (i) they satisfy the addition formulae of the functions $\cos x$ and $\sin x$; (ii) they are differentiable at $x = 0$, with $c'(0) = 0$ and $s'(0) = 1$. Then*

$$(14.2) \qquad \qquad c(x) + is(x) = \exp(ix)$$

*and consequently*

$$(14.3') \qquad \qquad c(x) \;=\; 1 - x^{[2]} + x^{[4]} - \ldots,$$
$$(14.3'') \qquad \qquad s(x) \;=\; x - x^{[3]} + x^{[5]} - \ldots$$

Put $e(x) = c(x) + is(x)$. The addition formulae show, by a trivial calculation, not even assuming that $c(x)$ and $s(x)$ are real-valued, that $e(x + y) = e(x)e(y)$. Moreover, the function $e(x)$ has derivative $e'(0) = i$ at $x = 0$ by hypothesis (ii). Corollary 1 of Theorem 11 immediately finishes the proof, the series (3') and (3") being obtained by separating the real and imaginary parts of the exponential series.

Theorem 12 leads us to *define* the functions $\cos x$ and $\sin x$ by the formulae (3), *for x complex as well.* We then obtain, as we shall see, all the elementary properties of the trigonometric functions.

(i) *Numerical calculation.* The series $\cos x$ and $\sin x$ converge with great rapidity. If one calculates $\exp(z)$ from its $n^{\text{th}}$ partial sum, the error committed is less than

$$|z|^{[n]} \left(1 + |z|/(n+1) + |z|^2/(n+1)(n+2) + \ldots +\right) \leq |z|^n . \exp(|z|)/n!$$

with a bound of the same kind for the trigonometric parallels. For reasonable values of $|z|$ – for the trigonometric functions, it is enough to do the calculations for $0 < z < \pi/2 < 1.6$ –, the denominator $n!$, about $4.10^7$ for $n = 11$, the sixth term of the series $\sin x$, rapidly becomes enormous with respect to the factor $\exp(|z|)|z|^n$, so that the approximation is excellent even for moderate values of $n$.

(ii) *Euler's Relations.* As we have already seen, the series expansions (3) trivially imply the relations

$$\exp(iz) = \cos z + i \sin z, \qquad \exp(-iz) = \cos z - i \sin z.$$

From these we deduce that

$$(14.4) \qquad \cos z = \left(e^{iz} + e^{-iz}\right)/2, \qquad \sin z = \left(e^{iz} - e^{-iz}\right)/2i,$$

the famous and fundamental formulae which frequently facilitate trigonometric calculations.

(iii) *Addition formulae.* For $x$ and $y$ real, we may write

$$\exp(ix) \exp(iy) = \exp(i(x + y))$$

and separate the real and imaginary parts using (2). Immediately we obtain

$$(14.5') \qquad \cos(x + y) = \cos x \cos y - \sin x \sin y,$$
$$(14.5") \qquad \sin(x + y) = \sin x \cos y + \cos x \sin y.$$

For $x$ and $y$ complex, we have to use the formulae (4) and write a few lines of the kind $(a + b)(c + d) = $ etc.

(iv) *Parity.* The formulae

$$\cos(-x) = \cos x, \qquad \sin(-x) = -\sin x$$

follow from the power series (3') and (3"). It is even more obvious that

$$\cos 0 = 1, \qquad \sin 0 = 0.$$

(v) *The Pythagorean Relation*

(14.6) $$\cos^2 z + \sin^2 z = 1.$$

Even for $z$ complex, the left hand side is equal to

$$(\cos z + i \sin z)(\cos z - i \sin z) = \exp(iz)\exp(-iz) = \exp(0) = 1.$$

For $z = x + iy$ with $x, y \in \mathbb{R}$, we have $e^z = e^x e^{iy} = e^x(\cos y + i \sin y)$, and since $|\cos y + i \sin y| = 1$ for $y \in \mathbb{R}$, by (6), we conclude that

$$|e^z| = e^{\mathrm{Re}(z)}$$

and more generally that

(14.7) $$|a^z| = a^{\mathrm{Re}(z)} \qquad \text{for } a > 0, \ z \in \mathbb{C}$$

since $a^z = e^{z \log a}$ with $\log a$ real; this formula is in constant use.

*Corollary:* the series $\zeta(s) = \sum 1/n^s$ converges in the half plane $\mathrm{Re}(s) > 1$, which completes Theorem 5 of Chap. II, n° 12. Its sum is holomorphic in this half plane since the series of derivatives

$$\zeta'(s) = -\sum \log(n)/n^s$$

converges normally on $\mathrm{Re}(s) \geq \sigma$ for any $\sigma > 1$ (see Chap. III, n° 17, example 3, the arguments extending immediately to the case where $s$ is complex). The same results for multiple series such as $\sum (m^2 + n^2)^{-s/2}$.

(vi) *Derivatives.* We have

(14.8) $$\cos' z = -\sin z, \qquad \sin' z = \cos z.$$

Simplest is to differentiate the power series (3') and (3") by straightforwardly applying the general rule of Chap. II, n° 19. We can iterate the formulae (8) and find the successive derivatives, for example $\cos'' z = -\cos z$, $\sin''' z = -\cos z$, etc. These formulae are *a fortiori* valid in $\mathbb{R}$.

(vii) *The number* $\pi$. Throughout this section of the text, the least easy, we work in $\mathbb{R}$. The psychological difficulty is to look for the number $\pi$ while appearing not to have found it already; there is a telling phrase of Pascal's on this theme. The situation is less ridiculous than it might appear: there are many "special functions" much more complicated than the circular functions in order to locate whose roots one is forced to use arguments similar to those which follow.

Consider the series

$$1 - \cos x = x^2/2 - x^4/2.3.4 + x^6/2.3.4.5.6 - \ldots$$

It is alternating, with decreasing terms for $0 \leq x \leq 3$: indeed one passes from one term to the next by multiplying by $x^2/3.4$, $x^2/5.6$, etc., i.e. by a factor $< 1$ since $x^2 < 3.4$. Its sum therefore lies between $x^2/2$ and $x^2/2 - x^4/2.3.4$, whence

$$(14.9) \qquad 1 - x^2/2 \leq \cos x \leq 1 - x^2/2 + x^4/24 \qquad \text{for } 0 \leq x \leq 3.$$

In particular we deduce that $\cos 2 < -1/3 < 0$ and that

$$-1 < \cos x < 1 \quad \text{for } 0 < x < 2,$$

as we see by examining the graphs of $1 - x^2/2$ and of $1 - x^2/2 + x^4/24$ between 0 and 2.

Since the function $\cos x$ is continuous and equal to 1 for $x = 0$, the intermediate value theorem of Chapter III shows that *the function* $\cos x$ *vanishes somewhere between* 0 *and* 2. Every limit of solutions of the equation $f(x) = 0$ being again a solution for every continuous function $f$, the set of roots $\geq 0$ of $\cos x = 0$ contains its greatest lower bound $a$; this is $> 0$ since $\cos 0 = 1$. *By definition*

$$(14.10) \qquad \pi/2 = \text{ least number } a > 0 \text{ such that } \cos a = 0.$$

One must have $\sin a = +1$ or $-1$; but the series defining $\sin x$ is also alternating with decreasing terms for $0 < x < 2$, whence

$$(14.11) \qquad x > \sin x > x - x^3/6 > 0 \quad \text{for } 0 < x < 2,$$

and consequently

$$(14.12) \qquad \cos a = 0, \qquad \sin a = 1.$$

The addition formulae now show immediately that

$$(14.13') \qquad \cos 2a \;=\; -1, \quad \cos 3a = 0, \quad \cos 4a = 1,$$
$$(14.13'') \qquad \sin 2a \;=\; 0, \quad \sin 3a = -1, \quad \sin 4a = 0$$

and more generally that

(14.14) $$\cos(x + a) = -\sin x, \qquad \sin(x + a) = \cos x$$

for any $x$ real or complex.

These relations show that, on $\mathbb{C}$, all multiples of $4a$, i.e. of $2\pi$, are periods of the trigonometric functions. In fact, they have no other in $\mathbb{R}$ (nor, as we shall see later, in $\mathbb{C}$, leaving tan aside). For let $4b \in \mathbb{R}$ be a period. By adding a number of the form $4ka$, with $k \in \mathbb{Z}$, we can assume $0 \leq b < a$, and so we need to show that $b = 0$.

Now, since $\cos 4b = \cos 0 = 1$, the relation $\cos x = 2\cos^2(x/2) - 1$ shows that either $\cos 2b = 1$, or $\cos 2b = -1$. The second hypothesis implies $\cos b = 0$ by virtue of the same relation; since $b < a$, least root $\geq 0$ of the cosine, this is absurd. If on the other hand $\cos 2b = 1$, then $\cos b = 1$ or $-1$; but the inequalities (9) show that $-1 < \cos x < 1$ for $0 < x < 2$ as we saw above, with strict inequalities; thus $b = 0$, qed.

This result also shows that $a$ is the only root of $\cos x$ between 0 and 2. Indeed, let $a'$ be another root between 0 and 2. Since $\sin a' > 0$ by (11), we have $\sin a' = 1$, and since $\cos a' = 0$ we see (addition formulae) that $4a'$ is a period like $4a$. So $a' = ka$ for an integer $k > 0$. If $k = 1$, we have finished. If $k > 1$, then $a \leq a'/2 < 1$ then, by (9), $\cos x$ is obviously $> 1/2$ between 0 and 1, qed.

We can even, better than nothing, verify that $2a (= \pi) > 3$ i.e. that the number $a$ introduced above is $> 3/2$. Since $\cos a = 0$, (8) shows that $1 - a^2/2 < 0$, i.e. $a > \sqrt{2}$; foiled. But the argument leading to (8) shows as well that
$$1 - x^2/2 + x^4/4! - x^6/6! < \cos x \qquad (0 < x < 2),$$

whence, by a short calculation, $\cos 3/2 > 679/2560 > 0$. Since we have seen that $\cos 2 < 0$, the function $\cos x$ must have a root between $3/2$ and 2, and since its only root between 0 and 2 is $a$, we obtain the required inequality.

To conclude this section we ought to rewrite the formulae (12), (13) and (14), replacing $a$ by $\pi/2$, but one may assume that the reader would then say, like a Parisian humorist à propos a best seller: "I haven't read it, I haven't seen it, though I've heard it mentioned".

(viii) *Arguments of a complex number.* Every *non zero* complex number can be put in the form

(14.15) $$z = r(u + iv)$$

in a unique way, with $r = |z|$ real $> 0$, and $u$ and $v$ real satisfying

(14.16) $$u^2 + v^2 = 1,$$

whence $|u| \leq 1$. Since the function cosine takes the values 1 and $-1$ and is continuous, the intermediate value theorem, already invoked, ensures the existence of a $t \in \mathbb{R}$ such that $u = \cos t$. Then $\sin^2 t = v^2$, so either $v = \sin t$, or $v = -\sin t = \sin(-t)$. On replacing $t$ by $-t$ if need be, we see that we can always put $z$ in the form

(14.17) $$z = r(\cos t + i\sin t) = |z|\exp(it).$$

The number $t$ is *unique up to a multiple of* $2\pi$. If indeed $t + c$ is another solution, we have $\exp(it) = \exp(it + ic)$, whence $\exp(ic) = 1$, and then $\exp(i(x + c)) = \exp(ix)$ for any $x \in \mathbb{R}$, and $c$ is a period of the trigonometric functions, so of the form $2k\pi$ as we saw above. "The" number $t$ is called the *argument* of the complex number $z \neq 0$. We shall study it in more detail in § 4 of this Chapter.

As one can always put $r$ in the form $\exp s$ with $s \in \mathbb{R}$, and in fact $s = \log r$, the relation (17) can again be rewritten, as $z = \exp(s + it)$. In other words, *the exponential function maps* $\mathbb{C}$ *onto* $\mathbb{C}^* = \mathbb{C} - \{0\}$. But in contrast to what happens in the real domain, the map is not injective, as we have just seen.

(ix) *Periodicity of* $\exp z$. The calculation of the values of $\cos x$ and $\sin x$ for $x = \pi$ or $x = 2\pi$ shows that

(14.18) $$e^{i\pi} = \exp(i\pi) = -1, \qquad e^{2i\pi} = \exp(2i\pi) = 1,$$

whence more generally

(14.19) $$\exp(z + k\pi i) = (-1)^k \exp z.$$

The function $\exp z$ thus has complex periods, namely all the multiples of $2\pi i$, and in particular is no more injective than the trigonometric functions. It has no other periods: the relation $\exp z = 1$, with $z = x + iy$, implies $|\exp z| = \exp x = 1$, whence $x = 0$, and $\exp(iy) = 1$, i.e. $\cos y = 1$ and $\sin y = 0$, whence $y = 2k\pi$ as we showed above. In other words,

$$\exp z = 1 \Longleftrightarrow z = 2ki\pi.$$

(x) *Logarithms of a complex number.* One defines "the" logarithm of a complex number $z \neq 0$ by agreeing that

$$u = \mathcal{L}og(z) \Longleftrightarrow z = \exp u,$$

but the pretend function $\mathcal{L}og$ is not a function: to each value of $z$ there correspond infinitely many values of $u$, differing one from the other by a multiple of $2i\pi$. This is a difficulty of the same kind as arises in defining "functions" such as $\arcsin x$ in the real domain. In other words, handle with caution. More exactly, put $z = |z|\exp(it)$ with $t = \arg(z)$. Then

$$z = \exp(\log|z| + it)$$

and so

$$(14.20) \qquad Log\, z = \log|z| + i.\arg z$$

where $\log|z|$ is the usual logarithm on $\mathbb{R}_+^*$. The ambiguity in the definition
of $\arg z$ causes exactly the same problem for the complex $Log$. We will find
it again *à propos* the primitive of the function $1/(x - a)$ for $a \in \mathbb{C}$, and
this is how Johann Bernoulli introduced it in 1702 (Cantor, p. 362, *à propos*
arctan). Calculations with complex logarithms or, what comes to the same,
with arguments of complex numbers, remains, three centuries later, one of
the most notorious sources of errors.

We saw in n° 11, Theorem 8 bis, that, as in the real case, the series
$u = z - z^2/2 + z^3/3 - \ldots$ satisfies $\exp u = 1 + z$ for $z \in \mathbb{C}$, $|z| < 1$. We deduce
that

$$(14.21) \qquad Log(1 + z) = 2ki\pi + z - z^2/2 + z^3/3 - \ldots$$
$$\text{for } z \in \mathbb{C},\ |z| < 1.$$

It is tempting to write Newton's binomial formula in the form

$$(14.22) \qquad (1 + z)^s = \exp[s.Log(1 + z)]$$

for $z$ and $s$ complex. Here again, handle with caution. If indeed one *defines*

$$(14.23) \qquad a^s = \exp(s.Log\, a)$$

for $a$ and $s$ complex, $a \neq 0$, the ambiguity inherent in the definition of the
complex log carries over to the definition of $a^s$, which will be defined only up
to a factor of the form $\exp(2k\pi is) = \exp(2\pi is)^k$. If $s$ is a rational number,
one obtains a finite number of possible values, but in the general case the
map $k \mapsto \exp(2k\pi is)^k$ of $\mathbb{Z}$ into $\mathbb{C}$ is injective, whence a countable infinity of
"determinations" of $a^s$. Here again, a notorious source of errors.

Formula (21), which lets one calculate $Log$ on a neighbourhood of 1,
extends to all other points $a \in \mathbb{C}^*$: every $z$ sufficiently close to $a$ is of the
form $z = a(1 + u)$ with $|u| < 1$, whence[25]

$$(14.21') \qquad Log\, z = Log\, a - \sum (1 - z/a)^n/n,$$

a power series converging for $|z - a| < |a|$. There are therefore, *on a neigh-
bourhood of a*, analytic functions whose values at every point $z$ form a subset
of the set of possible values of $Log\, z$. Every analytic function which possesses
this property on an open subset $U \subset \mathbb{C}^*$ is called an *analytic* or *uniform
branch* of $Log$ on $U$. The existence of such functions imposes serious restrictions on the "topology" of $U$, as we shall see in § 4.

---

[25] This means that all the possible values of $Log\, z$ are obtained by adding to the
series any value of $Log\, a$.

(xi) *Complex roots of the trigonometric functions.* The relations (4) show that $\sin z = 0$ is equivalent to $\exp(iz) = \exp(-iz)$, i.e. to $\exp(2iz) = 1$, i.e. to $2iz = 2ki\pi$, so that it has no other complex roots than the obvious real roots. Same result for $\cos z$.

(xii) *Expansions of $\pi$ in series.* There are as many as one wants. Let us start for example from the function $\sin x$ on $I =\,]-\pi/2, \pi/2[$. Since $\pi/2$ is the least root $> 0$ of $\cos x = \cos(-x)$, the function $\cos x$, which is $> 0$ for $x = 0$, is $> 0$ on $I$ (intermediate value theorem). So we have $\sin' x > 0$ on $I$, so that $\sin x$ is strictly increasing (Chap. III, n° 16), thus maps $I$ onto $J =\,]-1,1[$. The inverse map

$$\arcsin : J \longrightarrow I$$

exists and is differentiable, with

$$\arcsin' y = 1/\cos x = (1-y^2)^{-1/2}$$

as we have already remarked in Chap. III, formula (15.7). Since $|y| < 1$, Newton's binomial series gives us

$$(1-y^2)^{-1/2} = 1 + y^2/2 + 1.3y^4/2.4 + 1.3.5y^6/2.4.6 + \ldots;$$

see (11.12). The primitive series of the latter is

$$F(y) = y + y^3/3.2 + 1.3y^5/5.2.4 + 1.3.5y^7/7.2.4.6 + \ldots$$

and represents a differentiable function on $J$ such that

$$F'(y) = \arcsin' y$$

by the general theorem on term-by-term differentiation of power series. Chap. III, n° 16 shows then that $\arcsin y = F(y)$ up to an additive constant; it is zero since the two functions are zero at $y = 0$. In consequence,

$$(14.24) \quad \arcsin y = y + y^3/3.2 + 1.3y^5/5.2.4 + 1.3.5y^7/7.2.4.6 + \ldots$$

for $|y| < 1$; since $\arcsin(1/2) = \pi/6$ we obtain a series for $\pi$ which converges quite quickly.

Formula (24) is due to Newton, who had a proof very like ours: he *integrated* the binomial series term-by-term, which we shall not be able to do until Chap. V. Newton obtained the series for $\sin z$ by inverting (24), an easy exercise if one wants only the first terms.

Another possibility: use the function

$$\tan x = \sin x / \cos x,$$

still on the interval $I$. We have $\tan' x = 1/\cos^2 x = 1/(1 + \tan^2 x)$. The function is again strictly increasing and this time maps $I$ onto $\mathbb{R}$ in view of its limit values at the end points of $I$. Whence a function

$$\arctan y : \mathbb{R} \longrightarrow I = ]-\pi/2,\ \pi/2[,$$

with

$$\arctan' y = (1 + y^2)^{-1} = 1 - y^2 + y^4 - \cdots;$$

the argument used for the function arcsin shows here that

(14.25)                    $\arctan y = y - y^3/3 + y^5/5 - \cdots.$

For $y = 1$, we obtain

$$\pi/4 = 1 - 1/3 + 1/5 - \cdots,$$

Leibniz' series; for numerical calculations one should choose something else, as Newton observed at the time. Moreover he himself had a method based on his binomial series[26]. To do this he considered a circle, with equation $x^2 + y^2 - x = 0$, centre $(1/2, 0)$, and radius $1/2$, whence

(14.26)    $\begin{aligned} y = x^{1/2}(1-x)^{1/2} &= x^{1/2} - x^{3/2}/2 - x^{5/2}/8 - \\ &\quad - x^{7/2}/16 - 5x^{9/2}/128 - \cdots \end{aligned}$

by (11.19). Now he knew that the area of the curve $y = x^m$ contained between the verticals $0$ and $x$ is $x^{m+1}/(m+1)$ and "thus" that the analogous area of the curve (26) is

(14.27)    $z = x^{1/2}(2x/3 - x^2/5 - x^3/28 - x^4/72 - 5x^5/704 - \cdots).$

He took $x = 1/4$ and found, without any difficulty, $z = 0.07677\ 31061\ 63047\ 3$, the area of the curved triangle $AdB$ in the figure 2. Since the angle $ACd$ is equal to $\pi/3$, and since the area of the triangle $BdC$ is equal to $\sqrt{3}/32$, the area of the circular sector $ACd$ (namely $\pi/24$ since the circle considered has radius $1/2$) is equal to $0.07677 \ldots + \sqrt{3}/32$, whence Newton deduced

---

[26] See Vol. III of *Mathematical Papers* edited by D. T. Whiteside for the "tract" *De methodis serierum et fluxionum* of the winter of 1670–1671 and its English translation, pp. 223–227. And since Newton clearly did not care to squander his energy, he calculated simultaneously the areas of the equilateral hyperbola $x^2 - y^2 + x = 0$, the expansions of $y$ in series being identical to that for the circle up to changes of signs. The essential point is that if a curve is given by a power series in $x$ (with maybe non-integral rational exponents), the area bounded by the curve and the verticals $a$ and $b$ is $F(b) - F(a)$ where $F$ is the primitive series of $y$. In fact, Newton showed (in terms of fluents and of fluxions) that the derivative of the area with respect to $x$ is $y$ and so considered the result in question as obvious.

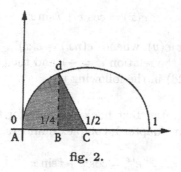

fig. 2.

$$\pi = 3.14159\ 26535\ 89792\ 8.$$

These methods set others to think. A little after 1700, the astronomer John Machin calculated a hundred decimal places of $\pi$ with the help of the formula

$$\pi/4 = 4\arctan(1/5) - \arctan(1/239);$$

one proves this[27] by using the addition formula $\arctan x + \arctan y = \ldots$ successively for $x = y = 1/5$, then $x = y = 5/12$, finally for $x = 120/119$ and $y = -1/239$, so as to reach $\arctan 1$. In 1717, the Frenchman de Lagny calculated some 250 decimal places using the formula $\tan \pi/6 = 1/\sqrt{3}$, whence

$$\pi = 2\sqrt{3}(1 - 1/3.3 + 1/5.3^2 - \ldots)$$

by applying (25). Euler used other similar formulae, and Gauss found the formula

$$\pi/4 = 12\arctan(1/38) + 20\arctan(1/57) + 7\arctan(1/239) + 24\arctan(1/268),$$

but one would not advise the reader to try to establish it if he is pressed for time; Gauss himself discovered it by chance in a context quite different from the numerical calculation of $\pi$.

(xiii) *Multiplication formulae.* The relation $\exp(ix) = \cos x + i\sin x$ shows that

(14.28)     $(\cos x + i\sin x)^n = \exp(nix) = \cos nx + i\sin nx;$

this is the famous *formula* of (Abraham) *de Moivre* of 1730, a French protestant refugee in England and author of a famous treatise on the calculus of probabilities; he proved it without referring to the exponential function, which is easy once one has the idea, since conversely the addition formulae for the trigonometric functions show directly, as we saw *à propos* Theorem 11, that the function

---

[27] See for example Hairer and Wanner, *Analysis by Its History*, pp. 52–53.

$$e(x) = \cos x + i \sin x$$

satisfies $e(x + y) = e(x)e(y)$, whence $e(nx) = e(x)^n$. There is nothing in this calculation other than the relation $i^2 = -1$ and the classical addition formulae. We shall clarify (28) in the following n°.

In a recent mathematics text book designed for the final year of French high schools there is, in Chapter I, complex numbers, a place where they put, *by definition*,

$$e^{ix} = \cos x + i \sin x$$

and where they show, with the help of the addition formulae, that $e^{i(x+y)} = e^{ix}e^{iy}$. No explanation is offered to the reader as to what the mysterious letter $e$ might signify, nor of a nonreal exponent. It would be difficult to imagine a more aberrant conception of mathematics: Euler's most famous formula degenerating into a pure and simple *notation*, and furthermore incomprehensible!

Since the books in use in high-schools conform strictly to the directives of the national Ministry of Education one is forced to conclude that this eminently original version of mathematics – I never met it before the year 2000 – is due to the committees who decide the programmes. It is difficult to imagine that mathematicians, even "applied", could ever have advocated it. But then, who is responsible? Electricians? In America, at the end of the XIX$^{\text{th}}$ century, engineers at General Electric were taught to use Euler's formula by Georg Steinmetz, a young German immigrant with a German Ph.D. in Physics who knew Mathematics; he soon became one of the most prominent electrical engineers in the USA, as famous as Edison and Sperry. See Ronald R. Kline, *Steinmetz: Engineer and Socialist* (Johns Hopkins UP, 1992).

(xiv) *Roots of a complex number.* The relation (28) shows in particular that

(14.29)    $(\cos x + i. \sin x)^n = 1$    for $x = 2k\pi/n$.

This shows in particular that the number 1 has $n$ $n^{\text{th}}$ roots in $\mathbb{C}$, situated in the complex plane at the vertices of the regular polygon of $n$ sides inscribed in the unit circle $|z| = 1$ and having a vertex at $z = 1$, namely the *roots of unity*

(14.30)    $\exp(2k\pi i/n)$    with $k = 0, \ldots, n - 1$;

adding a multiple of $n$ to $k$ clearly does not change the result. Since the equation $z^n - 1 = 0$ can never have more than $n$ roots, one obtains them all in this way, whence, using a theorem of algebra,

(14.31)
$$z^n - 1 = \prod_{0 \le k \le n-1} \left( z - e^{2ki\pi/n} \right).$$

More generally,

(14.32)
$$z^n - a = \prod \left( z - |a|^{1/n} e^{(it+2ki\pi)/n} \right)$$

for any nonzero complex number $a = |a|e^{it}$, where $|a|^{1/n}$ is the usual positive root. In these formulae, $k \in \mathbb{Z}$ can vary "modulo $n$". On replacing $z$ by $x/y$ in (31), one finds the identity

(14.33)
$$x^n - y^n = (x - y)(x - \omega y)\dots(x - \omega^{n-1}y)$$

where $\omega = \exp(2\pi i/n)$.

## 15 – Euler's relation *chez* Euler

In his *Introductio in Analysin Infinitorum* of 1748 which served as a Bible to mathematicians up at least to Cauchy's time, Euler did not establish the relation $\exp(ix) = \cos x + i \sin x$ by means of power series. He deduced it from de Moivre's formula (which did not presuppose these series) by an argument as always very ingenious and, as always, a little false. He wrote

$$\cos x + i \sin x = [\cos(x/n) + i \sin(x/n)]^n$$

and made $n$ tend to infinity. Then "evidently" $\cos(x/n) = 1$ and $\sin(x/n) = x/n$, so that the right hand side is "clearly" equal to $(1 + ix/n)^n$, "qed", if we know that $e^z = \lim(1 + z/n)^n$.

To justify this argument, we have to put

$$\cos(x/n) + i \sin(x/n) = 1 + z_n/n$$

and use Theorem 10: essentially, we have to show that $z_n$ tends to $ix$, in other words that

$$\lim\{n[\cos(x/n) - 1] + in\sin(x/n)\} = ix.$$

Since $\cos(x/n) - 1$ lies between $-x^2/2n^2$ and $0$ for $n$ large, by (14.9), the product over $n$ tends to 0. It remains to prove that $n\sin(x/n)$ tends to $x$, i.e. that the derivative of $t \longrightarrow \sin t$ is equal to 1 for $t = 0$. Qed, but (14.9) rests on the *series* for cosine.

The interest of (14.28) and of the analogous formula deduced from it on replacing $i$ by $-i$ (or, what comes to the same, $x$ by $-x$) is to provide the no less famous relations

(15.1')
$$\cos nx = \cos^n x - \binom{n}{2} \cos^{n-2} x \cdot \sin^2 x +$$
$$+ \binom{n}{4} \cos^{n-4} x \cdot \sin^4 x - \dots$$

(15.1")       $\sin nx = \binom{n}{1} \cos^{n-1} x . \sin x -$

$$- \binom{n}{3} \cos^{n-3} x . \sin^3 x + \dots ;$$

to obtain these one expands $(\cos x \pm i \sin x)^n$ by the binomial formula and adds or subtracts the two formulae so obtained. It seems that they are not there literally in de Moivre, and it was Euler who, by this method, wrote them in his *Introductio*. And when one is called Euler, one deduces, why not, the expansions in power series of $\cos x$ and $\sin x$, whence, starting from these, another proof of the relations (1') and (1"). The method, we see, is again of genius, – and, again, finesses the essential point.

Here it is: in (1'), for $x$ given, you replace $x$ by $x/n$ with $n$ infinitely large and so $x/n$ infinitely small; the left hand side becomes $\cos x$; on the right hand side you replace $\cos(x/n)$ and its powers by 1 since $\cos 0 = 1$, and $\sin(x/n)$ by $x/n$ since $\sin u \sim u$ for $u$ infinitely small; $\sin^k(x/n)$ then becomes $x^k/n^k$ and third term of (1') for example becomes

$$n(n-1)(n-2)(n-3)x^4/n^4.4! = 1(1-1/n)(1-2/n)(1-3/n)x^4/4!.$$

But since $n$ is infinitely large, $1/n$, $2/n$, etc. are zero, and the preceding expression reduces to $x^4/4!$; and so one (re)finds Newton's series $\cos x = 1 - x^2/2! + x^4/4! - \dots$

Once again we have to deal with a passage to the limit over $n$ in a sum whose number of terms is certainly finite, but increases indefinitely with $n$. This suggests using the dominated convergence theorem of n° 12, interchanging the letters $n$ and $p$ since here we sum over $p$ while passing to the limit over $n$.

In the general term

$$u_n(p) = n(n-1)\dots(n-2p) \cos^{n-2p-1}(x/n) . \sin^{2p+1}(x/n)/(2p+1)!,$$

let us put a factor $n$ in each of $2p+1$ terms of the form $n-k$ and gather to the sine the product of these factors $n$; we find

(15.2)  $(1-1/n)\dots(1-2p/n) \cos^{n-2p-1}(x/n)[n.\sin(x/n)]^{2p+1}/(2p+1)!.$

By Theorem 11, we need only find a convergent series $\sum v(p)$ with positive terms such that $v(p)$ majorises the expression above for any $n$, and then show that this tends to $x^{2p+1}/(2p+1)!$ as $n$ increases indefinitely.

It is easy to find $v(p)$. Clearly

$$|u_n(p)| \le [n.\sin(x/n)]^{2p+1}/(2p+1)!.$$

As we have seen above, $n.\sin(x/n)$ tends to $x$ as $n$ increases, so is, for $x$ given, majorised in modulus by a constant $M(x)$ whose value is not imporant because we then obtain the inequality

$$|u_n(p)| \leq M(x)^{2p+1}/(2p+1)! = v(p)$$

which suffices to prove normal convergence, since $\sum v(p) < +\infty$.

It thus remains to show that $u_n(p)$ tends to $x^{2p+1}/(2p+1)!$. The factors $1 - 1/n, \ldots, 1 - 2p/n$ present no problem: their product tends to 1. The denominator $(2p+1)!$ does not change and $n . \sin(x/n)$ tends to $x$ as we saw above; the product of the two last factors of (2) thus tends to $x^{2p+1}/(2p+1)!$. Since on the other hand $\cos(x/n)$ tends to 1, so does its $-2p-1$ th power for all $p$.

We have almost reached our aim: it remains to prove that

(15.3)                    $$\lim \cos^n(x/n) = 1$$

and this is the crucial point. It is clear that $\cos(x/n)$ tends to 1, but there is no theorem to say that if $u_n$ tends to 1, so similarly does $u_n^n$: this general statement is grossly false and Euler was particularly well placed to know this, since it very happily does not apply to the sequence $u_n = 1 + x/n$. He found, miraculously, that $\cos^n(x/n)$ really does tend to 1, but to prove this we have to use a sharper result than $\cos 0 = 1$, for example the inequality

(15.4)                    $$1 - x^2/2n^2 < \cos(x/n) < 1,$$

valid for $|x/n| < 2$, so for $n$ large; this shows that the difference between 1 and $\cos x/n$ is of the order of magnitude of $1/n^2$, in other words that $\cos(x/n)$ tends to 1 much more rapidly than, for example, $1 + x/n$. Still this argument remains to be completed; the inequality

$$(1 - x^2/2n^2)^n < \cos^n(x/n) < 1$$

will not yield the desired result unless one can show that the left hand side tends to 1; on putting $x = y\sqrt{2}$ this can be written $(1 - y/n)^n(1 + y/n)^n$ and so tends to $\exp(-y)\exp(y) = 1$; one might also use Theorem 21 with $z_p = -x^2/2p$.

All this, one sees, is like a high-wire act (netless *chez* Euler) and even breaks the vicious circle. The relation (14.9) which legitimates (4) reminds one very strongly of the beginning of the power series of $\cos x$, i.e. the result that one hopes to establish! We could deduce it from the relation

$$\cos x = (1 - \sin^2 x)^{1/2} \geq 1 - \sin^2(x)/2 \geq 1 - x^2/2$$

so long as we know that $|\sin x| \leq |x|$, which can be taken as geometrically obvious. Euler did not worry over these subtleties.

We said above that if a sequence $u_n$ tends to 1, the sequence $u_n^n$ need not do so too. The formula

$$\exp x = \lim(1 + x/n)^n$$

is witness to this. On pushing the false "argument" to the limit without having the same intuition as Euler for correct formulae, one might even claim that $1 + x/n = 1$ for $n$ infinitely large, and so that $(1 + x/n)^n = 1^n = 1$ – one sometimes meets this in students' work –, which would yield the relation $\exp(x) = 1$ for any $x$.

Its social impact would be immense as one sees from an argument due to Jakob Bernoulli, 1690, in an age when financial mathematics – interest, insurance, games of chance, etc. – interested many other people, including de Moivre, Leibniz and the English, waiting for Euler himself and his "tontines".

Suppose that you deposit your fortune for a year with a banker who, rather than calculating at the end of the year only the interest that he owes you the rate of $x\%$ per annum, instead allows *at each instant* for what he owes at the preceding instant. If, a little less generously, he does this $n$ times each year, your balance would be multiplied by $1 + x/n$ after the first period of $365/n$ days, then by $(1+x/n)^2$ after the second and so on until your fortune, at the end of the year, is multiplied by $(1 + x/n)^n$. This has been known for as long as moneylenders have existed – one in finds it already appearing in Babylonians examples! If you pass to the limit you will find that it has been multiplied by the factor $\exp(x)$ – or by 1 if you believe that $\lim(1+x/n)^n = 1$.

This fantasy formula allows you to prove that if you deposited your fortune with an infinitely honest banker, it would not increase even by a penny at the end of the year, and this at any rate of interest generously offered by the philanthropist. This would be the Triumph of Virtue: total honesty would cost strictly nothing to those who practiced it, and would bring strictly nothing to those who benefit from it. The same argument would show conversely that, if your banker lends you money at 1000% per annum and calculates at each instant the compound interest that you owe him, you will have to give him strictly no more, at the end of the year, than the sum he gave you at the beginning; this time, the Bankruptcy of the Sharks. One sees revolution.

In fact, the discoverer of the power series of $\cos x$ and $\sin x$, namely Newton, proceeded in quite another way: as we saw in n° 14, formula (24), an integration gave him directly, *via* his binomial formula, the power series for $\arcsin x$ and it was by inverting it that Newton found that of $\sin x$. The problem is to find a power series

$$(15.5) \qquad\qquad y = x + a_3 x^3 + a_5 x^5 + \dots$$

satisfying the relation

$$(15.6) \qquad x = y + y^3/3.2 + 1.3y^5/5.2.4 + \dots \quad (= \arcsin y),$$

which he did by the method expounded in Chap. II, n° 22: he calculated the successive powers in the series $y$, substituted the results in (6) and wrote that the total coefficient of each monomial $x^3$, $x^5$, etc. is zero, whence obtaining relations between the coefficients of (5) enabling him to calculate them one-by-one. For example

$$x = (x + a_3 x^3 + \ldots) + (x^3 + \ldots)/3.2 + \ldots$$

where the ... contain no further $x$ or $x^3$, whence $a_3 = -1/2.3$. It follows that

$$
\begin{aligned}
x &= (x - x^3/6 + a_5 x^5 + \ldots) + (x - x^3/6 + \ldots)^3/6 + \\
&\quad + 3(x + \ldots)^5/40 + \ldots = (x - x^3/6 + a_5 x^5 + \ldots) + \\
&\quad + (x^3 - 3x^2.x^3/6 + \ldots)/6 + 3(x^5 + \ldots)/40 + \ldots = \\
&= x + (a_5 - 1/12 + 3/40)x^5 + \ldots
\end{aligned}
$$

where the ... contain no further $x^5$, whence

$$a_5 = 1/12 - 3/40 = 1/120 = 1/2.3.4.5,$$

etc.

Further, integration provided him the series for $\log(1+x)$ which, inverted, produced the exponential series. We have seen that

$$y = \log(1 + x) = x - x^2/2 + x^3/3 - \ldots;$$

let us try to extract $x$ in the form

$$x = y + a_2 y^2 + a_3 y^3 + \ldots;$$

first we will have

$$y = y + a_2 y^2 + \ldots - (y + \ldots)^2/2 + \ldots$$

where the unwritten terms are of degree $\geq 3$ in $y$. So we must have $a_2 = 1/2$ to eliminate the terms in $y^2$. This done,

$$y = y + y^2/2 + a_3 y^3 + \ldots - (y + y^2/2 + \ldots)^2/2 + (y + \ldots)^3/3 + \ldots;$$

the terms in $y$ and $y^2$ cancel, as they must, and to eliminate the terms in $y^3$ we must have $a_3 - 1/2 + 1/3 = 0$, whence $a_3 = 1/6 = 1/3!$, etc.

This rudimentary method does not provide general formulae, which Newton confines himself to extrapolating from the first coefficients – if you believe in the perfection of Creation, you can go on to the end –, nor did it prove that the series obtained converged; but of course Weierstrass and his successors traversed the same path, including for power series in several variables. Given a *convergent* power series of the form

(15.7)                $x = y + b_2 y^2 + \ldots = g(y), \qquad |y| < R,$

(a case to which one reduces immediately if $b_1 \neq 0$), there exists one and only one *convergent* power series

(15.8)                $y = x + a_2 x^2 + \ldots = f(x), \qquad |x| < R',$

such that $g[f(x)] = x, f[g(y)] = y$. In terms of analytic functions: (7) represents an analytic function in a disc $|y| < R$, with $g(0) = 0$ and $g'(0) \neq 0$, and we have to prove the existence of one and only one analytic function $f$ on a neighbourhood of 0, such that $f(0) = 0$ and which, for $|x|$ sufficiently small, is the inverse map of $g$. There are analogous theorems in the differential calculus of several real variables as we have seen in Chap. III, n° 24, and they immediately set the problem into the framework of the theory of *holomorphic* functions, so also analytic *if* we know that the words "holomorphic" and "analytic" are synonymous.

But if one looks for a direct proof, the situation becomes complicated. One first adopts the point of view of formal series of Chap. II, n° 22: on substituting (8) into (7), and expanding the powers of $y$ and identifying the result with $x$, one obtains algebraic relations between the $a_n$ which enable one, theoretically, to calculate them one-by-one; this is clearly what Newton did for the first terms. This done, which is quite easy, the crucial problem remains: to show that if the coefficients of (7) are $O(q^n)$ for a number $q > 0$ [a necessary and sufficient condition for the radius of convergence of (7) to be $> 0$], one similarly has $b_n = O(q'^n)$ for some $q' > 0$. This demands non-obvious estimates of $b_n$ as a function of $a_p$, estimates analogous to, and more difficult than, those we used in Chap. II, n° 22, Theorem 17 *à propos* compositions of analytic functions. Neither Dieudonné, nor Remmert, serious people, have dared to expound the subject in their books; but see Serge Lang, *Complex Analysis* (Springer, 1999), Chap. II.

## 16 – Hyperbolic functions

When we have a reasonably civilised curve in the plane $\mathbb{R}^2$ it is generally possible to find an interval $I \subset \mathbb{R}$ and two continuous real functions $x(t)$ and $y(t)$ on $I$ so that, as $t$ traverses $I$, the point $(x(t), y(t))$ traverses all the curve, maybe several times (or even infinitely many times in the case of Peano); in other words: the function $t \mapsto (x(t), y(t))$ maps $I$ onto the set of points of the curve. This is called a *parametric representation* of the given curve. The level of civilisation of the curve will be higher as one can further require the functions $x$ and $y$ to have continuous derivatives. Peano's curve which passes through all the points of a square is at level zero.

The trigonometric or circular functions allow one to parametrise a circumference: as $t \in \mathbb{R}$ varies, the point $(\cos t, \sin t)$ describes the circle $x^2 + y^2 = 1$. If $a$ and $b$ are real nonzero constants, the point $(a \cos t, b \sin t)$ describes the curve $x^2/a^2 + y^2/b^2 = 1$, i.e. an ellipse with centre 0. The parabola $y = 3x^2$ is parametrised by the map $(t, 3t^2)$ of $\mathbb{R}$ into $\mathbb{R}^2$.

To parametrise the equilateral hyperbola $x^2 - y^2 = 1$, we need functions such that $x(t)^2 - y(t)^2 = 1$. This equation can also be written as $x(t)^2 + [iy(t)]^2 = 1$, so one should choose $x(t) = \cos t$, $y(t) = i \sin t$. This brings us into $\mathbb{C}^2$, but not into $\mathbb{R}^2$ if $t$ is real. If, on the other hand, $t$ is pure imaginary, it is clear from the series expansions that $\cos t$ and $i \sin t$ become real. One is thus led to introduce the functions

$$(16.1) \qquad \cosh t \;=\; \cos it = (e^t + e^{-t})/2 =$$
$$= \; 1 + t^2/2! + t^4/4! + \ldots = \sum t^{[2n]},$$
$$(16.2) \qquad \sinh t \;=\; -i \sin it = (e^t - e^{-t})/2 =$$
$$= \; t + t^3/3! + t^5/5! + \ldots = \sum t^{[2n+1]}.$$

These hyperbolic cosine and sine, as one calls them, have properties analogous to those of the circular functions, so similar that their statements hardly require formal proofs; in fact, one could even deduce them from the properties of circular functions from the formulae (1) and (2), clearly valid for $t \in \mathbb{C}$.

$$\cosh(-x) = \cosh x, \qquad \sinh(-x) = -\sinh x.$$
$$\cosh 0 = 1, \qquad \sinh 0 = 0.$$
$$\cosh' x = \sinh x, \qquad \sinh' x = \cosh x.$$
$$\cosh^2 x - \sinh^2 x = 1.$$
$$\cosh(x + y) = \cosh x . \cosh y + \sinh x . \sinh y.$$
$$\sinh(x + y) = \sinh x . \cosh y + \sinh y . \cosh x.$$

The difference from the circular functions relates to their behaviour for $x$ real. We have

$$\cosh x - 1 = (e^x - 1 + e^{-x})/2 = (e^{x/2} - e^{-x/2})^2/2 = 2 \cosh^2 (x/2) \geq 0,$$

whence $\cosh x \geq 1$. It is obvious that $\sinh x$ is $> 0$ for $x > 0$ and $< 0$ for $x < 0$. The function $\sinh x$ is strictly increasing since $\sinh' x \geq 1$. The function $\cosh x$ is strictly increasing for $x > 0$ and decreasing for $x < 0$.

As $x \to +\infty$, we clearly have $\cosh x \sim e^x/2$, with the same result for $\sinh x$; the two functions increase at an exponential rate. As $x \to -\infty$, we have $\cosh x \sim e^{-x}/2$ and $\sinh x \sim -e^{-x}/2$, so that $\cosh x$ tends to $+\infty$ and $\sinh x$ to $-\infty$.

We can pursue the analogy by introducing the functions

$$\tanh x \;=\; \sinh x/\cosh x = (e^{2x} - 1)/(e^{2x} + 1) = 1 - 2/(e^{2x} + 1) = -i \tan ix,$$
$$\coth x \;=\; \cosh x/\sinh x = (e^{2x} + 1)/(e^{2x} - 1) = 1 + 2/(e^{2x} - 1) = i \cot ix.$$

For $x \in \mathbb{R}$, the second formula assumes $x \neq 0$. For $x \in \mathbb{C}$, we leave to the reader the effort of finding the values of $x$ to be avoided in one or the other case. We have

$$\lim_{x \to +\infty} \tanh x = 1, \qquad \lim_{x \to -\infty} \tanh x = -1$$

since $e^{2x}$ tends to $+\infty$ in the first case and to $0$ in the second. We have the same results for $\coth x$ and, further,

$$\lim_{x \to -0} \coth x = -\infty, \qquad \lim_{x \to +0} \coth x = +\infty.$$

The derivatives of the functions $\cosh$ and $\sinh$ are immediately provided by the formulae

$$\tanh' x = 1/\cosh^2 x = 1 - \tanh^2 x,$$
$$\coth' x = -1/\sinh^2 x = 1 - \coth^2 x,$$

whence their directions of increase: the function $\tanh$ increases strictly on $-1$ to $+1$ between $-\infty$ and $+\infty$, the function $\coth x$ decreases strictly on $-1$ to $-\infty$ between $-\infty$ and $0$, and from $+\infty$ to $1$ between $0$ and $+\infty$. The graphs of these functions can be found in all the best textbooks.

In what concerns the expansions in power series of the functions $\tanh$ and $\coth$, the situation is necessarily the same as for the functions $\tan$ and $\cot$: Newton would have explained to you how to calculate the first terms, but the general formula is not obvious and involves the Bernoulli numbers which will appear in Chap. VI. Since moreover the change of variable $x = iy$ transforms the circular functions into the hyperbolic functions, it is not worth doing the same calculations twice.

The preceding functions admit inverse maps, at least partially. The function $y = \cosh x$ maps $[0, +\infty[$ onto $[1, +\infty[$ by the general theorems of Chap. III, whence a function

$$x = \arg\cosh y : [1, +\infty[ \longrightarrow [0, +\infty[;$$

it is differentiable at every point $y$ where $\cosh' y \neq 0$, i.e. for $y > 1$, and then

$$\arg\cosh' y = 1/\cosh' x = 1/\sinh x = 1/(\cosh^2 x - 1)^{1/2}$$

whence, since $\cosh x = y$,

(16.3)    $$\arg\cosh' y = (y^2 - 1)^{-1/2} \qquad \text{for } y > 1.$$

We can deduce a power series expansion for $\arg\cosh y$ from this. Formula (11.21) shows that

$$(16.4)(y^2 - 1)^{-1/2} = y^{-1}(1 - y^{-2})^{-1/2} = y^{-1}\left(1 + \frac{1}{2}y^{-2} + \frac{1}{2}\frac{3}{4}y^{-4} + \dots\right)$$

a power series in $y^{-1}$ converging for $y > 1$; on passing to the series of primitives as if dealing with a power series in $y$ we obtain the relation

$$\operatorname{arg\,cosh} y = \log y + C - \sum_{p \geq 1} \frac{1.3 \dots (2p - 1)}{2.4. \dots 2p} y^{-2p}/2p$$

where $C$ is a constant. This formal calculation, "integrating term-by-term" a series in $y^{-1}$ and not in $y$, is justified (while we do not yet have Chap. V at our disposal) by checking that if one differentiates the series obtained term-by-term, which produces the initial series (4), we obtain a series which converges normally on every compact interval $K \subset ]1, +\infty[$ (Chap. III, § 6. n° 17, Corollary of Theorem 19). Indeed, since the radius of convergence of Newton's series is equal to 1, the series (4) converges normally on the interval $[a, +\infty[$ for any $a > 1$ (Chap. III, § 4, n° 17, example 1), a more than sufficient result.

It remains to calculate the constant $C$. For this we use the relation

$$\operatorname{arg\,cosh} y = \log(y + (y^2 - 1)^{\frac{1}{2}}) = \log y + \log(1 + (1 - y^{-2})^{\frac{1}{2}})$$

below, and lets $y$ tend to $+\infty$; the last term tends to $\log 2$, whence

$$\operatorname{arg\,cosh} y = \log y + \log 2 + o(1) \quad \text{as} \quad y \to +\infty;$$

But when $y \to +\infty$, i.e. when $y^{-1}$ tends to 0, the sum of the power series in $y^{-1}$ obtained above tends to its constant term, namely 0. Hence we find that $\operatorname{arg\,cosh} y = \log y + C + o(1)$, whence $C = \log 2$.

The formula

$$(16.5) \qquad \operatorname{arg\,cosh} y = \log y + \log 2 - \sum_{p \geq 1} \frac{1.3 \dots (2p - 1)}{2.4. \dots 2p} y^{-2p}/2p$$

yields an asymptotic expansion (Chap. VI) of the function when $y$ is very large.

As $y$ tends to 1 the general term $u_p(y)$ of the series is positive and tends to $u_p(1)$ remaining $\leq u_p(1)$. If one shows that $\sum u_p(1)$ converges one can then pass to the limit under the $\sum$ sign (Chap. III, n° 8, Theorem 9). Now

$$u_{p+1}(1)/u_p(1) = (2p + 1)2p/(2p + 2)^2 = (1 + 1/2p)(1 + 1/p)^{-2}$$

whence

$$(16.6) \qquad u_{p+1}(1)/u_p(1) = 1 - s/p + O(1/p^2)$$

with $s = 3/2 > 1$. Now (Vol. II, Chap. VI, n° 5: Gauss' criterion, easy to prove) any series of positive terms satisfying a relation (16.6) is convergent if $s > 1$. This is the case here, and since $\arg\cosh y$ and $\log y$ tend to 0 when $y$ tends to 1, the formula (5) shows that $\sum u_p(1) = \log 2$.

*Exercise.* Show that, for $y$ close to 1, there is an expansion of the form

$$\arg\cosh y = \sum a_n(y - 1)^{n+\frac{1}{2}}.$$

On what interval with left end point 1 is it valid?

The function $y = \sinh x$ maps $\mathbb{R}$ onto $\mathbb{R}$, whence an everywhere differentiable inverse

$$x = \arg\sinh y : \mathbb{R} \longrightarrow \mathbb{R}$$

with

$$\arg\sinh' y = 1/\sinh' x = 1/\cosh x = (y^2 + 1)^{-1/2}.$$

The function $y = \tanh x$ maps $\mathbb{R}$ onto $]-1, 1[$, whence

$$\arg\tanh y : ]-1, 1[ \longrightarrow \mathbb{R},$$

with

$$\arg\tanh' y = 1/\tanh' y = 1/(1 - \tanh^2 x) = (1 - y^2)^{-1} \qquad (|y| < 1).$$

Finally, the function $\coth$ maps $\mathbb{R} - [-1, 1]$, the (obvious) union of two intervals, bijectively onto the same set, whence a function $x = \arg\coth y$ defined for $|y| > 1$ with

$$\arg\coth' y = 1/\coth' x = 1/(1 - \coth^2 x) = (1 - y^2)^{-1} \qquad (|y| > 1).$$

These two last functions thus provide primitives of the function $1/(1 - y^2)$ in each of three open intervals where it is defined, the two first serving to integrate $(y^2 \pm 1)^{1/2}$.

It remains to show how these inverse functions can be expressed in terms of the function $\log$. The formula

$$2y = 2\cosh x = e^x + e^{-x} = e^x + 1/e^x$$

shows that $e^{2x} - 2ye^x + 1 = 0$, an equation of the second degree in $e^x$. If one assumes $x > 0$ and so $e^x > 1$, it follows that $e^x = y + (y^2 - 1)^{1/2}$ since the other root is $< 1$. In consequence,

$$\arg\cosh y = \log\left[y + (y^2 - 1)^{1/2}\right] \qquad \text{for } y > 1.$$

An analogous argument shows that

$$\arg\sinh y = \log\left[y + (y^2 + 1)^{1/2}\right] \qquad \text{for } y \in \mathbb{R}$$

and that

$$\arg\tanh y \;=\; \frac{1}{2}\log\frac{1+y}{1-y} \qquad \text{for } -1 < y < 1.$$

$$\arg\coth y \;=\; \frac{1}{2}\log\frac{y+1}{y-1} \qquad \text{for } |y| > 1.$$

Clearly there are analogous formulae for the circular functions, but they involve the complex $\mathcal{L}$og and, because of this fact, serve mainly as sources of errors.

## §3. Infinite products

### 17 – Absolutely convergent infinite products

In analysis we meet *infinite products*

$$a_1 a_2 \ldots a_n \ldots = \prod a_n$$

to which we have to give a meaning. The only reasonable solution is to suppose that the partial products

$$p_n = a_1 \ldots a_n$$

tend to a limit $p$, which by convention will be the value of the product. Taken literally, the definition is of no interest if some of the $a_n$ are zero; so we agree to omit them from the product.

Experience shows that, with exceptions, the problem is again uninteresting (or that we do not know how to treat it ...) if the limit $p$ is zero. Suppose therefore that $p \neq 0$. Then $a_n = p_n/p_{n-1}$ tends to $p/p = 1$, which allows us to put $a_n = 1 + u_n$ where $u_n$ tends to 0. Then

$$\begin{aligned} \log |p_n| &= \log |1 + u_1| + \ldots + \log |1 + u_n| \leq \\ &\leq \log(1 + |u_1|) + \ldots + \log(1 + |u_n|) \end{aligned}$$

since the function log is increasing. Now we have shown in Chap. II, n° 10, that $\log x \leq x - 1$ for all $x > 0$, i.e.

$$\log(1 + x) \leq x \quad \text{for all } x > -1.$$

Thus

$$\log |p_n| \leq |u_1| + \ldots + |u_n|.$$

Suppose now that the series $\sum u_n$ is *absolutely* convergent. The preceding inequality shows that the sequence $(\log |p_n|)$ is bounded above by a number $M > 0$. The relation $\log |p_n| \leq M$ implies $|p_n| \leq e^M = M'$, so the sequence $(p_n)$ is bounded.

The inequality

$$|p_n - p_{n-1}| = |p_{n-1} u_n| \leq M' |u_n|$$

now shows that the series with general term $p_n - p_{n-1}$ is absolutely convergent, so convergent. Whence the existence of $p = \lim p_n$.

It remains to check that $p \neq 0$. For this, one replaces the product of the $a_n$ by that of their reciprocals $a_n^{-1} = 1 + v_n$, which replaces $p_n$ by $q_n = 1/p_n$. The relation $(1 + u_n)(1 + v_n) = 1$ shows that

$$|v_n| = |u_n|/|1 + u_n| \leq 2|u_n| \quad \text{for } n \text{ large}$$

since $|1 + u_n|$, which tends to 1, is $\geq 1/2$ for $n$ large. The series $v_n$ therefore converges absolutely. Consequently $q_n$ tends to a limit $q$, and since $p_n q_n = 1$ for all $n$, it follows that $pq = 1$, whence $p \neq 0$. In conclusion:

**Theorem 13.** *Every infinite product $p = \prod(1 + u_n)$ for which the series $u_n$ is absolutely convergent is itself convergent, with $p \neq 0$ if none of factors of the product is zero.*

One says that such an infinite product is *absolutely convergent*. This, for example, is the case of the infinite product

$$\sin x = x \prod_{n=1}^{\infty} (1 - x^2/n^2\pi^2)$$

for the function $\sin x$ which we have mentioned in Chap. II, n° 6, valid for all $x \in \mathbb{C}$. On replacing $x$ by $ix$ one finds an analogous product for the function $\sinh x$. The proof – we have already mentioned Euler's first, with his "algebraic equation of infinite degree", at the end of n° 21 of Chap. II – will be the object of the following n°.

The condition $\sum |u_n| < +\infty$, sufficient to ensure convergence of the product in the general case, is also necessary if the $u_n$ are *real and all of the same sign* for $n$ large. If $u_n \geq 0$ for all $n$, it is clear that $u_1 + \ldots + u_n < p_n \leq p$ for all $n$, whence the result. If $u_n \leq 0$ for $n$ large, then also $0 < 1 + u_n \leq 1$ since $u_n \to 0$; we may therefore assume these inequalities valid for all $n$, so that $1 \geq p_n \geq p_{n-1} > 0$ for all $n$; the relation $u_n = (p_n - p_{n-1})/p_{n-1}$ now shows that $u_n \sim (p_n - p_{n-1})/p$; since the sequence $(p_n)$ is decreasing and converges, the right hand side is the general term of a convergent series of negative terms, whence the convergence of $\sum u_n$ in this case.

*Example 1 (infinite product for the Riemann $\zeta(s)$ function).* Let $p$ be a prime number and consider

$$(1 - 1/p^s)\zeta(s) = \sum 1/n^s - \sum 1/(pn)^s.$$

Clearly this eliminates all the terms whose index $n$ is divisible by $p$ from the zeta series. On multiplying the result by $1 - 1/q^s$, where $q$ is another prime number, one removes from the remaining series all the terms whose indices are divisible by $q$ and so all the terms of the initial series whose indices are divisible by $p$ or $q$ (or by both). If one denotes the sequence

$$2, \ 3, \ 5, \ 7, \ 11, \ 13, \ 17, \ldots$$

of prime numbers by $(p_n)$, one sees that

(17.1)     $$(1 - 1/p_1^s) \ldots (1 - p_k^s) \, \zeta(s) = \sum_{\substack{n \text{ not divisible} \\ \text{by } p_1 \text{ or } \ldots \text{ or } p_k}} 1/n^s.$$

The integers $n$ figuring in the remaining series are, leaving $n = 1$ aside, all $\geq p_k$; any integer $n$ is a product of prime factors, which must be $\leq n$ and so $\leq p_k$ if $n \leq p_k$. Since the sequence $(p_k)$ is unending if one believes Euclid and other Greek geniuses, and even if you don't, the right hand side is thus, apart from the term $n = 1$, arbitrarily small for $n$ large so long as the series $\zeta(s)$ converges, i.e. for $\mathrm{Re}(s) > 1$. We conclude that

$$(17.2) \qquad \lim (1 - 1/p_1^s) \ldots (1 - p_k^s)\,\zeta(s) = 1.$$

We have thus almost established the following result:

**Theorem 14.** [28] *We have $\zeta(s) \neq 0$ for $\mathrm{Re}(s) > 1$ and*

$$(17.3) \qquad\qquad 1/\zeta(s) = \prod_{p \text{ prime}} (1 - 1/p^s),$$

*the infinite product being absolutely convergent.*

The absolute convergence of the infinite product reduces to that of the series $\sum |1/p^s|$; obvious, since it is a subseries of the series $\sum |1/n^s|$.

The preceding theorem explains the rôle that the function has played, since Riemann's famous memoir, in the problem of the distribution of prime numbers. It has given rise to an immense literature. "Analytic number theory" consists of using the methods of analysis to establish results in arithmetic.

The preceding formula, for example, *implies* the existence of infinitely many prime numbers. If such were not the case, the function $\zeta$ would be the restriction to $\mathrm{Re}(s) > 1$ of a function analytic on $\mathbb{C}$ except at the points $s = 2ki\pi/\log p$ which annul one of the factors, finite in number, of the product. In particular, it would tend to a finite limit when $s \in \mathbb{R}$ tends to 1, the limit of convergence of the series.

Now the function $\zeta(s)$ is decreasing for $s > 1$, like the functions $1/n^s$. When $s > 1$ tends to 1, the number $\zeta(s)$ thus tends, increasing, to a limit $M \leq +\infty$. If the latter were finite, all the partial sums of the series with positive terms $\zeta(s)$ would be $\leq M$ for all $s > 1$. But if the relation

$$1 + 1/2^s + \ldots + 1/n^s \leq M$$

is valid for all $s > 1$, it remains valid at the limit for $s = 1$. The hypothesis $M < +\infty$ would thus imply the convergence of the harmonic series. [In fact, we shall prove later that the product $(s - 1)\zeta(s)$ tends to 1 with $s$.]

This result, known for more than twenty centuries, is not very impressive. But by examining combinations of partial series of the form $\sum 1/(an + b)^s$, where $a$ and $b$ are given *relatively prime* integers, Dirichlet was able to prove by the same method, with the help of arithmetic calculations that the XX[th] century has widely generalised, that there are infinitely many prime numbers in the "arithmetic progression" $an + b$, as had been long conjectured. Experimental verification can surely be programmed on a computer and surely

---

[28] Euler, *Introductio* ..., I, Chap. XV.

has been – a lot more of this kind has been done[29] –, but the machine which will be capable of *proving* Dirichlet's Theorem is probably not for tomorrow.

## 18 – The infinite product for the sine function

Let us start from the identity

(18.1) $$x^n - y^n = (x - y)(x - \omega y) \ldots (x - \omega^{n-1} y)$$

where $\omega = \exp(2\pi i/n)$, cf. (14.33). On replacing $x$ and $y$ by $\exp(z)$ and $\exp(-z)$, the general term of the right hand side becomes

$$\exp(z) - \exp(-z + 2ki\pi/n) = \exp(ki\pi/n)[\exp(z - ki\pi/n) - \exp(-z + ki\pi/n)].$$

Since $1 + 2 + \ldots + n - 1 = n(n-1)/2$, the product of the exponential factors is equal to $\exp[(n-1)i\pi/2] = \exp(i\pi/2)^{n-1} = i^{n-1}$, whence

(18.2) $$\exp(nz) - \exp(-nz) =$$
$$= i^{n-1} \prod [\exp(z - ki\pi/n) - \exp(-z + ki\pi/n)].$$

On replacing $z$ by $iz$, using the formula

$$2i.\sin z = e^{iz} - e^{-iz},$$

and regrouping the factors $-2i$ which appear everywhere on the right hand side, we find

---

[29] Since 1946 we have seen number theorist specialists in the USA use the first electronic calculator, the ENIAC of Eckert and Mauchly, for arithmetic calculations; these clearly did not interest the "serious" users, but enabled them to check the functioning of the machine and to invent much more useful methods of calculation, as related by Herman H. Goldstine, *The Computer from Pascal to von Neumann* (Princeton UP, 1972), particularly pp. 233 and 273 à propos D. H. Lehmer; in particular they calculated $2,000$ decimal places of $\pi$. When the first programmable computer constructed by John von Neumann (there were also Eckert and Mauchly and their UNIVAC of 1950) at the Institute for Advanced Study at Princeton between 1945 and 1952 became operational, the exploit was celebrated in the course of an inauguration ceremony, where they exhibited calculations performed for Emil Artin, one of the "fathers" of "abstract" algebra, to show that a famous arithmetic conjecture of Kummer's, a century old, was probably incorrect. For the course of the story it may be better to rely on von Neumann himself: "*As far as the Institute is concerned, and the people who were there are concerned, this computer came into operation in 1952, after which the first large problem that was done on it, and which was quite large and took even under these conditions half a year, was for the thermonuclear problem. Previous to that I had spent a lot of time on calculations on other computers for the thermonuclear problem.*" (*In the Matter of J. Robert Oppenheimer*, United States Atomic Energy Commission, 1954, reprint MIT Press, 1971, p. 655, testimony of von Neumann to the Oppenheimer "trial".) Not mentioned in Goldstine.

(18.3)     $\sin nz = 2^{n-1} \sin z . \sin(z + \pi/n) \ldots \sin[z + (n-1)\pi/n]$

or again

(18.4)     $\sin \pi z = 2^{n-1} \prod \sin[(z + k)\pi/n].$

There are analogous formulae for $\cos nz$ (replace $z$ by $z + \pi/2$) and, on division, for $\tan nz$.

Suppose that $n = 2m+1$ is odd and, to simplify the notation, temporarily put

$$S(z) = \exp \pi i z - \exp(-\pi i z) = 2i \sin \pi z,$$

which will free us from having to carry around the factors $2i$ that would otherwise appear. Since, in these products extended over the $n^{\text{th}}$ roots of unity, the exponent $k$ can vary mod $n$, one can, in (3), let it take the values between $-m$ and $m$, whence

$$S(nz) = i^{n-1} \prod_{-m \le k \le m} S(z + k/n).$$

Since a simple calculation shows that

$$S(x + y)S(x - y) = S(x)^2 - S(y)^2,$$

we have to group the terms $k$ and $-k$ for $1 \le k \le m$; not forgetting the term $k = 0$ and taking account of the fact that $i^{n-1} = i^{2m} = (-1)^m$, we get

$$
\begin{aligned}
S(nz) &= (-1)^m S(z) \prod_{1 \le k \le m} \left[ S(z)^2 - S(k/n)^2 \right] \\
&= S(z) \prod \left[ S(k/n)^2 - S(z)^2 \right] = \\
&= S(z) \prod S(k/n)^2 \prod \left( 1 - \frac{S(z)^2}{S(k/n)^2} \right);
\end{aligned}
$$

note that $S(k/n) = 2i . \sin(k\pi/n)$ does not vanish for $1 \le k \le m$ since $n = 2m + 1$. Replacing $z$ by $z/n$, finally we find

(18.5)     $S(z) = S(z/n) \prod S(k/n)^2 \prod \left( 1 - \dfrac{S(z/n)^2}{S(k/n)^2} \right).$

But let us divide (5) by $S(z/n)$ and let $z$ tend to 0. Since $S(z)$ is equivalent to $2\pi i z$ and $S(z/n)$ to $2\pi i z/n$, the left hand side tends to $n$. On the right hand side, the terms $S(z/n)$ tend to 0, so that the second product tends to 1. Thus, in the limit,

(18.6)     $\prod S(k/n)^2 = n,$

which allows us to write (5) in the form

$$(18.7) \qquad S(z) = nS(z/n) \prod_{1 \le k \le m} \left(1 - \frac{S(z/n)^2}{S(k/n)^2}\right)$$

where, we recall, $n = 2m + 1$.

Now let $n$ tend to infinity. Since $S(z/n) \sim 2\pi iz/n$, the factor preceding the product tends to $2\pi iz$. For $k$ given, the ratio appearing in the general term of the product is equivalent to

$$(2\pi iz/n)^2/(2ki\pi/n)^2 = z^2/k^2.$$

If one is called Euler one deduces that

$$(18.8) \qquad S(z)/2i = \sin \pi z = \pi z \prod_{1}^{\infty} \left(1 - z^2/k^2\right),$$

and, on replacing $z$ by $iz$,

$$(18.9) \qquad \sinh \pi z = \pi z \prod \left(1 + z^2/k^2\right).$$

Since the series $\sum 1/k^2$ converges, the infinite product is absolutely convergent, which is a good sign ...

It remains to justify the passage to the limit from (7) to (8). The problem is analogous to that of passing to the limit in a sequence of series (Theorem 9 of n° 12).

For this, put

$$(18.10) \quad u_n(k) \;=\; -S(z/n)^2/S(k/n)^2 \quad \text{if } k \le m, \; = 0 \text{ if not,}$$
$$(18.11) \quad u(k) \;=\; -z^2/k^2 \quad \text{for all } k.$$

We have $\lim_{n \to \infty} u_n(k) = u(k)$ for all $k$ and need to deduce that

$$(18.12) \qquad \lim_{n \to \infty} \prod [1 + u_n(k)] = \prod [1 + u(k)].$$

For this, let us introduce the partial products

$$
\begin{aligned}
p_n(k) &= [1 + u_n(1)] \ldots [1 + u_n(k)], \\
p(k) &= [1 + u(1)] \ldots [1 + u(k)]
\end{aligned}
$$

and denote the two sides of (12) by $p_n = \lim p_n(k)$ and $p = \lim p(k)$. On agreeing to put $p_n(0) = p(0) = 0$ we have

$$(18.13') \quad p_n = \sum [p_n(k+1) - p_n(k)] = \sum u_n(k+1)p_n(k) = \sum w_n(k),$$

$$(18.13'') \quad p = \sum [p(k+1) - p(k)] = \sum u(k+1)p(k) = \sum w(k),$$

where obviously $\lim w_n(k) = w(k)$ since one is passing to the limit in a product of a fixed number $k$ of terms. To show that the infinite *product* $p_n$ tends to the infinite product $p$ when $n \to +\infty$, i.e. that the sum of the *series* $\sum w_n(k)$ tends to that of the series $\sum w(k)$, it thus suffices (Chap. III, n° 13, Theorem 17) to establish that there exists a convergent series $\sum v(k)$ with positive terms such that

$$v(k) \geq |w_n(k)| = |u_n(k+1)p_n(k)|$$

for any $k$ and $n$.

By (10) we must therefore first estimate the

$$|u_n(k)| = |S(z/n)^2/S(k/n)^2| = |\sin^2(\pi z/n)|/\sin^2(\pi k/n)$$

for $1 \leq k \leq m$, $n = 2m + 1$. Now we have $\sin t \geq 2t/\pi$ for $0 \leq t \leq \pi/2$ (examine the graphs of the two sides), whence

$$\sin^2(\pi k/n) \geq 4k^2/n^2.$$

On the other hand, $|\sin z| \leq |z|[1 + |z|^2/3! + \ldots]$, whence

$$|\sin(\pi z/n)| \leq |\pi z/n|[1 + |\pi z/n|^2/3! + \ldots] \leq M(z)/n$$

where $M(z) = \pi|z|[1 + |\pi z|^2/3! + \ldots]$ does not depend on $n$. It follows that

(18.14)  $$|u_n(k)| \leq M(z)^2/4k^2 = a(k),$$

the general term of a convergent series independent of $n$.

It follows that, for any $n$ and $k$, we have

$$|p_n(k)| \leq (1 + |u_1(k)|) \ldots (1 + |u_n(k)|) \leq \prod [1 + a(k)] = p'$$

the final result since the last product converges absolutely; whence, from (13') and (14), a bound

(18.15)  $$|w_n(k)| \leq p'a(k) = v(k)$$

by the general term of a convergent series independent of $n$, which, at long last, justifies the expansion of the sine function as an infinite product.

This proof, directly inspired by Euler in his *Introductio* (apart, of course, from convergence ...), is not, by far, the simplest[30], but it is surely the most spectacular. One may clearly generalise the argument:

---

[30] See other proofs in Remmert, *Funktionentheorie 2*, pp. 10–16.

**Theorem 15.** *Let $\sum u_n(p)$ be a series depending on an integer n. Suppose that (i) $\lim_n u_n(p) = u(p)$ exists for all p, (ii) there exists an absolutely convergent series dominating all the series considered. Then*

$$\lim_{n \to \infty} \prod_{p=1}^{\infty} [1 + u_n(p)] = \prod_{p=1}^{\infty} [1 + u(p)]$$

A variant of this result, with the same proof, is obtained on replacing the "discrete" variable $n$ by a continuous variable:

**Theorem 15 bis.** *Let $X$ be a set and $\sum f_p(x)$ a series of scalar functions defined on $X$ converging normally on $X$. Then the infinite product $\prod [1 + f_p(x)]$ converges absolutely for all $x \in X$ and its partial products converge uniformly on $X$.*

In the case where $X \subset \mathbb{C}$, we deduce that if the $f_p$ are continuous, then so is the product. If, when $x$ tends to a point $a$ adherent to $X$ (or to infinity), the $f_p(x)$ tend to limits $u_p$, then $p(x) = \prod[1 + f_p(x)]$ tends to the product of $1 + u_p$, etc.

In general, the series $\sum f_p$ will converge normally only on all compact subsets of $X$, in which case, of course, the same holds for the partial products. To deduce that, if the $f_p$ are continuous on $X$, so similarly is the product, one needs to know that, for all $a \in X$ and all sufficiently small $r > 0$, the set of $x \in X$ such that $d(a, x) \leq r$ is compact, in other words that $X$ is *locally compact*.

In these statements, one should pay more attention than we have done to the possibility that some factors of the products considered might be zero. In the situation of Theorem 15 bis, we have $|f_p(x)| < \frac{1}{2}$ for all $x$ once $p > N$; one should therefore suppress the first terms $N$ of the product to obtain a meaningful statement[31].

From the infinite product

(18.16) $$\sin z = z \prod (1 - z^2/n^2\pi^2)$$

one can deduce a slightly more general formula. On replacing $z$ by $z + u$, the general term becomes, after a line of calculation,

$$(1 - u^2/n^2\pi^2) \left(1 + \frac{z}{u - n\pi}\right) \left(1 + \frac{z}{u + n\pi}\right);$$

now the product of the factors $1 - u^2/n^2\pi^2$ is equal to $\sin u/u$. One deduces that

(18.17) $$\frac{\sin(z + u)}{\sin u} = (1 + z/u) \prod_{n=1}^{\infty} \left(1 + \frac{z}{u - n\pi}\right) \left(1 + \frac{z}{u + n\pi}\right);$$

---

[31] See the ultrarapid and ultraprecise exposition of R. Remmert, *Funktionentheorie 2*, pp. 2–10.

the right hand side can be written in the more seductive form

$$(18.18) \qquad \frac{\sin(z+u)}{\sin u} = \prod_{n \in \mathbb{Z}} \left(1 + \frac{z}{u - n\pi}\right),$$

though, taken literally, this is not meaningful since the product extended over $n > 0$ or over $n < 0$ is clearly divergent; if one uses (18) one has to combine the terms for $n$ and $-n$ to obtain a convergent product.

However this may be, one can take $u = \pi/2$, whence, after easy calculations,

$$(18.19) \qquad \cos z = (1 - 4z^2/\pi^2)(1 - 4z^2/9\pi^2)(1 - 4z^2/25\pi^2) \ldots$$

with the squares of the odd integers in the denominators. We can also deduce "easily" from (18) the expansion of $\cot u$ as a series of rational fractions. To do this, Euler wrote (18) in the form

$$\cos z + \sin z . \cot u =$$
$$= \left(1 + \frac{z}{u}\right)\left(1 + \frac{z}{u - \pi}\right)\left(1 + \frac{z}{u + \pi}\right)\left(1 + \frac{z}{u - 2\pi}\right)\left(1 + \frac{z}{u + 2\pi}\right)\ldots$$

and expanded the right hand side as a power series in $z$ multiplying the terms as if dealing with a finite product; first he found 1, then "evidently" the sum

$$z \sum_{n \in \mathbb{Z}} 1/(u - n\pi),$$

which he transformed into the product of $z$ by a more orthodox series by grouping the terms $n$ and $-n$, whence, on identifying with the term in $z$ of the expansion in power series of the left hand side of (18):

$$(18.20) \qquad \cot u = \frac{1}{u} + \sum_{n=1}^{\infty} \frac{2u}{u^2 - n^2\pi^2}.$$

In this same cycle of ideas, the expansion (16) can be written (change $z$ to $iz$)

$$1 + z^2/3! + z^4/5! + z^6/7! + \ldots = (1 + z^2/\pi^2)(1 + z^2/4\pi^2)(1 + z^2/9\pi^2)\ldots$$

or, on writing $z$ instead of $z^2/\pi^2$,

$$1 + \frac{\pi^2}{1.2.3} z + \frac{\pi^4}{1.2.3.4.5} z^2 + \frac{\pi^6}{1.2.3.4.5.6.7} z^3 + \ldots = (1+z)(1+z/4)(1+z/9)\ldots$$

(the notation $n!$ had not yet been invented, no more the $\sum$ and $\prod$). Here again, Euler expanded the right hand side as if it were a finite product and found

$$\sum 1/p^2 = \pi^2/3!, \quad \sum 1/p^2 q^2 = \pi^4/5!, \quad \sum 1/p^2 q^2 r^2 = \pi^6/7!,$$

etc. In the first series, one sums over the $p > 0$, in the second over all pairs $p, q$ such that $0 < p < q$, in third over the triplets such that $0 < p < q < r$, etc., as one does in ordinary algebra. Now this also tells you for example that

$$\left(\sum 1/p^2\right)^2 = \sum 1/p^4 + 2 \sum 1/p^2 q^2,$$

whence the immortal formulae

$$\sum 1/p^4 = (\pi^2/3!)^2 - 2\pi^4/5! = \pi^4/90,$$

etc.

## 19 – Expansion of an infinite product in series

In Chapter 10 of his *Introductio* Euler dealt with a much more general formula that he wrote as

$$(19.1) \quad 1 + Ax + Bx^2 + Cx^3 + \ldots = (1 + ax)(1 + bx)(1 + cx)\ldots;$$

and sought to calculate $A, B, \ldots$ as functions of $a, b, \ldots$. For him, "it is necessary, as we know in Algebra", that

> $A$ be equal to the sum of all the magnitudes $a, b, \ldots$, so equal to $a + b + \ldots$,
> $B$ be equal to the sum of the pairwise products of these magnitudes, so to $ab + ac + ad + bc + bd + cd + \ldots$,
> $C$ be equal to the sum of their products in threes, so to $abc + abd + bcd + acd + \ldots$,

etc. Euler then shows how to deduce from this the sum of the squares, or of the cubes, etc. ... of the coefficients $a, b, \ldots$ This is a rather difficult exercise in the manipulation of multiple series; the reader can ignore it without worry, as also the following n°.

The problem, in modern notation, is to justify the formula

$$(19.2) \qquad \prod (1 + u_n) = 1 + \sum A_n$$

where, for example,

$$A_3 = \sum_{i<j<k} u_i u_j u_k$$

and more generally

$$(19.3) \qquad A_n = \sum_{0 < i_1 < \ldots < i_n} u_{i_1} \ldots u_{i_n}.$$

Since it is a beautiful justification of the general theorems of Chap. II on unconditional convergence, we shall develop it. Of course we assume that $\sum |u_n| < +\infty$.

The proof is in several parts.

First we show that the series (3) converges (unconditionally, as in all that follows). If one does not impose any condition to the summation indices one clearly obtains the product series $(\sum u_i)^n$, which converges (Chap. II, n° 18, example 2 and n° 22). For this series, the set of indices is the Cartesian product $\mathbb{N}^n = \mathbb{N} \times \ldots \times \mathbb{N}$; the series (3), taken over a subset $J_n$ of $\mathbb{N}^n$, is therefore convergent (Chap. II, n° 18).

Let us write $J$ for the union of all the $J_n$ and put

$$(19.4) \qquad v_j = u_{i_1} \ldots u_{i_n} \quad \text{if } j = (i_1, \ldots, i_n) \in J_n \subset J.$$

The series $\sum v_j$ extended over $J$ is convergent too.

Now let us write $H_k$ for the set, finite, of $j = (i_1, \ldots, i_n) \in J$ with $n$ a priori arbitrary but $i_n \leq k$, whence $i_1 < \ldots < i_n \leq k$ and $n \leq k$. The $H_k$ form an increasing sequence whose union is all of $J$. We need only to show that the partial sums $S_k = \sum |v_j|$ extended over the $j \in H_k$ are bounded above. Now

$$1 + S_k = 1 + \sum_{j \in H_k} |v_j| = 1 + \sum_{0 < i_1 < \ldots < i_n \leq k} |u_{i_1} \ldots u_{i_n}|$$

where $n$ takes all values $\leq k$; this sum is obtained by expanding the product $(1 + |u_1|) \ldots (1 + |u_k|)$ "as in Algebra" (and there we are since $H_k$ is finite). Since $\sum |u_n| < +\infty$, these products are bounded above as we saw in n° 17 in proving Theorem 13. Similarly for $S_k$, whence the unconditional convergence of the series $\sum v_j$ extended over all the $j \in J$.

This calculation shows further that

$$(19.5) \quad \sum_{j \in J} v_j = \lim_{k \to \infty} \sum_{j \in H_k} v_j = \lim(1 + u_1) \ldots (1 + u_k) = \prod_{n=1}^{\infty} (1 + u_n).$$

But as we saw at the beginning of the proof, $J$ is the union of the pairwise disjoint sets

$$J_n : \quad \text{sequences } (i_1, \ldots, i_n) \text{ such that } 0 < i_1 < \ldots < i_n.$$

The associativity theorem then shows not only that the partial sums taken over the $J_n$ converge – these are the $A_n$ which interest us, see (3) –, but also that the series $\sum A_n$ is absolutely convergent and its sum is the total sum $\sum v_j$, i.e., by (5), the infinite product of $1 + u_n$. Euler was right:

**Theorem 16.** Let $\prod_{n \geq 1}(1 + u_n)$ be an infinite product where $\sum |u_n| < +\infty$. Then the series

$$(19.6) \qquad A_n = \sum_{0 < i_1 < \ldots < i_n} u_{i_1} \ldots u_{i_n}$$

*converges unconditionally for all* $n \geq 1$, *the series* $\sum A_n$ *converges absolutely, and*

$$(19.7) \qquad\qquad 1 + \sum A_n = \prod (1 + u_n).$$

The proof in fact shows a little more. Suppose that the $u_n$ are functions $u_n(x)$ defined on a set $X$ and that the series $u_n(x)$ is *normally* convergent on $X$: $|u_n(x)| \leq v_n$ with $\sum v_n < +\infty$. It is then clear that all the series featuring in the proof are dominated by the analogous series relative to the product $\prod(1 + v_n)$. In consequence, the sums $A_n(x)$ converge normally as well as the series $\sum A_n(x)$. If for example $X$ is an open subset $G$ of $\mathbb{C}$, if the $u_n(z)$ are analytic on $\mathbb{C}$ and if the series $\sum u_n(z)$ converges normally on all compact subsets $K \subset G$, one can conclude that all the functions appearing in the proof are analytic, on condition, as always, that we know that a normally convergent series of analytic functions is again analytic (Chap. VII).

We can even, in such cases, sometimes obtain an expansion of the infinite product in power series if we assume that the $u_n$ are themselves power series without constant term

$$u_n(z) = \sum_{p \geq 1} a_n(p) z^p$$

converging on a disc $|z| < R$ and that the series $\sum u_n(z)$ converges normally on every disc $|z| \leq r < R$. The general term of the series (6) is then a product of absolutely convergent series

$$(19.8) \quad v_j(z) = u_{i_1}(z) \ldots u_{i_n}(z) = \sum_{p_1, \ldots, p_n \geq 1} a_{i_1}(p_1) \ldots a_{i_n}(p_n) z^{p_1 + \ldots + p_n}$$

by the formula for multiplying absolutely convergent series, where we put $j = (i_1, \ldots, i_n) \in J_n$ as above. Calculating *formally*, the sum of $A_n(z)$ can be written

$$(19.9) \quad \sum_{n \geq 1} \sum_{0 < i_1 < \ldots < i_n} \sum_{p_1, \ldots, p_n \geq 1} a_{i_1}(p_1) \ldots a_{i_n}(p_n) z^{p_1 + \ldots + p_n} ;$$

for this to converge unconditionally it is necessary and sufficient that the analogous sum obtained on replacing the $a_n(p)$ and $z$ by their absolute values should converge. This amounts to substituting the product $\prod[1 + w_n(z)]$ for the given infinite product $\prod[1 + u_n(z)]$, where we have put

$$(19.10) \qquad\qquad w_n(z) = \sum_p |a_n(p) z^p|,$$

a convergent series for $|z| < R$. We are allowed to regroup the terms, in order to test the convergence of the new sum (9), which has positive terms (Chap. II, n° 18, Theorem 13). The sum over $p_1, \ldots, p_n$ is equal to the product of the $w_i(z)$ for $i = i_1, \ldots, i_n$ (multiplication of series). So it remains to prove

$$(19.11) \qquad \sum_{n \geq 1} \sum_{0 < i_1 < \ldots < i_n} w_{i_1}(z) \ldots w_{i_n}(z) < +\infty;$$

which is what Theorem 16 affirms *if we know* that

$$(19.12) \qquad \sum w_n(z) < +\infty, \quad \text{i.e. if} \sum_{p,n} |a_n(p) z^p| < +\infty$$

for $|z| < R$ (associativity: sum first over $p$, whence $w_n(z)$, then over $n$). This condition is realised if, for all $n$, the coefficents $a_n(p)$ are *real and of the same sign* (i.e. all $> 0$ or all $< 0$, the sign maybe depending on $n$), or, another important case, if the power series $u_n(z)$ each contain *only one nonzero term*, since $w_n(z) = |u_n(|z|)|$ in both these cases. This is also the case for a product such as

$$\prod 1/(1 + q^n z) = \sum (1 - q^n z + q^{2n} z^2 - \ldots)$$

with $|q| < 1$, since here $w_n(z) = \sum |q^n z|^p = |q^n z|/(1 - |q^n z|)$ and the series $w_n$ converges. The three cases, curiously, will be found again in Euler, as we shall see in the following n°.

In a case of this kind, one can calculate formally with the sum (9) without risk of falling into divergent series; one can perform arbitrary groupings of terms and, in particular, group all the monomials of the same degree in $z$, whence a power series whose coefficients are calculated "as in Algebra", but with the help of absolutely convergent series.

In the general case we know only that

$$\sum_n |u_n(z)| = \sum_n \left| \sum_p a_n(p) z^p \right| < +\infty,$$

which is not enough to ensure (12). Another way of proceeding consists of replacing (8) by

$$(19.13) \qquad v_j(z) = u_{i_1}(z) \ldots u_{i_n}(z) = \sum_p b_j(p) z^p$$

where

$$(19.14) \quad b_j(p) = \sum_{p_1 + \ldots + p_n = p} a_{i_1}(p_1) \ldots a_{i_n}(p_n) \quad \text{if} \quad j = (i_1, \ldots, i_n).$$

Then, by Theorem 16,

$$\prod [1 + u_n(z)] = 1 + \sum_{n \geq 1} \sum_{j \in J_n} v_j(z) = 1 + \sum_{j \in J} v_j(z) = v(z),$$

a series of analytic functions which converges (unconditionally) normally on every disc $|z| \leq r < R$, i.e. on all compact subsets of $|z| < R$. Now we will show in Chap. VII that if such a series converges normally on all compacta then (i) its sum is analytic, (ii) all the derived series converge normally on all compacta, (iii) the derivatives of the sum are the sums of the derived series. The function $v$ is therefore analytic on $|z| < R$, so (another general theorem) is the sum of its Maclaurin series

$$v(z) = \sum v^{(p)}(0)z^p/p! = \sum b(p)z^p,$$

on *all* the disc $|z| < R$, and

$$(19.15) \qquad b(p) = v^{(p)}(0)/p! = \sum_{j \in J} v_j^{(p)}(0)/p! = \sum_{j \in J} b_j(p)$$

with an absolutely convergent series. This result can be written

$$b(p) = \sum_{0 < i_1 < \ldots < i_n} \quad \sum_{p_1 + \ldots + p_n = p} a_{i_1}(p_1) \ldots a_{i_n}(p_n),$$

which is again the "obvious" result, on condition that one sums over the $p_k$ *before* summing over the $i_k$ and $n$, failing which one will not obtain a sum converging unconditionally: on replacing the $u_i(p)$ by their moduli one can certainly obtain a divergent series , as above. One is here at the limit of the capability of unconditional convergence, but since one never meets this general case in situations where one has to calculate the coefficients of the power series $u(z)$, this is not the place to linger, other than to put the reader on guard against the risks of excessive confidence.

The simplest case is that where $u_n(z) = a_n z$, treated (without considerations of convergence) by Euler, immediately providing him the expansions

$$(19.16) \qquad \prod_1^{\infty}(1 + a_n z) = 1 + \sum_1^{\infty} A_n z^n$$

with

$$(19.17) \qquad A_n = \sum_{0 < i_1 < \ldots < i_n} a_{i_1} \ldots a_{i_n}.$$

The power series obtained converges for any $z$ provided that $\sum |a_n| < +\infty$.

## 20 – Strange identities

Euler, as we have said, studied infinite products of arithmetic interest. For example he proved the formula

$$(20.1) \qquad \prod_{n \geq 1} (1 - z^n)^{-1} = \sum_{n \geq 0} p(n) z^n$$

where one puts $p(0) = 1$ and where, for all $n > 0$, $p(n)$ denotes the number of *partitions* of $n$, i.e. the number of possible ways of writing

$$(20.2) \qquad n = n_1 + \ldots + n_k$$

as a sum of any number of integers satisfying

$$(20.3) \qquad 0 < n_1 \leq \ldots \leq n_k.$$

The set $P$ of solutions of (3) for all values of $k$ is obtained as above: for $k$ given, the solutions of (3) form a subset of the Cartesian product $\mathbb{N}^k$ and $P$ is the union of the pairwise disjoint sets thus defined. If one considers the map

$$\pi : (n_1, \ldots, n_k) \longmapsto n_1 + \ldots + n_k$$

of $P$ into $\mathbb{N}$, then, to express it as ultra modern maths, we have $p(n) = $ Card $[\pi^{-1}(\{n\})]$ for all $n$, to clarify the rather vague classical definition.

*Example:* $5 = 1 + 1 + 1 + 1 + 1 = 1 + 1 + 1 + 2 = 1 + 1 + 3 = 1 + 2 + 2 = 1 + 4 = 2 + 3 = 5$, whence $p(5) = 7$ (errors and omissions excepted); the calculation rapidly becomes exasperating as the heroic reader may confirm by checking that $p(10) = 42$, though he is not obliged also to check that[32]

$$p(200) = 3\ 972\ 999\ 029\ 388.$$

If in a partition (2) one groups together the terms repeated several times one may also write it as

$$(20.4) \qquad n = h_1 i_1 + \ldots + h_r i_r$$

---

[32] I transcribe from Remmert, *Funktionentheorie 2*, p. 17, whose Chap. 1 expounds all the subjects treated in the present § by express methods, and even more, but assumes the elements of the theory of analytic functions as known. We would also dissuade the reader from searching for a general formula for $p(n)$; Euler gave a recurrence formula

$$p(n) = p(n-1) + p(n-2) - p(n-5) - p(n-7) + \ldots$$

in which the "pentagonal numbers" $\frac{1}{2}(3k^2 - k)$ appear, see Remmert; on the other hand there are formulae, whose proofs are very difficult, which provide asymptotic expansions in the sense of Chap. VI for $p(n)$ when $n$ is large.

The book of Hans Rademacher, *Topics in Analytic Number Theory* (Springer, 1973), covers this subject and many other domains, both simpler (Bernoulli numbers, gamma function, etc.) and more difficult, but assumes at least a good knowledge of the theory of analytic functions and, above all, a taste for "calculations" and "formulae" of the Eulerian kind.

More accessible: the classic of G. H. Hardy & E. M. Wright, *An Introduction to the Theory of Numbers* (OUP) and André Weil, *Number Theory. An Approach through History. From Hammurapi to Legendre* (Birkhäuser, 1984).

with integers $h_1, \ldots \geq 1$, arbitrary in number, and, this time,

(20.5) $$0 < i_1 < \ldots < i_r.$$

This said, the convergence of the product (5) is obvious for $|z| < 1$ by Theorem 13. To prove (1), we write that

$$\prod (1 - z^n)^{-1} = \prod (1 + z^n + z^{2n} + \ldots) = \prod [1 + u_n(z)]$$

with power series $u_n(z)$ without constant term, and, very happily, with all coefficients *positive*. As we saw in the preceding n°, we can then expand the product of these series as if dealing with a finite product of polynomials. In the series (19.9), i.e.

(20.6) $$\sum_{r \geq 1} \sum_{0 < i_1 < \ldots < i_r} \sum_{p_1, \ldots, p_r \geq 1} a_{i_1}(p_1) \ldots a_{i_r}(p_r) z^{p_1 + \cdots p_r},$$

we observe that $a_i(p)$, the coefficient of $z^p$ in the geometric series $(1 - z^i)^{-1}$, is equal to 1 if $p$ is a multiple of $i$ and to 0 if not. We can thus, in (6), restrict to summing over exponents which are multiples of $i_k$, which leads to the sum

(20.7) $$\sum_{r \geq 1} \sum_{0 < i_1 < \ldots < i_r} \sum_{h_1, \ldots, h_r \geq 1} z^{h_1 i_1 + \ldots + h_r i_r}.$$

To obtain $z^n$ in the general term, one must choose $r$, the $i_k$ and the $h_k$ so that $h_1 i_1 + \ldots + h_r i_r = n$; in view of the conditions imposed in the sum (7) on these numbers, such a choice corresponds exactly to a partition (4) of $n$. The term $z^n$ occurs $p(n)$ times, whence (1). Note in passing that, in view of the apparently vertiginous increase of $p(n)$, convergence (apart from at 0) of the series $\sum p(n) z^n$ is not obvious.

Euler's greatest success on this track – here one sees his great virtuosity ... – is the identity

$$\prod (1 - q^n) = \sum_{m \in \mathbb{Z}} (-1)^m q^{(3m^2 + m)/2} \qquad (|q| < 1)$$

which Jacobi, in his theory of elliptic functions, would present as a particular case of the formula (see Chap. XII, n° 6)

(20.14) $$\sum_{n \in \mathbb{Z}} q^{n^2} x^n = \prod_0^{\infty} (1 - q^{2n+2})(1 + q^{2n+1} x)(1 + q^{2n+1} x^{-1});$$

it is enough to replace $q$ by $q^{3/2}$ and $x$ by $-q^{1/2}$ and to observe that any positive integer is of the form $3n$, or $3n + 1$, or $3n + 2$, to obtain Euler's identity; all this converges for $|q| < 1$ and $x \neq 0$: the infinite product by Theorem 13, and the series because

$$\left| q^{n^2} x^n \right| = e^{n^2 \log |q| + n \log |x|},$$

with $\log |q| < 0$, so that for $n$ large the exponent is $< -n$, a bound which, lamentably weak as it may seem, is more than sufficient to prove the convergence of the series. For $x = 1$, Jacobi's identity shows that

$$\sum q^{n^2} = \prod \left(1 - q^{2n+2}\right) \left(1 + q^{2n+1}\right)^2.$$

If on the other hand one replaces $q$ by $q^{1/2}$ and $x$ by $-xq^{1/2}$, one obtains

$$\sum_{n \in \mathbb{Z}} (-x)^n q^{n(n+1)/2} = \prod_{n \in \mathbb{N}} \left(1 - q^{n+1}\right) \left(1 - q^{n+1}x\right) \left(1 - q^n x^{-1}\right);$$

on isolating the term $1 - x^{-1}$, which corresponds to $n = 0$, in the right hand side, and replacing $n$ by $n - 1$, one obtains

$$\frac{1}{x-1} \sum_{n \in \mathbb{Z}} (-x)^n q^{n(n+1)/2} = \frac{1}{x} \prod_{n \geq 1} (1 - q^n)(1 - q^n x)(1 - q^n x^{-1});$$

when $x$ tends to 1, the right hand side tends – Theorem 15 applies to each of the infinite products on the right hand side – to the product of the expressions $(1 - q^n)^3$; on the left hand side one has a power series in $x$, everywhere convergent because of the rapid decrease of the coefficients in $q$; this series is zero for $x = 1$, the terms in $n$ and in $-n-1$ cancel mutually; in consequence, the left hand side tends to the derivative at $x = 1$ of this power series. Hence, by grouping the terms $n$ and $-n-1$ of the series, a new miraculous formula

$$\prod_{n \geq 1} (1 - q^n)^3 = \sum_{n \in \mathbb{N}} (-1)^n (2n+1) q^{n(n+1)/2} = 1 - 3q + 5q^3 - 7q^6 + 9q^{10} - \cdots.$$

One of the curious aspects of these formulae, which Jacobi published in 1829, is that Gauss had found the majority of them in 1808, then kept them to himself as he had done so often, and, when Jacobi published his results, let him know that he knew them. Coming from the greatest mathematician of the age, and even, some claim, of all time[33], one can imagine the effect on a 25 year old "débutant" whose first great success this was. Jacobi told it to Legendre, who was indignant and spoke of an "excess of impudence" on Gauss' part[34]. If the Prince of Mathematicians wanted to preserve his own

---

[33] Dieudonné once suggested the use the "gauss" as the unit of measurement of the production of mathematical geniuses by humanity; he found, if I remember well, a dozen, as a very large maximum, for all the XIX[th] century and a half per year in our time, which may appear optimistic, supposing this "calculation" meaningful. One trembles at the idea of the 3.14 gauss that a China of two milliard inhabitants in the frontline of scientific progress might produce every week in 2197.

[34] For all this, see the historical remarks and the bibliography in Chap. 1 of Remmert already mentioned.

reputation, sufficiently established by then, even forgetting the first operational telegraph (it connected his office to that of the astronomer Wilhelm
Weber, at Göttingen), he failed, since everyone has always attributed these
results to Jacobi: whoever publishes first has the proper title, is the winner.
Present-day Americans, who, in all subjects, are under the influence of the
savage Darwinism of the XIX$^{th}$ century, have found the "apt" formulation:
*Publish or perish.*

The reader will maybe think that this is all the mode of 1750 or 1830 and
is no longer of interest; but no. This kind of formula continued to impassion
respectable mathematicians like Dirichlet, Hermite, Dedekind, Leopold Kro-
necker, Felix Klein, Henri Poincaré, Erich Hecke, Carl Ludwig Siegel, etc.,
up to about 1940. They have since been generalised formidably over about
twenty years in the ultra modern framework of the theory of semi-simple Lie
algebras, mainly by I. G. Macdonald. Even independently of these generalisations *unthinkable* before the second half of the XX$^{th}$ century, the Dedekind
function

$$\eta(z) = q^{1/12} \prod (1 - q^{2n}) = e^{\pi i z/12} \prod (1 - e^{2\pi i n z})$$

[we have put $q = e^{\pi i z}$ with $\text{Im}(z) > 0$ so that we have $|q| < 1$ and convergence
of the product] satisfies two functional equations

$$\eta(z + 1) = e^{\pi i/12}\eta(z), \qquad \eta(-1/z) = (z/i)^{1/2}\eta(z)$$

of which the first is trivial, though not the second. On raising it to the power
24 one obtains a function

$$\Delta(z) = \eta(z)^{24} = e^{2\pi i z} \prod (1 - e^{2\pi i n z})^{24} =$$
$$= q^2 - 24q^4 + 252q^6 - 1472q^8 + 4830q^{10} - 6048q^{12} - 16744q^{14} + \ldots$$
$$= \sum \tau(n)q^{2n}$$

which possesses the property of being multiplied by 1 or by $z^{12}$ when one
replaces $z$ by $z+1$ or by $-1/z$ and so, as one can show without much trouble,
satisfies the relation

$$\Delta\left(\frac{az + b}{cz + d}\right) = (cz + d)^{12}\Delta(z)$$

where $a, \ldots, d$ are rational integers satisfying $ad - bc = 1$. In this way one finds
an example of a "modular function"; all this is connected to the theory of
elliptic functions. The series $\sum \tau(n)/n^s$ constructed starting from $\Delta$ has, for
its part, properties analogous to those of the $\zeta$ function; they were conjectured
at the time of the Great War by an Indian autodidact, son of a minor employee
in Madras, who had learned classical analysis from the English textbooks of

1850 and was unknown until he introduced himself to G. H. Hardy, sending him several dozen identities *à la* Euler of which half were not only new, but very difficult to prove (Hardy had him come to Cambridge; he died of tuberculosis in Madras a dozen years later). About twenty years ago the Ramanujan conjecture (among other results) brought a Fields Medal, the mathematical equivalent of a Nobel Prize, to Pierre Deligne, who proved it, using all the machinery of geometry and algebraic topology erected after 1960 by Alexandre Grothendieck and his tribe: like Fermat's Last "Theorem" proved by Andrew Wiles (Annals of Mathematics 1995), this is an example of a seemingly totally "classical" problem, solved by "modern maths" thanks to an interpretation of the coefficients $\tau(n)$ of the infinite product $\Delta(z)$, which carry much more information than their mere definition.

Those who believe that "the age of formulae" is over deceive themselves greatly: the contemporary "formulae" certainly lie, in general, though not always, at a mathematical level incomparably more elevated than those of Euler; they involve functions, series, infinite products or integrals whose terms reflect the "structure" of some arithmetic, algebraic or analytic theory or other.

This said, there is much activity in other domains of mathematics, as important as this blend of Eulerian calculations, of elliptic or modular functions, and of theory of numbers, or of algebraic varieties, which fascinate some and repel others, despite, or because, of the antiquity of the tradition on which they rest ...

To return to Euler, after examining the products (15) he passed on to the more difficult identity that he wrote

$$(20.15) \quad 1/(1 - ax)(1 - bx)(1 - cx) \ldots = 1 + Ax + Bx^2 + Cx^3 + \ldots,$$

which amounts to forming the product of all the geometric progressions $\sum a^n x^n$, $\sum b^n x^n$, $\sum c^n x^n$, etc. The coefficient of $x^n$ on the right hand side is then clearly the sum of the products

$$(20.16) \qquad\qquad a^{n_1} b^{n_2} c^{n_3} \ldots$$

associated to each decomposition $n = n_1 + n_2 + n_3 + \ldots$ of $n$ into integers $> 0$; these are not the same partitions as the preceding: the $n_i$ are not subjected to the condition (3) since, to form a product of series, one chooses a term in each sum at random and adds the products obtained. We can for example choose for $a, b, c, \ldots$ the reciprocals of the *prime* numbers, and put $x = 1$ in the result; in this way we find the sum of all the reciprocals of products of any number of any powers of the prime numbers $2, 3, 5, 7, 11, \ldots$; since any integer $> 1$ can be written in a unique way in the form

$$2^{n_1} 3^{n_2} 5^{n_3} 7^{n_4} 11^{n_5} \ldots$$

with exponents $n_i \geq 0$ almost all zero, Euler obtained a supernatural identity that he wrote

$$\frac{1}{\left(1 - \frac{1}{2}\right)\left(1 - \frac{1}{3}\right)\left(1 - \frac{1}{5}\right)\left(1 - \frac{1}{7}\right)\left(1 - \frac{1}{11}\right)\cdots} = 1 + \frac{1}{2} + \frac{1}{3} + \frac{1}{4} + \frac{1}{5} + \cdots;$$

he also obtained more generally the infinite product of $\zeta(s)$ for $s$ an integer, which is more, or again less, reasonable according to the value of $s$. For him it was obviously just a formal calculation.

The partition series is obtained similarly on replacing $a$, $b$, $c$, ... in (15) by $z$, $z^2$, $z^3$, etc. and putting $x = 1$ in the product of all the geometric series obtained. It is useful to remark that if one forms the product of the first $n$ series, the coefficient of $z^p$ in the result is the same for all the $n > p$, since, after the first $p$ multiplications, one multiplies by series containing only terms of degree $> p$ in $z$ and starting with 1. We can thus obtain the coefficient of $z^p$ in the final result by calculating the product of the first $p$ progressions. We can even obtain the result keeping, in these $p$ geometric progressions, only the terms of degree $\leq p$, as Newton might have explained to you. To calculate $p(n)$ for $n \leq 10$ for example, it is enough to calculate the coefficients of $z, z^2, \ldots, z^{10}$ in the product

$$(1 + z + z^2 + \ldots + z^{10})(1 + z^2 + z^4 + \ldots + z^{10})(1 + z^3 + z^6 + z^9)$$
$$(1 + z^4 + z^8)(1 + z^5 + z^{10})(1 + z^6)(1 + z^7)(1 + z^8)(1 + z^9)(1 + z^{10});$$

it would be better to confide the task to a computer for the "large" values of $n$, which, in this context, manifestly occur very early, as we saw at the beginning of this n°. But since everything was possible to people like Euler, it is possible that he would have delivered himself to this amusing exercise since electronics and computer science were not yet, in his age, the two breasts at which advanced Humanity nourished itself, nor the milliard of breasts at which these two industries nourish themselves.

## §4. The topology of the functions $\mathscr{A}rg(z)$ and $\mathscr{L}og\,z$

**21** – In n° 14 we defined the argument of a complex number $z \neq 0$ as being any real number $t = \arg(z)$ such that

$$(21.1) \qquad\qquad z = |z|\exp(it)$$

and showed that the problem admitted infinitely many solutions, differing one from the other by arbitrary multiples of $2\pi$. The standard notation arg does not denote a true function; it is a *correspondence*, in the sense of Chap. I, between elements of the set $\mathbb{C}^*$ and elements of $\mathbb{R}$; for any $z \in \mathbb{C}^*$, the symbol $\arg(z)$ should therefore denote the *set* of $t \in \mathbb{R}$ such that $z = |z|\exp(it)$ even if, in practice, it almost always denotes one of these possible values. To avoid confusion, we shall write $\mathscr{A}rg(z)$ for this set, so that

$$(21.2) \qquad\qquad t \in \mathscr{A}rg(z) \Longleftrightarrow z = |z|\exp(it).$$

Similarly we agree that the notation $\mathscr{L}og\,z$ denotes the *set* of possible values of the complex log:

$$(21.2') \qquad\qquad \zeta \in \mathscr{L}og\,z \Longleftrightarrow z = \exp(\zeta).$$

Everyone who has taught the theory of holomorphic functions knows that arguments and, what comes to the same, logarithms of complex numbers, are one of the most frequent sources of theoretical incomprehension and of errors in calculation. We may therefore be pardoned for developing the subject quite untraditionally. This will also be an occasion to familiarise the reader with some topological techniques of much more general relevance.

   (i) *Graph of $z \mapsto \mathscr{A}rg(z)$.* A correspondence is defined by exhibiting its graph, namely, in this case, the set of pairs $(z, t) \in \mathbb{C}^* \times \mathbb{R}$ such that $z = |z|\exp(it)$. The Cartesian product can be represented in the usual three-dimensional space by choosing a system of rectangular coordinates $(x, y, t)$, identifying $\mathbb{C}^*$ with the horizontal plane $Oxy$ with the origin deleted, and $\mathbb{R}$ with the vertical $t$ axis; the product is then $\mathbb{R}^3$ less the vertical axis. The graph $\Gamma$ of the correspondence $\mathscr{A}rg$ is then the set of points $(x, y, t) \in \mathbb{R}^3$ such that

$$(21.3) \qquad x = r\cos t, \quad y = r\sin t \qquad \text{where } r = (x^2 + y^2)^{1/2} \neq 0;$$

the vertical through a point $(x, y)$ meets it in infinitely many points, each distant one from the other by a multiple of $2\pi$. This is a helicoidal surface analogous to an inclined ramp of infinite size rotating at constant angular speed about the $t$ axis and prolonged to infinity above and below; one could probably feature it in a science-fiction novel. The intersection of $\Gamma$ with the surface of a circular cylinder with axis $Ot$ is a helix of pitch $2\pi$ winding indefinitely around the cylinder.

On a surface such as $\Gamma$ one can do topology as on the plane: it supports a distance, namely the standard metric on $\mathbb{R}^3$. We can therefore speak of open sets, closed sets, continuous functions, convergent sequences, etc. in $\Gamma$, as well as of continuous maps with values in $\Gamma$; we shall do this in what follows.

One could also consider the graph of the correspondence $Log$; this would be the set of points $(z, \zeta)$ of $\mathbb{C}^* \times \mathbb{C}$ such that $z = \exp(\zeta)$. One has to work in $\mathbb{R}^4$ to represent it geometrically.

(ii) *Multiform functions and uniform branches.* Instead of speaking of "correspondences", mathematicians spoke until recently, and many still speak, of *multiform functions*[35], objects whose first peculiarity is not to be functions; the adjective "multiform" makes allusion to the fact that these pseudo functions are authorised to take several values at each point, like, for example, $(z^2 - 1)^{1/2}$ on $\mathbb{C}$ and, more generally, the "algebraic functions" obtained by choosing a polynomial $P$ with complex coefficients in two variables and associating to each $x \in \mathbb{C}$ the roots $y \in \mathbb{C}$ of the equation $P(x, y) = 0$; see the algebraic curve $x^3 - ax^2 + axy - y^3 = 0$ treated by Newton, who restricted himself to real variables (Chap. III, n° 14).

Now Newton had already remarked that on the neighbourhood of a point $(x_0, y_0)$ of the curve one can generally find a power series

$$y = y_0 + \sum a_n (x - x_0)^n$$

which satisfies the given equation $P(x, y) = 0$ identically. Although the equation $P(x, y) = 0$ does not allow one to consider $y$ as a true function of $x$, there nevertheless exist excellent functions $y = f(x)$ which satisfy it. In the simplest case of the correspondence $x^2 + y^2 = 1$ between points of $\mathbb{R}$, a correspondence whose graph is the circle with centre $O$ and radius 1, the formulae $y = (1 - x^2)^{1/2}$ and $y = -(1 - x^2)^{1/2}$ define two such functions on $[-1, 1]$.

One is thus led to define a *uniform branch* of a multiform function or correspondence: any true function, with only one value, whose graph is contained in that of the given correspondence. In the case of the argument, a uniform branch on a set $G \subset \mathbb{C}^*$ is thus a function $A : G \longrightarrow \mathbb{R}$ satisfying

(21.4)  $\qquad\qquad z = |z| \exp[i.A(z)] \qquad$ for all $z \in G$;

---

[35] Instead of considering $Log\,z$ as a (false) function defined on $\mathbb{C}^*$, one can save the situation by considering it as a true function defined on its own graph, namely $(z, \zeta) \mapsto \zeta$, since, on the latter, we have $\exp(\zeta) = z$. In the theory of analytic functions this artifice is the origin of the invention of Riemann surfaces of algebraic functions. Instead of considering, for example, $(z^4 - 1)^{1/7}$ as a function of $z \in \mathbb{C}$, which in the strict sense it is not, one works on the surface $G$ in $\mathbb{C}^2$ defined by the equation $\zeta^7 = z^4 - 1$ and studies the function $(z, \zeta) \mapsto \zeta$ thereon, and, more generally, the functions $(z, \zeta) \mapsto f(z, \zeta)$ where $f$ is a rational function of two variables. The true Riemann surface is a little more complicated, but is all the same the initial motivation for the construction and the theory of algebraic functions of a complex variable. See Chap. XI in Vol. III.

a uniform branch of $\mathcal{L}og$ will similarly be a function $L : G \longrightarrow \mathbb{C}$ satisfying

(21.4')                         $z = \exp[L(z)]$      for all $z \in G$.

The existence of such branches is obvious if one does not impose any supplementary condition on them: choose at random a point of the graph of
the given correspondence lying on the vertical through each $z$ ("axiom of
choice"). But to apply this procedure to the correspondence $\mathcal{A}rg\, z$ amounts
to attributing an argument to each complex number $z \neq 0$ by drawing lots;
utility zero. In this very case, and in all similar cases – the theory of holomorphic functions provides them *ad libitum* –, one looks to construct uniform
branches which are at least *continuous* functions of the variable, and even *analytic* when that is possible. In what follows, we always assume, implicitly,
and often explicitly, that the uniform branches in question are continuous.

(iii) *Uniqueness up to $2k\pi$ of uniform branches.* Given an arbitrary subset
$G$ of $\mathbb{C}^*$, do there exist on $G$ continuous uniform branches $z \mapsto A(z)$ of the
correspondence $\mathcal{A}rg$ or, what amounts to the same on putting

(21.5)                         $L(z) = \log |z| + i.A(z),$

of $\mathcal{L}og$? The reply depends on $G$ and is by no means obvious, as we shall
see. Another problem, which we can resolve now: how does one pass from one
uniform branch to another?

The answer depends on the fact that if $A'$ and $A''$ are two continuous
uniform branches of the argument in $G$ then the function $f(z) = [A'(z) -
A''(z)]/2\pi$ is continuous *and* has values in $\mathbb{Z}$. It will not escape anybody that
this kind of object is rarely met outside the case of *constant* functions. This
would be obvious if $G$ were an interval of $\mathbb{R}$, since the image $f(G)$ would then
be an interval (intermediate value theorem) contained in $\mathbb{Z}$, so reducing to a
single point.

In our present case, it is the same if $G$ is *connected*. Suppose, for simplicity
that $G$ is *arcwise connected*, the only useful case in practice; this means that,
for all $u, v \in G$, there exists a *continuous path* in $G$ joining $u$ to $v$, i.e.
a continuous map $\gamma : I = [a, b] \longrightarrow G$ of an interval of $\mathbb{R}$ into $G$ such
that $\gamma(a) = u$ and $\gamma(b) = v$ (for the case of an open subset of $\mathbb{C}$, look
again at n° 20 of Chap. II, where piecewise linear paths suffice). Now let $A'$
and $A''$ be two uniform branches of the argument in $G$, and let us suppose
them equal at a particular point $u \in G$ – if not, subtract a suitably chosen
multiple of $2\pi$ from $A''(z)$ – and let us show that $A'(v) = A''(v)$ for all
$v \in G$. To do this, let us choose as above a path in $G$ joining $u$ to $v$ and
put $f(t) = \{A'[\gamma(t)] - A''[\gamma(t)]\}/2\pi$. We obtain a continuous function $f$ on
$I$ with values in $\mathbb{Z}$, so constant; since it vanishes at $t = a$, it does so also at
$t = b$, qed[36]. In conclusion, *on a connected set $G \subset \mathbb{C}^*$ two uniform branches*

---

[36] A better proof would be to extend the intermediate value theorem as follows:
if $f$ is a continuous real function defined on a connected subset $G$ of $\mathbb{C}$, then

of the argument differ from one another by a constant multiple of $2\pi$ and two branches of $Log$ by a constant multiple of $2\pi i$.

It goes without saying that the connectedness of $G$ is essential: else take for $G$ the union of two disjoint open discs.

(iv) The answer to the above existence question is negative if $G = \mathbb{C}^*$. Suppose that indeed we have found on $\mathbb{C}^*$ a *continuous* real valued function $A(z)$ such that $z/|z| = \exp[iA(z)]$ for all $z$ and restrict to the $z \in \mathbb{T}$, the unit circle in $\mathbb{C}$. The map

$$t \mapsto \exp(it) = \cos t + i.\sin t$$

of $I = [0, 2\pi]$ into $\mathbb{T}$ is continuous and, as we saw in n° 14, surjective; the composite function $f(t) = A[\exp(it)]$ is therefore continuous on $I$. But the function $g(t) = t$ is also continuous and also satisfies $g(t) \in Arg[\exp(it)]$ for any $t$. Since $f(t) - t$ is, for all $t \in I$, an integer multiple of $2\pi$, we must have $f(t) = t + 2k\pi$ with an integer $k$ independent of $t$. Now $\exp(it) = 1$ for $t = 0$ or $2\pi$ and so $f(0) = f(1)$ since $f(t) = A[\exp(it)]$ depends, by definition, only on the point $\exp(it)$; absurd since $f(t) = t + 2k\pi$. *It is therefore impossible to choose an argument for every nonzero complex number $z$ so that it is a continuous function of $z$ on all of $\mathbb{C}^*$ or even only on the unit circle $\mathbb{T}$.* Same result for $Log\ z$.

(v) Let us show that, on the contrary, *it is possible to find a uniform branch of the argument* (or of the logarithm) *on the open set $G$ obtained by excising any half-line of origin $O$ from $\mathbb{C}^*$.* Modulo a rotation $z \mapsto e^{i\alpha}z$ about the origin, we may restrict to showing this for the open set $G = \mathbb{C} - \mathbb{R}_-$, where $\mathbb{R}_-$ is the set of real numbers $\leq 0$. Since the real negative numbers are characterised by the fact that their arguments are of the form $(2k + 1)\pi$, it is natural, on $G$, to choose for $A(z)$ *the* value of the argument of $z$ which satisfies

(21.6)                              $|A(z)| < \pi,$

determined unambiguously, and yielding, for $Log$, the function

(21.7)                         $L(z) = \log |z| + i.A(z)$

---

the image $f(G)$ is an interval. Consider two points $f(u)$ and $f(v)$ of $f(G)$, with $u, v \in G$ and, for example, $f(u) < f(v)$, and consider a number $c \in ]f(u), f(v)[$ not belonging to $f(G)$. The subsets $U$ and $V$ of $G$ defined by the inequalities $f(z) < c$ and $f(z) > c$ are then disjoint, satisfy $G = U \cup V$, $u \in U$, $v \in V$, and, finally, are open in $G$ since $f$ is continuous. But, by definition, one cannot decompose a connected space as two open disjoint nonempty sets: contradiction. The image $f(G)$ thus satisfies Theorem 5 of Chap. III, n° 4 which characterises the intervals. Still better: the image of a connected space under a continuous map is again a connected space.

on the same open set $G$. We shall show that $A(z)$ and then $L(z)$ are *continuous* functions of $z$ on $G$, which will establish the result. Here again this is "geometrically obvious", but a rigorous, i.e. analytic, proof, rests on a property of the complex logarithm which no sketch can provide.

Let us work on a neighbourhood of an arbitrary $a \in G$; we have shown above [n° 14, section (x)] that, on the disc $D(a) : |z - a| < |a|$, the values of $\mathcal{L}og\, z$ are given by

$$(21.8) \qquad \mathcal{L}og\, z = \mathcal{L}og\, a - \sum (1 - z/a)^n / n.$$

If we choose $\mathcal{L}og\, a = L(a)$ in (8), where $L(a)$ is given by (7), the right hand side is a continuous function of $z$ on $D(a)$, so its imaginary part is too; the latter being equal to $A(a)$ for $z = a$ and $|A(a)|$ being $< \pi$, the imaginary part of the right hand side of (8) is again, in absolute value, $< \pi$ on a neighbourhood of $a$: it is thus $A(z)$, whence it follows that

$$(21.9) \qquad L(z) = L(a) - \sum (1 - z/a)^n / n$$

for $|z - a|$ sufficiently small. Hence the continuity of the function (7) and thus of $A(z)$ on the open set considered. This argument even shows much more: *every uniform branch of $\mathcal{L}og$ is analytic.*

(vi) This result shows *a fortiori* that *there exist uniform branches of the argument on every open disc $D \subset \mathbb{C}^*$*: remove from $\mathbb{C}$ a half-line not meeting $D$, for example $\mathbb{R}_-$ if $D$ is contained in the half plane $\operatorname{Re}(z) > 0$, a case to which one can always reduce by a rotation around the origin. In this particular case, it is clear that, for $z \in D$, one can choose $A(z)$ in the interval $]-\pi/2, \pi/2[$. In the general case, one sees that if $A(z)$ is a uniform branch of the argument in $D$, then the set $A(D)$ of values taken by $A(z)$ for $z \in D$ is an interval of length $< \pi$.

(vii) *Liftings of a path.* By definition, a continuous path in $\mathbb{C}$ is a continuous map of a compact interval $I = [a, b]$ of $\mathbb{R}$ into $\mathbb{C}$; this definition extends trivially to any metric space, and in particular to the graph $\Gamma$ of the correspondence $\mathcal{A}rg$ introduced above. If $t \mapsto \gamma(t)$ is a path in $\Gamma$, we have

$$(21.10) \qquad \gamma(t) = (\gamma_0(t), A(t))$$

where $\gamma_0(t) \in \mathbb{C}^*$ and $A(t) \in \mathbb{R}$ are continuous functions of $t$; it is clear that $\gamma_0 : I \longrightarrow \mathbb{C}^*$ is a continuous path in $\mathbb{C}^*$ and that, for all $t \in I$, one must have $A(t) \in \mathcal{A}rg[\gamma_0(t)]$, the set of possible arguments of $\gamma_0(t)$.

If, conversely, one has a continuous path $\gamma$ in $\mathbb{C}^*$ and chooses a number $A(t) \in \mathcal{A}rg[\gamma(t)]$ for all $t \in I$ so that the function $A(t)$ is continuous, i.e. is a *uniform branch of the argument along $\gamma$* (similar definition to the case of $\mathcal{L}og$), then $t \mapsto (\gamma(t), A(t))$ is a continuous path in $\Gamma$ which projects horizontally onto $\gamma$; this is what one calls a *lifting* of $\gamma$ to $\varGamma$. One should pay attention to the fact that in general $A(t)$ depends on $t$ itself and not only on the point $\gamma(t)$, as shown by the map $t \mapsto \exp(it)$ of $[0, 4\pi]$ in $\mathbb{C}^*$ with $A(t) = t$.

It is clear, once again thanks to the intermediate value theorem, that the liftings to $\varGamma$ of a given path $\gamma$ (resp. the uniform branches of the argument along $\gamma$) can be deduced from one another by a vertical translation by (resp. by addition of) a multiple of $2\pi$ independent of the parameter $t$. In other words, the liftings to $\varGamma$ of a continuous path traced in $\mathbb{C}^*$ are entirely determined by their initial points, and *a uniform branch of the argument or of $Log$ along a path in $\mathbb{C}^*$ is entirely determined by its value at the origin of the path*. If, for example, $\gamma$ is $t \mapsto \exp(it)$, $0 \le t \le 6\pi$, the liftings of $\gamma$ to $\varGamma$ are arcs of the helix that makes three complete turns around the axis of $\varGamma$, the trajectories of a person using a staircase to mount three storeys.

(viii) *Liftings of a path: existence.* Let us now show that every continuous path $\gamma : I \longrightarrow \mathbb{C}^*$ possesses a lifting to $\varGamma$, or, what amounts to the same, that there always exists a uniform branch of the argument along $\gamma$.

First, such a function exists on every sufficiently small neighbourhood of any point $t_0$ of $I$. Choose an open disc $D \subset \mathbb{C}^*$ with centre $\gamma(t_0)$ and, on $D$, a uniform branch $A(z)$ of the argument in accordance with what we saw above; since the function $\gamma$ is continuous, $\gamma(t) \in D$ for all $t$ sufficiently close to $t_0$; the function $t \mapsto A[\gamma(t)]$, defined on a neighbourhood of $t_0$ and continuous like $z \mapsto A(z)$, clearly answers the question.

This done, put $I = [a, b]$, choose a number $\alpha \in Arg[\gamma(a)]$, and consider all pairs $(J, A_J)$ formed by an interval $J = [a, u[$ of origin $a$, with $u < b$, and of a continuous uniform branch $A_J(t)$ of the argument in $J$ satisfying $A_J(a) = \alpha$. The function $A_J$ satisfying these conditions is unique as we saw above. If now $(J', A_{J'})$ and $(J'', A_{J''})$ are two such pairs, we have $A_{J'}(t) = A_{J''}(t)$ on $J' \cap J''$. Writing $K = [a, v[$ for the union of all these intervals $J$, we can then unambiguously define a function $A_K(t)$ on $K$ by putting $A_K(t) = A_J(t)$ for all $J$ such that $t \in J$. The function $A_K(t)$ trivially satisfies $A_K(a) = \alpha$ and $A_K(t) \in Arg[\gamma(t)]$ for all $t \in K$. Since every $t \in K$ is interior to one of the intervals $J$, the function $A_K$ is continuous on $K$, so that $K$, which satisfies the conditions imposed on the intervals $J$, is the largest of these intervals.

Now there exists on a neighbourhood of $v$ in $I$ a continuous function $A(t)$ satisfying $A(t) \in Arg[\gamma(t)]$ on this neighbourhood. For $t < v$ sufficiently close to $v$, the two functions $A_K(t)$ and $A(t)$ are simultaneously defined and continuous; they are therefore equal, up to a (constant) multiple of $2\pi$; after subtracting it from $A(t)$ we may assume that $A(t) = A_K(t)$ for all $t < v$ sufficiently close to $v$. This allows us to extend the function $A_K(t)$ up to the point $v$, and beyond $v$ if $v < b$. The second eventuality is excluded since $K$ is the largest of the intervals $J$ considered above, qed.

In conclusion, *for any continuous path $t \mapsto \gamma(t)$ traced in $\mathbb{C}^*$ it is possible to choose the argument of $\gamma(t)$ so that it is a continuous function of $t$.* This choice is unique up to the addition of a constant multiple of $2\pi$, but, once again, the argument of $\gamma(t)$ thus chosen depends on $t$ and not only on $\gamma(t)$.

(ix) *Uniform branches and closed paths.* Let $G$ be an open connected subset of $\mathbb{C}^*$ and *assume* that there exists a uniform branch $A(z)$ of the argument on $G$. Choose an $a \in G$, put $A(a) = \alpha$ and, for $z \in G$ given, join $a$ to $z$ by a continuous path $\gamma(t)$ where $t$ varies in a compact interval $I$. The function $t \mapsto A[\gamma(t)]$ is clearly a uniform branch of the argument along $\gamma$, the branch whose initial value is $\alpha$ and final value is $A(z)$. In particular, *the final value does not depend on the path* $\gamma$ *chosen*, and since every other uniform branch of the argument along $\gamma$ differs from the preceding by a constant multiple of $2\pi$, the result persists in the following form: if one "follows by continuity" the value of the argument along a path joining $a$ to $z$ in $G$, the value obtained at $z$ depends only on the value chosen at $a$; in other words, the "variation of the argument" along a path in $G$ depends only on its origin and on its end point and, in particular, is zero along every *closed path*, i.e. one whose end points are equal.

If, conversely, this condition is satisfied, one can construct a uniform branch $A(z)$ of the argument in $G$ directly. Choose one of the possible of values $A(a)$ at a point $a \in G$ arbitrarily, then determine $A(z)$ by joining $a$ to $z$ by a continuous path in $G$: there exists a uniform branch of the argument along this path, one can assume that its initial value is $A(a)$ and $A(z)$ is then, by definition, its final value. The result being by hypothesis independent of the path chosen *in* $G$, this is a good definition. To show that the result is a *continuous* function of $z$, one chooses (figure 3) an open disc $D \subset G$ with centre $z$ and, on $D$, a uniform branch $A'$ of the argument according to the result obtained in (vi); one can clearly assume $A'(z) = A(z)$. This said, let $z'$ be a point of $D$; to join $a$ to $z'$, one can join $a$ to $z$ by a path $\gamma : [u, v] \longrightarrow G$ and follow this path by a path $\gamma' : [v, w] \longrightarrow D$ joining $z$ to $z'$; one clearly obtains a uniform (i.e. continuous) branch of the argument along the path joining $a$ to $z'$ by insisting that it be equal to $A[\gamma(t)]$ along $\gamma$ and to $A'[\gamma'(t)]$ along $\gamma'$. The final value being $A'(z')$, one thus has $A(z') = A'(z')$ on $D$, whence the continuity of $A$ on $G$.

In consequence, *for there to exist a uniform branch of the argument in $G$ it is necessary and sufficient that, for every $a, b \in G$ and path $\gamma$ joining $a$ to $b$ in $G$, the value at $b$ of the argument of $z$ obtained with the help of a uniform branch of the argument along $\gamma$ depends only on its value at $a$, and not on the path followed; or again: that the "variation of the argument along any closed path" in $G$ be zero.*

It is "geometrically obvious" that, for this to be so, it is necessary and sufficient that no closed path traced in $G$ surrounds the point $O$; no matter what mechanic or physicist explains to you that if you "follow the argument by continuity" of a point $z$ whose trajectory "surrounds" (in one sense or the other) the origin once or twice, at the end you obtain a value of the argu-

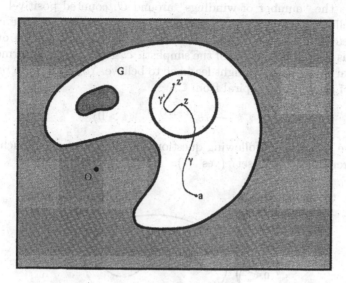

fig. 3.

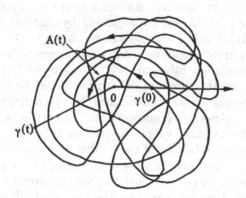

fig. 4.

ment equal to the value at departure increased by $2k\pi$, where the integer $k$ indicates the "number of windings" around $O$, counted positively or negatively following the "sense of rotation". Figure 4, where we have charitably abstained from using a Peano curve, shows that the calculation of $k$ is not always as simple as a sketch of the simplistic case of a circle (deformed a little to appear more serious) might lead one to believe. You could also remove the points of an unlimited spiral from $\mathbb{C}^*$

$$t \longrightarrow te^{i(t+1/t)} \qquad (t>0)$$

and pose yourself the following question: are there uniform branches of $\mathcal{L}og$ on the remaining open set? (yes ...).

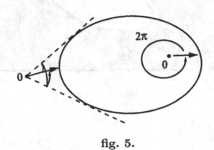

fig. 5.

The preceding short intuitive discourse contains several rather vague terms or expressions. "To follow the argument by continuity" means to construct a uniform branch of the argument along the trajectory $\gamma$, a topic which, in contrast to the physicists, we have explained properly. But what does one understand *mathematically* by a path which "surrounds" the origin? by the "number of rotations" that it makes around the latter? by the "sense of rotation"? The only correct answer is to declare that, *by definition*, the "number of rotations" that a closed path makes around the origin is, up to the factor $2\pi$, the difference between the values taken at the origin and at the end point of the path by a uniform branch of the argument along the given path. The physicists' argument then becomes not only correct, but tautologous.

It is not necessary to go searching for Peano curves to be brought to a halt before serious obstacles. Consider the unit circle $\mathbb{T}$ and a continuous and injective map $\varphi : \mathbb{T} \longrightarrow \mathbb{C}^*$; the formula $\gamma(t) = \varphi[\exp(it)]$, $t \in [0, 2\pi]$, transforms it into a closed path as simple as possible since its image is a "curve" $C$ homeomorphic to a circle; further, $\gamma$ is bijective apart from the fact that $\gamma(0) = \gamma(2\pi)$. It is "geometrically obvious" that one can distinguish "the interior" of the curve $C$ from its "exterior": these are the two connected components of the open set obtained by deleting the curve from $\mathbb{C}$. It is no less "obvious" that the interior is bounded and that the exterior is not: see figure 5. If one assumes that $O$ is interior to $C$, it is "obvious" that the trajectory of the point $\gamma(t)$, which describes $C$ once and only once, "winds once" around $O$, so that the variation of the argument along $\gamma$ must "evidently" be, according to the physicists, equal to $\pm 2\pi$. It is also "obvious" that, if $O$ is exterior to the curve, this variation must be zero.

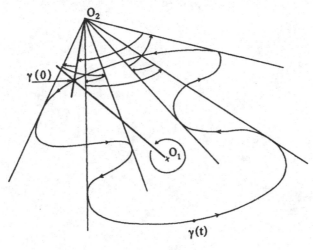

fig. 6.

All this is correct: it is Jordan's famous theorem, though he did not prove it perfectly, a quasi-impossible enterprise a century ago. But for the *evidence*, see the appendix in Chap. IX of Vol. 1 of Dieudonné's *Treatise on Analysis*: ten pages of ultra-condensed arguments due essentially to one of the greatest of contemporary topologists.

We will find these problems again in Vol. III, for example *à propos* looking for primitives of analytic functions. We know (Chap. II, § 3, n° 19) that if we have an analytic function $f$ on an open subset $G$ of $\mathbb{C}$, then on a neighbourhood of each $a \in G$ there exists a power series in $z - a$ having derivative $f$, but the real problem is to decide whether there exists an analytic function $g$ defined on *all of $G$* and such that $g'(z) = f(z)$ for all $z \in G$. The case of the function $f(z) = 1/z$ on $G = \mathbb{C}^*$ shows that not always is there a solution.

# Index

# Universitext